Planting the World

Planting the World

Joseph Banks and His Collectors: An Adventurous History of Botany

Jordan Goodman

WILLIAM COLLINS

William Collins
An imprint of HarperCollins*Publishers*
1 London Bridge Street
London SE1 9GF

WilliamCollinsBooks.com

First published in Great Britain in 2020 by William Collins

1

Maps by Martin Brown

Typeset in Garamond by Palimpsest Book Production Limited,
Falkirk, Stirlingshire
Printed and bound in Great Britain by CPI Group (UK) Ltd, Croydon

MIX
Paper from
responsible sources
FSC C007454
www.fsc.org

This book is produced from independently certified FSC™ paper
to ensure responsible forest management.

For more information visit: www.harpercollins.co.uk/green

For Cordelia

'I envy you your situation within two miles of an Erupting Volcano you will easily guess I read your Letters with that Kind of Fidgetty anziety which continuously upbraids me for not being in a similar Situation I envy you I pity myself I blame myself & then begin to tumble over my Dried Plants in hopes to put such wishes out of my head which now I am tied by the leg to an armchair I must with diligence suppress'

Joseph Banks to William Hamilton, 4 December 1778

CONTENTS

Maps

Key to Maps

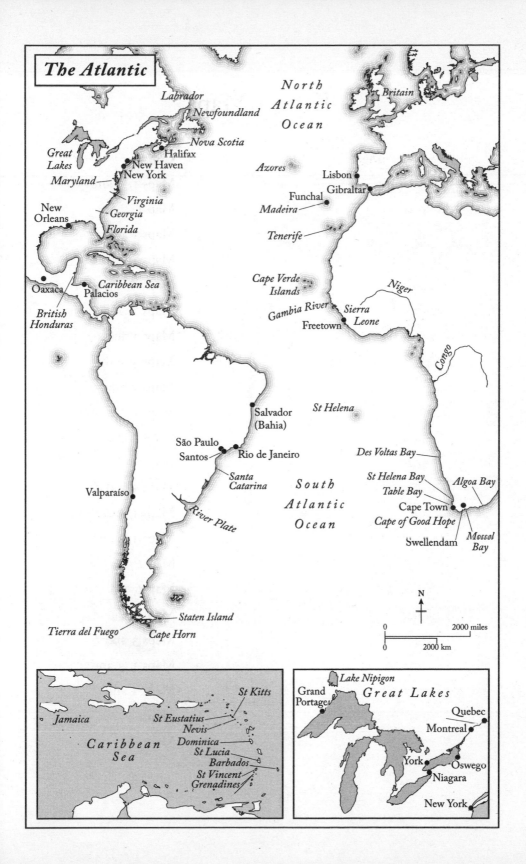

The Atlantic

Labrador

Newfoundland

Nova Scotia

Halifax

Great Lakes

Maryland

New Haven
New York

North Atlantic Ocean

Britain

Azores

Lisbon

Gibraltar

Funchal

Madeira

Virginia

New Orleans

Georgia

Florida

Tenerife

Oaxaca

Caribbean Sea

Palacios

British Honduras

Cape Verde Islands

Gambia River

Niger

Sierra Leone

Freetown

Congo

St Helena

Salvador (Bahia)

São Paulo
Santos

Rio de Janeiro

Des Voltas Bay

Santa Catarina

St Helena Bay

Table Bay

Algoa Bay

South Atlantic Ocean

Cape Town

Valparaíso

River Plate

Cape of Good Hope

Swellendam

Mossel Bay

N

0   2000 miles

0   2000 km

Staten Island

Tierra del Fuego

Cape Horn

Jamaica

St Kitts

St Eustatius

Nevis

Dominica

St Lucia

Barbados

St Vincent
Grenadines

Caribbean Sea

Lake Nipigon

Grand Portage

Great Lakes

Quebec

Montreal

York

Oswego

Niagara

New York

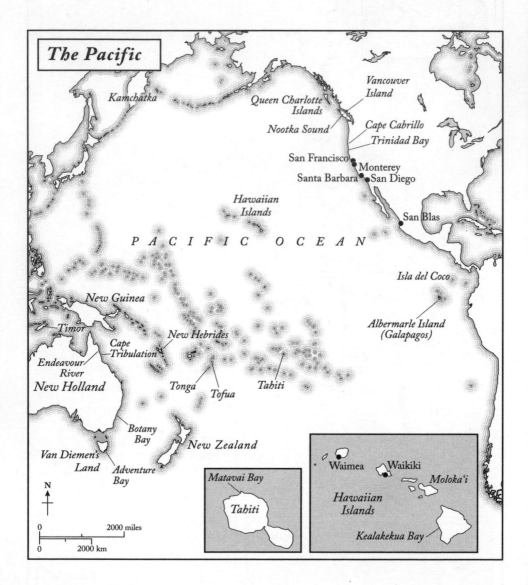

The Pacific

Kamchatka

Queen Charlotte
Islands

Vancouver
Island

Nootka Sound

Cape Cabrillo

Trinidad Bay

San Francisco • Monterey

Santa Barbara • San Diego

San Blas

Hawaiian
Islands

P A C I F I C   O C E A N

Isla del Coco

New Guinea

Albermarle Island
(Galapagos)

Timor

Cape
Tribulation

New Hebrides

Endeavour
River

New Holland

Tonga   Tofua

Tahiti

Botany
Bay

New Zealand

Van Diemen's
Land

Adventure
Bay

N

0   2000 miles

0   2000 km

Matavai Bay

Tahiti

Waimea   Waikiki

Moloka'i

Hawaiian
Islands

Kealakekua Bay

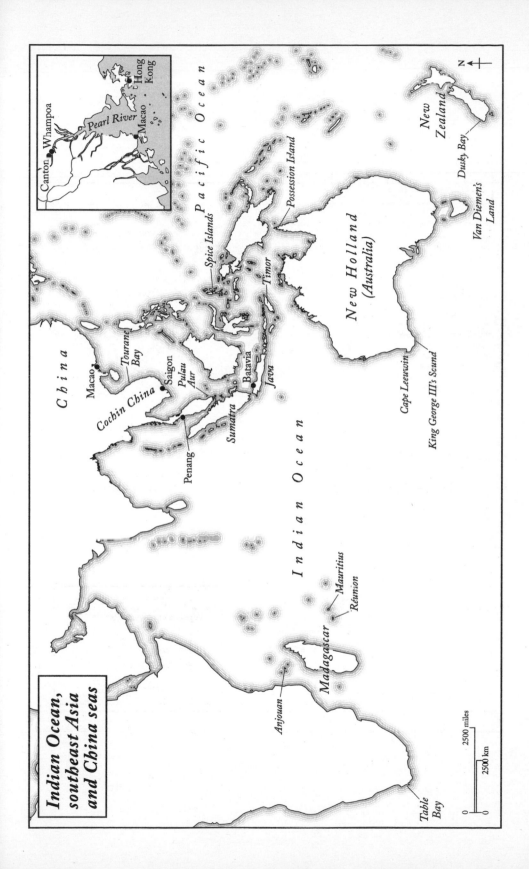

Indian Ocean, southeast Asia and China seas

Canton • Whampoa
Pearl River
• Hong Kong
Macao •

Pacific Ocean

New Zealand

Dusky Bay

Van Diemen's Land

Possession Island

New Holland (Australia)

Spice Islands

Timor

China

Touranc Bay

Macao •

Cochin China

Saigon
Pulau Aur

Penang •

Sumatra

Batavia •

Java

Indian Ocean

Cape Leeuwin

King George III's Sound

Mauritius
Réunion

Madagascar

Anjouan

Table Bay

N

2500 miles
2500 km
0
0

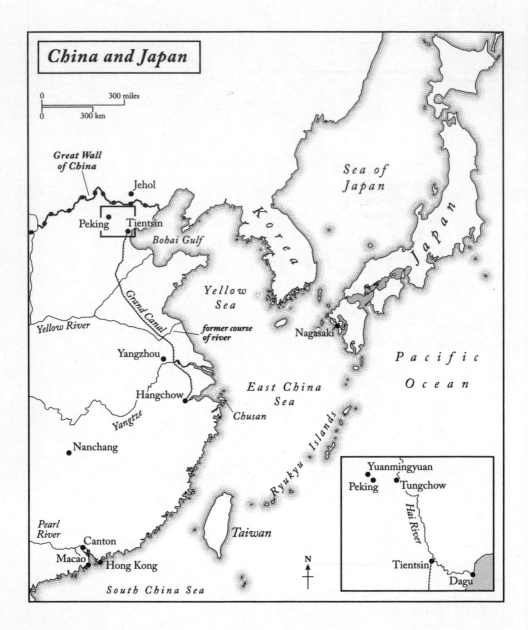

China and Japan

0   300 miles

0   300 km

Great Wall of China

Jehol

Peking   Tientsin

Bohai Gulf

Yellow River

Grand Canal

Yellow Sea

former course of river

Yangzhou

Hangchow

Chusan

Yangtze

Nanchang

Pearl River

Canton

Macao   Hong Kong

South China Sea

East China Sea

Taiwan

Ryukyu Islands

K o r e a

Sea of Japan

J a p a n

Nagasaki

P a c i f i c O c e a n

N

Yuanmingyuan

Peking   Tungchow

Hai River

Tientsin

Dagu

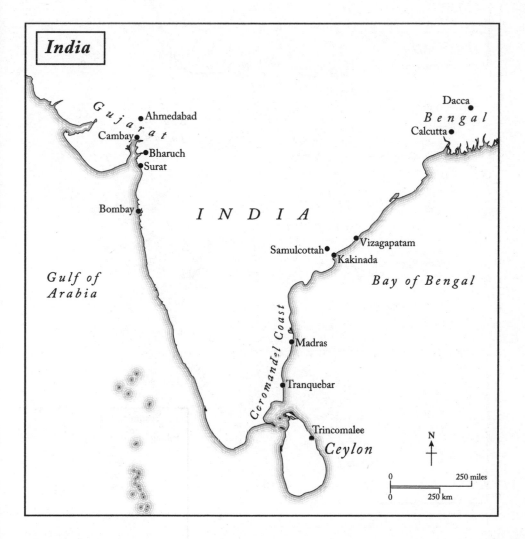

India

Gujarat

Ahmedabad

Cambay

Bharuch

Surat

Bombay

I N D I A

*Gulf of
Arabia*

Samulcottah

Vizagapatam

Kakinada

Bay of Bengal

Ceromandel Coast

Madras

Tranquebar

Trincomalee

Ceylon

Dacca

B e n g a l

Calcutta

N

0 _____ 250 miles

0 _____ 250 km

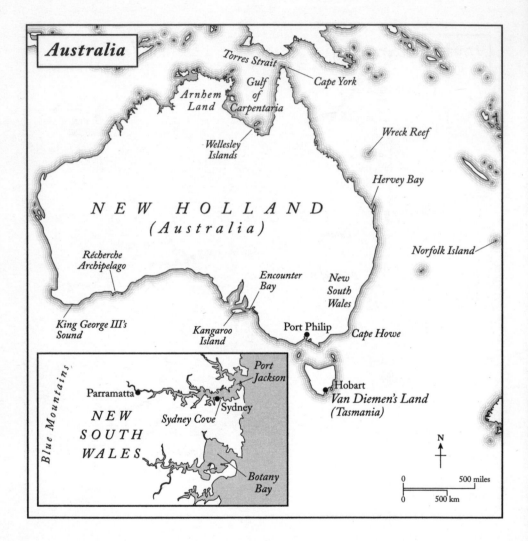

Australia

Torres Strait

Cape York

Arnhem
Land

Gulf
of
Carpentaria

Wellesley
Islands

Wreck Reef

Hervey Bay

NEW HOLLAND
(Australia)

Récherche
Archipelago

Norfolk Island

Encounter
Bay

New
South
Wales

King George III's
Sound

Kangaroo
Island

Port Philip

Cape Howe

Blue Mountains

Port
Jackson

Parramatta

NEW
SOUTH
WALES

Sydney

Sydney Cove

Hobart

Van Diemen's Land
(Tasmania)

Botany
Bay

N

0   500 miles

0   500 km

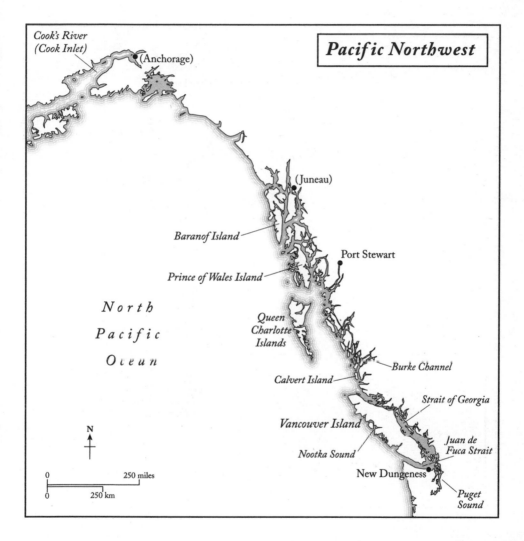

Cook's River
(Cook Inlet)

(Anchorage)

Pacific Northwest

(Juneau)

Baranof Island

Port Stewart

Prince of Wales Island

North

Pacific

Ocean

Queen
Charlotte
Islands

Burke Channel

Calvert Island

Strait of Georgia

N

Vancouver Island

Juan de
Fuca Strait

Nootka Sound

New Dungeness

0   250 miles

Puget
Sound

0   250 km

Illustrations

PEOPLE

Joseph Banks, oil portrait by Sir Thomas Lawrence, after 1795. (© *Trustees of the British Museum*)

Archibald Menzies, oil portrait by E.U. Eddis, 1836. (*Linnean Society of London*)

Francis Masson, oil portrait by George Garrard, 1887. (*Linnean Society of London*)

John and Alexander Duncan. (*Courtesy of Cathy and Andrew Duncan*)

Allan Cunningham, oil portrait, c.1835. (*National Library of Australia PIC Screen 51 #R88*)

Portrait of Sir John Barrow (1764–1848). (*Wellcome Collection. Attribution 4.0 International CC BY 4.0*)

Matthew Flinders, c.1800, watercolour miniature portrait. (*Mitchell Library, State Library of New South Wales*)

Captain William Bligh. (*GL Archive/Alamy Stock Photo*)

Robert Brown. Line engraving by C. Fox, 1837, after H. W. Pickersgill. (*Wellcome Collection. Attribution 4.0 International CC BY 4.0*)

George Macartney and George Staunton. Pencil, pen and ink, wash and watercolours. Images taken from Album of 372 Drawings of Landscapes, Coastlines, Costumes and Everyday Life Made during Lord Macartney's Embassy to the Emperor of China. Originally published/produced in 1792–1794. (© *British Library Board. All rights reserved/Bridgeman Images*)

Daniel Charles Solander. Lithograph by Miss Turner after J. Zoffany (*Wellcome Collection. Attribution 4.0 International CC BY 4.0*)

Arthur Phillip, oil portrait, 1786, by Francis Wheatley. (*Mitchell Library, State Library of New South Wales*)

THINGS

Drawing of a flat peach made in Canton by John Bradby Blake and Mok Sau c.1773. (*Oak Spring Garden Foundation, Upperville, Virginia*)

Sketch by William Bligh of the plant nursery on HMS *Bounty*, 1787. (*Sir Joseph Banks Collection, SS 1:1, Sutro Library, California State Library*)

Sketch of plant cabin on HMS *Guardian*, 1789. (*Sir Joseph Banks Collection, G 1:5, Sutro Library, California State Library*)

The mutineers turning Lieutenant Bligh and part of the officers and crew adrift from HMS *Bounty*, painted and engraved by Robert Dodd, 1796. (*Dixson Library, State Library of New South Wales*)

HMS *Providence* and HMS *Assistant* at the Lizard, 1791, watercolour by Lieutenant George Tobin. (*State Library of New South Wales, SAFE/PXA 563*)

William Alexander's watercolour drawing of an *opuntia* and cochineal from Rio de Janeiro, 1792. (*Sir Joseph Banks Collection, En 1:20, Sutro Library, California State Library*)

Breadfruit, 1792, watercolour by Lieutenant George Tobin. (*State Library of New South Wales, SAFE/PXA 563*)

White peony by an unknown Chinese artist, c.1794–6. (*Courtesy of Cathy and Andrew Duncan*)

A view of Funchal, Madeira, oil painting by William Hodges (*1744– 97*). (*Art Collection 2/Alamy Stock Photo*)

Transplanting of the bread fruit trees from Otaheite, painting by Thomas Gosse, 1796. (*Historic Images/Alamy Stock Photo*)

PLACES

The library at the home of Joseph Banks, Soho Square, undated drawing attributed to Ralph Lucas (1796–1874). (*Natural History Museum*)

Kew Gardens: the Pagoda and Bridge, 1762, oil painting by Richard Wilson. (*Yale Center for British Art, B1976.7.172, Paul Mellon Collection*)

Cape Town, 1791, watercolour by Lieutenant George Tobin. (*State Library of New South Wales, SAFE/PXA 563*)

Adventure Bay, 1792, watercolour by Lieutenant George Tobin. (*State Library of New South Wales, SAFE/PXA 563*)

Point Venus, Matavai Bay, 1792, watercolour by Lieutenant George Tobin. (*State Library of New South Wales, SAFE/PXA 563*)

Presidio at Monterey, c.1792, watercolour by William Alexander (after John Sykes). (*Edward E. Ayer Digital Collection, Newberry Library*)

Friendly Cove, Nootka Sound, c.1792, watercolour by William Alexander. (*Edward E. Ayer Digital Collection, Newberry Library*)

Qianlong's court (showing George Thomas Staunton on one knee in front of the Emperor), 1793, watercolour by William Alexander. (*Huntington Library, Art Museum and Botanical Gardens. Gilbert Davis Collection*)

A scene at Tientsin, 1793, engraving by William Alexander, *The Costume of China*, 1805.

View of Sir Edward Pellew's Group, Gulf of Carpentaria, c.1802, oil painting by William Westall. (*ZBA7944, © National Maritime Museum, Greenwich, London*)

The European factories, Canton, 1806, oil painting by William Daniell. (*Yale Center for British Art, B1981.25.210, Paul Mellon Collection*)

The road to the home of Baron von Langsdorff on the outskirts of Rio de Janeiro, 1817–18, watercolour by Thomas Ender. (*HZ 13845, Graphic Collection Academy of Fine Arts Vienna*)

A panorama of São Paulo, 1817–18, watercolour by Thomas Ender. (*HZ 13309, Graphic Collection, Academy of Fine Arts, Vienna*)

Dramatis Personae

Clarke Abel (1780–1826). Surgeon to the Amherst embassy to China 1816–17

Adam Afzelius (1750–1837). Botanist, student of Linnaeus, Swedenborgian, hired by Sierra Leone Company and Banks's plant collector in Sierra Leone 1792–6

William Aiton (1731–93). Head Gardener Kew 1759–93, Banks's main contact at the royal garden

William Townsend Aiton (1766–1849). Son of the above and Head Gardener Kew 1793–1841, taking over his father's duties and contact with Banks

William Alexander (1767–1816). Artist to the Macartney embassy to China 1792–4

John Allen (1773–1806). Lead miner, Banks's appointed mineralogist on HMS *Investigator* 1801–3

William Amherst (1773–1857) 1st Earl Amherst of Arracan. Leader of the third British embassy to China (Amherst embassy) 1816–17, Governor-General of India 1823–8

James Anderson (1738–1809). Physician General East India Company Madras, naturalist, first person to bring cochineal to Banks's attention in 1786

George Austin (?–1790). Banks's appointed gardener on HMS *Guardian* 1789

Joseph Banks (1743–1820). Botanist, President of the Royal Society

Sarah Sophia Banks (1744–1818). Sister of Joseph Banks

(Lady) Dorothea Banks (1758–1828). Wife of Joseph Banks

Francis Baring (1740–1810). Banker, head of Baring and Co. London's chief merchant bank, Director 1779–1810 and Chairman East India Company 1792–3, responsible for the Company's role in the Macartney embassy to China 1792–4

Francis Louis Barrillier (1773–1853). Army officer, surveyor in New South Wales 1800–3

John Barrow (1764–1848). Comptroller on Macartney embassy to China 1792–4, Secretary to George Macartney Cape Colony 1797–1804, Second Secretary to the Admiralty 1804–45

Henry Bathurst (1762–1834) 3rd Earl Bathurst. Politician, Secretary of State for War and the Colonies 1812–27

Nicolas Thomas Baudin (1754–1803). Naval officer, commander French exploration vessel *Le Géographe* on its intended circumnavigation of Australia 1800–3, met Matthew Flinders in present-day South Australia in April 1802

Franz Bauer (1758–1840). Botanical artist active at Kew

Ferdinand Bauer (1760–1826). Botanical artist, brother of the above, Banks's appointed artist on HMS *Investigator* 1801–3

Lord Beauchamp. See Francis Seymour Conway

Andreas Berlin (1746–73). Botanist, student of Linnaeus, plant collector in Sierra Leone alongside Henry Smeathman 1773

Charles Blagden (1748–1820). Physician, Secretary to the Royal Society 1784–97, Banks's confidant

John Bradby Blake (1745–73). Supercargo East India Company Canton 1770–3, co-producer of hybrid Chinese-British botanical drawings

William Bligh (1754–1817). Naval officer, commander HMS *Bounty* and HMS *Providence* 1787–9 and 1791–3 respectively, Fourth Governor of New South Wales 1806–8

Essex Henry Bond (1762–1819). Naval officer, commander East India Company ship *Royal Admiral*

Jacob Bosanquet (1755–1828). Director from 1782, Deputy-Chairman Court of Directors 1797, 1802, 1810, Chairman Court of Directors 1798, 1803, 1811 East India Company

James Bowie (c.1789–1869). Gardener, Banks's plant collector in Brazil 1814–16 and at the Cape 1816–23

William Broughton (1762–1821). Naval officer, commander HMS *Chatham* 1791–3

Robert Brown (1773–1858). Botanist, Banks's plant collector in Australia on HMS *Investigator* 1801–3, Banks's librarian 1810–20

Earl of Buckinghamshire. See Robert Hobart

Charles Bunbury (1740–1821). Politician, Chairman Bunbury Committee on crime and punishment 1778–9

David Burton (?–1792). Gardener, Banks's plant collector in Australia

George Caley (1770–1829). Gardener, botanist, Banks's plant collector in Australia 1800–10, Superintendent St Vincent Botanic Garden 1816–22

John Call (1731–1801). Engineer, politician, promoted transportation and the establishment of penal colonies

Duncan Campbell (1726–1803). Merchant, shipowner, contractor for transportation and for the provision of convict hulks

Lord Castlereagh (1769–1822). Politician, President of the Board of Control 1802–6

Charles Cathcart (1759–88). Military officer, politician, leader of the first British embassy to China (Cathcart embassy) 1787–8

Henry Chamberlain (1773–1829). Diplomat, British consul-general in Brazil, an important contact for James Bowie and Allan Cunningham while plant collecting in Brazil

John Clarkson (1764–1828). Naval officer, abolitionist, Governor of Sierra Leone 1792, responsible for bringing the first settlers from Nova Scotia

David Collins (1756–1810). Military officer, judge advocate of New South Wales 1788–96

James Colnett (1753–1806). Naval officer, fur trader, commander merchant vessel *Prince of Wales* 1786–8

Francis Seymour Conway (1743–1822) Lord Beauchamp. Politician, headed the Beauchamp Committee hearings into transportation 1785

James Cook (1728–79). Naval officer, explorer, commander HMS *Endeavour* 1768–71, HMS *Resolution* 1772–5, 1776–9

John Cranch (1758–1816). Naturalist on HMS *Congo* 1816

John Crosley (1762–1817). Astronomer on HMS *Investigator* 1801–3

Allan Cunningham (1791–1839). Gardener, Banks's plant collector in Brazil 1814–16 and Australia 1816–31

Alexander Dalrymple (1737–1808). Cartographer, First Hydrographer of the British Admiralty 1795–1808

Nathaniel Dance (1748–1827). Naval officer, commander East India Company ship *Earl Camden* 1802–4 engaging French fleet in the Battle of Pulo Aura 1804

William Devaynes (1730–1809). Politician, merchant, banker, Chairman East India Company 1770–1807 *passim*

James Dickson (c.1738–1822). Botanist, nurseryman in Covent Garden

George Dixon (1748–95). Naval officer, fur trader, commander merchant vessel *Queen Charlotte* 1785–8

Jonas Dryander (1748–1810). Botanist, Banks's librarian 1782–1810

Alexander Duncan (1758–1832). Surgeon to the East India Company Canton 1788–97, Banks's plant collector in Canton during this period

John Duncan (1751–1831). Brother of the above, Surgeon to the East India Company Canton 1783–8, Banks's plant collector in Canton during this period

Henry Dundas (1742–1811) 1st Viscount Melville. Politician, Treasurer of the Navy 1782–1800, Home Secretary 1791–4, President Board of Control 1793–1801, Secretary of State for War 1794–1801, First Lord of the Admiralty 1804–5

Hinton East (?–1792). Planter, botanist, Receiver-General of Jamaica 1779, Banks's important contact in Jamaica

John Ellis (c.1710–76). Naturalist, early proponent and supporter of the Linnaean classificatory system

Richard Cadman Etches (1744–1817). Merchant, entrepreneur, formed the King George's Sound Company the first British company to trade for furs in the Pacific Northwest 1785

Matthew Flinders (1774–1814). Naval officer, commander HMS *Investigator* 1801–3

George III (1738–1820). King of the United Kingdom 1760–1820

Hugh Gillan (?1745–98). Physician to the Macartney embassy to China 1792–4, plant collector in Brazil and China on this mission

Peter Good (?–1803). Gardener, Banks's plant collector in Calcutta 1794, Banks's appointed gardener and plant collector on HMS *Investigator*

William Goulburn (1784–1826). Politician, Under-Secretary of State for War and the Colonies 1812–21

Erasmus Gower (1742–1814). Naval officer, commander HMS *Lion* conveying members of the Macartney embassy to and from China 1792–4

Charles Green (1734–71). Astronomer on HMS *Endeavour* 1768–71

William Grenville (1759–1834) 1st Baron Grenville. Politician, Home Secretary 1789–91

Charles Francis Greville (1749–1809). Politician, antiquarian, collector

William Hamilton (1730–1803). Antiquarian, collector, British Ambassador to Naples 1764–1800

George Harrison (1767–1841). Civil servant, Assistant Secretary of the Treasury 1805–26

Lord Hawkesbury. See Charles Jenkinson

Robert Hobart (1760–1816) 4th Earl of Buckinghamshire. Politician, President of the Board of Control 1812–16

James Hooper (?–1830/1). Gardener, Banks's plant collector in China on the Amherst embassy 1816–17, head gardener Buitenzorg (now Bogor) Botanic Gardens 1817–30

John Hope (1725–86). Physician, Professor of Botany and King's Botanist University of Edinburgh 1761–86

Anthony Pantaleon Hove (?–1830). Gardener, Banks's plant collector in southwest Africa 1786, Gujarat 1787–8

John Hunter (1737–1821). Naval officer, Second Governor of New South Wales 1795–1800

Charles Jenkinson (1729–1808) Lord Hawkesbury. Politician, President of the Board of Trade 1786–1804

Robert Banks Jenkinson (1770–1828) 2nd Earl of Liverpool. Politician, Prime Minister of the United Kingdom 1812–28

William Jones (1746–94). Jurist, philologist, orientalist, judge Supreme Court Calcutta 1783–94

Engelbert Kaempfer (1651–1716). Naturalist, physician to the Dutch East India Company Nagasaki 1690–2

Paul Ke (c.1752–?). Missionary, Chinese interpreter to Macartney embassy to China 1792–4

William Kerr (1779–1814). Gardener, Banks's plant collector in Canton 1803–10, Superintendent Ceylon Botanic Garden 1810–14

James King (1750–84). Naval officer, commander HMS *Discovery* on James Cook's Third Voyage 1779–80

John King (1759–1830). Civil servant, Permanent Under-Secretary of State at the Home Department 1791–1806

Philip Gidley King (1758–1808). Naval officer, Third Governor of New South Wales 1800–6

Johan Gerhard Koenig (1728–85). Pharmacist, physician, botanist, plant collector on India's Coromandel Coast 1768–85

Robert Kyd (1746–93). Military officer, Superintendent Calcutta Botanic Garden 1787–93

David Lance (c.1755–1820). Supercargo East India Company in Canton 1775–89 and 1803–4

Baron Georg Langsdorff (1774–1852). Physician, naturalist, diplomat in Brazil 1813–30

William Elford Leach (1791–1836). Zoologist, assistant librarian British Museum 1813–22

James Lee (1715–95). Nurseryman, joint owner Vineyard Nursery Hammersmith 1745–95

Jacob Li (c.1752–?). Missionary, Chinese interpreter to Macartney embassy to China 1792–4

James Lind (1736–1812). Astronomer, physician, accompanied Banks to Iceland 1772

Carl Linnaeus (1707–78). Botanist, physician, Professor of Medicine Uppsala University 1741–72, Rector Uppsala University 1750–71, established modern botanical nomenclature

Charles-Alexandre Linois (1761–1848). Naval officer, commander of the *Marengo* and the French fleet during the Battle of Pulo Aura 1804

Lord Liverpool. See Robert Banks Jenkinson

David Lockhart (1786–1845). Gardener, Banks's plant collector on HMS *Congo* 1816, Superintendent Trinidad Botanic Gardens 1818–45

Israel Lyons (1739–75). Botanist, mathematician, astronomer, an early exponent of Linnaean taxonomy and nomenclature, taught (1764) and inspired Banks

George Macartney (1737–1806). Diplomat, Governor of Madras 1781–5, leader second British embassy to China (Macartney embassy) 1792–4, Governor of the Cape Colony 1796–8

James Main (c.1765–1846). Gardener, plant collector in China 1790s

William Marsden (1754–1836). Orientalist, First Secretary of the Admiralty 1804–7

Nevil Maskelyne (1732–1811). Mathematician, Astronomer Royal 1765–1811

Francis Masson (1741–1805). Kew gardener, plant collector at the Cape, the Atlantic and Caribbean, the Mediterranean and North America.

James Matra (1746–1806). Sailor, diplomat, promoter of settlement in New South Wales 1783–5

Murray Maxwell (1775–1831). Naval officer, commander HMS *Alceste* on the Amherst embassy to China 1816–17

Archibald Menzies (1754–1842). Assistant surgeon Royal Navy, gardener, botanist, Banks's plant collector in the Pacific on the *Prince of Wales* 1786–9 and on HMS *Discovery* 1791–4

Anne Monson (c.1727–76). Botanist and plant collector

John Montagu (1718–92) 4th Earl of Sandwich. First Lord of the Admiralty 1748–51, 1763, 1771–82

Dan Moowattin (c.1790–1816). Dharug guide with George Caley in Australia 1805–10

Thomas Morton (?–1792). Secretary to Court of Directors East India Company 1783–92

David Nelson (?–1789). Gardener, Banks's plant collector on HMS *Discovery* 1776–80 and on HMS *Bounty* 1787

Evan Nepean (1752–1822). Administrator, Under-Secretary of State for the Home Department 1782–94, First Secretary to the Admiralty 1795–1804

August Nordenskjöld (1754–92). Chemist, Swedenborgian, hired by Sierra Leone Company

Franz Pehr Oldenburg (1740–74). Soldier, plant collector in the Cape meeting Banks, Solander and Masson 1771–3

Mungo Park (1771–1806). Surgeon, Banks's sponsored explorer in West Africa 1795–7

Sydney Parkinson (c.1745–71). Banks's appointed botanical artist on HMS *Endeavour* 1768–71

William Paterson (1755–1810). Military officer, botanist and explorer in Australia 1791–1810, Lieutenant Governor of New South Wales 1801–10

Robert Peel (1788–1850). Politician, Under-Secretary of State for War and the Colonies 1810–12

James Pendergrass (1766–1851). Naval officer, commander East India Company ship *Hope* to and from Canton 1805–6

Arthur Phillip (1738–1814). Naval officer, commander HMS *Sirius* First Fleet to New South Wales 1787–8, First Governor of New South Wales 1788–90

Constantine Phipps (1744–92) 2nd Baron Mulgrave. Naval officer, explorer, Banks's close friend for over thirty years

William Henry Pigou (fl.1775–83). Supercargo East India Company in Canton

William Pitcairn (1712–91). Physician, botanist, President Royal College of Physicians 1775–85, owner of substantial garden in Islington

William Pitt (1759–1806). Politician, Prime Minister of Great Britain 1783–1801, Prime Minister of the United Kingdom, 1801, 1804–6

Home Popham (1762–1820). Naval officer, Rear Admiral 1814–20, politician 1804–12 *passim*

Nathaniel Portlock (c.1747–1817). Naval officer, commander merchant vessel *King George* 1785–8, commander HMS *Assistant* 1791–3 to Tahiti

John Pringle (1707–82). Physician, President of the Royal Society 1772–8

Puankhequa I (1714–88). Principal Hong merchant in Canton

Puankhequa II (1755–1820). Son of the above and principal Hong merchant in Canton

Qianlong Emperor (1711–99). Sixth Emperor of the Qing dynasty China 1735–96

William Ramsay (fl.1771–1814). Secretary, Court of Directors, East India Company c.1792–1806

John Reeves (1774–1856). Assistant inspector of teas East India Company in Canton, Banks's plant collector in Canton 1812–20

John Rennie (1761–1821). Civil engineer involved in assessing the steamboat HMS *Congo* 1815–16

Matthew Riches (1755–?, fl. 1791–1814). Naval officer, commander of the East India Company ship *Thames* to and from Canton 1806–7

Edward Riou (1762–1801). Naval officer, commander HMS *Guardian* 1789–90

Henry Roberts (1756–96). Naval officer, Lieutenant HMS *Discovery* 1776–80

William Roxburgh (1751–1815). Surgeon, botanist, Superintendent Calcutta Botanic Garden 1793–1814

Patrick Russell (1727–1805). Physician, naturalist to the East India Company in the Carnatic 1785–91

Friedrich Sello var. Sellow (1789–1831). Gardener, botanist, plant collector in Brazil 1814–31

Granville Sharp (1735–1813). Lawyer, abolitionist, proponent of the settlement of Sierra Leone 1787–9

Henry Smeathman (1742–86). Naturalist, entomologist, plant collector in Sierra Leone for a consortium including Banks 1771–5

Christen Smith (1785–1816). Banks's botanist on HMS *Congo* 1816

Christopher Smith (?–1807). Banks's appointed gardener on HMS *Providence* 1791–3, Banks's appointed gardener on the *Royal Admiral* 1794, Superintendent Penang Botanic Garden 1805–7

James Smith (?–1789). Banks's appointed gardener on HMS *Guardian* 1789

James Edward Smith (1759–1828). Botanist, President of the Linnean Society of London 1788–1828

Percy Smythe (1780–1855) 6th Viscount Strangford. Diplomat, British Envoy to Portugal (Brazil) 1806–15

Daniel Solander (1733–82). Botanist, student of Linnaeus, Keeper Natural History Department British Museum 1773–82

Anders Sparrman (1748–1820). Botanist, student of Linnaeus, naturalist on HMS *Resolution* 1772–5

George Spencer (1758–1834) 2nd Earl Spencer. Politician, First Lord of the Admiralty 1794–1801

George Leonard Staunton (1737–1801). Physician, diplomat, Secretary to George Macartney on the Macartney embassy to China 1792–4

George Thomas Staunton (1781–1859). Son of the above, sinologist, East India Company employee 1798–1816

Philip Stephens (1723–1809). Politician, First Secretary of the Admiralty 1763–95

John Stuart (1713–92) 3rd Earl of Bute. Politician, Prime Minister of Great Britain 1782–3, adviser to Augusta Dowager Princess of Wales concerning Kew 1751–72

George Suttor (1774–1858). Market gardener, plant collector in Australia 1800–20 *passim*

Olof Swartz (1760–1817). Botanist, student of Carl Linnaeus the younger, Banks's close friend

Emmanuel Swedenborg (1688–1772). Mathematician, engineer, visionary

Lord Sydney. See Thomas Townshend

Edward Thompson (c.1738–86). Naval officer, commadore of the Africa Station Royal Navy 1782–6, commander HMS *Grampus* 1785–6

Thomas Boulden Thompson (1766–1828). Naval officer, commander HMS *Nautilus* 1785–6

Henry Thornton (1760–1815). Politician, banker, abolitionist, Chairman Sierra Leone Company 1791–1808

Carl Peter Thunberg (1743–1826). Botanist, student of Linnaeus, surgeon to the Dutch East India Company Nagasaki 1775–6

Thomas Townshend (1733–1800) 1st Viscount (Lord) Sydney. Politician, Home Secretary 1783–9

George Tripp (1752–1830). Naval officer, commander HMS *Grampus* 1786, commander HMS *Sphinx* 1790–1

James Kingston Tuckey (1776–1816). Naval officer, commander of HMS *Congo* 1815–16

William Tudor (?–1816). Comparative anatomist on HMS *Congo* 1816

George Vancouver (1757–98). Naval officer, commander HMS *Discovery* 1791–5

Gerard De Visme (1726–97). Merchant based in Lisbon 1746–91

Carl Bernhard Wadström (1746–99). Industrialist, abolitionist, promoter of the colonisation of West Africa

Samuel Wallis (1728–95). Naval officer, commander HMS *Dolphin* 1766–8

James Watt Jr. (1769–1848). Mechanical engineer, businessman, consultant to the navy regarding the steam engine on HMS *Congo*

William Westall (1781–1850). Banks's appointed landscape artist on HMS *Investigator* 1801–3

Whang at Tong (fl.1774–96). Linguist, scholar and businessman in

Canton, visited England 1774–?81 and accompanied Banks in and around London

William Wilberforce (1759–1833). Politician, abolitionist, Director Sierra Leone Company

James Wiles (1768–1851). Banks's appointed gardener on HMS *Providence* 1791–3

George Yonge (1731–1812). Politician, Secretary at War 1782–94

Prologue

When I first thought of writing this book, I had in my mind an image of Joseph Banks that I had pieced together from the various biographies that had been written about him in the twentieth century and supplemented by other accounts in which he appeared as a major character. The image I formed was of a man who was an autocratic administrator; a man of great personal wealth who manipulated his powerful connections at the highest level in order to make Britain the greatest nation on earth and its empire second to none; a man who used science for these ends; a man who could be seen and caricatured as a powerful spider at the centre of an enormous web.

This was the image I brought to my research, but as I began reading through Banks's voluminous extant correspondence – letters to and from him – and the endless amount of scribbled-on paper, in the form of notes, lists, draft reports and memoranda – the image of the powerful spider at the centre of the web began to dissolve, to be replaced by a radically different impression, that of someone who was less authoritarian, less one-dimensional and less in control; who was more responsive and less manipulative; someone whose relationships were more personal and informal; and who was more appealing, in short, more human. Ultimately, Banks was a figure who represented the culmination of an era that was on the wane, an era when an individual could still, by amassing a great library and mastering its contents, provide critical aid and information to a great variety of projects in the wider world.

This book is not a biography although I do provide a biographical sketch in the introduction. I am focusing on just one of Banks's many interests, albeit the central and most fervently held, his love for plants. This passion combined the scientific study of plants and his personal dedication to supplying King George III's garden in Kew with the finest examples of the world's botanical bounty.

* * *

As a young man, Banks had gone as a naturalist on three important scientific voyages, one of which was an unprecedented circumnavigation of the globe lasting almost three years. Yet, after he was thirty, he never went to sea again; it is tempting to see his enthusiasm for sending other men to collect and move plants across the seas on his behalf as a vicarious reliving of his earlier experiences.

Many of these collectors were gardeners who had worked at Kew, but others, mostly from Scotland and the European continent, were trained botanists; a few were ship's surgeons and government officials, and others who having no specifically relevant occupation had become fascinated by natural history. Banks chose some of these men to travel and collect on his behalf; but many others contacted him to volunteer their services.

The twenty-two chapters in this book, based primarily on Banks's correspondence and jottings, tell the stories of these collectors: the different ways in which they entered Banks's life, their practices and problems across the world – how they interacted with their shipboard companions, especially the ship's captain; how they decided which plants to gather; how they found them and how they tried to move them by ship across the oceans. Banks referred to these activities as his favourite project and to himself as a projector: his role, he once explained, was no more than to facilitate their execution. Though the projects did not always begin with botany in mind, nor for that matter were they necessarily even initiated by Banks, once he became involved, either by being invited or by just joining in, he gave them a botanical twist.

Building chronologically and across the world, layer by layer, each chapter reveals how and when Banks became involved; what he did to give the projects a botanical direction, and with what consequences. At the same time, it recounts, where possible in their own words, how collectors pursued their instructions, how they conducted themselves far away from home, how they interacted with their environment, local materials and particularly local people. The results were not always satisfactory: collectors, like their plants, often did not survive the ordeals that beset them.

Banks, like his contemporaries, took it for granted that plants were there for the taking. They did not concern themselves with questions of ownership; at the same time, they were also generally unaware of the consequences of transplantation and the introduction of foreign species into new habitats. They assumed it was beneficial.

Banks was immensely curious, and it is his desire for knowledge that drives the narrative of the book. People and plants were the motivating force for his actions and sent his mind and his collectors exploring the world and its oceans.

Introduction

Joseph Banks and Kew

'If I am to do all to write all to direct all & to pay for all & no human being feel inclined to thank me I shall I fear in due time feel as sulky as an old sow who has lost her scrubbing Post.'[1]

Who was Joseph Banks?[2] He was born on 13 February 1743 in London to William and Sarah Banks, the couple's first and only son.[3] William entered Middle Temple, London, but is not known to have pursued any legal career. Though he and his wife, who was from Derbyshire, spent some of their time in the St James's area of London at their Argyll Street home, most of their life was lived at their country estate at Revesby Abbey, Lincolnshire. Here, William, now also a Member of Parliament for a Cornwall constituency, devoted himself to the Lincolnshire estates which he had inherited from his father in 1741.

Joseph's formal education began at the age of nine at Harrow School but four years later he was transferred to Eton, where he followed a curriculum primarily in Latin, Greek and English literature, which he studied dutifully but without much enthusiasm or success. Towards the end of his stay, he was also instructed in algebra, geography and French.

Next was Christ Church, Oxford, which he entered as a gentleman-commoner on 16 December 1760, aged seventeen. During his first year at the university, Joseph's father William died aged forty-one. When he came of age in 1764, Banks inherited his father's Lincolnshire estates and thus became an extremely wealthy young man.

Banks continued to attend university and during his time there he became intensely interested in natural history. As a boy, he had collected plants and insects like many others of his age and class but now he wanted to study botany seriously. To get some kind of formal instruction in botany – the subject was not on the curriculum at Oxford – Banks, who now had money, and with the support of Humphrey Sibthorp, the Sherardian Professor of Botany (who did not teach the subject), paid to be present at a set of lectures delivered during the summer of 1764 to a group of sixty enthusiastic students by Israel Lyons, a Cambridge botanist and astronomer.[4]

Lyons was one of the earliest exponents of the new Linnaean system in Britain and shared his understanding of and passion for it with Banks. In 1735, Carl Linnaeus, who was born in southern Sweden in 1707, published a new, simple and radical classification system based on the sexual characteristics of plants (the number of stamens and pistils).[5] The nomenclature he devised consisted of a binomial description of genus followed by species. At Uppsala University, where he taught medicine and botany, Linnaeus educated a large number of devoted students, many of whom travelled abroad searching for plants and spreading knowledge of his botanical system.[6] Although Banks wanted to meet Linnaeus, he never did (Linnaeus died in 1778) but his close associates, Daniel Solander and Jonas Dryander, were both educated by Linnaeus, as was one of his collectors and many others in his circle.

Lyons and Banks became close friends and their association worked to both men's benefit over the coming years until 1775, when Lyons died at the age of thirty-six.

Banks left Oxford without a degree shortly after Lyons's lectures. He could have done anything he liked – gone into the law, Church, or the City – or simply enjoyed himself as a wealthy young man about town, but his great enthusiasm was for plants. As soon as he could he moved to Bloomsbury close by the newly opened British Museum, to which he obtained a reader's ticket on 3 August 1764, and there threw himself into the study of botany, helped by its world-famous herbaria, illustrations and texts.

At that time, the British Museum was the only public space in London where natural history could be studied. While he was there, Banks became acquainted with others like himself, and through his new contacts and friendships, he was elected in his absence at age twenty-three a Fellow of the Royal Society on 1 May 1766.

Like every enthusiastic naturalist, Banks went out and about botanising, observing and collecting living specimens in their habitat. A rare chance to botanise beyond Europe came Banks's way in April 1766, when an old school friend, now Lieutenant Constantine Phipps, invited him to join HMS *Niger* bound for Newfoundland and Labrador, on fisheries protection duty.[7] Banks eagerly accepted the opportunity.

The *Niger*, with Banks aboard, was away from England for nine months, from 22 April 1766. Six of those months were spent in and around Newfoundland and Labrador. Coming home by way of Lisbon on 26 January 1767, Banks landed with a substantial haul of new natural specimens – plants, birds, insects and fishes – all of which needed to be classified and some of which were illustrated as well, for which task he principally employed the great and highly established Linnaean artist, Georg Ehret, and the young Scottish artist, Sydney Parkinson, to whom Banks had just been introduced by James Lee, the part owner of the famous Vineyard Nursery in Hammersmith.[8]

Exciting as this adventure was, it paled into insignificance when compared to the next one. Not only was it a longer expedition and to a part of the world that only a few Europeans had ever been to, but it would be partly sponsored by the Royal Society, of which he would become one of its most famous Fellows. This epic voyage to the South Pacific would become a defining moment in Banks's life and interests.

There are many beginnings to the voyage of HMS *Endeavour* but a significant one took place on 15 February 1768 when King George III received a 'Memorial for Improving Natural Knowledge' from the Royal Society.[9] The Society was appealing to the King for his financial support to send two men to the South Pacific somewhere in a rectangle bounded by a latitude 'not exceeding 30 degrees [south] and between the 140th and 180th degrees of longitude west', as defined by Nevil Maskelyne, the Astronomer Royal, to observe the transit of Venus, predicted to be visible there on 3 June 1769 – an event that would not occur again until 1874.[10] The Royal Society had been discussing since at least June 1766 how they would contribute to observing and recording this rare but crucially important astronomical occasion, from which it was hoped they could calculate the size of the solar system. On 19 November 1767, the Society's newly constituted Committee for the Transit had agreed the general plan of sending observers on a ship to the Pacific that would need to be rounding Cape Horn no later than January 1769.[11] Time was running out for adequate preparations to be made. 'The Royal Society', the memorial pleaded, 'was in no condition to defray this Expence [which they had estimated at £4000, not including the cost of the ship], their Annual Income being scarcely sufficient to carry on the necessary business of the Society.' Time was of the essence. Several other European powers (the memorial pointed to France, Spain, Denmark and Russia) were already making their own preparations for the event and Britain, in the forefront of astronomical science, simply could not afford to be a bystander. The memorial was signed by James Douglas, Earl of Morton, the Society's president, and fourteen Fellows including Benjamin Franklin and Nevil Maskelyne.[12]

By late February 1768, the King had consented to defray the costs of sending observers to the southern hemisphere and, at the same time, he ordered the Admiralty to provide a suitable ship to take them to their destination.[13] By the end of March, the Admiralty had agreed which ship to purchase. It was called the *Earl of Pembroke*. It had been built in Whitby a little more than three years earlier and was currently lying unused in the Thames to the east of the present location of Tower Bridge. Just over a week later, on 5 April, the Admiralty informed the Navy Board, who were responsible for the day-to-day administration of the Royal Navy, that the ship, now renamed HMS *Endeavour*, and at the relatively small size of 32 metres long and 9 metres wide, should be prepared and armed as necessary for 'conveying to the southward such persons as shall be thought proper for making observations on the passage of the planet Venus over the sun's disc'.[14]

For the time being, the ship had no commander and, more importantly, no specific destination in the southern hemisphere for observing the astronomical event. In the discussions leading to the drafting of the memorial, however, suggestions were made that one of the Marquesas Islands, which had been sighted by Álvaro Mendaña in 1595, or one of two islands in Tonga (then named Rotterdam and Amsterdam Islands and last seen by Abel Tasman in 1643), might be suitable, but no one was certain precisely where in the ocean these likely candidates were.[15]

While the issue of the ship's destination remained unresolved, that of the *Endeavour*'s commander was moving swiftly along. Sometime during the week after the order to get the ship ready for sea, that is before mid April 1768, the Admiralty found in James Cook the man that they wanted to appoint to command. Cook was not a young man and, as far as the Royal Navy was concerned, fairly inexperienced, but he certainly had talent.[16] He was born in Yorkshire and first went to sea when he was eighteen years old working for a Whitby company involved in carrying coal from the northeast of England to London. Once Cook's apprenticeship was over he sailed on ships throughout the North Sea, from Holland in the south to Norway in the north. He did well, and he was promoted, but then, and for no reason that has come down to us, he volunteered, in 1755, to join the Royal Navy in Wapping, East London. Two years later he became a master, qualified, therefore, to sail naval ships. By then, however, war had erupted between Britain and France and their respective allies, and Cook was sent on a naval squadron to North America where he participated in several battles in and around the St Lawrence River. When, in 1763, peace came to bring an end to what was then referred to as the Seven Years War, Cook, who had by now distinguished himself in surveying and cartography, in addition to navigation, was appointed to be the surveyor on a naval expedition to Newfoundland over which Britain had been given sovereignty under the terms of the peace treaty. There he remained, apart from short spells back in London, until the early part of November 1767, when he returned bearing a cache of elegantly produced maps and hydrographic surveys of the coasts of this geographically complicated island.[17]

Cook intended to return to Newfoundland in the spring of 1768, once he was satisfied that the engravers were competently handling his manuscript maps, but this didn't happen. His requisitions to the Admiralty to prepare his surveying ship for the next season coincided precisely with their search for someone to command HMS *Endeavour*. Cook never crossed the North Atlantic again. Instead, from now until his murder in Hawaii in February 1779, his life was bound up wholly with the Pacific.

It was now May 1768. The *Endeavour* was being prepared but where was it heading? Cook, and Charles Green, whom the Royal Society had already appointed as the expedition's astronomer, were the designated observers. They needed an island on which they could erect their observatory. Would Cook be able to find the Marquesas or Amsterdam and Rotterdam island? And if he could, would the ship be welcomed or attacked?

While these questions were being discussed in the Admiralty and the Royal Society, something wholly unexpected happened.[18] On 20 May 1768, Samuel Wallis, a naval commander, arrived in London with incredible news. In August 1766, Wallis had been given command of HMS *Dolphin* whose objective was to sail into the Pacific in search of *Terra Australis Incognita*, the substantial land mass that was supposed to exist in southern latitudes – Alexander Dalrymple, the noted hydrographer and cartographer, preferred to use the term 'Southern Continent'

and many followed his example.[19] Wallis reported that high land had
been seen in the distance during the voyage but what caught his and
everyone else's imagination was his discovery of an extraordinary island
and civilisation, which he named, in honour of his sovereign, 'King
George the Third's Island'.[20] Wallis was an excellent navigator and
equipped with the latest instruments to calculate that most elusive of
navigational parameters – longitude. He reported that this island,
which had abundant food and water, a healthy climate, a good
anchorage and welcoming people, and which we now know as Tahiti,
lay at 17 degrees 30 minutes latitude and 150 degrees longitude, west
of London, precisely within Maskelyne's oceanic rectangle.

Wallis knew nothing about the Royal Society's interest in tracking
Venus and the Admiralty had not expected him to arrive back in
London for at least a year, that is sometime in 1769. As it happened,
because of widespread illness among his crew, his own weakness and
serious doubts that his ship could stand any more wear and tear, Wallis
had decided to abandon a part of his surveying objectives and hurry
home by way of the Cape of Good Hope (in spite of his instructions
to return by way of Cape Horn).[21] History would have been very
different had he carried out his instructions to the letter.

Wallis's discovery of the island and of an excellent anchorage in the
very north of the island, at a place he named Port Royal, or Matavai
Bay in Tahitian, where he had anchored on 23 June 1767, could not
have been better news for the Royal Society. The vague destination of
the Marquesas and Tonga was now replaced by a firm, precise and,
therefore, perfectly findable location. The predicted date of the transit
was almost the same as the date of Wallis's anchorage so that what he
described then, especially the weather, would equally apply to the
Endeavour's stay. On 9 June 1768, a fortnight after Cook had officially
taken charge of the *Endeavour*, the Council of the Royal Society
endorsed the choice of the island discovered by Wallis as the expedition's destination.[22] In the following month, the Admiralty reaffirmed
the Society's decision of where to observe the track of Venus when
they presented their instructions to James Cook, who had, in the
meantime, been promoted to the rank of lieutenant.[23] To guide him
to Tahiti, the Admiralty presented Cook with copies of 'such Surveys,
plans and Views of the Island and Harbour as were taken by Capt
Wallis, and the Officers of the Dolphin when she was there'.[24]

The Royal Society Council meeting minute of 9 June 1768 recorded
the important decisions that had been taken since the 'Memorial' of

mid February: the observers, Cook and Green, had been chosen and their salaries agreed; the ship and its commander had been commissioned; and the location pinpointed in Maskelyne's rectangle of southern sea.

At this point, the scientific aspects of Cook's expedition to the Pacific were astronomical and geographic. The minute of the Royal Society's Council meeting, which recorded Cook and Green's appointment, also had a small note to the effect that the Society's secretary would be asking the Admiralty that 'Joseph Banks . . . being desirous of undertaking the same voyage . . . for the Advancement of useful knowledge . . . He . . . together with his Suite . . . be received on board of the Ship, under the Command of Captain Cook.'[25]

Banks attended his first meeting at the Royal Society on 12 February 1767 shortly after his return from Newfoundland and Labrador.[26] Though he was not in London when, in November 1767, the Committee of the Transit recorded its decisions about how the Society wished to have Venus's track observed, it is very possible that he knew about it shortly afterwards, and certainly by the time of the 'Memorial' to the King on 15 February 1768, Banks had made up his mind to try and join the expedition.[27] Over the next few months, by dint of careful negotiations and relationships, especially with Philip Stephens, the First Secretary of the Admiralty, whom he had met at the British Museum, Banks convinced those in authority that he should go to the Pacific.[28] The Royal Society Council minute of 7 June 1768, requested the Admiralty to accept Banks, accompanied by seven others, including two artists (Sydney Parkinson and Alexander Buchan), a secretary (Herman Spöring) and four assistants and servants (James Roberts, Peter Briscoe, Thomas Richmond and George Dorlton), all paid for by him, to join the ship.[29]

More than a month later, on 22 July, the Admiralty informed Cook that the Royal Society's request had been accepted. Instead of seven in Banks's accompanying suite, they now stipulated that eight, in addition to Banks, would be going.[30] The eighth person was Daniel Solander, probably the most important person in Banks's intellectual life since Israel Lyons.

Solander was Linnaeus's best and most favourite student, and had been invited to England from Sweden, especially by the botanist John Ellis, to expound his teacher's new system of classification. Since 1763 he had been busily working on cataloguing the Museum's natural-history collections, primarily those that Hans Sloane had bequeathed. In the following year he was made a Fellow of the Royal Society.[31]

Solander, ten years Banks's senior, probably met Banks when he first used the British Museum's Reading Room, and soon after this meeting he took over Banks's botanical education where Lyons had left off. He had prepared Banks for his Newfoundland voyage and, on his return, helped him catalogue the plants that had been collected.[32] It is not surprising then that Banks confided in Solander that he was planning to join the *Endeavour*. Solander was 'very excited by my plans, and immediately offered to furnish me with information on every part of natural history which might be encountered on such an ambitious and unparalleled mission'. Banks later explained that several days later, when they were dining at the home of a mutual friend, the topic of the *Endeavour* came up. Solander jumped to his feet and asked Banks if he wanted a companion to join him. Banks replied, 'Someone like you would be a constant benefit and pleasure to me!' Solander did not hesitate. 'I want to go with you,' he exclaimed.[33]

On 24 June 1768 Solander wrote to the Trustees of the British Museum to tell them about Banks's offer, and that the Archbishop of Canterbury, who had the power to grant leaves of absence, had agreed he should go. Solander added that this unique opportunity would allow him to collect for the museum.[34] Banks may have been well known in the Royal Society, especially its exclusive dining club, which he frequented increasingly after the 'Memorial' had been sent to the King, and in the British Museum's Reading Room, but in the world of botany, it was Solander who was the more famous. He was a great addition to the voyage.

This was now quite a different expedition from what had been planned by the Royal Society when they petitioned the King for financial help. It wasn't just advances in astronomy and geography that they hoped would gain from the expedition. Now natural history, and botany in particular, had a leading role. There were also two Fellows of the Royal Society on board.

John Ellis, who had known Banks since 1764, wrote to Linnaeus excitedly, telling him about the forthcoming voyage.[35] Ellis's main news for Linnaeus was that his student, Daniel Solander, was accompanying Joseph Banks, whom he described as a 'very wealthy man', to the Pacific. Ellis added that they were very well-equipped, with a fine library and all of the tools necessary to collect and preserve natural history specimens; or, in Ellis's own words: 'No people ever went to Sea better fitted out for the purpose of Natural History, nor more elegantly.' What Ellis did not mention was the huge quantity of cases and book shelving that Banks was taking on board – 'such a Collection . . . as almost frighten me', Banks remarked.[36]

Banks and his suite were given rooms next to Cook's. The 'scientific gentlemen' would be sharing his great cabin: specimens in bottles and in presses, nets and hooks, and sheets of drawing paper were jammed up next to maps and mathematical instruments.[37] Deferentially, Ellis concluded his letter to Linnaeus by saying that 'All this is owing to you and your writings'.

On 30 July 1768, Cook received his instructions. He was to take the ship to Port Royal Harbour by way of Cape Horn. On the way, the Lords of the Admiralty remarked, 'You are at Liberty to touch upon the Coast of Brazil, or at Port Egmont in Falkland Isles, or at both in your way thither.' The first stop though was Madeira, where Cook was ordered to 'take on board such a Quantity of Wine as you can Conveniently stow for the use of the [Ship's] company'.[38]

So, on 25 August, the *Endeavour*, with almost one hundred men on board, ten of whom had already been to the Pacific on the two previous voyages of HMS *Dolphin*, left Plymouth for the Pacific Ocean.

Following his instructions, Cook took the *Endeavour* to Madeira where he stocked up with 14,000 litres of wine. Banks and Solander had been collecting specimens from the sea as the *Endeavour* made its way south, but Madeira now gave them the first opportunity to try out their methods for collecting on land and for recording and drawing botanical specimens, in the ship's great cabin.[39] With the generous assistance of the English Consul and the resident English physician (himself a naturalist) and despite it not being the best time of the year for botanising, by the end of their five days' stay, over three hundred species of plants had been collected – Solander reported to Linnaeus that of these fifty or sixty were new species.[40]

On 18 September, Cook set sail for Rio de Janeiro on the other side of the Atlantic. The stay in the city, from 14 November until 7 December, was generally a frustrating time for Banks and his entourage. Their welcome from the authorities was frosty, and they were not given permission to land. It was a bitter disappointment, especially when compared to their warm reception in Madeira. Surreptitiously evading the restrictions, Banks and Solander managed a few precious hours on shore and, in the end, either by their own means or by bribing locals to bring plants to the ship, they managed to collect about three hundred specimens: Parkinson drew about 10 per cent of them. The ship's company hurriedly wrote letters home as they did not know when they would get another chance to send them. Soon they would be entering a part of the Pacific where there would be no passing European ships to which they could entrust their letters. They did not even know at this stage by what route they would be returning home, or when.

For about five weeks, the *Endeavour* made its way south through the Atlantic until 14 January 1769, when the ship anchored in a sheltered bay near the tip of Tierra del Fuego. Solander and Banks rushed to collect as much as they could. Banks was anxious to go into the interior. The local people seemed friendly and the naturalists' activities were not made difficult as in Rio de Janeiro. But it was here that the first tragedy of the voyage struck.

When they were climbing a part of the interior that resembled the Alps, the weather suddenly turned cold, with snow and icy winds. They were too far from the ship to make it back before nightfall and two of Banks's black servants, George Dorlton and Thomas Richmond, having drunk too much, literally froze to death.

Banks continued to collect but stayed closer to the ship. On 21 January, the *Endeavour* left its anchorage and headed for Cape Horn where the ship left the Atlantic and entered the Pacific Ocean. The botanic haul was small but with about a hundred specimens, it was respectable nevertheless.

For the next two months and more, Cook took a northwest course, making straight for Tahiti and for the *Endeavour*'s planned anchorage in Matavai Bay, which they reached on 13 April, well in time for the rendezvous with Venus's track across the sun, and strictly within the instructions laid down by the Admiralty.

How much anyone on the ship knew about Tahiti is unclear. Not long after Wallis's arrival in London from the Pacific, some London newspapers carried reports of the discovery of a 'large, fertile, and extremely populous' island. Descriptions of the people were included such as the following: 'The first day they came along-side with a number of canoes . . . there were too [sic] divisions, one filled with men, and the other with women; these last endeavoured to engage the attention of our sailors, by exposing their beauties to their view.'[41]

The *Endeavour*'s men, including Banks, were mostly young and hungry for experiences, and they were very impressed by the beauties on view. They soon discovered how different Tahitian society was from what they were accustomed to at home. Banks spent as much time learning Tahitian ways, particularly their uninhibited sexual practices – what he called 'enjoying free liberty in love' – as he did botanising.[42]

However, shortly after their arrival, Banks suffered yet another tragedy. Alexander Buchan, the landscape artist, died suddenly on 17 April. Banks was devastated in more ways than one as he explained: 'I sincerely regret him as an ingenious and good young man, but his Loss to me is irretrievable, my airy dreams of entertaining my friends in England with the scenes I am about to see here are vanished. No account of the figures and dresses of men can be satisfactory unless illustrated with figures: had providence spard him a month longer what an advantage would it have been to my undertaking but I must submit.'[43]

The transit observations were made as planned. Banks continued to explore the island accompanied, at various times, by Cook, by John Gore, the third lieutenant, and by William Monkhouse, the ship's surgeon, who had been with Banks on HMS *Niger* in Newfoundland and Labrador. On 4 July Banks did something he had never done before but which would become part of his botanical practices: in and around the encampment of what was called Point Venus, Banks planted seeds of watermelons, oranges, lemons, limes and other varieties he had brought with him from Rio de Janeiro and distributed large quantities of the same to the local people.[44]

Banks and Solander crisscrossed the island but on many days they collected little if anything. The botanical haul, at just over three hundred plants, was on the small side and about the same as they had collected in Madeira in a much shorter period of time and at a less opportune time of the year. On the other hand, Banks was particularly impressed by the Tahitian agricultural accomplishments, especially their cultivation of the breadfruit tree, which provided the population with its main source of nourishment.

On 13 July 1769, three months after arriving, Cook and the ship's company bade farewell to Tahiti. Cook had carried out all but one of his instructions and now he turned to this final one. 'When this Service is perform'd', the Admiralty had written, 'you are to put to Sea without Loss of Time, and carry into execution the Additional Instructions contained in the inclosed Sealed Packet.'[45]

These additional instructions told Cook that he was now to look for *Terra Australis Incognita*, the southern land mass that Wallis and some of his men thought they had seen in the distance when they were in the area.[46] The Admiralty told Cook that he should first look for land by sailing south to latitude 40 degrees; if nothing was found, then he should turn westward and search again in between latitudes 40 and 35 degrees until he met the eastern side of New Zealand.

Cook did what he was told and found no land mass in the great ocean, until 6 October 1769, after being at sea for almost three months, land was spotted at last. Was this the edge of the sought-after 'Southern Continent'? Cook decided that the only way to know for certain was to follow the coast and see where it went. He did just that. For almost six months, the *Endeavour* sailed in and around the coast until, at the end of March, Cook confidently concluded that New Zealand was, in fact, made up of two major islands and, therefore, unrelated to *Terra Australis Incognita*. There was certainly no southern land mass in this part of the ocean and with that recognition, as Banks put it, came 'the total demolition of our aerial fabric called continent'.

Banks and Solander botanised whenever they could though the circumstances were not as pleasant as they had been on Tahiti. The Maori people, who had not seen any Europeans since Abel Tasman's minimal contact with them on South Island in 1642, were variously curious, friendly and outright hostile and warlike. But the plant collection, consisting entirely of plants new to European science, was significant and outnumbered those from the places that the ship had already visited.

Together with his officers, Cook now decided to sail back to England by way of the East Indies and the Cape of Good Hope, because it was too risky at this time of year to go back by Cape Horn.[47] This meant that Cook was hoping to meet the land which Abel Tasman had discovered in 1642, to which he gave the name Van Diemen's Land, and to follow its coast northward until reaching its northern extremity.[48] Van Diemen's Land was shown on one of the maps Cook had with him, which had been drawn by Alexander Dalrymple, and which he had initially given to Banks.[49] So, on 1 April 1770, the *Endeavour* sailed westward towards the eastern coast of New Holland.

A little over a fortnight later, at a place Cook called Point Hicks (near the present border of Victoria and New South Wales), land was spotted extending to the northeast and to the west but nothing was seen to the south where Van Diemen's Land was supposed to be. Cook continued on the course he had decided upon when he was about to leave New Zealand. He turned the *Endeavour* to face north and began sailing along the coast of New Holland.

On 27 April, Cook, Solander and Banks, with four rowers, attempted to land but the surf beat them back. The next day, 28 April, Cook took the ship a little further up the coast, where there appeared to be an opening like a harbour, in which he anchored the ship. It had initially been called Stingray Bay but a little over a week later, Cook, in recognition of Banks and Solander, renamed it Botany Bay.

The first encounter with the local people was tense. They carried spears and the British fired a few shots into the air. Fortunately, no one was hurt. After that, the locals were wary of contact. Banks and Solander saw no one when they were ashore collecting although they suspected there were people about. After a few days they had collected so much Banks was afraid their haul would spoil before they had time to dry and press it in their collection books.

Banks and Solander already had enough natural history specimens, not just plants, but birds and insects as well, to keep them busy for a very long time. Cook had no reason to remain longer than it took to replenish the ship's water supplies. On 6 May 1770, the *Endeavour* left its anchorage and began its voyage northward in the Tasman Sea.

Cook sailed close by the coast, close enough for Banks to see the land with the help of his telescope, and even make out the kind of trees that grew and the birds that populated the shore. Though he saw fires, he didn't see any people. On 23 May, Banks got his first opportunity since leaving Botany Bay to collect. While he was away, those still on board spotted nearly twenty local people gathering on the beach though they soon retreated into the surrounding forest.

For the next few weeks the ship made its way up the coast, stopping infrequently and giving little time for Banks and Solander to collect much. Nothing remarkable was noted but then, 'scarce were we warm in our beds when we were calld up with the alarming news of the ship being fast ashore upon a rock'.[50] The ship, now inside the Great Barrier Reef, to the northeast of a point of land Cook named Cape Tribulation, had 'struck and stuck fast', and was being cut into by coral.[51]

The pumps were worked to their limits and everything was done, including throwing overboard much of the ballast and all the guns on the deck, to float the ship. Banks confessed that he was on the point of packing everything up he could save and 'prepared myself for the worst'.[52] By a combination of luck and skill, the ship, still leaking, was made to sail and Cook, carefully avoiding shoals and shallow water, began looking for somewhere on the shore where he could repair it. On 17 June, Cook, having spotted a likely place, was finally able to moor the ship in the mouth of a river.

It took seven weeks for the repairs to be made in the inlet of what Cook came to call Endeavour River (present-day Cooktown, Queensland). Banks and Solander did very well, even better than at Botany Bay: altogether they described almost 1000 species.[53] Cook had his mind on other matters. The ship was almost repaired and it was time to leave. On 4 August, Cook moved the ship from its mooring. For almost three weeks, he gingerly steered it northward, mostly between the coast and the Great Barrier Reef, avoiding shoals and visible coral formations until, on 20 August, the *Endeavour* reached the northernmost point of land, which Cook named York Cape (now Cape York).

Cook was close to waters that had been well charted by earlier European explorers. One important question remained unresolved, however: was the northern part of New Holland, where the ship now stood, attached to the southern coast of New Guinea as shown on several contemporary maps; or were these two land masses separated by a channel or a strait, supposedly discovered by the Spanish explorer Luis Vaez de Torres in 1606, the track of which was shown on Dalrymple's map and which Cook believed existed?[54] As he sailed around the Cape, clinging to the coast, there was always water to starboard: the only land he saw was the coast on his port side. Cook concluded that he was in a strait, to which he gave the name Endeavour and which formed a part of the track Torres had taken more than one hundred and fifty years earlier. As he kept sailing in a westerly direction, the strait widened and led directly into the Arafura Sea and eventually to the heavily populated island of Java.

Batavia (present-day Jakarta), where the *Endeavour* anchored on 9 October 1770, had been the centre of the Dutch East India Company's Asian trade network since the early seventeenth century, its harbour frequently teeming with European ships on their way to and from the East Indies.[55] Here, Cook could ensure that the ship would be expertly repaired in order for it to make it back safely to England. He could send his first despatches and a copy of his journal to the Admiralty, and the ship's company could, for the first time since they were in Rio de Janeiro almost two years earlier, write precious letters home with some certainty that they would get to their destinations – it was from letters written here that Londoners, reading reports in the news-papers, first learned of the *Endeavour*'s safe arrival in Batavia.[56] Unfortunately, for the ship's company, they were now exposed to a range of tropical diseases against which they had no protection. Many became ill, including Banks and Solander. The surgeon, William Monkhouse, was one of the first to die, followed quickly by his mate; then Charles Green's servant and three more men.

On 25 December 1770, the *Endeavour* was ready to resume its voyage home. 'There was not I believe a man in the ship but gave his utmost aid to getting up the Anchor, so completely tird was every one of the unwholesome air of this place', wrote Banks.[57]

The worst fatalities, however, happened when the ship was back at sea heading for the Cape of Good Hope. On 24 January 1771 Herman Spöring died, who had acted as Banks's secretary and also produced some fine drawings; two days later, Sydney Parkinson died and two days after that it was Charles Green's turn. Banks's accompanying suite, which had already been reduced by the earlier deaths of Richmond, Dorlton and Buchan, was reduced to three. Solander, Briscoe and Roberts were all that remained of the original eight. There were also deaths among the ship's company and these continued as the ship made its way through the Atlantic.

On 14 March the ship anchored in the harbour of Cape Town. A month later they were on their way again and after a short stay at the British East India Company's island of St Helena, Cook set a course for the English coast which he hoped to reach without stopping en route. The survivors were desperate to get home.

At three o'clock on 12 July 1771, a little short of three years on its circumnavigation, the *Endeavour*, the first British scientific voyage of its kind, landed on the coast of southern England at Deal.

Cook's achievements were many. He was both a skilled navigator and a superb surveyor and cartographer. He not only discovered that New Zealand was formed of two islands and that the east coast of New Holland, from Point Hicks in the south to Cape York in the north, was continuous, but he surveyed the coasts and produced the first charts of both places. During the voyage of the *Endeavour*, besides producing these entirely new charts, he improved upon and corrected those already existing of Tahiti and the area around Cape York.[58]

But it was not Cook who was fêted on the *Endeavour*'s return.

Banks, cutting a more dashing figure, and Solander, depicted as fatherly and studious, were immediately taken into the nation's heart as heroes. They were the talk of the town and their company was much sought after.

The most eminent person eager to meet Banks was King George III. The meeting happened on Friday, 2 August 1771, at St James's. Francis Seymour Conway, Lord Beauchamp, who knew Banks from Eton and Oxford, and whose father was the Lord Chamberlain, performed the introduction (Banks and Beauchamp would meet again many years later under very different circumstances).[59] The newspaper articles described nothing of what happened that day between the King and Banks apart from commenting that '[Banks] was received very graciously.'

The son of Frederick, the Prince of Wales, who had died in 1751, George ascended the throne in October 1760, on the death of his grandfather George II. George III was five years older than Banks and he had taken a keen interest in the voyage of HMS *Endeavour*, supporting it ardently and committing £4000 of his own money to it.[60] So he had a stake in knowing what had been collected. London's botanic community was certainly aware that the King was anxious to see Banks and Solander, even before they arrived.[61] Less than a fortnight after their first meeting, the King requested that both Banks and Solander, accompanied by Sir John Pringle – who at the time was Queen Charlotte's personal physician, a leading member of the Council of the Royal Society and a friend of Banks and Solander's – should meet him at his summer home in Richmond on Saturday for 'a private conference . . . on the discoveries they made on their last voyage'.[62] As a member of the Council of the Royal Society, Pringle was very interested in the voyage of the *Endeavour* and would have been involved in aspects of the planning for the observation of the transit of Venus.[63] After the ship returned, Pringle met Banks and Solander, both separately and together, sometimes at his home and other times at Banks's home in New Burlington Street, and on many other occasions, finding out about those aspects of the voyage that interested him most.[64]

At their meeting with the King, Solander and Banks no doubt brought examples of the plants they had collected.[65] Newspaper articles at the time report that Solander had already been to the royal gardens at Kew and had planted some samples from the voyage – 'they have been set in the Royal Gardens . . . and thrive as well as in their natural soil', commented one article.[66] For the rest of that month, Banks and

Solander made frequent visits to Richmond during which time the King also examined their collection of plant drawings.[67]

When, in the year following the triumphant return of the *Endeavour*, the government decided on a second voyage to the Pacific, Banks began planning it as if it were his own. He convinced the First Lord of the Admiralty to let him radically alter the structure of HMS *Resolution*, the main ship, to accommodate him and his substantial entourage and equipment. He had gone too far. The ship was deemed unseaworthy on its first trial. Cook agreed, so did the Admiralty and the Navy Board, and the ship was ordered to be returned to its original state. Banks was devastated and angry at this turn of events and removed himself, his entourage and equipment from the ship and instead chartered a vessel, the *Sir Lawrence*, for his own scientific expedition to Iceland by way of the Hebrides.[68]

The *Resolution* debacle was undoubtedly a great disappointment to Banks but from it he learned a valuable lesson. He no longer tried to impose his will on others but sought instead to influence and persuade. In time he managed to restore his friendly relations with the Admiralty and Cook.

After the expedition to Iceland, and apart from a brief trip to Holland in 1773, Banks never went to sea again, but by then he had already spent four and a half years on ships sailing over much of the globe and collecting natural-history specimens. In this respect, Banks's experiences set him apart from most naturalists of the time, but there was more to it than that, for during the time he was at sea he learned about how ships worked; about shipboard spaces, and how they might be altered for global botanical projects; and about how naval careers advanced and how much commanders mattered. Though Banks never went to sea again after he was twenty-nine, ships and the sea shaped the rest of his adult life.

Instead of travelling Banks established himself at home in England. In the summer of 1777, he moved from his accommodations in New Burlington Street to a grand house in Soho Square where his sister Sarah Sophia joined him. In the following year, at the age of thirty-five, he was elected President of the Royal Society, a post he occupied until his death in 1820. In 1779, Banks married Dorothea Hugessen and she joined the Soho Square household. A pattern of life was laid down: most of the year was spent in London, with a couple of months in the autumn at his country estates.[69]

Soho Square was much more than just a family home. It housed

Banks's personal library of books (in many languages and exceeding 20,000 titles at his death) and an unknown quantity of pamphlets and drawings;[70] as well as a vast herbarium, and zoological and mineral collections.[71] From 1773, Solander, Linnaeus's disciple, was always near Banks, helping him with his collections, especially those gathered in the Pacific and on the Iceland expedition.[72] Soho Square's international scholarly resources – including Solander: in himself, a major attraction – were made freely available to interested visitors from all parts of the world. Eventually, a five-volume catalogue of the library's holdings was published and made public, under the guidance of Jonas Dryander, a Linnaean-trained Swede like Solander, who became Banks's librarian following the latter's death in 1782.[73]

In the study, close by the library and herbarium, were volumes of letters, both incoming and copies of those going out. It is estimated that at his death these volumes contained 100,000 letters and represented a global correspondence network of several thousand people from all walks of life, many of whom became life-long friends.[74]

Banks clearly lived a busy and satisfying life in England and yet, as he confided to his friend Sir William Hamilton, the British Ambassador to the Neapolitan court, he sorely missed the excitement of the scientific adventures of his youth.[75]

Although Banks and Solander were initially presented to the King at his summer home in Richmond, he soon moved to Kew, in a property that then stood in the gardens. The story goes back to the early 1730s when Frederick, the Prince of Wales, George's father, leased a house opposite to what is now called Kew Palace.[76] This house, which soon came to be called the White House, was designed to be a royal residence and act as a family retreat during the summer months. Not long after signing the lease, the beginnings of a garden were laid out and plants were brought into cultivation. Over the following years the gardens were expanded as more land was bought.

On 20 March 1751, Frederick died unexpectedly aged forty-four and the house and gardens passed to his wife, Augusta, the Dowager Princess of Wales. With her close botanical advisers, especially John Stuart, the 3rd Earl of Bute, who had his own magnificent garden at Luton Hoo, Augusta, who spent part of each year at the White House, was able to expand Kew gardens. Many exotic plants were donated by Bute himself and leading London botanists, many of whom he knew well – other plants were purchased from London nurserymen.[77]

A key moment in the history of the gardens for the next few decades was the appointment in 1759 of William Aiton to be in charge of the physic garden.[78] Aiton was born in 1733 in Lanarkshire, Scotland and came to London in 1754, where he found work at the Chelsea Physic Garden, then under the direction of Philip Miller. Between Aiton and Bute, Kew's stock of plants became large and diverse. By the end of the 1760s, with plants from many parts of the world, notably North America, Kew was, according to many contemporary observers, Britain's best-stocked garden and rivalled similar royal gardens in Europe, especially the Jardin du Roi in Paris and the Schönbrunn in Vienna.[79]

Then another tragedy struck the royal family. Augusta died in February 1772. George III, who had been spending the summer months at Richmond Lodge, now removed his family to Augusta's White House, which Queen Charlotte, his wife, began to redesign for their occupancy.[80] Bute disappeared from the scene and the royal gardens at Kew now came under the King's direct control.

Banks's relationship with Kew certainly went back to 1764, for it was then that he met Aiton, possibly through Daniel Solander or James Lee at the Vineyard Nursery, Hammersmith.[81]

For the next few years Banks had little more to do with Kew than to visit and observe.

By the end of 1776, however, Banks's relationship with Kew had changed significantly.[82] Banks, himself, had trouble defining his new role: all he could say, as he tried to describe it in a letter to the Spanish Ambassador in London in 1796, was that for many years he 'exercis[ed] a kind of superintendence over His Royal botanic gardens'.[83]

The 'superintendence', as he called it, might have been the single most important part of Banks's exceedingly busy life. It was always on his mind and, whenever the opportunity presented itself for a naturalist or gardener to accompany a voyage, Banks tried to ensure that Kew's needs were not forgotten.

This was the beginning of a very long relationship between Banks and the King. It would last for almost forty years and ended only in 1810 when George's debilitating illness made contact impossible. Banks would meet the King whenever possible on a Saturday, usually at the royal gardens at Kew, and they would spend several hours walking, talking about plants and other topics of mutual interest, particularly about the development of the gardens at Kew.[84]

The stakes were very high. Plants mattered. The greater the splendour, the finer and rarer the visual and sensual experience they offered, the better. For the first few decades of Kew's existence, its stock of plants had been the result of donations and exchanges with similar gardens. Many new varieties from all over the world found themselves at Kew by this route but there had not yet been any attempt at a systematic collection in the wild. By the time of the change in Banks's relationship with the royal garden, however, this had altered dramatically. Kew's first plant collector was already abroad, and over the following thirty years, he and Banks became very close.

PART I

To Every Corner of the Earth

Preface

Between 1777 and 1779 Joseph Banks's life changed in several important ways that set a pattern for the rest of his days. In 1777, he moved
with his sister, Sarah Sophia, into his permanent home in Soho Square.
It was a large house with enough space to accommodate his domestic
life and his professional interests. His great library focused principally
on natural history and its associated texts, manuscripts, drawings and
specimens. In 1778, Banks was elected President of the Royal Society
after having been a Fellow for just over a decade; he was only thirty-five
years old. In 1779, he married Dorothea Huggesen, who moved into
Soho Square with him and Sarah Sophia. In the same year, Banks
leased and subsequently bought Spring Grove, a property with extensive grounds in Heston, Middlesex. Over time Banks had gardens laid
out, and greenhouses and hothouses built. Produce grown there was
sent to Soho Square and this was where Banks carried out a number
of important horticultural experiments. Banks also began what became
an annual pilgrimage to manage his Lincolnshire estates, centred on
his country home at Revesby Abbey, where he, his wife and sister
would spend every September and October.

In contrast to the English focus of his domestic life in London and
Lincolnshire, Banks was being drawn in other, more outward-looking
and global directions. In 1776, Francis Masson, Kew's first official
collector since 1772, became Banks's direct responsibility. Masson had
already been to the Cape of Good Hope collecting for the royal gardens
at Kew, and he was on his way to Madeira to continue his assignment
when Banks took over his direction. During the next three decades,
while Masson travelled through the Atlantic region, Banks handled
all his preparations, telling him where and what to collect, arranging
his finances and managing the receipt of his specimens for Kew.

Banks, already a wealthy man, now settled down as a county notable
and the President of the Royal Society. His contact with Masson offered

Banks something entirely different, something unpredictable that must have reminded him of his experiences on the *Endeavour*. Banks could share vicariously in the excitement of finding new plants to send to Kew, of making Kew a place where plants from all over the globe could thrive, far from their native habitats. It gave Banks, as he said himself, the greatest of pleasures, to harness the intellectual resources of Soho Square, its library and herbarium, to the practical horticultural experience and knowledge of Kew, all for the benefit of the King and his garden. Banks would continue pushing these projects into new geographic regions, whenever the opportunities arose.

These early collectors, including Masson, were mostly Scots – they were generally better trained and more knowledgeable about botany than their English counterparts. Most of these men sought out Banks rather than the other way round. They expanded their own and Banks's geographical horizon, collecting plants in parts of the world – the Pacific Northwest, China, southwest Africa and the Coromandel Coast – whose botany was hardly known in Europe.

1

1772: Masson Roams the Atlantic

Joseph Banks did not choose his first collector himself. Francis Masson had been appointed as Kew Garden's first plant collector by Sir John Pringle. Sir John had been a close friend of the royal family even before 1764 when he was made Physician in Ordinary to Queen Charlotte. Liked and trusted by the King, he had replaced the Earl of Bute as adviser to the royal garden at Kew. Though, as he admitted, 'I myself am so little a Botanist', he was very well connected in cosmopolitan scientific circles, and would have acted as the King's agent in selecting Masson, no doubt taking the advice of the head gardener at Kew, William Aiton.[1]

Masson had been working under Aiton, as a gardener at Kew, and had made a good impression. He was a fellow Scot, born in Aberdeen in 1741, but little is known about his life before Kew.[2] Although Banks took no credit for selecting Masson, saying Pringle did it all, he does seem to have had a hand in deciding where he was sent.[3] According to Masson, writing in 1796, it was Banks who 'suggested to his Majesty the idea of sending a person, professionally a gardener, to the Cape'.[4]

Aside from recommending the destination, it's unlikely Banks had anything to do with any instructions for Masson. Banks was busy planning a second voyage to the Pacific with Cook. This, and the intense pressure of classifying the huge botanical collection from the first voyage, and preparing the botanical drawings made on the *Endeavour*, took up most of his time.[5]

The choice of the Cape as the destination for Kew's first plant collector, may not seem obvious; it was under Dutch rule for one thing. However, other circumstances did recommend it. Banks and Solander had spent some time there, from 14 March to 16 April in 1771, when the *Endeavour* made its last substantial stop before returning to England.[6] The plant collecting had not been as productive as they had expected, because, for almost half their stay, Solander had been confined to bed suffering from a fever. Referring to what possible botanical treasures might be found beyond the port, Banks commented 'I can say but little . . . not having had an opportunity of making even one excursion owing in great measure to Dr Solanders illness.'[7] Even so, in the vicinity of the ship's anchorage, they managed to collect more than three hundred varieties of plants, including a gardenia, an acacia and a heather.[8]

Observing the plants being cultivated by Dutch farmers in the fields around, and in the Dutch East India Company's botanic garden, Banks concluded that though the climate was milder than that of England, the food crops, at least, were pretty much the same. This would have led him to conclude that the Cape area might be ideal for collecting plants that would be easy to grow at Kew, unlike the tropical plants that needed a protective habitat and artificial heat.

This observation would have been confirmed by the fact that Kew was already growing plants from the Cape, many of which had been introduced to the garden in the 1730s by Philip Miller, head gardener of the Chelsea Physic Garden.[9] Not only had there been these living plants for Masson to see but Hans Sloane's herbarium was then at the British Museum, which contained an impressive collection of Cape plants that had come into Sloane's possession from other collections and collectors.[10] Also, since the early years of the seventeenth century, Cape plants figured in specialised texts, such as the famous *Hortus Cliffortianus*, compiled by Linnaeus, and many of these publications were at the British Museum or in Banks's home.[11]

These factors alone recommended the Cape as a collecting destination but also important was the fact that maritime contact between it and Europe was excellent. Table Bay, the Cape's harbour, was always full of foreign ships, primarily from the Dutch, Swedish and English East India Companies either heading into or returning from the Indian Ocean.[12] When the *Endeavour* arrived in Table Bay on 14 March 1770, Cook noted that there were already sixteen ships at anchor; over the following month he reported that four British East India Company and seven Dutch East India Company ships left for Europe.[13] With so many ships bound for Europe and with a sailing time of less than three months, living plants would have their best chance of survival at sea if shipped from the Cape.

What may have scaled the decision was that when Banks and Solander were at the Cape, they had met a Swedish soldier, Franz Pehr Oldenburg, who was working for the Dutch East India Company at the Cape and who was very interested in natural history, having already amassed a personal herbarium.[14] Before Banks and Solander left the Cape for England, they had made an agreement with Oldenburg to collect specimens for them after their departure. The specimens arrived in London sometime in 1772.[15] If Oldenburg was still at the Cape when Masson arrived, the benefits would be substantial, since, in addition to his local botanical knowledge, he spoke Dutch, which Masson did not.

It was decided that Masson would travel to the Cape by the beginning of April 1772 at the latest.[16] On 5 May Cook received an order from John Montagu, 4th Earl of Sandwich and First Lord of the Admiralty, informing him that Francis Masson would be joining the ship for a passage to the Cape of Good Hope.[17] He would thus be sailing with Banks for part of the way, but five days after the order from Sandwich, the *Resolution* was given its first trial for its intended Pacific voyage. At the mouth of the Thames, Robert Cooper, the ship's first lieutenant, declared to Cook that he would not risk the ship at sea in its present top-heavy state. The *Resolution* was ordered back to Sheerness, and the superstructure, which Banks had designed and which the Admiralty had built for him, was removed.[18] Learning of this Banks withdrew from the voyage and took his personal, and rather large, entourage with him.

In its new slimmed down version, the *Resolution*, in company with HMS *Adventure*, left Plymouth on 13 July 1772. Masson had new scientific companions – Johann Forster and his son Georg Forster, the two naturalists who were hurriedly assigned to the ship following Banks's departure. After two short stops, the first at Madeira and the second in the Cape Verde Islands, the ships anchored in Table Bay. Masson stepped into a bustling town at the foot of Table Mountain, a Dutch settlement with a population of around 5000 people, about half of whom were white: the black population was mostly composed of slaves owned by the Dutch East India Company and imported from East Africa and the Indian Ocean region.[19]

What instructions he carried with him we don't know but within less than two months, on 10 December 1772, Masson headed to the interior, in an easterly direction, accompanied by Franz Pehr Oldenburg, who was still at the Cape, and an unnamed Khoikhoi, who was in charge of the wagon driven by eight oxen.[20]

Masson's report of his expedition, written to Sir John Pringle, was read to the Fellows of the Royal Society in London. It was short and, apart from providing some information about where the party went, it contained few details. What we do know is that they got as far as the town of Swellendam in the Western Cape, about one hundred and fifty miles to the east of the Cape. (When Cook was at the Cape he reported that the Dutch had settled an area which, at its greatest extent, exceeded 900 miles and 28 days' journey time.)[21] On 20 January 1773, after a two-day stay in Swellendam, they set off to return to the Cape by the same route they had taken out. Masson mentioned that he had collected and sent seeds of several species of heather back to Kew, which had germinated and been successfully grown there while he was still at the Cape.[22]

On his return to Cape Town, probably in February 1773, Masson was in for an unexpected treat: a fellow plant collector and a trained botanist had arrived in town.[23] Carl Peter Thunberg, who was born in the same year as Banks but in southern Sweden, had studied under Linnaeus at Uppsala University and had taken a medical degree. Like other disciples of Linnaeus, he left Sweden for foreign shores. Thunberg arrived in Amsterdam in August 1770 where he worked on plant collections from other parts of the Dutch Empire. In December 1771, he began his long journey to the East via the Cape where he arrived on 16 April 1772 in order to collect plants and learn Dutch. After acquainting himself with Cape Town and its local botany, including day trips into the surrounding countryside, on 7 September, Thunberg set out with three European companions, two Khoikhois and an oxen-driven cart for the interior: first to the north and then eventually in the direction of Swellendam and beyond. They were back in Cape Town on 2 January 1773.

At some point between February and September 1773, Masson and Thunberg met and decided to travel together on a botanical expedition. On 11 September they set out from Cape Town and headed in a northerly direction for about 100 miles from their point of departure. Masson could see that the decision to go to the Cape to collect was the right one as he looked down on the scene unfolding before him, on the sea coast from St Helena Bay back to the Cape: 'The whole country', he wrote, 'affords a fine field for botany, being enamelled with the greatest number of flowers I ever saw, of exquisite beauty and fragrance.'[24] Having reached their most northerly point, the party turned southeast, heading again to Swellendam, and finally reached the ocean at Mossel Bay on 16 November. There they turned inland and made their way again in a southeasterly direction until they met the ocean for a second time at Algoa Bay (near present-day Port Elizabeth) on 14 December, almost five hundred miles from Cape Town. They had been away for months. They pushed on a little further to the east until they were on the banks of the Sundays River, which flows into Algoa Bay. But the Khoikhoi people they employed refused to go any further. The oxen were sick and the carts near collapse.[25] Reluctantly, they turned back and, retracing their steps, they arrived in Cape Town on 29 January 1774.

Masson referred on many occasions to the beauty of the plants he and Thunberg encountered but, apart from mentioning that he obtained specimens of what he called geraniums, stapelias and ericas near the end of the journey, he failed to specify what else he had collected.[26]

While they were in Cape Town during 1774, Thunberg and Masson met Lady Anne Monson, the highly accomplished botanist and collector, who was on her way from England to Calcutta with her husband, a colonel who was returning to Bengal to serve on the supreme council.[27] She, Thunberg and Masson frequently went botanising together in the surrounding countryside.[28]

After only a few weeks these delightful excursions came to an end. Lady Monson continued her voyage to Calcutta on the *Pacific*, an East India Company ship. On 29 September 1774, Thunberg and Masson decided to strike out into the interior again. They headed out of town in a northeastern direction to an area they had never visited before. This time they took in mountains and desert, plotting a circular route that took them to a point around three hundred and fifty miles north of Cape Town before swinging to the southeast and eventually to the southwest back to Cape Town, which they reached on 29 December 1774.

Thunberg must have been a real inspiration for Masson: highly educated and one of Linnaeus's own students – a mine of up-to-date botanical information and an intrepid traveller, not just an accomplished gardener like Aiton. For his education as a travelling botanist, collecting in unfamiliar terrain, Masson could have had no better teacher than Thunberg.

Thunberg's departure for Batavia, his ultimate destination, on 2 March 1775, may have convinced Masson to leave as well. He left the Cape sometime towards the end of March but before that Masson met Anders Sparrman. Sparrman, born in Sweden in 1748, was another student of Linnaeus's, who had first arrived at the Cape on 12 April 1772 to study the region's natural history.[29] His plans, however, were put on hold when he met the Forsters and was invited to join them on the *Resolution* on its Pacific voyage.[30] The *Resolution*, having dropped Masson off at the end of October, now collected Sparrman before setting off for the Pacific less than a month later on 22 November 1772. There is no evidence that Masson and Sparrman met at this time.[31] It was only when Sparrman came back to the Cape on 22 March 1775 on the *Resolution*'s return voyage to England from the Pacific that the pair are known to have met.[32]

Masson was back in London in either July or August 1775.[33] He had been sending seeds and living plants to Kew from the Cape, but now on his return, he had much more to show.[34] Solander was at Kew, where Masson had gone back to live, on Saturday, 26 August and reported to John Ellis, his close friend and fellow botanist, that he had just seen Masson 'with a great cargo of new plants . . . a glorious collection'.[35]

Solander did not give many details about the collection but the antiquary Richard Gough alerted his friend and fellow antiquary the Reverend Michael Tyson that Masson had brought home a great variety of new species of plants from the Cape of Good Hope, 'which grew and blowed at sea'.[36] Tyson, himself, then went to Kew to see Masson and was amazed by what he found: totally new species of heather, protea, geraniums, and an undisclosed number of dried varieties.[37] Tyson spoke of more than two hundred examples of just two species but when Masson wrote to Linnaeus at the end of December 1775, his first letter to 'the Father of Botany', introducing himself as having been employed 'some years past, by the King of great Brittain, in collecting of Plants for the Royal Gardens at Kew', and remarking that he had made two trips into the interior with his student Thunberg and had met Sparrman, he announced that he had 'added upwards of 400 new species to his Majesties collection of living plants'.[38]

What no one mentioned was that Masson had also collected a magnificent plant known by its popular name as the bird-of-paradise flower but which, when it arrived at Kew in 1773, was given the name *Strelitzia reginae*, in honour of Queen Charlotte, Duchess of

Mecklenburg-Strelitz.[39] It is likely that Thunberg directed him to this plant since he had already come across it on his first journey to the interior.[40] This may have been Banks's favourite plant:[41] certainly, when, in 1795, it came to putting together a substantial gift of plants for Catherine II, Empress of Russia, among many Cape plants collected by Masson, *Strelitzia reginae*, which Banks described as 'one of the most rare, & certainly . . . one of the most beautiful plants in Europe', headed the list of plants chosen.[42] Even more remarkable, perhaps, is that Masson collected the first specimen of the cycad *Encephalartos altensteinii*, which was at Kew in 1775, and which, in 2009, was re-potted: it is believed to be the oldest living pot plant in the world.

Banks, too, was delighted with Masson's collection, 'a profusion of Plants unknown till that time to the Botanical gardens in Europe'.[43] In his estimation, 'by means of these, Kew Gardens has in great measure attained to that acknowlegd superiority which it now holds over every similar Establishment in Europe; some of which, as Trianon, Paris, Upsala, etc., till lately vyed with each other for pre-eminence, without admitting even a competition from any English Garden.'

Masson's second assignment, also authorised by Sir John Pringle, was to collect for Kew on the Atlantic islands of Madeira, and the Canaries, and in the British West Indies.[44] It is likely that Banks and Solander once again had a hand in deciding where Masson should go: they had both been to Madeira in 1768 on the *Endeavour*'s outward voyage and, during their stay from 12 to 18 September, had collected many plants that they brought back with them in 1771.[45]

Masson left for Madeira on 19 May and arrived there on 5 June 1776.[46] Around this time, Pringle, who was by now George III's Physician Extraordinary, in addition to being President of the Royal Society, no longer had time for Kew.[47] Banks took over. He arranged Masson's finances, sent him instructions and supplies, and, in return, Masson sent Banks plants and seeds and letters.[48]

In what might have been his first letter to Banks, dated 28 July 1776, Masson noted that he hadn't yet explored very much of the island but he believed he had already found some new plants. He had about fifty plants for Banks and some seeds for Aiton.[49] Apart from that he said little: he remained a man of few words for the next thirty years.

Masson would have arrived well equipped to explore the botany of the Atlantic islands, especially Madeira and the Canaries. The islands' plants were well represented in Sloane's herbarium at the British Museum.[50] Both island groups were a common stopover for British ships, both naval and East India Company vessels, on their way into the South Atlantic. It was where they stocked up with wine and fresh fruits for the next part of the voyage. Those who were interested in botany helped themselves to the local flora. There was a good, if small, selection of plants from the area already growing at Kew.[51]

From the summer of 1776 until the summer of 1779, Masson stayed and collected in the Atlantic region – seeds for Aiton and plants for himself and Banks.[52] Madeira was the base where he spent most of his time when he wasn't travelling. The itinerary he followed during this period began with a stay in Madeira which lasted for almost a year until May 1777; this was followed by a trip to the Azores, not planned originally but which Charles Murray, the British Consul in Funchal, who found Masson very convivial, convinced him he should visit.[53] Masson stayed in the Azores from late May 1777 to early December 1777, during which time he visited the islands of São Miguel and Faial – Johann and Georg Forster had collected on Faial between 14 and 19 July 1775 on the final homeward stretch of Cook's second voyage, and might have discussed this with Masson – and São Jorge;[54] Masson left the Azores in December for the Canaries, via Madeira, and remained there until early October 1778, and from there he travelled back to Madeira, where he stayed until May 1779.[55]

Because of a lack of plant lists, it is not possible to give a definite account of the shipments, but about one thousand plants, many of them new to European science, would probably be close to the mark. Collecting was not a problem. Masson went well off the beaten track where there was plenty to be found.[56] What was difficult was shipping his collections back to England. The problem, as he said himself, was that while he was in the islands the American revolutionary war was in progress on the other side of the Atlantic. American cruisers were present but not British naval vessels. He had to use whatever shipping was available: a Dutch ship on one occasion; on another a French brig on its way to Dunkirk with a stop at Portsmouth, and even a British privateer making its way back to Plymouth.

However the difficulties of this trip were as nothing when compared to those of his next destination, the West Indies.[57] Masson left Madeira sometime in the early part of 1779 for Barbados. What happened then is unknown, but at some time he went from there to Grenada, where he became embroiled in the war between Britain and France, France by this time fighting on the American side. The French assaulted and captured the island from the British in the early part of July 1779.[58] Masson was, in Banks's words, 'call'd upon to bear arms in its defence, which he did, & was taken prisoner'.[59] When he was finally released, he made his way to St Eustatius, a Dutch Leeward Island, between St Kitts and St Martin.

There he found some respite from the conflict and island-hopped, taking in St Kitts and Nevis but found that the people there were unsupportive of his efforts: 'a philosopher', he wrote, 'is [a] monster of nature to the people in this part of the world'.[60] He then headed to Saint Lucia, where rains and an impenetrable forest kept his bota-nising to a minimum, but he was still able to put together a small collection of thirty plants for Banks. Then things went from bad to worse: in the early days of October 1780, a deadly hurricane hit the island and Masson lost the entire collection he had with him at the time, his clothes and papers, and almost his life.[61]

Masson never gave up. Certainly the hurricane did not stop him. His next port of call was Jamaica. He had already been to six islands. We know that Masson was in Jamaica in March 1781 but not much more than that. He originally planned to try to get into central America from there but was unable to, presumably because of the continuing conflict. Instead, he finally made up his mind to return home and was back in London in November 1781.[62]

Where to go next was always the question for Masson. In December 1778, while he was still in Madeira, Masson had written to Linnaeus's son, also called Carl (the great Linnaeus had died in early January of the same year) mentioning that he had been in touch with 'Mr. De Visme'.[63] Indeed, just before he left Madeira for the West Indies, Masson had informed Banks that he was sending the Canary Islands collection to Lisbon to that same De Visme who would then forward it to Banks.[64] So, it is not surprising that in March 1783, Masson went to Lisbon where he met several of Banks's acquaintances, including Gerard de Visme, who was a central figure in the English merchant community, and whom Banks had met in November 1766 on his way back to England from Newfoundland and Labrador on HMS *Niger*.[65]

De Visme had a garden in what is now Benfica, a suburb of Lisbon, and it would seem that he invited Masson to write up a catalogue of the plants that were growing there.[66] When he saw the garden, Masson was not particularly impressed by it or its owner – 'he knows little or nothing of Botany or natural history', Masson wrote to Banks, 'but possessed a very great share of vanity and conceit.'[67] Masson declined to do the catalogue. For his part de Visme was very proud of his gardening achievements, reminding Banks that he had seen his garden in its infancy when he visited Lisbon almost twenty years earlier. De Visme claimed to Banks that the reason Masson did not undertake the catalogue was that 'he has worn himself out with fatigue, in roving over wild grounds, & sands, in search of new productions . . . I expect he will add one, to ye martyrs of Botany.'[68]

If he was worn out, as de Visme wrote, Masson certainly hadn't finished travelling for not only did he visit Lisbon, but also southern Spain, Gibraltar, North Africa and Madeira again, though the order is not clear. He was back in London by late 1784 or early 1785.[69]

Whenever he returned, Masson, as we have come to expect, was not one to stay put for very long and so, on 16 October 1785, he boarded the *Earl of Talbot* in Portsmouth, an East India Company vessel heading to the Cape. He arrived at Table Bay on 10 January 1786.[70] Though his reception does not seem to have been as warm as it had been the first time, and though he expressed a wish to be sent to Madras, where there was a vacancy as the East India Company's naturalist (see Chapter 5), he remained at the Cape.[71]

Masson had been sent to the Cape a second time, as he told Thunberg, to collect living plants for Kew, as opposed to herbarium samples and seeds, from the mountains in the environs of the Cape.[72] He did not always follow his orders, as he confessed to Thunberg, but ventured further into the interior. Banks knew this as he paid Masson's bills and though he disapproved of him not fulfilling his instructions, there was little he could do about it.[73] Masson also failed to follow his instructions exactly as to what he collected: he often sent seeds, herbarium plants and bulbs, as well as the living plants he had been sent to get.

It isn't clear how long Masson intended to remain at the Cape but it appears that no time limit was placed on his stay. On several occasions he complained to Banks that he hadn't heard from him in almost two years, and he feared that the letters destined for him had ended up in India or Canton as the East India Company convoys often decided to bypass the Cape. Masson carried on as if he had been instructed to stay. He had no problems collecting new and unusual plants – to Thunberg he mentioned heathers, amaryllis and protea, plant varieties he had collected before, but emphasised that he believed he had found more than twenty new species of *Stapelia*, which he was going to include in a monograph he was planning.[74] The problems that he did encounter were that he couldn't always find a ship on which to send his plants back to Kew. It wasn't that there were no ships in harbour but rather that the ones that came in on their way to England, especially from India, were often so crowded with passengers, that there was no room left for plants.

William Aiton, for whom Masson had been working for maybe thirty years, died at the beginning of February 1793 (William Townsend, his son, took over the duties at Kew) and, from this point on, the tone of Masson's letters to Banks changed. He now began referring increasingly to wanting to return to England and by October 1794, he started settling his affairs in preparation for a return voyage. On 17 March 1795, fearing that he might get caught up in a possible conflict between Britain and the Netherlands, he found passage on a returning East India Company ship, taking his collection of living plants that he had been growing in his garden at the Cape with him, and was back in England in August of the same year.[75]

As far as one can tell from the few references he made to it in his correspondence with Banks, Masson's botanical haul was substantial – about one half of the number of geraniums growing at Kew by 1795 had been introduced by Masson.[76] It is also clear that of all the plants he collected he was particularly proud of the *Stapelia*. As he told Thunberg, he intended to publish on this group and during the quiet periods at the Cape, Masson arranged his collection and began to draw the individual plants. He brought this work to fruition in 1796 when his monograph *Stapeliae Novae*, describing new species of *Stapelia*, was published in London. The book was dedicated to the King and contained descriptions of forty-one species, all laid out in proper Linnaean fashion, in Latin, and illustrated in colour.

'Still enjoying, though in the afternoon of life, a reasonable share of health and vigour' was how Masson described himself, aged mid fifties, in the dedication to his monograph.[77] He also added that he was 'anxious to recommence my employment as a collector'.

'I am ready to proceed to any part of the globe, to which your Majesty's commands shall direct me. Many are the portions of it that have not yet been fully explore by Botanists.' But where to? He had already been to the Cape twice, the islands of Madeira, the Canaries, the Azores and a number of West Indian islands.

Banks recognised that Masson was not done with travelling yet. 'Masson . . . is Fat & Cheerfull', he wrote to a mutual friend in Lisbon, 'he has already got his Travelling notions Returned into his head he says he does not mean to Settle yet & if a good climate is Proposed to him will readily undertake a new voyage.'[78]

Banks knew just the place for Masson. It was Buenos Aires and its hinterland in which Banks was very interested. On 10 April 1796, Banks wrote to the Spanish Ambassador in London asking for permission to send Masson to collect seeds for Kew.[79] Masson, Banks emphasised, would keep himself to himself, not enquiring into the local political situation: he had been collecting for the King for near thirty years and he had conducted himself in an exemplary fashion.

There was no such trip, probably because several months later Spain sided with France and declared war on Britain. If Banks had set his sights in a general way on the far side of the Atlantic, any place there associated with Spain was now out of the question.

By the end of July 1797, he had decided to send Masson to North America, particularly Upper Canada. At the time Upper Canada, part of British Canada, consisted of what is now southern Ontario, and land which stretched westward following the shores of Lake Huron and Lake Superior until Lake Nipigon (west and north of here was the territory, called Rupert's Land, ceded to the Hudson's Bay Company in 1670): its European population, mostly recently arrived, stood at around 25,000.

Banks arranged the necessary paperwork.[80] A set of instructions (now sadly lost) was drafted, and letters of introduction from the Duke of Portland, the then Home Secretary, to the several governors in the British part of North America (Lower Canada, New Brunswick and Nova Scotia) were drawn up in case Masson ventured into these jurisdictions. A few days later, Rufus King, the American Ambassador in London, provided letters of introduction for his new country, the United States, giving Masson a clear run to venture wherever his collecting instincts took him, even down the Mississippi River.

The canvas was enormous. Masson decided to start at New York and proceed from there, but the voyage there did not start well.[81] He left Gravesend in September 1797. Somewhere in the Atlantic near the Azores, the ship was intercepted by a French privateer, then shot at by a French pirate ship, which then took and kept the ship: Masson and several crew were put on a ship heading for Baltimore and then transferred to another ship which finally landed them at New Haven, Connecticut.

Masson arrived in New York during the final week of 1797, after a gruelling four months at sea.[82] He spent the winter in New York and then in the spring of 1798, following Banks's instructions, he worked his way up the Hudson and the Mohawk River until he reached the American settlement of Oswego on the south shore of Lake Ontario. He was heading westward along the southern shore of the lake, for Niagara, in Upper Canada, one of the first places to be settled by those British colonists who had remained loyal to the Crown during the American revolution.[83] Masson arrived at the small town, which had at the beginning of July lost its status as Upper Canada's capital to the town of York (soon to be renamed Toronto), directly opposite on the north shore of Lake Ontario. He botanised in the area but stayed in nearby Queenstown in the house of a 'Mr Hamilton' (Robert Hamilton, merchant and judge) who took him in as his guest because 'being a new place there was no decent lodging to be got'.[84] Masson wanted to travel to Detroit, at the western end of Lake Erie, but got no further than Fort Erie at its eastern end where he waited in vain for favourable winds to take him down the lake. Summer was quickly drawing to an end. He took the boxes of seeds he had collected and made his way across Lake Ontario to York and from there eastward to the much larger, older and more-established town of Montreal, in Lower Canada, where he arrived on 16 October 1798.[85]

Within a day or so of arriving, Masson learned that a vessel was ready to sail to Quebec, further down the St Lawrence River and, from there, he hoped to find another one scheduled to cross the Atlantic to Europe. It was probably his last chance to send samples of his collecting activities before winter set in and the ports froze. Masson sent a box of plants and a bag of 'Zizania', wild rice, especially for Banks.[86]

For the next few years, Montreal was Masson's base. It was a good choice. Not only did it send shipping to England either directly or by way of Quebec, but it was the headquarters of the North West Company, a fur trading business founded some fifteen years before Masson's arrival by a group of enterprising Scots living in Montreal.[87] Each year some of the partners fanned out into what is now western Canada from their wintering base at Grand Portage (now in northern Minnesota), on Lake Superior, in search of beaver, and then in July they would return to their base where the partners from Montreal would be waiting for them to freight the cargo of furs back to the St Lawrence River.[88] Masson hoped to take advantage of this arrangement for his botanical purposes.

His hopes were realised.[89] During the summer of 1799, he travelled with the Montreal partners through southern Ontario to Georgian Bay, Lake Huron and finally Lake Superior to Grand Portage, Upper Canada, where the two sets of partners met. Here and on his travels Masson collected more than one hundred different varieties of seed. For his return trip to Montreal, he went back to Niagara where he collected examples of living plants, in order to send them to Banks from Quebec before winter closed in again.

In the following summer, Masson wanted to travel to Pittsburgh by way of Lake Erie but for some reason this did not happen.[90] Instead he returned to the northern Great Lakes and came back with another collection of living plants, including purple trillium.[91]

From what we can gather, Banks appeared to have been pleased with the collections he had received from Masson since his arrival in North America.[92] But then, two years later, Banks's attitude to Masson appeared to have changed significantly. On 2 February 1802, Banks wrote to him saying that his collections had become disappointing, and that he was not venturing far enough to find new species. Banks had by then already received several hundred specimens, but many of them were not new.[93]

Then there is a three-year gap in the surviving records. It seems there must have been some discussion of Masson returning to London but he was intent on staying, remarking that he was in very good health and was hoping to travel further down the Gulf of St Lawrence to collect.[94] This was in May 1805.

By the end of the summer, Masson was clear that he was being recalled and though he continued collecting, including living plants during the summer, he did not seem to be disappointed to be leaving.[95] He intended to spend the winter in Montreal and then make his way home via New York in the following spring.

In the event, Masson did not come home. He died in Montreal on 23 December 1805 aged sixty-four. According to John Gray, a friend who was a Montreal attorney and merchant, with very close ties to the North West Company, Masson left few possessions: clothing, especially for the winter months, two maps, two journals, a few drawings, a tea pot and caddy. Gray specifically noted that alongside a small number of authoritative, Linnaean, botanical texts, was a copy, no doubt well-thumbed and annotated, of William Aiton's *Hortus Kewensis*, published sixteen years before, which told Masson precisely which plants were growing in the royal gardens.[96]

Masson went further and collected more, and over a longer period of time (thirty years) than any other collector associated with Banks. Much of what he did was on his own initiative. Banks described him as 'indefatigable': no one who met Masson had a bad word to say about him.

2

1779: Return to Botany Bay by Way of Southwest Africa

As Banks sifted through the vast collections made during the *Endeavour*'s Pacific voyage, several thousand specimens in all, he would have been reminded of the time he had spent in New Holland, at Botany Bay, and on the Endeavour River, in the period from late April to early August 1770, but few other people were interested. New Holland hardly got a mention in the press and when it did it was described as a hostile place, unlike Tahiti, which was described as paradise.[1]

More information about New Holland became available at the end of September 1771, a little over two months after the return of the *Endeavour*, when an anonymous publication, *A Journal of a Voyage Around the World*, appeared in London, and told the story of Cook's three-year voyage. In it New Holland came across as a very disagreeable place.[2] John Hawkesworth's officially sanctioned publication in 1773, of what he called 'Lieutenants Cook's Voyage Round the World', devoted many more pages to New Holland than the earlier publication: it painted a much more attractive picture of the country's resources, but, it, too, had little positive to say of its inhabitants.[3] Two days later, the journal of Sydney Parkinson, the *Endeavour*'s artist, who had died on 26 January 1771 (as the ship was making its way back from Batavia to the Cape of Good Hope), was published – and it confirmed most of the previous descriptions.[4]

After that New Holland disappeared from the scene.

Seven years went by.

* * *

On 30 November 1778, following the resignation of Sir John Pringle, Banks, aged thirty-five, was elected, almost unanimously, President of the Royal Society.[5] In the early part of 1779, a few months after his election, Banks received an invitation from Sir Charles Bunbury, MP for Suffolk and Steward of the Jockey Club, to appear before his House of Commons committee to give expert evidence. This was Bunbury's second committee investigating the general issue of crime and punishment in England.

The first, which convened on 23 March 1778, was set up to assess the effectiveness of an Act of Parliament, which was given Royal Assent on 23 May 1776 and which authorised for the first time the use of hulks, disused naval warships and other vessels, to house prisoners on the River Thames.[6] The Act, the result of desperate government action on a very serious and growing problem of where to house prisoners, and discussed by prison reformers such as John Howard, parliamentarians and religious leaders alike, was passed as a temporary measure and would be in force for two years in the first instance.

Why the desperation and why was it temporary? In 1718, Parliament had passed the Transportation Act which authorised foreign banishment, specifically to the American colonies, as a punishment for serious crime, the term length, depending on the kind of crime, being either seven or fourteen years.[7] Between 1718 and the outbreak of the American War of Independence in 1776, about 50,000, men, women and children had already been disposed of in this way, ending up, for the most part, working on tobacco plantations in Virginia and Maryland.[8] But now, with war raging in the American colonies, there was nowhere to transport the prisoners. Hulks were considered a reasonable solution, especially as they provided the opportunity for hard labour, preparing ballast for ships; and the two years that the Act ran was viewed as a trial period. They interviewed several witnesses. These included Duncan Campbell, who ran the government contract for hulks, and Daniel Solander, who had paid an unannounced visit to the *Justitia*, a retired East India Company ship and the first ship supplied by Campbell as a prison, to inspect the accommodation. Bunbury's committee concluded that the system was working well and should continue.[9] Once again the Act was passed and Royal Assent was granted on 22 May 1778.[10]

Not everyone in Parliament and among the public was satisfied with this conclusion – hulks were only absorbing a small proportion of the prison population and many believed the conditions on board, often fatal, were worse than those in the prisons. In response, on 16 December 1778, Bunbury was asked to convene a second committee to seek more detailed information on the number and condition of prisoners convicted of serious crimes in London and the southern English counties jails; to provide more information concerning conditions on the hulks; and, finally, and perhaps most surprisingly, to provide 'A short Account of the Acts relating to Transportation, and of the Proposals which have been stated to your Committee, for recurring in some Degree to that Mode of Punishment'.[11]

Transportation was back on the agenda. Duncan Campbell was called again, this time to talk about transportation. For twenty years he had been taking British prisoners to Virginia and Maryland. He was asked for his opinion as to where he would send prisoners now. He answered that Georgia and Florida would be suitable but not for large numbers.[12]

On hearing this, the Committee 'thought proper, therefore, to examine how far, Transportation might be practicable to other Parts of the World'.

It was time to call Banks. He was no expert on transportation but he was one of the few people in London who could speak authoritatively on another part of the world that might be suitable as a destination for those whose punishment, in the eyes of the Judiciary, was transportation.[13]

The report of Banks's appearance began thus: '*Joseph Banks*, [was] requested, in case it should be thought expedient to establish a colony of convicted Felons in any distant Part of the Globe, from whence their escape might be difficult, and where, from the Fertility of the soil, they might be able to maintain themselves, after the First Year, with little or no aid from the Mother Country, to give his Opinion What Place would be most eligible for such Settlement?' Banks answered straight away that that place was '*Botany Bay*, on the Coast of *New Holland*, in the *Indian Ocean*, which was about Seven Months Voyage from *England*'.[14]

Then Banks explained his choice further. He spoke glowingly of Botany Bay. The climate was right – like Toulouse, he said – though he admitted he had only been there for a week when the weather was 'mild and moderate'; the soil would be able to support large numbers of settlers, and sheep and cattle would thrive – there were no predators; fishing was good, water plentiful, and there was an abundance of timber. The inhabitants, he added, were few in number – no more than fifty in the neighbourhood he reckoned; they were willing to share their land (but not their produce). There were no Europeans anywhere nearby – 'escape would be very difficult'. A settlement would begin to maintain itself after the first year, he asserted, though for the first year the settlers would need to bring all of their necessary provisions with them from England. Would the Mother Country reap benefit from this settlement, the Committee asked? Banks replied confidently that 'if the People formed among themselves a Civil Government', the population would grow, they would demand more European goods and, with a land mass exceeding that of the European continent, they 'would furnish Matter of Advantageous Return'.[15] Banks made no mention of the Endeavour River area where he had spent seven weeks, considerably longer than at Botany Bay.[16]

This was probably the first time that Banks spoke publicly about Botany Bay. He was certainly an illustrious witness but one suspects that what he said made little impression. The real focus of attention, as an alternative to the American colonies, was a place much closer to home and with connections that went back more than a century – West Africa. Bunbury had invited six witnesses to speak about their experiences in this part of the world. Between them they had more than thirty years' experience living and working on the coast. One of the witnesses, John Roberts, who had been in command of the West African trading and slaving forts for nine years, advocated an African penal colony – his preferred choice was a port, 400 miles upriver from the mouth of the Gambia River. He produced figures to back up his case that the overall cost of this penal colony would be less than the corresponding costs to keep prisoners on hulks. Two other witnesses were in broad agreement with Roberts while the three remaining witnesses were not in favour, citing especially the horrible mortality rates among Europeans. Transportation was not intended to be a death sentence.[17]

After taking expert evidence, Bunbury's recommendation was that the laws governing transportation, which specified the American colonies as the destination, should be altered to include 'any other Part of the Globe that may be found expedient'.[18]

As the revolutionary war across the Atlantic progressed, North America became increasingly closed as a destination for transportation: some convicts were sent to West Africa, typically to be enlisted there in the military, but there were few of them. Suggestions of where convicts should be sent were made from all quarters. Possible destinations included Nova Scotia and Newfoundland in Canada, the West

Indies, British Honduras, Gibraltar, Menorca and St Helena, but none of these came to anything.[19] It may have been that the government was hoping that the war would be won and that transportation to the American colonies could resume as before.

Nothing more was heard of Botany Bay.

Four years later, the subject of New South Wales, as it was now generally referred to, suddenly reappeared in Banks's life.

On 28 July 1783, from an address off Grosvenor Square in London's Mayfair, James Matra wrote a letter to Joseph Banks. This was not the first time Matra had written to Banks but the subject of this letter was out of the ordinary, very different from his former concerns. He was very agitated.

Matra, or Magra, his birth name (he changed it in 1775), met Banks in 1768 on HMS *Endeavour*. Born in New York in 1746, one of three sons of a wealthy Corsican couple who had come to New York by way of Ireland, James Matra was drawn to the sea and served on several British ships during the Seven Years War (1756–63). A few more naval appointments followed on ships that took him to Chatham, Dublin and finally London. There, on 25 July 1768, he signed on as Able Seaman on HMS *Endeavour*. One month later, the ship left Plymouth for the Pacific.[20]

Matra, who was promoted to full midshipman just before the ship arrived back in England, is hardly mentioned in the surviving journals. In his own journal, Banks acknowledges his presence but does not record his name, referring to him instead as 'one of our Midshipmen an American'.[21]

Less than a month following the return of the *Endeavour*, Matra was promoted but he chose not to pursue a naval career.[22] Instead he remained in London, probably spending some of the time in the company of Banks, Solander and those in their circle.[23] Six months later, though, a new career opportunity opened when he was appointed British Consul to the Canary Islands, based in Tenerife. He remained in this post until March 1775 when he returned to London, where, towards the end of the year he petitioned the King to have his name changed to Matra.[24] A short trip to New York followed and then in mid 1778, he accepted a return to consular life when he agreed to be Secretary to Sir Robert Ainslie, British Ambassador in Constantinople.[25]

Not particularly happy in this situation, Matra was back in London in July 1781. And it was here, as an exiled American Loyalist, a British subject who had remained loyal to the Crown during the revolutionary war, that he formed his grand design, the essence of which he shared with Banks in July 1783.

In his letter, Matra, who had been corresponding with Banks for several years, asked about rumours he had heard that there were plans being considered to settle New South Wales.[26] Matra had learned, he continued, that Banks was the guiding hand in one of these plans and added, as he had 'frequently revolved similar Plans', that he would 'prefer embarking in such a Scheme'.[27]

There is no surviving evidence that Banks was involved; but Matra was probably correct that there was a plan afoot, which he attributed to Sir George Young, a senior naval officer, and Sir George Jackson, Second Secretary to the Admiralty.[28] One month after writing to Banks, Matra offered his own proposal to the government to settle New South Wales with American Loyalists and other free settlers, 'an Object . . . which may in time, atone for the loss of our American colonies'.[29]

Matra's proposal was wide-ranging and covered all aspects of how the settlement would be provided for and what its future would hold. The key point is that it would pay for itself, as it could grow the same economically useful plants that were already growing nearby in the islands of the East Indies to the benefit of other European nations – sugar, coffee, tea, indigo and cotton received special mention. Matra also pointed out, and in some detail, the strategic role, both political and economic, that New South Wales could play in a new Pacific region. He had the backing of 'some of the most intelligent, & candid Americans' for whom this proposal held out their best opportunity of re-establishing their lives. And what's more, Matra had Banks's full support.[30]

No sooner had Matra laid his ideas before the government than matters took an unexpected turn. Lord Sydney, the Home Secretary, met with Matra and laid before him the idea that was brewing in the Cabinet that New South Wales might be suitable for convicts punished with transportation. With negotiations completed leading to the signing of the Treaty of Paris between Britain and the newly established United States, the government had to look for a new destination for transportation. Matra was very enthusiastic about Sydney's idea. 'I believe', he wrote in April 1784, 'that it will be found, that in this Idea, good Policy & Humanity are united.'[31]

Events moved ahead very swiftly, because the number of people convicted of crimes in England was growing rapidly.[32] This was not the time to be concerned with a free settlement. Pressure on local prisons was becoming intolerable and the chorus demanding the government resume transportation was increasing in volume.[33]

Even though, with Banks's backing, Matra spoke glowingly of New South Wales, it was not at this time the government's favoured option. North America was being reconsidered. On two occasions in 1783 and 1784, transportation was resumed clandestinely to Maryland but the experiment failed.[34] And then, at the end of 1784, West Africa, which had been discussed at the Bunbury Committee, became the government's favourite choice, though it did not publicise the fact.[35]

When word got out that the government was seriously considering West Africa, opposition to the idea, especially from Edmund Burke, gained momentum – he and other Members of Parliament were convinced that West Africa was a graveyard for white people. Transportation, they reiterated, was a punishment, not a death sentence. The government paused and agreed with the idea put forth by Francis Seymour Conway, Lord Beauchamp, MP for Orford, that a committee should evaluate the government's recommendation. Chaired by Beauchamp, the committee sought expert testimony on the suitability of West Africa as a penal colony. The government was looking for support to establish a settlement on Lemaine Island, four hundred miles from the mouth of the Gambia River. The specific legislation, the Transportation Act of 1784, was already in place. Unlike its earlier 1718 version, which specified the American colonies as the destination for transportation, this Act left it up to the government to decide on a suitable place of banishment. If everything went according to plan, then all the government would need to do would be to raise an Order-in-Council to specify Lemaine Island as the convict destination.

As it turned out, nothing went according to plan. In its first stage of hearings, held between 26 April and 3 May 1785, the Beauchamp Committee heard one witness after the other pour scorn on the idea of transportation to West Africa.[36] In its first report, on 9 May, Beauchamp's Committee confirmed that West Africa was unsuitable.[37]

In the second stage of its hearings, beginning on 9 May, the Committee widened its questioning beyond West Africa. One of the first to appear was James Matra. There was nothing surprising in this. Since he had proposed New South Wales as a settlement in August 1783, other influential people had added their voices to his, including

Sir George Young, a prominent naval officer, and his brother-in-law Sir John Call, a military engineer and wealthy MP with wide experience in Indian affairs.[38] Matra appeared twice and repeated many of the claims he had been making for the past two years.[39]

Banks, too, was invited to give his opinion. It was a Q and A. 'This Committee', Beauchamp began, 'would be glad to know where in your Voyage with Captain Cook it occurred to you that there were any places in the newly discovered Islands to which persons of such Description might be sent in a Situation where they might be able by Labour to support themselves?'[40]

Banks could have suggested a number of places where he had been: Tahiti, New Zealand, Endeavour River and Botany Bay. From this list of possibilities, he chose Botany Bay, in which he had spent the least amount of time. Why Botany Bay over the others? Banks never explained this. He may not have recommended New Zealand because the reception of the *Endeavour* by the local people was hardly friendly; he may have thought Tahiti too small and densely populated; and ruled out Endeavour River because its climate was too warm and too humid for the British constitution. In his answers to the committee's questions, Banks spoke very highly of Botany Bay in much the same way as he had six years earlier in front of the Bunbury Committee. He reiterated his key observations: the land was fertile and there was fresh water; the climate was similar to that of Europe; the natives were not hostile, and in any case they were armed only with spears made of fish bones; they would make way for the newcomers. When asked about the provision of women for the male convicts, Banks answered that he did not foresee a problem: 'I have no doubt', he said, 'that they might be obtained from the South Sea Islands at no other Expence than the Charge of fetching them.'[41]

Banks repeated his observation that there were no 'wild Beasts' – or predators – and that though he had seen quadrupeds, they were unlikely to be a food source, unlike the plentiful supply of birds. Banks summed up his opinion succinctly: 'From the Fertility of the Soil the timid Disposition of the Inhabitants and the Climate being so analogous to that of Europe I give this place the preference to all that I have seen.'[42]

Matra and Banks made their case eloquently and with conviction. After all, they had actually been there. The Committee at this point were sufficiently convinced to ask for costings from other witnesses. When these were put before the Committee, it was clear that settling

Botany Bay would be no more expensive than keeping convicts in hulks.[43] On the second to last day of the hearings, Matra was called back again – his third appearance – and this time the conversation turned to issues of procuring livestock for the fledgling settlement, and the competing claims to the territory from other European powers. Matra assured the Committee that livestock could easily be purchased from the Indonesian islands to the north and that Britain was alone in claiming the land.

Everything seemed to point to Botany Bay.[44] But then on 25 May 1785, the very last day of the proceedings, Sir John Call, who had previously authored a proposal to settle New South Wales, drew the Committee's attention to a part of the world no one had yet considered – Das Voltas Bay in present-day Namibia.[45] Call, an MP and member of the Committee, had previously given evidence in the first stage of the hearings. Based on his acquaintance with the West African coast going back to the 1750s, he categorically dismissed the area as being suitable for transportation. But now, and at the last moment, he offered a new suggestion.

Within government circles, the idea of the southwestern coast of Africa as a convict destination was not entirely new. Commodore Edward Thompson, who was in command of the Africa station of the Royal Navy, had advocated it almost two years earlier, in July 1783, and then again just before the Committee's proceedings began.[46]

The Committee never called anyone to testify on the suitability of this part of the world and the hearings came to an end as scheduled. Two months later, on 28 July 1785, the Committee reported its findings and recommendation to the House of Commons. Not a single word was devoted to Botany Bay. It is as though Banks's and Matra's exhaustive testimony had never taken place. Instead, the report concluded that on the basis of the evidence laid before them only Das Voltas Bay was suitable as a destination and eventually a settlement. As the report argued, both convicts and American Loyalists could be accommodated; the area had not been claimed by anyone else and lay in a good geopolitical space, with the Portuguese to the north and the Dutch to the south; Brazil, and its offshore whaling areas, was not far away; and the sailing time to England was reasonable.[47]

Why did the Das Voltas scheme trump that of Botany Bay? The documentary record is unfortunately silent on this question. Speculations have taken the place of solid facts but these need not detain us here.[48] What matters is that less than a month after the report was made, on

22 August 1785, Lord Sydney, the Home Secretary, asked the Admiralty to send a naval vessel to the area, to judge its suitability for settlement and to find the most likely location for it.[49]

It fell naturally to Edward Thompson to sail there and report back. Under the cover of a regular visit to inspect the British West African slaving stations, Thompson prepared to set sail in two ships. He carried secret orders that at an appropriate time he should send HMS *Nautilus*, the smaller of his two ships, southward towards Das Voltas Bay to survey the area.[50]

Banks was back in the picture, now advising the government about the new African destination. The government, Banks told William Aiton at Kew, in a letter dated 29 August 1785, wanted a naturalist to accompany Thompson to the coast of Africa, to a part 'unknown to us in the Article of Natural History'.[51] This man, Banks explained, should be an enthusiastic gardener; he should know some botany; he should know how to collect and dry specimens; and he should be able to distinguish between different kinds of soils. But most of all, he should be a keen observer and recorder. In Banks's own words: 'He should be ready to write down his opinion of every thing which occurs in his department & to enable him to do so I shall draw up full instructions of the things he is expected to notice.'

Banks had clearly been drawn into the government's confidence and he told Aiton not to waste a moment choosing 'the best who offers'. But there was more to this than satisfying the government's request. The voyage, Banks emphasised, would also result in entirely new specimens for him and seeds for Kew. He pressed Aiton. 'If it can be done well we have future opportunities . . . for God's sake be active and do not let such an opportunity slip you.' Banks was loath to lose any opportunity to collect for the royal garden at Kew.

As it was summer, the time of the year that Banks habitually spent at Revesby Abbey to tend to estate matters, he left it to Jonas Dryander, his Swedish librarian and a trained botanist, and to Dr Charles Blagden, his close friend and Secretary of the Royal Society, to make the final choice of best candidate. Aiton did not have a name at hand, but, fortunately, Johann Graefer, a German gardener working in London, went to see Blagden and brought with him a young Pole, a Mr Au, who was then working at Kew.[52] At some point Au had worked at a nursery garden in Islington owned partly by Graefer and this was how the two men had met.[53] Banks, apparently, also knew Au, having recommended him to Aiton at some point, and was pleased with the suggestion of his appointment.[54] There was no one else, but there was a problem that made Banks anxious.[55] Evan Nepean, Under-Secretary of State at the Home Office, had raised an objection to having a 'foreigner' entrusted with such a sensitive mission.[56] Nepean asked Au to attend an interview.[57] It was getting tense.

Several days passed and letters from Blagden, Dryander and Banks crisscrossed but finally, thanks to Blagden's personal intervention and Banks's arguments in Au's favour, Nepean withdrew his objection and Au was given the go-ahead on 15 September 1785 to join Thompson's convoy to Africa.[58]

Blagden seemed to know instinctively that he had got the right man. As he told Banks: 'He is literally the only person that offered; & yet fortunately much superior to any one we could have expected.' Almost without pause, Blagden listed his attributes: 'his education has been far beyond his present situation in life,' he had an excellent writing style, could draw, knew many languages and several sciences, 'attended the hospitals in London', knew surgery, 'is bold, active & animated with the most laudable ambition of being distinguished'.[59] Fine words, indeed.

Blagden spoke with Thompson who expected Au in Portsmouth – 'everything seems well settled,' Blagden assured Banks.[60] Banks now drew up his promised 'Instructions'. The document detailed every aspect of Au's substantial task: if he found signs of agriculture, he was to report on the nature of the soil, the tools and manure used, and the crops harvested; he was to collect specimens of everything that was cultivated; he was to describe the topography, the water quality, the forests, and take samples of every living botanical specimen, flowers, fruits, roots and seeds. And, without detracting from this objective, he was 'to collect such seeds as he shall find ripe or bulbous roots as he may dig up which on his return home may furnish the Royal Gardens at Kew with something valuable to that magnificent collection'.[61] There is no doubt that Banks had high hopes for this mission. The settlement of convicts was not his aim – that was the government's business: appropriating the natural history of somewhere new was, and he relished the anticipated results.

HMS *Grampus* and HMS *Nautilus* sailed for Africa on 28 September 1785. Au was on board but not under that name. Both the instructions drawn up for him and his entry in the ship's muster announced a change of name: the man whose education, according to Blagden, far surpassed his station in life was now known as Antoni, and then Anthony Pantaleon Howe – it wouldn't be the last time he changed his name.[62]

By the turn of the year the two ships were nearing the point where they would separate. Then disaster struck. On 16 January 1786, Commodore Thompson suddenly became very ill and died the very next day. George Tripp, the second-in-command, took over the *Grampus* and with it charge of the Africa station. The command of HMS *Nautilus* fell to the nineteen-year-old Thomas Boulden Thompson, the late Commodore's 'nephew' and heir.[63]

On 2 February, the two ships parted as planned. The voyage to southwest Africa now began and it proved to be long and dull but, finally, on 21 March 1786, the ship arrived in St Helena Bay, about 150 kilometres to the north of Cape Town. The plan was to survey the coast northward from this point, including finding Das Voltas Bay, whose precise location was disputed.

It didn't begin well and it continued badly. On 11 April they found Das Voltas Bay but the soil looked barren and dry. On 27 April they entered Walvis Bay and met four local people whose homes, located some way inland, looked poverty-stricken, more like huts than houses: they resembled, Thompson remarked, 'the halves of bee hives, with the backs to the wind'.[64] It was all too depressing and on 17 May, Thompson decided that he had seen enough and took the *Nautilus* to the west heading for the island of St Helena, leaving the African coast for good. His terse comments in his journal on that day said it all: 'completed the Survey of the Coast from Lat 32.47.47 S. to 16.0.0 S. without finding a Drop of fresh Water or seeing a Tree'.[65]

HMS *Nautilus* was back in Spithead on 23 July 1786. Immediately Thompson went to London to see Lord Sydney. Over a period of three weeks, the two men discussed the voyage and Thompson's conclusions.[66] Sydney would have found the disappointing news painful. The government had invested so much in the hope of a settlement in southwest Africa and now it was over.[67]

A settlement at Das Voltas was deemed impossible and a substitute place was desperately needed. There were at least 1300 felons on hulks who had been sentenced to transportation and they needed to be moved.[68] On 18 August, Lord Sydney wrote a letter to the Treasury. The report of the officers of the *Nautilus*, he remarked, concluded that 'the coast [was] sandy and barren, and from other Causes unfit for a settlement.' He then went on: 'His Majesty has thought it advisable to fix on Botany Bay situated on the coast of New South Wales.'[69] Banks was not mentioned by name as recommending the destination, but, incorrectly, Cook was.

There was no need for a survey. Cook had already done that when the *Endeavour* was on the coast. Events moved very quickly. By the end of the month Lord Sydney had informed the Admiralty of the decision and of the King's command that a warship and a tender should be prepared to accompany a fleet of ships carrying 750 convicts to Botany Bay. A month after that, the Admiralty confirmed that they had ordered two ships to be fitted and that the lead ship, HMS *Sirius*, would be under the command of Captain Arthur Phillip.[70]

The newspapers were soon on to the story.[71] Botany Bay and transportation became inseparably linked.[72] The Europeanisation of Australia would now begin.

Anthony Pantaleon Howe had returned with a collection of seeds from some of the places HMS *Nautilus* anchored.[73] They went to Kew. Though it was not a voluminous collection, Banks was pleased with the way Howe had conducted himself.

Howe would soon be sent on another, and much more, daring mission.

3

1780: The First Circumnavigation of Archibald Menzies

A little before Banks became involved in the potential location of penal colonies, he had been focusing on the Pacific Northwest. It began with the fur trade, in particular the pelt of the sea otter, but, as usual, Banks saw an opportunity for collecting plants.

On 7 October 1780, the *Resolution* and the *Discovery*, the ships of Cook's third Pacific voyage, anchored in Deptford, after a four-year voyage to find a passage linking the Atlantic and Pacific Oceans. The trip had been a disaster: no passage was found, Cook was killed in a skirmish on Hawaii on 14 February 1779, and his successor, Charles Clerke, had died on the return voyage.

The tragic news of Cook's death had travelled overland across Russia to reach London many months before the return of the ships.[1] John Montagu, the 4th Earl of Sandwich and First Lord of the Admiralty, was nominally responsible both for the expedition and for the publication of its narrative. Whatever enthusiasm he may have had for the expedition, he lacked any for the publication and, no sooner had the ships docked, than he was in touch with Banks hoping for help. Montagu had all the journals of the voyage to hand but, as he told Banks, 'I had so much trouble about the publication of the last two voyages, that I am cautious or rather unwilling to take upon me to decide in what manner & for whose emolument the work shall be undertaken.'[2] Banks immediately came to his aid and wrote to James King, who had taken command of HMS *Discovery* after Clerke died, asking him to write the narrative. King was delighted to accept and thereby came under the patronage of Banks, someone he admired and whom he referred to as 'the common Center of we discoverers'.[3]

Until the ships returned, no one in London had any detailed knowledge of the expedition's proceedings. It was all in the logs and journals, which had only arrived with the ships and would now be put at King's disposal and, by extension, at that of Banks, who was in overall charge of the publishing project.

At some point during the writing of the narrative Banks must have learned, either by reading it in draft form or hearing of it from King, that during April 1778, while the *Resolution* and the *Discovery* lay at anchor in Nootka Sound (the indigenous name which Cook renamed King George's Sound), Vancouver Island, the ships' company, officers and crew alike, were treated to an unusual sight. Cook noted in his journal that no sooner had they entered the bay than they were surrounded by canoes filled with furs of various animals but, in particular, those of what Cook called the 'sea beaver' (but which we know as the sea otter). The indigenous people were there to trade and seemed to know what they were about and what they wanted. 'For these things', Cook remarked, 'they took in exchange, Knives, chisels, pieces of iron & Tin, nails, Buttons, or any kind of metal. Beads they were not fond of and cloth of all kinds they rejected.'[4]

Banks would not have been surprised to learn this. Though he had never been to this part of the world he knew a lot about it. He would have known that the Russians, beginning with Vitus Bering's expedition to the North Pacific in 1741, had discovered the fur-bearing mammals in the area and, soon afterwards, began trading them to the Chinese. He would have known that Georg Steller, the expedition's naturalist, produced the first description of the sea otter and, particularly its pelt, the gloss of which, he reported, 'surpasses the blackest velvet'.[5] Banks would also have been familiar with Peter Simon Pallas's published account of his travels in Russia, which provided a detailed account of the trade in furs from Russia to China at the Siberian–Chinese border trading post of Kyakhta.[6] William Coxe's 1780 publication, *Account of the Russian Discoveries Between Asia and America*, the first of its kind in English, also spoke glowingly of the trade in furs, and was only the latest in the long line of publications that discussed the Pacific fur trade.[7] Banks would have concluded from all this information that there was no reason why the British could not follow the Russians into this lucrative space.

The possibilities of a British-controlled Pacific fur trade had, in fact, already been advocated by both James Matra and Banks when their proposal for settling Australia, dated 23 August 1783, was put before the government.[8] The plan, it will be recalled, proposed a community of free settlers with the addition of American Loyalists who would produce for export to Britain and re-export to Europe a number of crops suited to the climate and which could be imported for cultivation from India, the nearby Spice Islands, New Zealand and elsewhere. Because Matra and Banks were imagining a settlement that would not be an expense to Britain, but rather increase its wealth, they proposed that commerce should be opened between New South Wales and the fur-rich Pacific coast which Cook had surveyed. The conclusive element, Matra and Banks emphasised, was the fantastic return on investment that the fur trade guaranteed. 'The skins which they procured', Matra and Banks wrote referring to Cook's third voyage, 'then sold in China at 400 hard Dollars each, though for the few they brought home, of the same quality, they only received about Ten Pounds each.' And there was more. 'As our situation in New South Wales, would enable us to carry on this Trade with the utmost facility, we should no longer be under the necessity of sending such immense quantities of Silver for the different Articles we import from the Chinese Empire'.

The government, as seen in a previous chapter, did not take up Matra's proposal for settling Australia with free settlers and American Loyalists, and the commercial possibilities of a Pacific fur connection were not pursued. But the idea of British ships plying a trade between Nootka Sound and China remained attractive.

It got a tremendous boost when James King's authoritative telling of Cook's final voyage was published on 4 June 1784. King laid it out simply and starkly in a chapter in volume three which described the journey of the *Resolution* and the *Discovery* from Macao, the Portuguese settlement at the mouth of the Pearl River in southern China, to the main port city of Canton, almost 100 kilometres further up river. The whole stock of sea otter pelts that were sold in Canton fetched just under £2000. Considering that by the time the ships got to Canton, some of the pelts had already been sold on the Kamchatka Peninsula and others had been spoiled, this figure represented an enormous profit margin.[9] King also mentioned that some of the ship's company were so obsessed by the potential wealth they were on the point of mutiny, intending to force the ships to return to the Pacific coast of America. Indeed, as it turned out, while they were in Canton two seamen from the *Resolution* commandeered the ship's cutter. A search was made but to no avail. 'It was supposed', King wrote, 'that these people had been seduced by the prevailing notion of making a fortune, by returning to the fur islands.'[10]

King remarked that the British fur trade should be an extension of the East India Company's trade in Canton and, with two ships making the return voyage to Nootka Sound from Canton annually, they could bring as many sea-otter skins as possible to this lucrative market. At the time the East India Company held a monopoly of all British trade in most of the Indian and Pacific Oceans. King added that in order to have something with which to trade at Nootka Sound, each ship would need to carry five tons of unwrought iron, 'a forge, and an expert smith, with a journeyman and apprentice, who might be ready to forge such tools, as it should appear the Indians were most desirous of . . . iron is the only sure commodity for their market.'[11]

At a price of just under £5, King's three-volume account of Cook's last voyage was not cheap. The entire print run, however, sold out in three days and a contemporary reviewer commented that second-hand sets were changing hands for more than twice the original price.[12] No doubt much of the attraction was the account of Cook's death but some readers must have been keen to learn about the fur bonanza that was the Pacific Northwest.

Richard Cadman Etches was just the sort of person to be intrigued by the promise of the fur trade. Born in Derbyshire, Etches came to London in 1779, the same year as Cook died, and seems to have established himself quickly in the City's tea and wine business, going into partnership with Robert Hanning Brooks.[13] Details about his activities are at best sketchy but on 13 March 1785, Etches met with Banks to discuss his proposal: to send two ships to the Pacific Northwest to obtain furs for trade in Canton as King had recommended.

Banks thought that Etches should be more ambitious. Japan, for example, should be included as a possible market.[14] Banks had been led to think about Japan as a commercial possibility probably as early as 1774 when John Blankett, a British naval officer, wrote a report to Lord Sandwich at the Admiralty on the potential of Japan and the nearby islands as a market.[15] But what impressed Banks the most was most likely the visit to Soho Square by Carl Peter Thunberg, who had been with Francis Masson collecting plants at the Cape, and who had been at the Dutch East India Company settlement in the Bay of Nagasaki for more than a year in 1775 and 1776. Thunberg had stayed at Banks's home in late 1778 and early 1779 on his way back to Sweden from Japan, and while he was there he had spoken to Banks about the potential of Japan as a market.[16]

Banks agreed to write to Thunberg, who by then was in Uppsala, asking him to expand on the Japanese market and, specifically, his opinion of the likelihood of British ships being allowed to trade in the 'northern Provinces', something which Thunberg had said was possible.[17]

Banks also suggested that Etches should think bigger in terms of how he wanted to finance the venture. Etches agreed with him and, in addition to the £20,000 subscription he had arranged to fit out the two ships, he would try to increase the nominal capital to £200,000 if the first venture were a success – which he didn't doubt it would be.[18]

Not long after the meeting with Banks, Etches floated his new business, which he appropriately called 'The King George's Sound Company'. It had nine partners: Richard Cadman Etches, his brother John, a number of other merchants, and two naval officers from Cook's third voyage, George Dixon, who served as an armourer on HMS *Resolution*, and Nathaniel Portlock, who was appointed on HMS *Discovery* as master's mate and then transferred to HMS *Resolution* following Cook's death.

Portlock may already have worked for Etches but apparently he didn't know Dixon. Banks, however, did know him. Dixon had been in contact with him in late August 1784 to propose an expedition across America starting in Quebec. Dixon wanted it to be as much a scientific expedition as a commercial one. He volunteered to undertake the astronomical work and to assign David Nelson, whom he knew from the *Resolution*, to be responsible for natural history. As Banks had recommended Nelson to be the naturalist on the *Resolution*, he would have been pleased with this suggestion.[19] All the other personnel, thirty or forty men, Dixon thought he could get in Quebec where there was a greater likelihood of finding men who knew the trade and even some indigenous languages.[20] In his reply Banks welcomed the idea but thought the time was not right, since a new nation, the United States, had just been declared on that continent.[21]

As a consolation, Banks may have suggested to Dixon that he should attach himself to a commercial venture to the Pacific, and it is very likely that it was Banks who told Dixon about Etches.[22] However the connections were made, what was important was that by including two veterans of Cook's final voyage, Etches was drawing a line directly from Cook to himself, coupling London, again, to Nootka Sound.

Thunberg's reply to Banks has not survived, but it is highly likely that he repeated his positive opinion of Japan. In Etches' proposal to the East India Company to allow him to trade at Canton, he referred to his having received 'the most flattering encouragement from conversation of Gentlemen of the greatest eminence and abilities (for knowledge of Japan)'.[23] The East India Company granted approval in the early part of May for the venture to proceed and to open up commercial relations with Japan.[24]

Etches then bought two ships and began fitting them out for the voyage. By the beginning of June the ships were being provisioned and then a problem was discovered. Etches also needed the permission of the South Sea Company, whose charter included the Northwest Coast of the Pacific, to trade. Negotiations, conducted by George Rose, Secretary to the Treasury, were quickly begun and by 4 August, all the paperwork was in order.[25]

On 29 August, a distinguished party assembled at Deptford to inspect the ships. In the group were George Rose and Joseph Banks; his schoolfriend Constantine Phipps, now Baron Mulgrave, a member of the Board of Control for India and the Board of Trade; and Sir John Dick, British Consul at Leghorn from 1754 to 1776, a civil servant who had taken an early and keen interest in the venture.[26] Though the ships were ready they had still not been named. George Rose named the larger of the two the *King George*, and Banks, charged with naming the smaller ship, christened it *Queen Charlotte*.[27]

And so, on 31 August 1785, the first two ships of the King George's Sound Company left Gravesend, but it took two more weeks, until 16 September, before they could clear the English coast for the long voyage to the Pacific Northwest by way of Cape Horn and the Hawaii Islands. Portlock was in overall command of the expedition, and captained the *King George*, while Dixon took responsibility for the *Queen Charlotte*.[28]

Too impatient to wait for news of the first expedition, which was somewhere in the Pacific, Etches began to plan a second.[29] He found another veteran of Cook's voyages to command it – James Colnett, who had served as a midshipman for three and a half years under Cook on HMS *Resolution* on his second voyage to the Pacific.[30]

In the summer of 1786, Colnett, now a first lieutenant, took leave from the Royal Navy (it was peacetime and many naval officers were on half-pay), and, in his own words: 'Having been recommended to a Company of Merchants trading to the NW side of America, the beginning of July 1786 I receiv'd a Letter from their Secretary offering me the command of a Cutter to perform a Voyage to KING GEORGE'S or NOOTKA SOUND . . .'[31] That ship was named the *Prince of Wales*. Its tender was the *Princess Royal*, under the command of Charles Duncan, a naval man like Colnett.

Preparations for the voyage were quick and even though they missed their sailing date of early August, on 23 September 1786 the ships left Deptford for the Pacific. The second Etches expedition was on its way and this time Banks had a botanical collector on board   Archibald Menzies.

Menzies had first written to Banks at the end of May 1784, when he sent him a letter and a small parcel of seeds from Halifax, Nova Scotia. Menzies explained that he was doing this 'at the request of Dr. Hope, Professor of Botany at Edinburgh'.[32] While Banks did not know Menzies, he certainly knew John Hope well, having been corresponding with him since 1766 at least.[33] Hope, a physician and an early and ardent supporter of the Linnaean system, had been Regius Professor of Botany in the University of Edinburgh since 1768 and in charge of the Royal Gardens (forerunner of the Royal Botanic Gardens, Edinburgh).[34] Menzies had been born in Weem near Aberfeldy, Perthshire in 1754, the son of a tenant farmer. He had attended Hope's botany lectures while studying medicine at the university. He also worked for Hope as a gardener between 1775 and 1778.[35]

Menzies never got his medical degree but may have been granted a licence to practise surgery. This would have been enough for him to apply as a surgeon's mate in the navy, which he did in 1782, and joined HMS *Nonsuch* that year.[36] A year later he was on HMS *Assistance* as assistant surgeon and, when he wrote to Banks, he was stationed in Halifax, the base of the North American Station of the Royal Navy, on the same ship.

In 1784, Menzies, now thirty years old, returned to Halifax from cruising the waters of the Caribbean, as part of the duties of the British naval presence in the region.[37] The parcel for Banks contained seeds from where the ship had called: Barbados, Dominica, St Kitts and Nevis in the West Indies, and Sandy Hook, near New York. For those plants he could examine, Menzies remarked, he included their Linnaean names.[38] Menzies ended his letter by telling Banks that the land around Halifax promised excellent botanising and that he would 'wholly devote [his] vacant hours to natural history . . . I have no doubt that I shall be enabled to send you another parcel early in the autumn.'[39]

He was as good as his word. On 2 November 1784, Menzies sent Banks a second parcel – the first one he knew had arrived safely and was much appreciated – of seeds from Nova Scotia, several of which he could not find mentioned in the edition of Linnaeus he had with him.[40]

Over the next two years Menzies continued to remain in contact with Banks and sent him seeds, including some from the Bahamas. Unfortunately, as Menzies had to admit, he had been unable to find the species of plants that Banks had specifically asked for.[41]

Still, Banks must have been very pleased. Not only did he have specimens of plants from Nova Scotia but he was now in touch with a professional collector who might just be called a botanist – Menzies's command of the Linnaean system qualified him for that appellation.

On 12 July 1786, a day before the East India Company gave Etches permission to trade in the Pacific Northwest, Banks got a letter from Menzies that he would soon be in England.[42] A little over five weeks later, 21 August 1786, Menzies wrote to Banks from Chatham to announce his arrival. He said that he was forwarding another parcel of seeds to him and expected to be in London himself in a matter of a few days. He added that he had heard that 'there is a Ship, a private adventurer, now fitting out at Deptford to go round the World – Should I be so happy as to be appointed Surgeon of her, it will at least gratify one of my greatest earthly ambitions, & afford one of the best opportunities of collecting seeds & other objects of Natural History for you . . .'[43]

What happened next is unclear but, on 7 September, Menzies told Banks that he had been appointed surgeon to the expedition and was expecting to sail soon.[44] Menzies was the perfect candidate: he had served on naval vessels, had studied with John Hope, knew his plants and the Linnaean system well. He even had a highly glowing reference from his old teacher.[45]

Did Banks recommend him or was it Hope? Menzies's letter to Banks, unfortunately, reveals nothing.[46] All we can be certain of is that Menzies did ask Banks to intervene with Etches to allow him more freedom to collect on the voyage and that Etches agreed; and that Menzies visited Soho Square on 27 September while Banks was at Revesby to offer some more plants from Halifax (including an orchid, which Jonas Dryander, Banks's librarian, thought superb) and presumably to acquaint himself with Banks's Pacific collection.[47]

Though Menzies kept predicting an early departure, in fact, the *Prince of Wales* and the *Princess Royal* did not leave the English coast until 16 October 1786.[48] On 16 November, when the ships were in the Cape Verde Islands, Menzies wrote his first letter to Banks. He thanked him for writing to Etches to give him more freedom to collect than was originally stated because 'the west coast of N. America presents to me a new & an extensive field for Botanical researches . . . & I can assure you that I shall lose no Opportunity in collecting whatever is new, rare or useful.'[49] Menzies emphasised that his confidence was supported by experience: before he had left for the voyage, William Aiton at Kew had shown him a specimen of *Houstonia caerulea* (popular name little or azure bluet) which he had raised from seed sent to him from Nova Scotia, the implication being that if he could do it once he could do it again and again.[50]

The ships would be heading for Cape Horn before entering the Pacific. Colnett had planned to stop at what was then called Staten Island, but which now has the name of Isla de los Estados, and which lies about twenty miles off the southernmost tip of Tierra del Fuego. His instructions were to land Samuel Marshall, a naval lieutenant, and the son of Captain Samuel Marshall, whom Colnett had befriended in Cowes, with a party of fifteen men, to establish a sealing settlement.[51] Menzies was going to take this opportunity to collect specimens of *Drymis winteri* (then referred to as *Wintera aromatica* and popularly known as Winter's bark), from which an anti-scurvy medicament could be prepared – its properties were first noted by Captain John Winter, after whom the plant is named, in 1578, and reiterated forcefully by Hans Sloane in 1693. (Banks, who had seen the plant when he was in Tierra del Fuego on the *Endeavour*, had specifically asked for it to be collected.)[52]

On 17 November, the day after Menzies wrote his letter, the ships continued their voyage, passing from the hot and humid tropics to the squally and snowy South Atlantic. On 26 January 1787, the ships anchored in New Years Harbour, Staten Island, almost in the exact same spot as the *Resolution* when it was there – with Colnett – in 1775. While the main task – landing the sealing party – was proceeding, Menzies lost no time botanising.

As the ships were about to depart, Menzies wrote to Banks. The weather had not been kind 'in this wild and inhospitable clime' but he had managed to make several excursions and find many plants that were not listed in Linnaeus. Of equal, if not greater importance, Menzies was delighted to have found the *Wintera aromatica* growing everywhere and in flower – 'This beautiful tree . . . loads the circum-ambient Air with a most pleasing aromatic Odor.'[53] Menzies had collected and potted some twenty young plants and, as he had no possibility of keeping them alive during his voyage, he was assured by Lieutenant Marshall that he would take them back to England on the relief ship that was to be sent from London for the party, together with seeds which he would collect for Menzies.[54]

Menzies carefully told Banks that though the plants were addressed to him, he was to send some of them to John Hope in Edinburgh. Menzies did not know that two days before he arrived in Cape Verde, Hope had died in Edinburgh – when Menzies did finally learn of Hope's death, he referred to him as 'my best and only friend'.[55] The plants fared just as badly, as it turned out. The *Duke of York*, the relief ship owned by Etches, left on 21 April 1787 for Staten Island but, on 11 September, while at New Years Harbour it went down. The crew and the sealing party were all saved as they managed to leave in boats, but the plants went down with the ship.[56]

Though the intention was to sail to Hawaii, which Cook had visited in January 1778, Colnett decided to go straight for Nootka Sound.[57] The decision certainly meant that Colnett would be arriving at his destination sooner than anticipated but, on the other hand, the long sea voyage was bringing on scurvy. Everyone was thankful that on 5 July 1787, after about five months at sea, the coastline of Vancouver Island was sighted: the next day they were visited by several canoes bearing a few furs.[58]

From now on the season and the weather would dictate the expedition's plans. The general idea was to begin at Nootka Sound and move north-ward along the coast, trading as much as possible. But progress turned out to be slow as the ships were in poor shape. After sailing along Vancouver Island, they set course for the Queen Charlotte Islands where they anchored for two weeks during the second half of August 1787. Colnett, fearing more damage to the ships, sailed across the strait – Hecate Strait – that separated the islands from the coast, and he made base on an island which Menzies named for his London patron – Banks Island. There they remained for almost three months to repair the damage, by which time, as Colnett remarked, winter had definitely closed in: 'the Country half cover'd from the summits of its hills to the water's Edge with snow'. On 19 November, the *Prince of Wales* and the *Princess Royal* departed the Pacific Northwest to spend the winter in Hawaii.[59]

However, the time there did not go well: violent confrontations soured relations and trade. Colnett must have been anxious to leave and, on 13 March 1788, full of provisions and now with three Hawaiians on board, who wished to join the voyage, the *Prince of Wales* and the *Princess Royal* headed back to the Pacific Northwest.[60] They arrived safely in April.

For the next three or four months both ships traded, one southward and the other northward with the intention of meeting at any one of several predetermined points. This never happened, but thanks to the information that was left in letters in the safe-keeping of Haida chiefs on the coast, Colnett knew that Charles Duncan, in command of the *Princess Royal*, was safe. Within a day of each other, on 17 and 18 August 1788, both ships left the coast for Hawaii with their cargo of pelts, the *Princess Royal* having out-performed the *Prince of Wales*.[61]

All along, Menzies had been botanising whenever he could and had found many plants that were new to western science, including a variety of penstemon with large lavender to purple flowers, that was named in his honour.[62] He continued to collect specimens, whenever he could, some of them which were new to Europe such as the trailing raspberry (*Rubus pedatus*), and *Sanguisorba menziesii*, a plant with grey-green leaves and red flowers, in the shape of bottlebrushes.[63]

The first part of the expedition was now over. It was time to realise the proceeds of the venture, to sell the pelts in Canton during the trading season which would begin in October.

On 12 September, the two ships met at a familiar anchorage in Moloka'i. Once provisioning for the next leg of the voyage was completed, the ships left Hawaii on 30 September and arrived in Macao on 12 November 1788, a little over two years since leaving England.[64]

The furs, almost two thousand of them, did not sell as well in

Canton as Colnett had predicted, but the return, at just over £20,000, nevertheless justified the outlay.[65] Colnett did better than any of his predecessors in the Pacific Northwest, so much so that the original plans were altered substantially.[66] John Etches, Richard Cadman Etches's brother, and the *Prince of Wales* supercargo, the merchant responsible for the sale of the ship's cargo, decided to join forces with another fur-trading scheme headed by a group of private traders in Canton. The decision was taken that the *Princess Royal* would remain behind in Canton and join three other ships for the next season and that the *Prince of Wales* would return to England.[67]

Now under the command of James Johnstone, previously the ship's chief mate and a close friend of Menzies's (they had been together on HMS *Assistance* in the eastern Atlantic), the *Prince of Wales* set sail for England from Macao on 1 February 1789 – Charles Duncan having relinquished command of the *Princess Royal* because of ill health was returning home as a passenger, as was Menzies. After stopping briefly at Sumatra, where Menzies collected some fine plants, and St Helena, the *Prince of Wales* anchored off the Isle of Wight on 14 July 1789 and was back on the Thames several days later.[68]

What had Menzies accomplished? Without a plant list to consult, it is impossible to be precise, but it has been estimated that Menzies collected in the region of one hundred plants.[69] This may not seem a lot, but given that many of the specimens were new to European botany the figure is much more impressive than it sounds. Banks was, of course, a recipient of both dried plants and seeds, but so were other renowned botanists.[70] Professor Daniel Rutherford, John Hope's successor at Edinburgh, got seeds and dried plants from the ship's major calling places; as did James Edward Smith, the President of the newly founded Linnean Society of London.[71]

When he returned to London, Menzies spent a lot of time at Soho Square to try to make sense of his collection, comparing the specimens with those in Banks's herbarium and supported by the vast textual and illustrative material in the library.[72]

With the return of Menzies from his circumnavigation, Banks's interest in the Pacific fur trade faded. His relationship with Menzies, however, would soon enter a new and more extensive phase: a second and longer circumnavigation.

While it was a minor part of the Pacific Northwest story from Banks's point of view, during the time Menzies was in the Pacific, China moved centre stage into Banks's life.

4

1782: The Brothers Duncan in Canton

Banks had been thinking about China for some time. The country's botany was of enormous interest to the West. With a population of 300 million and a landmass the size of the European continent, China had a storehouse of botanical riches hardly known in Europe. By the mid eighteenth century, Chinese plants were growing in several royal and private gardens across Europe, but the numbers and varieties were insignificant when compared to what Europeans believed remained beyond their reach.[1]

As Banks knew, Europeans were not free to go where and when they wanted in China. The Qianlong Emperor, who ascended the throne in 1735, was responsible for many of the restrictions, especially as they applied to Europeans wishing to trade. In 1757, the so-called Canton System came into force, whereby most European merchants became confined to Canton.[2] They were allowed to remain there for only part of the year, known as the trading season, normally from October through to March – during the break between the end of the southwest monsoon and the beginning of the northeast monsoon.[3] In Canton, the European merchants, better known as supercargoes, traded under the umbrella of their respective national East India Companies, and lived in factories, buildings that were rented out to them by the resident Hong (Chinese) merchants.[4] More than thirteen such factories, sited on the banks of the Pearl River and separated from the rest of Canton, were allocated to the foreign merchants.[5] The Europeans and the Hong, who generally did not speak each other's languages, were obliged to do business with the help of a corps of Chinese middlemen, known as the linguists, who spoke a kind of pidgin English.[6] No women were allowed in the factories. When the trading season was over, the supercargoes typically moved to Macao, often joining their families, where they awaited the beginning of the next cycle.[7]

Restrictions on movement had a direct bearing on what Europeans knew of Chinese natural history. Banks was well aware of this. When, for example, in 1768, he was preparing for the epic voyage on HMS *Endeavour*, Thomas Falconer, a classical scholar who knew an impressive amount of natural history, warned him that he would learn hardly anything about Chinese botany in Canton. 'The Europeans', Falconer commented, 'have but little communication with the Natives, & none beyond the suburbs of Canton. You will have a better yield at Batavia if you stop there, as our China Ships sometimes do.'[8]

Banks's knowledge of Chinese natural history, especially its botany, was second to none.[9] In his own library he had virtually everything written by Europeans about China's natural history, much of it contained in compilations of letters sent home from Jesuit missionaries based at the imperial court in Peking.[10] The titles spanned the seventeenth and eighteenth centuries, from Michael Boym's early *Flora Sinensis*; through to the thirty-four-volume *Lettres édifiantes et curieuses*, compiled and published between 1703 and 1776 by Jean-Baptiste Du Halde, a French Jesuit historian, the contents of which provided the most comprehensive European understanding of China in the eighteenth century; and finally to the *Mémoires concernant l'histoire, etc. des Chinois*, a multi-volume work, a collaboration between French academicians, missionaries and two Chinese students who lived in France where they studied western science.[11] Banks particularly valued this compendium, especially its third volume, published in 1778, which contained the section on Chinese plants and trees.

Though highly esteemed, these compilations had one very important limiting factor: the information they contained was derived from secondary sources. But this was not true of Banks's primary source on far-eastern botany, Engelbert Kaempfer's *Amœnitatum exoticarum* (*Exotic Pleasures*), published in 1712, a first-hand botanical treatise of the oriental world by a European botanist. Kaempfer trained as both a physician and a naturalist. After several years travelling to Russia and modern-day Iran he joined the Dutch East India Company as physician, and, in 1690, he was posted to the Company's factory in Nagasaki, Japan, and began applying his training to the field of Japanese natural history.

For two years Kaempfer administered to his Dutch patients, finding as much time as possible to botanise locally: as a physician he was allowed on the mainland where he continued his botanical researches. When, in 1695, he returned to Europe, he settled down in his Westphalian hometown to organise his herbarium and his botanical notes from all of his travels. The result of the latter exercise was the publication in 1712 of the illustrated *Amœnitatum exoticarum*. Kaempfer died four years later.[12]

Banks's copy in the British Library still bears the annotations of Daniel Solander in which he noted the 'new' Linnaean names for Kaempfer's plants.[13] The fifth part of the book is dedicated to Japanese plants and Kaempfer describes each plant in a manner that would have warmed Banks's botanical heart: Kaempfer 'placed the various parts of the plant under the botanist's knife', beginning with the root and following through eventually to the seed.[14] While Kaempfer covered a wide range of plants – those bearing fruit and nuts and those valued for their flowers – he singled out four plants for special mention and attention: the paper mulberry tree, the musk seed, a creeper and, most importantly, *Camellia sinensis*, the tea plant. The plants in this section of Kaempfer's book, though Japanese, were indigenous to China and their names were written out in Chinese characters.

Besides printed books about Chinese natural history, Banks also had a vast stock of illustrated material. Some of it accompanied texts, such as Boym's, Kaempfer's and Pierre Joseph Buc'hoz's recently published *Herbier*, a collection of Chinese medicinal plants that were growing in China and in European gardens.[15] In addition Banks also had manuscript drawings of Chinese plants, without textual information, especially a collection made in Canton in the early 1770s by a Chinese artist working in collaboration with John Bradby Blake, one of the East India Company's supercargoes, and his English-speaking Chinese assistant, Whang at Tong.[16]

If, by the time Banks had finished poring over his books and drawings, he still wanted to know more Chinese botany, he could leave his library in Soho Square and stroll over to the British Museum in Bloomsbury, where, as President of the Royal Society, he held the office of Trustee. In this privileged position he could examine yet another and earlier collection of texts and images of Chinese and Japanese plants that had formerly belonged to Sir Hans Sloane.

Sloane, who was President of the Royal Society from 1727 to 1741, was an avid collector of natural history objects, books and manuscripts. When he died in 1752, he donated his vast collection, which also included the Chinese plant specimens in James Petiver's herbarium, and Kaempfer's entire natural history collection, to the British nation.[17] In the following year, Sloane's bequest became the first collection to form the British Museum.

When Banks had finally reached the limit of experiencing Chinese plants through text and image, in his own and Sloane's collection, he could always go to the royal gardens in Kew and see thriving examples of the real thing. Plants from China had been arriving in London from the 1720s, sometimes ending up in private gardens and nurseries and sometimes at the Chelsea Physic Garden. By 1789 when William Aiton, the head gardener at Kew, published his *Hortus Kewensis*, as many as one hundred different plants from China were growing there.

As he approached the gardens, Banks's mind would have been focused on China as soon as the 163-foot Pagoda slowly came into view. Built by the architect Sir William Chambers between 1761 and 1762 – Chambers also built Somerset House, where the Royal Society was housed from 1780 – it was designed both to be looked at and to be looked out from: 'from the top you command a very extensive view on all sides in some directions upwards of some forty miles distance, over a rich and variegated country.'[18]

All this was well and good. London's institutions and Banks's library and herbarium were exceptional when it came to information about China's natural history, but it was no substitute for collecting living plants from their natural habitat. How to do this was not obvious: only the East India Company, with limited personnel, had access to Canton. And then, out of the blue, a possibility appeared.

John Duncan was born in Brechin, Scotland in 1751 and after completing his medical studies he began working as a surgeon on board East India Company ships sailing to India and China. In late winter 1782, he was returning to China, this time as the East India Company's resident surgeon in Canton. Accompanying him on board the *Morse*, was William Henry Pigou, who was going back as the East India Company's Chief of the Council of Supercargoes.[19] Their destination was Canton, where the English were the major presence among the foreign traders.[20] It is not clear which of the two men knew Banks but they jointly wrote a letter to him from Rio de Janeiro dated 31 May 1782, en route to Canton. In it they informed Banks that they were sending him a few specimens of plants, birds, insects and other examples of natural history from Rio de Janeiro, where they had been since late April; and that once they got to Canton, for where they were leaving the following day, they would be happy to collect there on his behalf. Both Pigou and Duncan excused themselves for not being 'sufficiently acquainted with the Study of Botany' and, therefore, could do little more than collect.[21]

Banks wrote back to Pigou and Duncan on 10 August 1782 to thank both men for their collections. He was pleased with them and took up their offer of collecting for him in Canton.

Pigou and Duncan were in Canton in early October 1783 and began collecting immediately when they read Banks's letter. In early December, as the *Morse* was preparing to leave for England, they sent Banks a box of shells, some fish from Macao in a jar, a Chinese magnolia in a pot; and a 'Wanghee', a species of bamboo, which they told Captain Henry Wilson, who was returning on the ship as a passenger, that he should deliver 'Dead or Alive'.[22] Other plant specimens, including another Chinese magnolia in a pot, were despatched on the *Northumberland*, another East India Company ship leaving Canton around the same time.[23]

Duncan settled easily into Canton, even going into business with the Hong merchants.[24] Banks seemed to warm to him very quickly and lobbied successfully on his behalf with the directors of the East India Company to raise his salary by £200 per year. In return, Duncan did what he could to find the plants that Banks wanted sent to England. Some were easy to get – the *Yu Lan*, or Chinese magnolia, and water lilies, for example. Most likely Duncan got the specimens from one of the city's nursery gardens, three miles upstream from where the European and American trading companies had their factories. The nursery mostly supplied the local Chinese market but it also sold flowering plants in pots, fruit trees and seeds to foreign buyers.[25] A later visitor to the nursery left a vivid description of its layout and colours. 'Camellia with double white red & variegated flower, also a single variety with a rose Colord flower . . . Yulan, a species of Magnolia with White Red & variegated Flowers, the two last not [seen]; Azalea indica, Daphne indica . . . Chrysanthemum indicum double red, white, flesh Colord, & orange Colord. This plant is used by the Chinese Ladies as an ornament for the head.' There were also pots with fruit trees and pots planted with dwarf trees.[26]

Whether Duncan found anything new when he visited the nursery was a matter of luck. One of the plants which was on Banks's most wanted list – 'among our greatest desiderata' – the mountain or tree peony, *Mu dan* in Chinese, *Paeonia suffruticosa*, as it is known in western botany, both men knew had to be obtained through personal contacts, specifically among the Hong merchants, since the plant grew to the north of Canton where the climate was more like England.[27] On 4 April 1787, Duncan wrote to Banks to say that he had managed to get an example of it 'from a Merchant as a present'.[28]

Duncan didn't say which merchant but it is most likely to have been one of the Hong merchants. Duncan had good connections with these men, including Puankhequa I, the city's chief merchant, as he had had business dealings with them.[29] The *Mu dan* that Duncan had received was, he remarked, the true *Mu dan*: highly prized, very expensive and 'so difficult to procure' – only Hong merchants would have had access to these plants; they grew far to the north.[30] As it turned out, the *Mu dan* example survived the voyage to England, but succumbed to its first English winter in Kew.[31]

Duncan continued to collect for Banks and managed to get two more specimens of the *Yu lan*. These he sent with John Kincaid, a surgeon with the East India Company at Madras, who was returning home as a passenger via Canton.[32] When, at the end of June 1788, Kincaid was safely back in London, he had to tell Banks the sad news that the two plants had not survived the voyage; but that two fans, which Duncan was sending as presents to Dorothea and Sarah Sophia Banks, were as good as new.[33]

John Duncan's health was deteriorating and on 14 January 1788, less than a month after sending his final consignment to Banks with Kincaid, he was granted sick leave to return to London.[34] By then, Alexander Duncan, his younger brother, was already in Canton having been given permission to leave Calcutta, where he was serving in the East India Company's medical establishment, to visit his ailing brother. Alexander immediately completed his brother's outstanding consignment of more *Yu lans*, and promised to fulfil the remaining instructions.[35]

John Duncan was back in London several months later and contacted Banks at the earliest opportunity. He told him about his having had to leave Canton and that his brother, Alexander, was now acting as the factory's surgeon, the supercargoes there having insisted that he stay on. Banks agreed to use his influence with the East India Company to make Alexander's position a permanent one. Banks wrote to Thomas Morton, the Secretary to the Court of Directors of the East India Company, asking him 'to confer a favor of no small matter'.[36] This was not how appointments were normally made – the medical service in Calcutta decided on who would be the surgeon at Canton – but though the process was delayed, in the end, John Duncan and Joseph Banks got their way and Alexander Duncan got the appointment.[37]

Duncan started collecting directly for Banks in late 1788 and continued to do so for about eight years. In his very first letter to Banks, dated 6 December 1788, Duncan proudly told Banks that he had more *Yu lan*s – the last ones he had sent he was sure would not have made it – and seeds of the red and white water lily. All of the specimens were in good shape but to give them the best chance of surviving the voyage back he entrusted the consignments to two passengers, one of them probably Henry Lane, a supercargo, who was returning to London on the *Talbot*; and another, a Mr Turnly who was to be on the *Minerva*. Duncan was taking the precaution of spreading the risk and enhancing the chances of survival by trusting passengers to care for the plants, and using two different ships. In all of his attention to plants, Duncan did not forget a present – a 'Box of Souchon, called here Powchong Tea' – for Lady Banks.[38]

When he read the letter, Banks might well have concluded that Alexander Duncan was going to be more enterprising in collecting plants than his brother had been. He and another 'Gentleman' (a Mr Mieron), Duncan told Banks, had already 'engaged a Chinese' to bring him specimens of the much-sought-after lacquer tree from the interior. Banks would have been right in his assessment. In the next trading season, Banks learned that Duncan had befriended Consequa, one of the Hong merchants and a nephew of Puankhequa I, and had discussed aspects of Chinese botany with him.[39] Like many other Hong merchants, Consequa had a garden in Henan (Honam), the island on the other side of the Pearl River from the East India Company factories, but Consequa's would become one of the most elaborate in Canton.[40] With his connections both inside and outside Canton, Consequa would be the vital and desperately sought-for link to the Chinese countryside.

*   *   *

In all his dealings with John Duncan, Banks, though grateful for what he had collected and sent, must have been frustrated that he could not convey specific requests to him. Duncan had already excused himself for not knowing much botany and therefore would not have been able to identify any particular plant Banks might have wanted. It is for this reason that Duncan's consignments consisted of the easily obtainable magnolia, peony and water lily. Banks had a very clear idea of the kind of plants he wanted from China – those that were not already growing in Kew or anywhere else in England.

Even though Alexander Duncan was no better at botany than his brother, this ceased to matter so much because by the end of 1789 he had received from Banks a remarkable quarto, 'The Book of Chinese Plants'.[41]

The person responsible for it was Banks's old friend James Lind, with whom he had last travelled to Iceland in 1772 and who was now physician-in-ordinary to the royal household in Windsor.[42] Lind, like many medically trained men of the time, practised his profession on East India Company ships and, beginning in 1766, he went to Canton. He was back in Canton in late 1779 or early 1780.[43] By then, apparently, he had learned enough Chinese to able to write characters and to translate some words.[44] He also collected Chinese books.[45]

In 1789, Lind was visiting Soho Square and at some point began work on an extraordinary project, what he called 'A Catalogue of such Chinese and Japonese [sic] plants whose Chinese characters are known and are botanically described' and 'to shew which of them have been introduced into His Majesty's botanic Garden at Kew'.[46]

Lind had worked out that he could cross-reference the Chinese plants mentioned in Kaempfer's *Amœnitatum exoticarum* and in the *Bencao gangmu*, China's most important compendium of *materia medica* (Banks owned a copy), with the Chinese drawings in his library (probably those that came from John Bradby Blake) and the Chinese plants actually growing in Kew, which he would know thanks to William Aiton's newly published *Hortus Kewensis*.[47] The result of this exercise was an inventory of precisely those plants that were not growing in Kew and for which Lind, with the help of his sources, could supply Chinese characters and possibly even a drawing. Lind could identify around one hundred and fifty plants that fell into this category.

'The Book of Chinese Plants' that Duncan received from Banks in late 1789 was compiled from Lind's inventory. Each page had a coloured illustration of a Chinese plant that was not then growing in England together with its Chinese characters. In the corner of each page there were several crosses added to convey how much Banks wanted an individual plant: four, the highest number, signified that the plant was unknown in the West and was a must: one, the lowest number, signified that the plant was known but had never been seen in its living state.[48] The purpose of the Book was to make plant collecting in Canton more focused and efficient. It took the guesswork out of collecting and did not require the collector to know Chinese: the plant's Chinese name could be shown to its provider for identification; and the illustration would be used for verification.

Duncan immediately put the Book to work. The characters were not always accurate, but the drawings made all the difference.[49] In his next couple of letters to Banks, Duncan prided himself on getting seventeen plants, which appeared in the Book, including the Chinese flat peach.[50]

It didn't matter now that Duncan did not know Linnaean nomenclature because this book dispensed with it. Once he had a specimen, he simply described it in a letter, using Banks's own words, and then referred to it by its romanised Chinese name and its corresponding Chinese characters, which were inserted into the letter by someone who could write Chinese. One such plant, which Duncan called *Tsu-lung-tsow* and which he described as having 'Leaves in form of Tancard with a lid', was the pitcher plant, *Nepenthes mirabilis*.[51] Duncan entrusted it to the ship's commander Captain Brodie Hepworth, who had previously brought Banks botanical specimens from China and whom he recommended as someone who cared for plants. Duncan told Hepworth that even though it might look to him as though the plant was dead on arrival, it could be pruned and might revive. The plant survived the voyage and was one of the first carnivorous plants to grow at Kew.[52]

Banks had asked Duncan to return the book to him but Duncan insisted that he still needed it. On 23 January 1790, Duncan placed ten new plants, which he had found from the book, on the *General Elliott*, about to leave Canton for London. Robert Drummond, the ship's captain, was known to pay plants the attention they deserved and Duncan was hopeful that they would survive the voyage.

The book, Duncan told Banks, had done its magic. Twelve more plants from the book and a box containing flat peach trees, which Banks had especially wanted, were ready for shipment. This time, Duncan had entrusted part of the consignment to Captain Edward

Cumming of the *Britannia*, because, as he put it, 'he always keeps his Plants in the Balcony', a safer place than most on a ship.[53]

Captain Cumming was also returning the book to Banks. No sooner had it left, than Duncan asked Banks to send it back to him so that he could continue collecting from it. By the time the trading season opened in the autumn of 1792, it was back in Canton.[54] It would leave Canton again in March 1794. Up to this point, the book had crossed from London to Canton and back four times, a total distance of nearly one hundred thousand miles and about one and a half years at sea. But its voyages were not yet over.

Duncan was not a botanist, as he himself admitted; he didn't know the Linnaean system and it was Banks's book of Chinese plants that made any kind of sensible collecting possible in China.[55] Even so, the activity of collecting remained difficult. Though Alexander Duncan knew most of the Hong merchants, did business with them, visited their gardens and took botanical advice from them, he felt very much at their mercy. He often remarked to Banks that they did not deal with him squarely, that they delayed in getting the plants he asked for. He also accused them of 'chicanery', and of deliberately frustrating his collecting activities. The lacquer tree was a case in point. In early February 1790 Duncan wrote to Banks expressing his frustration that he would probably never get a specimen of it. As he remarked: 'Whether its from the policy of the Country, or other causes, I cannot find out, but we are always disappointed, by flattering evasive answers.'[56] Duncan's own explanation of the paucity of botanical specimens and information was that it was down to 'the jealous nature of the Chinese'.[57]

The interaction between the Chinese and Duncan may not have been easy or always fruitful but these problems were nothing compared to the anxiety he must have felt when the returning East India Company ships brought Banks's yearly letter. It would only be at this point, many months after the plants were sent, that Duncan would learn of their fate. He wouldn't have been any too hopeful when he opened Banks's letter. Duncan did everything he could to increase the plants' chances of survival: he tried to contract with sympathetic captains, those who knew some botany, or some, like Captain Cumming, who kept the plants in protected spaces; he used ship's surgeons as often as he could and even passengers; and he tried to spread the risk by dividing his collections among several ships. Even so, Duncan estimated that nine out of the ten plants sent from Canton died on the returning voyage.[58]

The sea was not kind to plants. They needed the right amount of light and water to survive anywhere, ships included. But on ships, where space was at a premium, there was no satisfactory place to keep them. In respect of access to light and the absence of sea spray – salt was the plant's real killer – the main deck was the best place. But this was also where the ship's animals roamed – dogs, cats, goats, sheep and rats – who particularly enjoyed digging up soil. Plants needed to be in special containers to protect them from these destructive visitors. Then there was the issue of water. This was the ship's scarcest resource and the quantity available was monitored to the last drop. It would take a courageous captain to water the plants before watering the crew. The plants, therefore, depended on rain water, not too much, not too little and at the right time.[59]

Added to this was the fact that the shipping routes in the Canton trade were some of the longest in the commercial world – six months from Canton to London was not unusual; and the ships passed through several climatic zones – temperature, rainfall, pressure and wind speed varying tremendously – as they worked their way from the South China Sea to the North Atlantic.

When, at the other end of the chain, Banks received bad news from an East India Company ship's captain, freshly returned from Canton, he would not have been surprised, but very frustrated. As a case in point, James Glegg, the surgeon of the East India Company ship, the *Earl of Mansfield*, to whom John Duncan entrusted two pots, one containing the 'Canton *Mou dan*' and another containing a water lily, explained to Banks what had gone wrong on the return voyage. On 4 December 1787, when he arrived in Portsmouth, Glegg wrote to Banks with the sad news that the water lily, which he had kept in his cabin, had died because water had been rationed to half a pint of water per day. The other plant, 'a beautiful shrub I had in my own Cabbin, and by begging and borrowing water, I persevered well, when it spread its Branches in a vine like manner, and with the Flowers on it was very pleasing; but alas! When off the Cape of Good Hope, a heavy Sea broke into the port of my Cabbin and entirely filled the jar in which your plant was, from which it faded fast.'[60] Glegg pointed out that the plant under the care of the ship's commander also died, even though it was well watered.[61] The sad truth was that most plants did not survive the voyage.

One of Duncan's largest and one of his last shipments to Banks came in March 1794. The plants were very beautiful, Duncan excitedly told

Banks, and he hoped that the 'tedious passage does not destroy' them. For this shipment, he took the usual precautions, and in addition he put the finest plants in charge of James Main, who was returning home on the *Triton* ship. Main had been in Canton since the *Triton* arrived in December 1793 collecting for Gilbert Slater, a wealthy gentleman, with strong East India Company connections, and with a great desire to provide Chinese plants for his large garden near Leyton, to the east of London.[62] Slater had already introduced several Chinese plants to England, including a semi-double crimson rose and the white-petalled *Camellia japonica*.

Duncan had managed to get four tree peonies from Henry Browne, chief of the East India Company supercargoes, from northern China. Main was now in charge not only of Slater's collection but also of two of the four tree peonies from northern China: Duncan was excited by the prospect of its arrival in London – 'I am in great hopes, the Moutan (tree paeony) will honor you, with some of its beautiful Flowers the ensuing season,' adding that, 'It is astonishing as any thing I have ever met, to see how that King of Plants, delights in a Cold Air – and on the contrary how it droops by the heat.'[63]

Though Main's track record in sending plants back to Slater was not unlike Duncan's own, Duncan felt, nevertheless, that having Main travel with the plants would lead to the very best outcome. It was not to be.

On 23 September 1794, Main was back in London, now working for Archibald Thomson's nursery in Mile End. He didn't go into details at this stage but simply reported that the news was not good.[64] Three of the four plants in his charge had died. Most of the damage, Main later explained to Banks, happened before the voyage was half over. He didn't know why. He had had ten years' experience as a gardener and had even taken with him a set of instructions about how to care for plants at sea that Banks had drawn up for a different voyage. As Main put it: 'Transporting plants from China, is Attended by difficulties, not generally thought of . . .'[65] The final *coup de grâce*, as far as the plants were concerned, was delivered in the Channel when the *Triton* was run down by a British naval ship. The ensuing crash, when the mast toppled onto the deck, flattened most of the plants that had survived the climatic ordeals.[66]

Miraculously one of the tree peonies did survive it all, though only just. It had been kept in the ship's stern gallery. As soon as possible, it was taken to Kew, becoming the first 'Moutan' in the King's Garden.[67] Duncan had tried to get this plant to Banks for some time and at last it was accomplished.

During the trading season of 1794/95, Duncan had no live plants to send but instead, and in the way of compensation, he managed to lay his hands on what he described as 'the most beautiful piece of workmanship I have ever sent home'. It was a present for Lady Banks – 'an elegant Ivory Fan, with a Moutan painted in the centre, and a red and white water Lily, on the sides'.[68]

Captain Andrew Patton of the East India Company ship, *Ocean*, was entrusted with this singularly valuable cargo and it arrived back in London in late July 1795. In what could be one of the last letters that Banks wrote to Duncan, on 29 May 1796, Lady Banks insisted that her thanks to Duncan should take precedence over everything else. As Banks wrote, it was 'the most beautifull Fan she has ever seen probably the most beautifull one that ever reached England at least it has been hitherto unrivalled wherever Lady Banks has exhibited it & that I can assure you was in as many places as she has visited'.[69]

In what reads like a farewell letter, Banks also assured Duncan of his great service to Kew – 'we have since you have favored us with your assistance gaind many very elegant additions to the collection'. The 'Moutan', Banks wrote, was doing well. 'We have I hope possessed ourselves of the proper mode of cultivating the Moutan . . . [it] has been kept cool all the winter & has this spring pushd out a head of leaves full two feet in diameter . . . the whole appearance of the Plant is far more healthy than it was when received.'[70] Knowing all about the problems of transporting plants by sea, Banks promised Duncan that in time for the next trading season he would send him a Kew-trained gardener to return with the plants he had collected. Banks commented that Main was over-confident and 'took very few instructions'. The gardeners at Kew, on the other hand, were beginning to understand how to care for plants aboard ship.[71]

The Kew gardener never went to Canton. At the end of October 1796, Duncan may have performed his final botanical act when he collected plants, both economic and ornamental, for the East India Company's botanic garden in Calcutta. He had still not learned the Linnaean system but he had learned about observing plants. His description of the plants he was sending was full and detailed.[72] Soon after, Duncan wound up his business affairs and left Canton, never to return.[73] In 1799 he bought a property near Arbroath, Scotland, but before that his services to botany were recognised when he was made a Fellow of the Royal Society on 19 April 1798.[74]

With Duncan's departure, Banks was left without anyone to collect Chinese plants for him. A few years later, however, he did find someone who knew botany to go to Canton; and by that time, he had also devised an ingenious solution to the problem of moving living plants across vast ocean distances.

5

1786: The Madras Naturalists and Dreams of Oaxaca

In May 1787, Banks received a letter from India that amazed and excited him. It was from James Anderson, Physician General to the East India Company's establishment in Madras. It told of a remarkable discovery he had made, which could substantially impact Britain's trade with the rest of the world.

This is the first surviving letter from Anderson to Banks, but it is unlikely it was the first he sent, as it is entirely lacking the rather formal protocols used in introductory letters at that time. It is likely that someone else in Madras had previously introduced them.

Banks was already acquainted with most of the important European naturalists based on India's Coromandel Coast, many of whom were sending him plant specimens. There was a flourishing natural-history network in the area, but the most likely candidate for making the connection was Johan Gerhard Koenig, who had become very close to Banks.

Koenig was born in what is now Latvia to German parents and trained as an apothecary in Riga. In 1748, at the age of twenty, he went to Denmark where he worked in several provincial pharmacies.[1] Though he learned something about plants during the time he studied *materia medica* in Riga, his serious education in botany began when he attended university in Uppsala during the years 1757–9, where he became a student of the already famous Carl Linnaeus. It was here that Koenig met and befriended Daniel Solander who had himself come under the spell of Linnaeus shortly after he had enrolled as a student in the university in Uppsala in 1750.[2]

In 1759, Koenig returned to Copenhagen where he took up an appointment as a hospital pharmacist.[3] Soon his career blossomed: he went on two botanical expeditions, the one to Iceland in 1764 being especially important; and he also obtained a medical degree.[4] No sooner had he earned his medical credentials in 1767, than Koenig's life took another and dramatic turn when he was appointed to be the next doctor and botanical collector for the Lutheran Danish-Halle mission station based in Tranquebar, on the Coromandel Coast, nearly 300 kilometres south of Madras, where he arrived in early July 1768. The Danish East India Company was established there in 1620 and continued, though now trading under the name of the Asiatic Company, to use the town as its base. The missionaries, mostly Germans from Halle an der Saale, first arrived in Tranquebar in 1706, under the patronage of Frederick IV, the King of Denmark and Norway.[5]

Though the Tranquebar mission valued natural history very highly, Koenig didn't get along with the missionaries: he found them too controlling and restrictive – for example, he was not allowed to send specimens anywhere other than Copenhagen.[6] Also, his salary was too low for him to travel much beyond Tranquebar and this obviously affected the range and quality of his botanical collections.

It seemed he was trapped but in 1774, and probably through the intervention of the Linnaean-minded community of naturalists in Copenhagen, he was able to leave Tranquebar and travel to Madras, the headquarters of the British East India Company on this stretch of the southeastern coast.

Madras was the Company's second trading post in India and in 1640, with the building of Fort St George, it became an increasingly powerful commercial and political force in the region.[7] When Koenig arrived there he found that there were several British residents and visitors interested in natural history, among them Lady Anne Monson, who knew Solander and who we've already heard of when she met another one of Linnaeus's students, Carl Peter Thunberg, at the Cape on her way to India.[8] At that time, Koenig could not have chosen a better place to pursue his interests in Indian botany.

One of the first things Koenig did, once free of the restrictions of the mission station, was to write a letter to Solander in which he included a sample of seeds he had collected between Tranquebar and Madras. Koenig had previously written to Solander without any botanical enclosures. Now he promised Solander that he could expect 'beautiful and better examples' in the future.[9]

In 1774, Solander, after returning from the voyage to Iceland with Banks, resumed his duties as the 'Keeper of Natural and Artificial Productions' at the British Museum. He was also busily working with Banks cataloguing the *Endeavour*'s natural-history collection and looking after his herbarium.[10] Koenig's seeds would find their home in this collection.

In this same letter in which he promised seeds, Koenig made it known that he wanted help from Solander and Banks in getting him an appointment with the East India Company and, in return, he offered his services to the two men – 'the approval of the Professor and Mr. Banks is desired and for this I offer my most beseeching prayers, and to gain this will be my chief effort.'[11]

In fact, Koenig did not go to work for the East India Company, but instead, in February 1775, the Nawab of Arcot, Muhammad Ali Khan Wallajah, the local Indian ruler who had a strong interest in natural history and who had already had a European as his personal physician, offered Koenig a job as a naturalist.[12]

Though he was freer to pursue his interests than he had been with the Tranquebarians, Koenig was financially no better off working for the Nawab, because it seems that the latter was in very serious debt to his creditors in the East India Company and could hardly manage to pay his naturalist at all. Koenig was naturally desperate to leave this situation, and the Nawab discharged Koenig from his employ at the end of 1777.[13] Relying entirely on the good will of friends and contacts back in England, Koenig continued collecting, but time was running out. In July 1778, he wrote to the East India Company begging them to hire him as their naturalist; or, as he put it, their 'Natural Historian' with a view to 'compile a Natural History of this Country' on Linnaean principles.[14]

Koenig's appeal was successful. It was the first time that the East India Company had hired a naturalist. Probably, unknown to Koenig, on the other side of the world, Banks, just recently elected President of the Royal Society, recommended his appointment.[15]

Koenig now had financial security and, even better, the opportunity to travel. Over the next few years, he visited present-day Thailand, the Spice Islands, Sri Lanka, Calcutta and Tranquebar. Throughout this whole period, he collected plant specimens and sent them to his many correspondents in Europe, particularly Solander and Banks.[16]

In early 1785, Koenig was transferred to Calcutta, the main site of the East India Company's interests in Bengal, a major upward step in his career, but while collecting his possessions from Tranquebar and Madras to take to his new destination, he began feeling unwell and, on 26 June, he died from dysentery in Kakinada, between Madras and Calcutta. On 6 June, knowing he was dying, Koenig drew up his will: he bequeathed all his papers and specimens to Banks.[17]

Koenig had been a key person in Banks's botanical world, the first to send him specimens from India, but there were others in Banks's inner circle of naturalists. One of these was Patrick Russell who had spent more than twenty years in Aleppo, mostly working for the Levant Company as its physician.[18] Bringing with him a large consignment of specimens from Syria, Russell arrived in London sometime after 1772 and soon after became acquainted with both Solander and Banks, who studied his collection with great interest. Early in 1782, Russell headed for India and in August, he arrived in Vizagapatam, north of Madras, accompanying his younger brother who had been appointed to the East India Company as a chief administrator. Earlier in June of the same year, as they were making their way up the coast of India, Russell had met Koenig in Tranquebar and they immediately became fast friends.[19] Not long after Koenig's death, Russell was offered and accepted the post of Company Naturalist in his place.[20]

Another naturalist was William Roxburgh. Born in 1751, Roxburgh had studied medicine and botany in Edinburgh.[21] His medical career began on East India Company ships but in early January 1776 he decided on a different path. After several months of travel, he arrived in Madras and at the end of May 1776, he was appointed assistant surgeon at Fort St George. At some point after this, certainly before early 1779, Roxburgh met Koenig in Madras and they, too, became close, closer even than Russell and Koenig.[22] Through Koenig, Roxburgh came indirectly under the influence of Linnaeus and he and Koenig botanised together whenever they could over the next few years. In fact, it was while visiting Roxburgh that Koenig died.

It is very likely that Koenig introduced Roxburgh to Banks for it was Banks who wrote to Roxburgh first asking him to collect plant specimens for him.[23] This letter initiated a long correspondence between Banks and Roxburgh which lasted almost thirty years, during which time Roxburgh kept Banks fully informed of the increasing European knowledge of Indian botany.

* * *

When, in July 1778, Koenig petitioned the Madras Council of the East India Company to appoint him as their first naturalist, he referred them to the two surgeons, Gilbert Pasley and James Anderson, of the Madras Medical Board, who 'from their own extensive knowledge are able to form a proper Judgement of my capacity'.[24]

Anderson was the junior of the two men. Born in Long Hermiston, Scotland, in 1738 and educated in medicine at the University of Edinburgh, Anderson, like so many other Scots, began his medical career serving on East India Company ships. He arrived in Madras in 1765 where he joined the Madras Medical Board as the Company's assistant surgeon.[25] He rose through the ranks and at the time of Koenig's petition he was in the post of Surgeon General.[26] He had also just begun work on a botanical garden on a piece of waste land that the Company had granted him.[27]

Anderson's career continued to blossom, and in April 1786 he reached the highest rank, that of Physician General. In the same year, on 3 December, he wrote to Joseph Banks from Madras, the letter which Banks received in May 1778, the receipt of which we learned of at the beginning of the chapter.[28]

Anderson was reporting that he had found cochineal insects attached to a grass which, in this part of India, was fed to horses. He had examined the insect with the help of a magnifying glass and it was consistent with its description in the scientific literature: Anderson was referring to the publications of Antonie van Leeuwenhoek, famous for his development of the microscope and the first to investigate the structure of the insect using the instrument; to Hans Sloane and the Abbé Guillaume Raynal, both of whom had written on cochineal; and especially to René Antoine de Réaumur, whose magisterial, encyclopaedic natural history of insects contained a definitive account of cochineal.[29]

Anderson claimed that he had observed the insects and noted that 'multitudes of the young are daily issuing forth of a red colour with six legs and two Antennae: some have wings and are said to be males': and that he had 'macerated them in water and spirits of Wine, and find it communicates to both a colour equal to the Cochineal of Mexico'.

Banks may not have been particularly interested in cochineal at that time but he would have instantly recognised the importance of Anderson's find.

Cochineal is a scale insect which, when crushed and treated, produces a brilliant red liquid. It had been used in Mexico and Peru as a dyestuff for almost two thousand years.[30] It first arrived in Spain in the 1520s as a consequence of the Spanish conquest of Mexico and by the end of the sixteenth century it was the most sought-after red dyestuff in Europe.[31] Producing a startlingly beautiful and vibrant colour never seen before, cochineal was worth its weight in gold and anyone who knew anything about it knew its value.[32]

Anderson impressed Banks even more when he revealed that not only had he planted a small piece of ground with the grass but that he had set out '1000 Opuntia Plants for the purpose of cultivating them in the Mexico way'.

This was an important point because it was well known, especially from Réaumur's extensive discussion of it, that in Mexico the female cochineal insect, which produced the red dye, fed and lived its entire life on a variety of *Opuntia*, a genus in the cactus family and commonly called prickly pear or nopal.[33] Réaumur wrote about how the Mexicans had domesticated the cochineal insect and were harvesting them from the nopal (cactus) plants which were grown in plantations, called nopalries. The work was intensive and strictly regulated by the seasons. The resulting dye was far superior to and more intense than the one derived from wild cochineal.[34]

Anderson ended his letter to Banks by saying that he would be sending him a sample of the dyestuff he derived from the insects. He feared that the sample he had sent was too damp and that he would take more care in drying the dyestuff in future.[35]

Anderson admitted that this was not going to be easy. He was not certain that what he had been calling cochineal was what Linnaeus had described – looking at it under the microscope, he was inclined to give the insect a new name – and the dyestuff was not so blood-red as he had hoped. Added to that, the cactus that grew in this part of India was much spinier than the one in Mexico: he was not sure he would be able to induce enough insects to live on it?

Banks received the first two of Anderson's letters sometime in May 1787 and replied to him, agreeing with his main points.[36] The insect was probably not cochineal but another species, so far undescribed in European sources; and it did not give a very good red. It was more like the dye that came from a scale insect species called Polish cochineal, which was now little used. Banks tried out the dye himself and concluded it did not give a good colour, and the results from two professional scarlet dyers, whom he had asked to test it, were so poor that he thought Anderson should abandon the whole project.

However, as Banks put it: 'Every evil in life has its accompanying benefit.' Had Anderson not brought the subject of the cochineal to his attention, he admitted, it would never have occurred to him that there was a real opportunity to break the Spanish monopoly on cochineal. India, he argued, had a climate in which the Mexican cochineal could survive, in which spikeless cactus could be cultivated, and an abundant enough labour force to make a success of the project. Banks was certain that the price of Indian cochineal would undercut Spanish sources and, by extension, make Britain if not self-sufficient then at least less dependent on Spain.[37] Anderson would be the perfect person to put the plan into action. 'If I find the great men at the India House inclined to hear reason on the subject, I shall not fail to propose it to them before the next ships sail,' Banks wrote.

Before 1785, Banks does not appear to have had any direct contact with the East India Company, despite knowing several of the naturalists associated with it. This changed in April 1785 when Banks received a letter from Thomas Morton, the Secretary to the Court of Directors of the Company, asking him to look over a sample of Chinese hemp that had come his way and to distribute it to those most qualified to judge its properties, in the hope that it might, in future, be grown in Britain.[38] Though nothing much came of this, it did signal that, in the wake of the loss of the American colonies following the War of Independence (1776–83), the East India Company and the government were looking for alternative and more secure sources of raw materials. This entailed the use of naturalists to make proper scientific assessment of the natural resources available.[39]

Banks now found himself being consulted by the Company on several issues and becoming friendly with its leading figures, and those in the government closely involved in its policies: men such as Sir George Yonge, the Secretary for War, Henry Dundas, the chief figure on the government's Board of Control (which sought to impose restrictions on the Company's expansionary territorial tendencies); and the Directors of the Company.[40] Banks was consulted, for example, on the plans to begin a Company botanic garden in Calcutta; and on the introduction of tea to India.[41] So when Banks told Anderson that he would share his ideas with the Company, he knew that he would receive a sympathetic hearing.

* * *

In 1787 a remarkable book published on the French island of St Domingue (now Haiti) told the story of Thiery de Menonville's daring journey in 1777 to Oaxaca in central Mexico where he procured a quantity of domesticated cochineals and their nopal habitat, and successfully brought back both the insect and the plant to the island.[42] It was no secret that the local people of Oaxaca and the surrounding area were cultivating cochineal and nopals on a large scale – in the first volume of his major work, *A Voyage to the Islands . . .*, 1707, Hans Sloane had a full-size engraving of a Oaxacan nopalry.[43]

Banks had a copy of de Menonville's book in his library and studied its contents very closely.[44] It wasn't so much de Menonville's travel narrative that would have caught his attention but his meticulous description, in the second volume of the book, on how to rear cochineal and cultivate nopal. It was the first published account of the details of cultivation of both insect and plant and, importantly, it was based on first-hand observation rather than hearsay.[45] So taken was Banks by the book that he shared it with Thomas Morton at the East India Company.[46] Anderson also had a copy of it in Madras.[47]

William Petrie, who worked for the East India Company in Madras and who had established the Company's first observatory there, was in Madras when Anderson began writing to Banks. In a letter to Banks, mostly concerned with the natural history of snakes, Petrie informed Banks that he knew Anderson was working on a cochineal project and added a few words about his character: 'His application is intense,' he wrote, 'Sparing neither fatigue of the Body or Mind, & a perseverance which no difficulties can check. In addition to this, he is a man of great humanity.'[48] Banks had never met Anderson but Petrie's description of him was certainly borne out by the intensive correspondence reaching Soho Square from Madras. Over a period of thirteen months Banks received more than one long letter per month from Anderson on the subject of cochineal.

However, Petrie's letter revealed something that Banks found extremely disturbing. Anderson, Petrie remarked, had published the first six letters he had written to Banks in the *Madras Courier*, the town's first newspaper, which had only begun publication two years earlier – and the same letters were promptly republished by the Company's own press – and Anderson had continued sending his letters to the press until fourteen letters were published by early 1788.[49]

Banks was furious. The last thing he wanted was publicity for the project – he'd hoped to keep it secret from the Spanish and the

Mexicans in particular. As soon as he learned that the letters had been published, Banks stopped writing to Anderson: 'I persevered in declining to answer your letters during the time you were printing your Correspondence . . . from a firm persuasion that no benefit could accrue, either to the company or to you, from that undertaking', Banks wrote.[50] That was a polite way of putting his displeasure. Anderson, despite Banks's objection, continued to make public what Banks thought should be kept secret. When, on one occasion, Banks learned that Anderson had printed a set of instructions based on Banks's advice, he was beside himself with anger.[51] To Thomas Morton at the East India Company in London, with whom he had by now become well acquainted, Banks referred to Anderson's 'incredible imprudence' and to him being 'actuated by a degree of absurdity wholly above comprehension'.[52]

Banks not only stopped writing to Anderson but, more importantly, stopped working with him on the cochineal project, bypassing him and dealing directly with the East India Company instead. Anderson was not deterred by Banks's decision and continued to work on his nopalry, which was being supervised by Dr Andrew Berry, his nephew.[53] As Petrie had observed, Anderson was tireless and he did get some yield of cochineal which was shipped back to England.[54] The quantities were modest but the real problem was that the Madras dye was less effective than the one from Oaxaca.[55]

Anderson seems not to have concerned himself too much about which species of insect produced the red dye; nor whether wild varieties of cochineal were inferior to the domesticated ones in the quality of the dye they produced. Banks, however, was bothered by this. Without the real cochineal being raised in India there would be no chance of undercutting the Spanish. Banks knew it had been domesticated in Oaxaca but could this be reproduced elsewhere? He embraced the texts that consistently spoke about the superiority of the domesticated cochineal; and he interrogated every text he could lay his hands on which included observations of dye-producing scale insects, concluding, finally, that these were all wild varieties.[56] Oaxaca was the only certain source of the real domesticated cochineal.

During the first wave of an interest in cochineal in 1788, Banks did what he could to help the East India Company develop a cochineal industry. One of the first things he did was to send samples of Mexican cactus from Kew to the Calcutta Botanic Garden where they were

successfully cultivated and from where they were successfully distributed to other parts of India.[57] At the same time and inspired by the
epic story of de Menonville's travels to and from Oaxaca, he suggested
that the Company should send someone to Mexico to get the real
cochineal. Banks even found someone who was prepared to do this.
At the time, however, the Company was not interested in spending
money on this venture and when they learned that the wild cochineal
and the cactus were plentiful and there for the taking in Rio de Janeiro,
they decided to follow that route.[58] Though Banks disagreed with them
he continued to advise them and produced instructions for managing
the insects and plants on a long sea voyage.[59]

The idea of a direct 'procurement' surfaced again in the Court of
Directors in 1790 and Banks, again, reiterated his original plan.[60] Banks
reminded them that when this issue was last raised, he had found
someone living on the Gulf of Honduras who would be prepared to
undertake the risky venture. His brief was to get the cochineal and
the cactus and deliver them to England for which he was to receive
the sum of £1000 (near £100,000 in today's value). Banks would ask
this person to accompany the precious cargo together with some local
people who knew how to cultivate the insect and plant and could be
induced to settle in India. Banks added that he would take care of all
the details concerning the transport and care of the insect and plant.

Nothing happened. Banks was frustrated.[61] Then, in mid 1792,
Francis Baring, the Chairman of the Court of Directors of the East
India Company, head of Baring and Co., one of the most important
banking concerns in the world, and a man who knew a bit about
cochineal, got in touch with Banks about reviving the idea of sending
someone to Oaxaca once again.[62] Banks indicated that he was not
willing to put any more effort into this scheme – '[I] Cannot but
seriously Lament the time [I have] spent in Promoting it – unless the
Company could guarantee that this time they were serious.'[63]

Baring's reply offered Banks the guarantee he sought but then
towards the end of his letter he raised the old topic of getting wild
cochineal from Brazil. Banks was clearly upset and irritated. 'I have
reported against Sylvester being brought from Brasil,' he noted in
exasperation and in his own hand at the end of Baring's letter.[64] Six
weeks later, Baring and several of his directors finally got Banks's point
about going to Oaxaca, doubled the premium to £2000 and told
Banks that he should go ahead with his plan.[65]

Banks did not at first warm to the idea but soon changed his mind

– 'it is a Favorite Plan of Mine,' he admitted.[66] Two years had, however, been wasted since he last pursued this and he was 'ignorant whether or not the Person on whom I meant to rely for the first attempt towards putting it into Execution is still alive or if he is Whether he still resides as he did'.[67]

The person Banks had in mind was James Bartlet. Banks had heard from Bartlet when he wrote from Honduras on 21 July 1791, telling Banks that a botanical tour of the area would prove very rewarding.[68] Bartlet was the brother of Alexander Bartlet who, in company with two others, ran a trading company between London and the West Indies and owned plantations on several islands.[69]

On 14 January 1793, Banks wrote to Bartlet. He could not hide his irritation at the Company's neglect of his plan to supply India with the true cochineal – four years in the making, he remarked – but the main thrust of the letter was to assure him that the East India Company now meant business and that the original plan stood. Banks noted that 'a Frenchman' [de Menonville] had proved that both the cochineal and the nopal tolerated a sea voyage well but that, in any case, once the cargo arrived in London, Banks would put the cochineal in his hothouse where he was already growing nopal.[70]

It all sounded promising but, in the event, it soon unravelled, as Banks explained to Hugh Inglis, the Deputy-Chairman to the Company's Court of Directors.[71] Banks had no reply from Bartlet; it had now been three and a half years since Banks had written to him.[72] Even if someone could be found to take Bartlet's place, Banks remarked, he doubted if the venture would be successful now because of Anderson's foolhardy decision to print and distribute Banks's directions for transporting cochineal and nopal by sea, including sending them to a botanist working for the King of Spain. This and the publication of de Menonville's book, Banks argued, meant that the Spanish were more vigilant than ever and the venture was now 'more likely to End in the Loss of Liberty or Even of Life, than the final attainment of the Object in Question'.

For these reasons Banks was bowing out of the plan which he had been enthusiastic about years before. Besides which, with Britain now at war with France, and rumours rife of a secret treaty between France and Spain, it would be foolhardy to try to enter Mexico in secret and steal the plants and insects.

Even though Banks said that he was no longer willing to be involved with the cochineal project, he did leave a little room for manoeuvre and ended the letter to Inglis with the promise that if something did come up, he 'would not decline to Proceed'.

Sure enough something did occur. On 23 June 1802, Banks received a letter from Robert Sproat, a Scottish physician.[73] The letter, dated 5 March 1802 and sent from present-day Honduras near the town of Palacios on the coast, reminded Banks that they had been introduced in 1794 when Sproat was temporarily in London. Sproat also told Banks that he had, as agreed, sent him part of a collection of plants in 1796, but, unfortunately, the ship carrying them was intercepted and ended up in New Orleans, then under Spanish control, where none of the contents survived; the other part of the collection, which remained with him, was destroyed during an attack by the local people. What Banks made of all this we don't know but a passage in the letter a little further on would have jumped off the page. It reads: 'I also had the Opuntia sent me from the city of Comayagua & having transplanted it in an adjoining plantation it flourished exceedingly. I was even flattered with the full hopes of getting in to my possession the Cochineal Insect, for conceiving it to be one of your especial desiderata, I used my utmost influence to obtain it from the interior of Guatemala; I at last received the Insect but to my great astonishment found that either through ignorance or design in the sender they were all bruised to death! A circumstance I relate not Sir to claim any merit for the exertions I made, but for your information respecting the existence of the Insect in that country.' Sproat concluded this part of his letter by commenting that with more care the operation might have succeeded and been sent to England – 'without trial it is but conjecture at best, which affords no satisfaction.'

In his 27 August 1802 reply, Banks confirmed that Sproat's remarks on cochineal gave him 'particular pleasure' because it seemed very likely that the real cochineal and the cactus could be procured from Guatemala and sent to India.[74] If Sproat was 'inclined to undertake this patriotic task' – which looked possible without 'any hazard of loss to yourself' – Banks would make him a generous settlement: £500 on receipt of the cactus plants in London; £1000 for the delivery of healthy cochineal; and a further £1000 for the delivery of both insect and plant to India. Banks provided Sproat with a condensed version of his instructions on how to keep the cochineal and the cactus alive on a sea voyage, with the added incentive of a gratuity for the ship's captain 'proportioned to the state of health in which either Plants or Insects arrive'. Finally, it was up to Sproat to try and entice one or two local people skilled in nopalry and cochineal cultivation to care for both all the way to India and then to stay or be sent home if they so wished.

The letter to Sproat showed Banks at his most excited. 'I need only add', he remarked towards the end of the letter, 'that as I who am constantly employed in the execution of projects of this nature without ever having accepted remuneration in any shape have not the slightest view to advantage in this.'

As it turned out, nothing further seems to have happened. Did Sproat receive Banks's letter? Was he alive? Did he decline the offer? We don't know. All we do know is that the East India Company kept offering a £2000 reward for the 'importation of the cochineal insect from South America', but there were no takers.

As time went on the Company lost interest, and the dream of Oaxacan cochineal in India died.[75]

PART II

Floating Gardens and the Cotton Club

Preface

By 1786 Joseph Banks had been President of the Royal Society for eight years – he would continue being elected annually until his death in 1820. It was a position which gave Banks enormous prestige not only in Britain's scientific community but also further afield. Over the following five years, Banks became a member of at least fifteen other scientific societies, some of them modelled on the Royal Society, some new, some well established, and stretching from Philadelphia to St Petersburg. Banks was now in touch with many of the world's scientific practitioners, not only those interested in natural history, but ranging over all fields of scientific enquiry. His correspondence files grew accordingly.

Enlightening as it was, and satisfying as it must have been to be fêted by so many scientific institutions, it was Banks's botanical projects which continued to dominate his attention. By 1786, his first collectors were either in the field or on their way. Francis Masson was at the Cape, on his second visit there, collecting living plants for Kew; Anthony Pantaleon Howe was in southwest Africa on HMS *Nautilus*; Archibald Menzies was on his way to the Pacific Northwest; John Duncan was already sending plants to Banks and presents to his wife Dorothea from Canton; and on the Coromandel Coast, James Anderson had written his first letter to Banks about his discovery of cochineal.

The tables and other surfaces in the library at Soho Square must have been covered with all the available texts, drawings and maps of the places where his collectors were, as Banks envisioned what they may have been experiencing. And, all the time, plant specimens were arriving, sometime at Soho Square, if, as seeds, they were in envelopes addressed to Banks; and, other times, at Customs House, if they were bulbs or living plants, needing clearance before being sent to Kew.

The geographic scope of these collections was already wide but over the next few years they spread out even further. Howe (now called Hove), who had returned from southwest Africa, was sent to western India on a secret mission; and Menzies, who had left Canton for London, would soon be on his way back to the Pacific as part of a naval convoy, commanded by George Vancouver, who had been on both Cook's second and third voyage, to keep the delicate recently negotiated peace between Britain and Spain and to continue Cook's survey of the Pacific coast. Menzies was mainly collecting living plants for Kew, but he also collected botanical specimens and planted seeds given to him in England and at the Cape whenever and wherever he could.

In addition to all this, Banks became involved with three other government-initiated projects before 1791. The first focused on establishing and sustaining Britain's new penal colony in New South Wales. The second and third projects were concerned with transplanting the breadfruit tree from Tahiti to the British West Indies on specially converted naval vessels.

Banks's experiences on the *Endeavour*, now more than a quarter of a century in the past, were still being called upon, and he took all the opportunities that the government projects presented him with to further his own personal project of supplying Kew with living plants from all over the world, on an unprecedented scale.

6

1786: The First and Second Fleet

The extremely disappointing report from HMS *Nautilus* put paid
to the idea of forming a convict settlement on the southwest coast
of Africa and paved the way for Botany Bay as the only feasible alternative. Banks, who had so far merely recommended it as a suitable
spot for a colony, soon found himself involved in planning the practical details of its day-to-day existence. This was not through the
government, but because of Banks's relationship with yet another of
his acquaintances who was interested in natural history, Arthur Phillip.
This was Captain Arthur Phillip who was to be the first Governor of
New South Wales. He was born in 1738, and so five years older than
Banks, and had been at sea since he was nine. First in merchant ships,
and then in the Royal Navy, he rose quickly through the ranks. Early
in the 1770s, Phillip went on half-pay, and spent some time in France,
possibly spying for the British.[1] By 1774 he was back in London, but
not for long. Soon he was sent to Lisbon. By mid January 1775 he
was a captain in the Portuguese Navy for service in Brazilian waters.
The Portuguese had asked the British for naval assistance and Phillip
had been recommended. He was described to the Portuguese authorities as highly intelligent, an astute and accurate observer, with
extensive naval and military expertise, fluent in French, and other
European languages. Phillip was the perfect man for the job.[2]

He was. He stayed with the Portuguese Navy for three years. During
this time in Brazil, he was surprised to learn that a colony of cochineal
insects had been found living on cactus plants on the Brazilian island
of Santa Catarina, the southern naval base between the towns of Santos
and Porto Alegre; and the Viceroy of Brazil, the Marquis de Lavradio,

had ordered production to be expanded to the point at which Brazilian cochineal could be exported and compete with the Spanish monopoly.[3]

Phillip already knew something about the making of cloth – his wife's first husband had been a very prosperous cloth merchant and Phillip himself had spent several years in and around the cloth cities of northern France.[4] The valuable cochineal naturally attracted his attention and he became very interested in understanding the insect's behaviour. While stationed at the island, he actually bred the insects in his cabin and observed their life cycle. Unfortunately, while on patrol in the River Plate, much colder than on Santa Catarina, all the insects died. Nevertheless, Phillip had learned enough about their behaviour to be confident that the insect and the cactus could be transplanted to one of the British islands in the West Indies where the tropical climate would provide the right conditions: and the insect could be 'bred to the great advantage of the Nation, as well as to the very great profit of the Planter.'[5] Phillip's cochineal experience would come in very useful later on.

In 1778 Phillip returned to his naval career in Britain and was eventually made Captain of his own Royal Navy ship. He apparently continued to spy on the French, reporting directly to Evan Nepean, Under-Secretary of State at the Home Office.[6]

It seems likely that it was Nepean who recommended Phillip to be Commander of the First Fleet. Not only had he employed him previously, but the two men had become close friends. He would have known Phillip's service record and maritime experience, which included sailing in both the Atlantic and Indian Oceans.[7]

If Nepean did recommend Phillip to the Home Office, it seems the Admiralty were not too keen on appointing him, but eventually they gave way. On 3 September 1786, Lord Howe, the First Lord of the Admiralty, told Lord Sydney, the Home Secretary, that he was satisfied as to Phillip's abilities, and that the Admiralty would not object further.[8] Phillip was on his way, as Commander of the First Fleet, to be Governor of the first penal colony in New South Wales.[9]

Although there is no record of Phillip meeting with Banks during the seven months of preparation for the voyage to New South Wales, Phillip would undoubtedly have consulted him. Phillip and the government intended that the colony should become self-sufficient in time.[10] Banks's unique knowledge and experience of the area would have been invaluable in achieving that end. However they collaborated, Banks did not just give advice in planning the agricultural future of the colony. He provided a generous array of cuttings of fruit trees and vegetables, including apples, pears, peaches, nectarines, oranges, lemons and soft fruits; artichokes, garlic and horseradish; and a number of herbs. But this was nothing when compared to what Banks supplied in seed: every conceivable vegetable; wheat, barley, rye and oats. Banks was taking no chances; many of the seeds were duplicated, even triplicated, and distributed over HMS *Sirius* and HMS *Supply*, the two naval vessels chosen for the voyage to New South Wales, and over two of the store ships as well.[11] For Banks, this was not just an opportunity to discover which European plants would grow in the southern latitudes: in Phillip, who he knew was interested in natural history, Banks now had a willing collector in a place which, apart from his own brief visit, was hardly known to natural history in the West.

The First Fleet – two naval vessels, three store ships and six convict ships – left Portsmouth on 13 May 1787. Just over two weeks later, on 3 June, it anchored in the harbour of Santa Cruz, the main town of Tenerife. It was from here that Phillip replied to the letter Banks had sent him at Portsmouth. In it Banks had asked him to collect specific plants and seeds from Rio de Janeiro. Phillip sent his letter of agreement in a packet ship that was returning to London.

From Rio de Janeiro, Phillip wrote to Banks that he had managed to get hold of four different types of ipecacuanha. Known in Europe since the middle of the seventeenth century as native to Brazil, ipecacuanha featured as an important element in European *materia medica*.[12] It was the root of this flowering plant that, when dried and ground into a powder, was used to treat a number of medical conditions. In small doses, ipecacuanha encouraged sweating and coughing; in larger doses, the plant acted as an emetic and cathartic and was used to treat forms of dysentery.[13]

Getting hold of the dried roots of ipecacuanha in Britain was not a problem and it featured in the apothecary's list of remedies. However, Banks was interested in seeing the whole plant which was, as he must have told Phillip, hardly known in Europe.[14] He sent Banks preserved in rum one of the specimens, which Phillip had been led to believe by a local surgeon to be the best.

In Rio de Janeiro Phillip loaded up with those plants that Banks had been unable to supply him with in London. Many of them were tropical – indigo, cotton, coffee, bananas, tamarind and cacao – (Banks believed that the climate of Botany Bay was similar to that of southern France), but he managed to get some vines and tobacco as well. He

also took three ipecacuanha plants potted in earth for cultivation in
Botany Bay. In addition to these plants, 'I have procured every Seed
I think likely to grow in NSW,' he told Banks. Now back in Rio de
Janeiro after more than a decade, Phillip did not forget his experience
with cochineal when he had successfully reared the insects in his cabin.
If only he could keep them alive en route to New South Wales, he
thought that they might thrive in their new habitat.[15]

Cape Town was the next destination, which the fleet reached on 13
October 1787. There to welcome Phillip was Francis Masson, Kew's
first dedicated collector, who was collecting for Banks in the Cape,
on his second trip there in fifteen years. Phillip loaded his ship with
local plants – figs, sugar, apples, oranges and lemons – flour, wheat,
barley and bread.[16] On 13 November, the day after the Fleet's departure,
Masson wrote to Banks saying that he had seen Phillip and that his
'Cabbin was like a Small Green House with plants from Brazil'.[17]
Phillip also wrote to Banks assuring him, not for the first time, that
he would be sending him seeds and plants from New South Wales.[18]

The ships, each 'like another Noah's Ark', left Cape Town on 11
November.[19] Botany Bay was the fleet's next and final destination.

Carrying in the region of 1300 people between them, the ships of the
First Fleet arrived there between 18 and 20 January 1788. The cold of the
southern ocean, disease and accidents had killed off many plants, though
some of those that had been put on board at Rio de Janeiro not only
survived but began to bloom as the ships approached warmer habitats.[20]

Almost immediately, Phillip decided that it was a mistake to attempt
a settlement in Botany Bay: the area was very swampy and there was
hardly any fresh water. After quickly investigating Port Jackson, the
next bay northward along the coast, which Cook had seen but had
not entered, and finding it would meet all of their requirements, Phillip
ordered the fleet to remove from Botany Bay and harbour in Sydney
Cove. Here, on 26 January 1788, Phillip formally took possession of
New South Wales, which was defined as 'extending from the northern
cape or extremity of the coast called Cape York, in the latitude of
10°37' south, to the southern extremity of the . . . South Cape, in the
latitude of 43°39' south, and all of the country inland to the westward
as far as the one hundred and thirty–fifth degree of longitude'.[21]

On 15 May, Phillip wrote his first despatch to Lord Sydney, in which
he recounted his proceedings from the Cape to the Australian coast
and explained why he had decamped from Botany Bay. He also sent

his first observations of the country around the new settlement; and of the indigenous people. This report, Phillip remarked, was put together over a period of time. He didn't know when he would get a chance to write such a long letter again, seeing, as he put it, that 'the canvas house I am under [is] neither wind nor water proof.'[22]

Two months later, on 2 July 1788, Phillip wrote his first letter from New South Wales to Banks. The information about Botany Bay was completely at odds with what Banks had given in evidence to both the Bunbury and the Beauchamp Committees. Phillip didn't bother going into any detail about why he had abandoned Botany Bay so quickly – he referred him to Evan Nepean, the under-secretary to Lord Sydney, the Home Secretary, for the reasons.

What he wanted Banks to know was that life was not going to be as easy in the settlement as Banks had suggested; that the 'natives' were far more numerous and not as welcoming as Banks had found them – three convicts had already been killed; and that food, particularly fish, was not as abundant as Banks had thought. The indigenous people, Phillip added, were having a hard time and there was little he could do to help them.

On the plus side, however, Phillip could now send Banks the first consignment of seeds; and he also threw in a stuffed kangaroo and reported that 'these Animals are very numerous, but after being fired at grow very shy'.[23]

Though Phillip's first duty was to the settlement, he did as much as he could to collect seeds and plants for Banks. He knew that this had to be done before the convict ships and the transports left Port Jackson for London. The first ships began to leave in May 1788, but Phillip had nothing for Banks by then and it would take him until the last ships sailed, the convict ship *Alexander* and the store ships *Fishburn* and *Golden Grove*, in November, before the collections were ready. What exactly was sent is not known but some of the 'Tubs of Plants' and seeds must have made it because several were introduced into Kew around the time that the ships returned – the consignment included several species of *Banksia*, orchids and a 'Kangaroo vine'.[24]

It shows how indebted Phillip felt to Banks that he tried so hard to get him the botanical specimens he'd asked for. Conditions were not good in the fledgling colony. While quite a few of the plants Phillip had brought from Rio de Janeiro and the Cape – including the cochineal insect on its prickly pear cactus – were doing well, these were what Phillip referred to as the luxuries; the basics, especially wheat, were doing very poorly. Added to that, most of the cattle that Phillip had bought at the Cape had not survived or had wandered off into the bush. In his despatches to Lord Sydney and Evan Nepean at the Home Office, and Philip Stephens at the Admiralty, Phillip pleaded that the next convict ships should bring provisions, food, clothing and medicines that would last at least two years.[25]

In the meantime, the colony was facing starvation. To save it, Phillip sent HMS *Sirius* to Cape Town for vital supplies. The ship left in early October 1788. Because of the winds at that time of the year, it had to go by way of Cape Horn. It didn't reach Cape Town until 2 January 1789. After filling its hold with enough supplies for a whole year, the *Sirius* arrived back in New South Wales on 8 May 1789.[26] The ship's supplies were only just in time to avert disaster.

By then, in London, the Second Fleet, taking more convicts and supplies to New South Wales, was preparing, and Sydney had received Phillip's letter begging him for more supplies.[27] He wrote to the Admiralty formally instructing that they should get a naval ship ready to sail to New South Wales, though it is clear that, informally, the preparations were already under way.[28]

The ship in question, HMS *Guardian*, was virtually new, having been no further than the Channel since 1784, when it was completed. Most of the guns were removed in order to make more room for supplies. The 26-year-old Lieutenant Edward Riou received his commission to command the ship on 21 April 1789.[29]

Riou had been a midshipman in 1776 on HMS *Discovery*, one of two ships on Cook's third voyage. When the ship returned from its intensive exploration of the Pacific in October 1780, Riou was promoted to the rank of lieutenant. For the next several years he was stationed in the West Indies, in the Channel and, finally, for two years, he was in Newfoundland as second lieutenant to Captain Erasmus Gower (who will crop up again in a later chapter).

Riou thus had experience in the world's main oceans and though he had not landed in Australia, he came very nearly in sight of Van Diemen's Land, having sailed across the Southern Ocean from the Cape towards New Zealand.

Within days of Riou's commission, Banks was involved in preparing the voyage. It seems that Banks saw the *Guardian* as a suitable vehicle

for sending new and interesting plants back from New South Wales
for Kew. Banks wrote to Nepean at the Home Office on this exact
point less than a week after Riou's commission. Banks suggested that
the area at the stern of the ship bounded by a rail – the taffrail – could
be adapted as a storage unit for plants in pots and should be glazed
over.[30]

At some point in the next few weeks, plans for the *Guardian* began
to change quite radically. Collecting for Kew in New South Wales
remained an objective, but the scope of the ship as a plant carrier was
substantially increased. Banks had been in conversation with William
Grenville, the new Home Secretary, who had succeeded Lord Sydney,
as well as other members of the government and the Admiralty. The
upshot of these meetings was the decision that the *Guardian* should
now take to New South Wales 'such trees and plants as are useful in
food or physic, and cannot conveniently be propagated by seed in
pots of earth'.[31]

The original idea of using the taffrail was now defunct. A much
larger space was needed. On Lieutenant Riou's invitation, Banks visited
the *Guardian* and had a look.[32] Riou had intended to be there to show
Banks around, but, for some reason, the two men missed each other.[33]

Banks began his inspection with the commander's cabin. It was far
too small for what Banks intended; and the volume of supplies the
ship was intending to carry meant that there was little spare room
below deck. Botanists like John Ellis in Britain and Henri-Louis
Duhamel du Monceau in France had recommended that the best
method for transporting living plants across oceans was for them to
be packed in organic matter, such as moss and earth, and placed in
specially designed containers which would allow air to circulate and
which could be moved around the ship to take advantage of the best
conditions.[34]

This method worked well for small numbers of plants but the
volume that Banks envisioned being moved meant that this time-
honoured technique wasn't appropriate. He needed a radically new
solution and he had one in mind. As he told Grenville, he, with the
assistance of the ship's master and a shipbuilder, 'caused the form of
a small coach to be chalked upon the deck in such a manner as they
both agreed would not be at all in the way of working the ship'.[35]
Coach was the technical term for such a construction but Banks liked
to call it an apartment because it was his intention that it would also
house a gardener and his equipment. Its dimensions were 16 feet by

12 feet and 5 feet high, and it could be constructed, Banks stated confidently, in little more than one week.[36] It would house 93 pots of plants. Flexibility was built into the design of the coach so that it could respond easily to changing conditions. It had sliding shutters and a canvas cover to be used when the weather was fine but direct sunlight was to be avoided. In cold weather, glazed units in fitted frames took the place of the open gratings and a stove with chimney was provided for even colder weather.[37] Banks knew a lot about greenhouses – he saw them at Kew and he had several of them in his own property at Spring Grove.[38]

On the day he completed his inspection of the ship, Banks left a letter on board for Riou seeking his agreement to the idea of a coach so that an order for construction could be submitted. In addition to laying out the coach's dimensions and the materials from which it would be made, Banks emphasised that the coach would also serve as a home for the plants Riou would be bringing back from New South Wales for the King.[39]

By return of post, Riou, not surprisingly, agreed to Banks's plans, both to the construction of the coach and to returning home with plants from New South Wales.[40]

Once that was done, Banks turned his attention to the gardeners who would accompany the plants. James Smith and George Austin had worked at Kew, were experienced and up to the task. They would be going to New South Wales as Superintendent of Convicts, in which capacity they would be supervising convicts in gardening and agricultural work in general with the ultimate aim of passing on their specific expertise to the colony.[41] Smith was going to stay on in New South Wales for three years and would be collecting seeds and plants for Kew whenever he could.[42]

Banks issued two sets of instructions concerning the transport of plants and the duties of the gardeners. The first, to Lieutenant Riou, outlining what he would need to do to accommodate the responsibilities of the gardeners and the health of the plants, both on the outward and return voyages. James Smith, Banks told Riou, was to be given overall charge of the coach and that was where he should live, ready, at a moment's notice, to open or close windows as necessary; and Riou was to give every possible assistance to Smith in collecting rain water whenever possible on the understanding that in hot conditions, the plants would need plenty of watering. Riou was also told to choose one of the ship's company to shadow Smith, and to learn from him how to care for plants so that he would become the gardener on the homeward voyage looking after the plants for the King, which would be stowed in the coach in the same manner as the plants on the outward voyage.[43]

The instructions to Smith dealt with the details of how best to keep plants healthy on their outward voyage; how much air and water to allow them and when; how to collect, keep and apportion water; and how to avoid, and to do so at all costs, sea water settling on the plants' leaves, a situation, which if not remedied immediately, would lead to the plant's death. As for the coach, Banks continued, 'plants on board a Ship, like cucumbers in February, require a constant attendance'. Circumstances at sea changed often and abruptly and the gardener had to be prepared to open and shut the coach several times a day. In order to keep a watchful eye and act quickly and decisively, Banks warned him to 'beware of Liquer, as one drunken bout may render the whole of your Care during the Course of the voyage useless & Put your Character in a very Questionable situation'.[44] Finally, as far as the voyage was concerned, Smith was to be constantly vigilant that shipboard animals – cats, dogs, mice and rats – and cockroaches did not get into the coach. Cat and dogs, Banks commented, could 'destroy the whole garden in half an hour'. Vermin, like mice and cockroaches, could be dealt with by keeping the coach perfectly clean. As for monkeys and goats, Banks hoped these would not be allowed on board at all.[45]

Banks repeated that in New South Wales Smith should collect plants for the King's Garden, and these he was to place in the same pots he went out with, and send them back with the *Guardian*, whenever the ship was ready to sail. For himself, he wanted Smith to look out for and collect seeds and dried specimens, with flowers and fruit, to dry them and place them between paper sewn into a book. Smith was asked to share all these instructions with George Austin, the other gardener.

On 15 July, Riou took Smith and Austin's instructions to the ship, which was waiting in Portsmouth. He told Banks that as far as he was concerned, the coach was the King's property and that he would be treating it as such.[46]

The plants, fruit trees, herbs and vines, many of them from Kew, some from Banks's own garden and some from the nursery of Ronalds in Brentford, began to be loaded onto the *Guardian* in early July and continued in anticipation of an early sailing date.[47] At the same time, all of the other non-perishable supplies for the colony – clothes, hats, shoes, needle and thread, cloths and blankets, sugar, currants, pearl barley and sundry medicines – were placed on board. 'We are excessively deep, nay too deep I fear to carry a Single Cow for the Cape,' Riou commented.[48]

On 7 September Smith wrote to Banks assuring him that all the plants were on board and in good shape but that Austin, who was suffering from a swelling in his legs, was not. The sailing date had been announced as the next day. Riou had assured Smith that he was taking the business of the plants as seriously as he did the ship itself. Banks heard from Riou that he was very pleased with Smith: 'he behaves in ye most attentive quiet, & but best manner, I wish only that all of the Superintendants [sic] had been men of his disposition.'[49]

It was looking good. As planned, on 8 September 1789, the *Guardian* left the anchorage in Spithead and headed towards the Cape. There were 124 people on board: 88 men in the ship's company, 9 Superintendents of Convicts, including Smith and Austin, who were travelling to New South Wales to fill various posts in the colony, 2 other passengers, and 25 convicts with sentences varying from 7 years to life and who had been selected because they had special practical skills.[50]

After a short stop at Tenerife, where Riou bought about 2000 gallons of wine for the colony, the *Guardian* anchored in Table Bay at the Cape, on 24 November. Riou had previously reported to Banks from Tenerife that despite the fact that most of the plants had already spent almost three months on board, they were doing well. Now, in the Cape, Riou asked Smith to prepare a report on how the plants had fared on the Atlantic part of the voyage. Despite the most attentive of care, almost 20 per cent of the botanical cargo had either perished or was expected to do so.[51] The herbs fared worst – 'the death of the herbs', Smith told Banks, 'is owing to the heat we had in crossing the line, as we was a week nearly becalm'd, and then it was exceeding hot, on the 13th of Octr the Thermometer ran to 104 degrees high, which was too hot for any English herbs to live in.'[52]

Riou, following Banks's orders in case of disaster, asked Smith to replace as many of the plants as he could and add to the list any plants which he could only get in Table Bay and which he thought would do well in New South Wales. Smith turned immediately for advice to Francis Masson, Banks's collector who was still at the Cape. Plants were bought and seeds too, mostly from the garden of Colonel Robert Jacob Gordon, commander of the Dutch garrison and an excellent botanist and gardener; and Riou went on his own spending spree for live animals – bulls, cows, stallions, fowl of all kinds, rams, ewes, boars and rabbits.[53] To accommodate this menagerie, Riou dismounted almost all of the guns and built stalls on both sides of the main deck and coops for all the fowl on the quarter deck. 'With [the livestock] and an addition to all culinary Fruits of this Country amounting to about 150 Trees in number we were in a situation to be a most comfortable sight to Governor Phillip,' Riou jotted down in his notebook.[54]

On 11 December 1789, the *Guardian* sailed out of Table Bay for the final leg of its voyage to Port Jackson.

Almost a fortnight later, by which time the *Guardian* had sailed more than 2000 kilometres to the southeast of the Cape, Riou spotted a large iceberg. Two boats were despatched from the ship to collect ice to supplement the fresh water taken on at the Cape. The animals were in need of hydration. The boats had hardly returned when a thick fog set in and the ship continued its course believing that the iceberg was being left behind. Instead, buffeted by winds from all directions, they were heading straight for it. Suddenly, out of the fog, Riou saw 'a body of ice full twice as high as our masthead, showing itself through the thickest fog I ever witnessed'. The ship seemed to stick to the ice and then the rudder tore away. Riou ordered the decks to be cleared of cattle, guns and gun-carriages; the spare anchors and everything else from below that could be thrown overboard were jettisoned. It was in vain. The water kept flooding the ship and the pumps were overwhelmed. Riou offered those on board the chance to take to the launch and four smaller boats to save themselves.[55] Forty men, including four convicts, went in the boats and fifteen others, led by Thomas Clements, the *Guardian*'s master, went in the launch.

Riou, the remaining crew, the five Superintendents of Convicts, and twenty convicts, sixty-one people in total, or about half of the total number who had sailed on the *Guardian* from England, stayed with the ship. Miraculously, they managed to keep it afloat and steer whatever was left of it northwest. Almost two months to the day after the *Guardian* hit the iceberg, a floating mass of timbers was spotted in the sea outside Table Bay. Whalers were sent out to help the wreck to safety.

Those who stayed with the *Guardian* were lucky to be alive.[56] The convicts, Riou told the Admiralty, had behaved so helpfully, working the pumps day and night, and so he had promised them he would do whatever was in his power to pardon them.[57] William Grenville, the Home Secretary, who had the power to do this, agreed with Riou. The twenty convicts were put on the *Neptune* and the *Scarborough*, two of the convict transport ships of the Second Fleet that arrived at the Cape on 13 April 1790. Six either died on the voyage to New South Wales or shortly after. Governor Arthur Phillip pardoned the remaining fourteen convicts – though they had to remain within the confines of the settlement until their sentences had expired, they were free men.[58]

Those who went on the launch, fifteen in total, also survived. A French merchant ship from Mauritius chanced upon the boat in the middle of the ocean on 3 January, and took the castaways to the Cape.[59] Some of the survivors went back to London taking passage on East India Company ships that were returning from China, and it was with their arrival in London on 23 April 1790 that the news of what had happened in the Southern Ocean on Christmas Eve was first made public.[60]

Smith and Austin were both on one of the smaller boats. They and their thirty-eight compatriots didn't survive. Nor did the ninety-three pots of plants.

Captain George Tripp had commanded HMS *Grampus* on the voyage along the African coast to search for a suitable site for a new penal colony with HMS *Nautilus*. He was instructed by the Admiralty on 8 October 1790 to sail HMS *Sphinx* as quickly as possible to the Cape.[61] He was ordered to pick up Riou and bring him back home, and what remained of HMS *Guardian* that had not been sold off. By 15 May 1791 Riou was back in London. In addition to bringing the ship's figurehead and some guns and shot, Riou had been given a consignment of plants from Francis Masson for Banks.[62] There were two boxes of seeds and bulbs and 'a large growing plant of *Strelitzia alba*', all for Kew.[63] Named after Augusta, Princess of Wales and mother of King George III, the person most responsible for bringing Kew into the Georgian era, this *Strelitzia* was the second species of this genus introduced by Masson, the first, in 1773, being the *Strelitzia reginae*. Banks was very pleased to hear of Riou and his courage, and thanked him for the plants, especially the *Strelitzia alba*.[64]

Though Riou's voyage was a disaster there was one piece of good news. The plant cabin Banks had had built for the *Guardian* had worked. Almost all of the fruit trees and plants, Riou told him, were still alive when the ship came to grief. This information encouraged Banks to believe that he had solved the problem of moving live plants across the globe.[65]

7

1787: Anthony Pantaleon Hove in Gujarat

We first met Anthony Pantaleon Hove in Part I as Mr Au. He returned from his voyage on the *Nautilus* as Mr Howe. Subsequently he added Anthony (sometimes Anton) Pantaleon and changed Howe to Hove.[1] We don't know why.

Hove, as he will now be called, did not come away empty-handed from his voyage on HMS *Nautilus* in 1785 and 1786, but he must have been as disappointed in the natural history of the southwestern African coast as the ship's commander was about its suitability for settlement. But the first leg of the voyage went well for Hove and he was able to send back almost one hundred species of seeds from places along the West African coast before the ships parted company.[2] At least three species of what Hove collected in southwestern Africa and called geraniums were listed in Aiton's *Hortus Kewensis* of 1789 as growing in the King's Garden.[3]

Though his African plant haul was not as large as expected, as far as Charles Blagden, Secretary to the Royal Society, and who was responsible for recommending Hove in the first place, was concerned, it showed that his assessment of Hove's capabilities was correct. Within a year, Hove would be ready to set sail for a much more daring adventure lined up for him in another part of the world by Joseph Banks.

The Board of Trade, or, full title, the Committee of Privy Council for Trade and Foreign Plantations, under the chairmanship of Charles Jenkinson, Lord Hawkesbury, had cotton on its mind. This was not surprising. For centuries, India had, as one recent book describes it, 'clothed the world'.[4] Now, that supremacy was being challenged by Britain. The British cotton industry was undergoing revolutionary change, both technical and organisational.[5] Located mostly in Lancashire and the adjoining counties in the northwest of England, the cotton industry epitomised what we now call the Industrial Revolution, incorporating the rapid application of mechanised production, the use of inanimate power (in the form of steam and coal), and factory organisation. Though it was not the only industry to undergo these changes, the rapidity with which it did set the cotton industry apart. It was not only the speed of change that impressed but also the degree to which it enlarged Britain's economy.[6]

In the midst of all this change and expansion there were problems. One of these surfaced in 1786. Early in that year, the Manchester fustian makers (fustians are heavy cotton cloths) informed the Board of Trade that they were becoming increasingly concerned that the West Indies, their principal source of raw cotton, was not producing adequate supplies of the fine cotton used to make their best-selling velvet-like cloths.[7] They insisted that the Board should do something to encourage the cultivation of fine cotton on the islands.[8] A month later, Lord Sydney, Home Secretary, sent a circular to the governors of the British islands telling them to encourage their planters to seek out the finest cotton growing in the Caribbean region and to dedicate themselves to its cultivation.[9]

Even if the planters could be convinced to grow cotton (instead of sugar, their traditional crop), there was still the question of quality: were the finer kinds of cotton actually growing in the region? The problem was sent back to Lord Hawkesbury and the Board of Trade.

On 23 February 1787, Hawkesbury invited John Hilton and other Manchester manufacturers to London to give their advice, 'when the Committee propose to take into consideration the most effectual mode of promoting the cultivation of the finer sorts of Cotton in the British West India Islands'.[10]

A few days later, on 26 February, Hilton, accompanied by Peter Drinkwater, another prominent manufacturer, appeared before the Board of Trade. Also invited was Henry Dundas, the Treasurer of the Navy and *de facto* President of the Board of Control, a body created by the East India Company Act of 1784 to manage the British government's interest in India.[11] Dundas had already questioned the East India Company about cotton growing and manufacture in India.[12]

The Mancunians had brought samples of raw cotton from ten different parts of the world: the cheapest was from Thessalonika (in present-day Greece); West Indian cotton prices were mid range and Brazil near the top; but the highest prices (nearly triple the lowest price) were being paid for cotton coming from Surat, in western India.[13]

The members of the Board of Trade were, of course, particularly interested in the latter source and questioned the witnesses specifically about it. 'If you could obtain the Surat cotton of the best quality, either from the East Indies or from any other part of the world at the average price above mentioned, are you of opinion that the manufacturers of this Country could equal the manufacturers of the East Indies in the perfection of the manufacture . . .?' a member asked. 'In the finest muslins they certainly exceed us,' came the answer, followed by the opinion that in time and with expected technical improvements, British manufacturers could produce as fine a yarn as India.[14] Then another question: 'If we could import the yarn of the finest and best quality could you weave it into muslins of as high a Quality as are made of the like sort of yarn in India?' 'We have no doubt we could,' the Mancunian replied.

Hilton and Drinkwater had been invited because they knew the Lancashire cotton scene intimately. What they didn't know was that Hawkesbury was already involved in the early stages of a daring scheme and that he was only waiting to hear confirmation from the manufacturers on certain specific points before putting his plan into operation.

It seems that while that Hawkesbury was getting the Board of Trade to find ways of encouraging the cultivation of fine cotton in the West Indies, discussions were already taking place behind the scenes between Banks, Hawkesbury and Evan Nepean, Under-Secretary at the Home Office, about finding a way of getting samples of the finest cotton from India. On 22 February 1787, the day before Hawkesbury invited Hilton and Drinkwater to give evidence to the Board, Nepean told Banks that Hawkesbury wanted to talk to him, in Nepean's presence, about the 'Cultivation of Cotton, and he means to solicit your opinion upon the best mode to be adopted for obtaining some Samples of that article from the Neighbourhood of Bombay where it is said the finest hitherto known has been produced.'[15] Nepean invited Banks to Whitehall to meet with him and Hawkesbury on the very same day that the Board of Trade would be meeting Hilton and Drinkwater, which was 26 February 1787. In his letter, Nepean added that he hoped 'my friend Mr. Ow might not object to a trip, if he should have a sufficient knowledge of Cotton to be able to distinguish the difference of the quality'.

This meeting with Hawkesbury might have been the first time that he and Banks actually met, although they had been in correspondence for the previous three years on a variety of topics, and their friendship would continue for almost another quarter of a century.[16]

No record of the meeting survives but the minutes of a subsequent meeting on 31 March 1787 of the Board of Trade with Hawkesbury in the chair, and to which the Prime Minister had been invited, do survive. They simply report that Hove had received the King's commission to collect plants and cotton seeds in the area around Bombay, and while there to find out more about how the local farmers cultivated fine cotton so that the information could be shared with the planters in the West Indies. Joseph Banks would be preparing Hove's instructions.[17]

Obviously this plan must have been discussed and devised at the meeting between Banks, Hawkesbury and Nepean. Later Nepean told Banks that Hawkesbury liked the plan and that he would be able 'to fix it'.[18] By the end of March, the plan was in the final stages of execution.[19]

Banks got straight to work and the instructions were ready in a few days and given to Hove on 2 April 1787 before he left London to join the *Warren Hastings*, the East India Company ship that would carry him to Bombay.[20] They were an example of the kind of instructions Banks would be writing to other collectors in the future. Hove was told that he was being sent out to collect for the royal gardens at Kew; and that he was to collect everything he found that he had not seen already growing at Kew, whether ugly or beautiful, small or large, useful to medicine and manufactures as much as to the botanist. Banks instructed him to explore the area of Gujarat, from the modern town of Bharuch north to the modern town of Ahmedabad, where, he remarked, there were many beautiful and valuable plants known only by the drawings that had been brought back from the area.

These instructions Banks referred to as 'Public'. At the same time, he drew up another and much more important set of instructions for Hove and which he called 'Private'.

These instructions began with, 'As the real object of your Mission is to procure for the West Indies seeds of the Finer sorts of Cotton with which the Ahmood [Ahmedabad] Country where you are ordered to reside abounds . . . you are continually to keep that in view as your main object, & consider the Collections for His Majesty's Garden as secondary to it.'[21]

Banks told Hove that as soon as he had identified which area grew the finest cotton he was to go and live there, learn the local language and ingratiate himself to the local people by offering them his medical services. Hove would be travelling in the guise of a physician: before leaving London, he had learned a little medicine from John Hunter, one of Britain's leading surgeons and anatomists who lectured to students at his large Leicester Square home.[22]

Hove was then to find someone growing cotton successfully and 'near him Fix yourself & pay a daily attendance to his manner of culture, writing down constantly your observations upon it, and communicating them to me on all opportunities'. Hove was to note the nature of the soil, the quantity and frequency of manuring, rains, sowing and harvesting. To gather all of this information, Banks estimated he would need a year. When at the end of this period, he was ready to leave, he was to obtain as large a quantity of the best seeds and samples of the cotton they produced that he could. 'These seeds', Banks emphasised, 'you must protect & preserve with the Strictest attention'.

All of this was to be done secretly. It seems that only half a dozen people knew Hove's real purpose in going to Gujarat. It was essential that the East India Company did not suspect what he was up to.[23] As far as the East India Company was concerned Hove was in western India on the King's business and they were to make his stay as easy as possible, giving him passage on their ships, to and from India, providing him with letters of introduction and allowing him to draw money from them in Bombay during his stay.[24] The East India Company had substantial interests in the Gujarat cotton growing area, purchasing raw cotton there and exporting it to Canton, in order to help finance the tea trade.[25] They were also the largest exporters of Indian cotton textiles to Europe, a trade which, at this time, accounted for around half of the company's total commodity trade, and was worth several million pounds annually.[26] They were, therefore, in direct competition with the Lancashire industry. Had they known Hove's real intentions, his presence would have been unwelcome, to say the least.

In order to keep Hove's work secret, Banks insisted that he should send all seeds and plants (including cotton seeds) to Soho Square, under the guise they were for Kew. To prevent anyone spying on his correspondence in India, Hove was to write in Polish to his brother in Poland, describing in detail all of the information he was able to get about cotton cultivation. This information, translated into English in Poland, should be sent to Banks in London.[27]

As a reward for his work, Banks promised Hove a plantation in the West Indies where he could grow the finest Gujarat cotton.

On 13 April 1787, Hove wrote to Banks from Deal assuring him that 'I will observe Minutely Every Transaction, which [I] shall met with in the East, and inform you of it with greatest exact'ness & fidelity.'[28] Hove hadn't had much time to prepare for his mission – Banks later remarked that Hove 'was dispatchd with such haste as precluded all extension of Enquiry on that subject'.[29] He did, however, find enough time to become acquainted with at least one publication that described the interior of India, and specifically the area to the north of Surat. Written by Pierre Marie François de Pagès, and published in 1782, the book relayed the adventures of the author as he travelled the world between 1767 and 1776. He was in Surat and the adjoining Maratha region for six months, travelling on foot and dressed as a local. Banks had a copy of Pagès's travel account in his library.[30] Hove also had time to speak to George Forster, a servant of the East India Company based in Madras, who had walked from Bengal to the Caspian Sea, and was at present in London.[31]

Banks informed Hawkesbury that Hove was on board the *Warren Hastings* at Deal and that he had everything he needed for his mission.[32] On 14 April 1787, the ship sailed out into the Channel and the Atlantic Ocean – three months later it was in the Indian Ocean and anchored in a bay of the island of Anjouan, one of the Comoros Islands, off the coast of Mozambique.[33] It stayed there for nearly a fortnight, giving Hove ample time to explore the island with a guide and to collect some of the more interesting plants he found for Kew. On 17 July 1787, the *Warren Hastings* left for Bombay, which it reached on 29 July.

Hove's first chance to write to Banks came in late September 1787 when he took advantage of the sailing of the East India Company ship, the *General Eliott*, from Bombay to London.[34] The letter had little to report apart from the fact that Hove had managed to collect some living plants, including a local magnolia, which grew at a cool altitude and which he thought would do well in England. A passenger on the ship had agreed to take care of this and some other living plants. Those he had collected on Anjouan he was leaving behind in Bombay and would deal with once he returned from visiting the cotton region to the north.

Hove explored as much of the area around Bombay, Surat and to the north as he could. In his next letter to Banks, written from Bombay in early February, Hove reported that he had reached Ahmedabad on 24 October 1787 and had begun studying the cotton industry as instructed. But then, after describing cotton growing in the area, he announced that he had only stayed there for one month because it was unsafe – he had been robbed twice and he felt that the locals were becoming suspicious of his presence.[35] He was going to Cambay (now known as Khambhat), on the coast, where he hoped to find even better cotton growing.

This letter may have raised some concerns in Banks's mind but Hove's removal to Cambay would have alleviated them. However, the next letter was even more alarming.[36] Hove revealed that his dealings with Company officials and local people were often fraught. He had been advised to present gifts to the local rulers as he passed through their territory but this did little to make him feel safer. The bottom line was that the sum allotted to his mission, £300, was gone in less than three months. His expenses were mounting and he was finding it increasingly difficult to find anyone to fund him. He'd had no luck with the Governor of Bombay (the chief of the East India Company establishment). The only person Hove had found to assist him financially was Dr Helenus Scott, a senior military surgeon attached to the East India Company in Bombay, who also had a lively interest in botany and who sent Banks botanical specimens. Hove had overspent wildly; on 17 March 1788, he had drawn a bill in Bombay on Hawkesbury to the tune of £2125.[37]

Banks was appalled. Although he was at Revesby Abbey attending to his property, he sent a copy of Hove's letter to Hawkesbury. In the covering letter, Banks insisted that Hove should be recalled before he incurred any further expenses.[38]

It was not just Hove's overspending that was bothering Banks. As he explained to Hawkesbury, it seemed that Hove had been sent to the wrong part of India, that 'the Cottons of Ahmood & its vicinity are by no means the Finest in India but on the contrary are of a medium Quality while the Extra Fine Cotton . . . is now found in the very heart of our Settlements in India.'[39]

Banks's source for this reappraisal was Langford Millington, a wealthy cotton planter with estates on Barbados and St Vincent, who had gone to see Banks sometime in late summer or early autumn of the previous year. He had brought a letter of introduction from Joshua Steele, a planter himself, founder of the Society of Arts of Barbados (modelled on the Society of Arts of London) and an old friend of Banks's. Steele referred to Millington's wealth but added that he was a 'scientific Man'.[40] Who better than a cotton planter to judge on the origin of the finest cotton. Banks listened to Millington. It was not the first time that Banks had heard that the fine quality of Bengal cotton exceeded that of any other part of India. Robert Kyd, the Superintendent of the East India Company's botanic garden in Calcutta, had said the same in his 1786 report.[41]

Not long after reading Kyd's report Banks had written to his old friend William Jones, the philologist and jurist who was living in Calcutta, to try and get him samples of cotton from Dacca which, he thought, was the fine cotton to which Millington referred.[42]

There was no time to lose. Hove was wasting money on what was now in Banks's opinion a misguided mission. Hawkesbury agreed and asked Banks to write to Hove recalling him.[43] Hawkesbury would contact the East India Company in London authorising them to make the necessary arrangements, including the advance of even more money to get Hove back from Bombay.

Banks wrote the letter to Hove and sent it overland. Though expensive, this was the quickest method of communicating with India – two months as opposed to four if the letter went entirely by sea. The ships of the current trading season had left and, though Hawkesbury noted that there was a ship departing in November, the delay was deemed unacceptable.[44]

By land it was. Banks wrote to Hove on 30 September 1788.[45] The tone of the letter was severe. 'I was surprised & very disagreeably so by the Contents of your last Letter & cannot help being of the opinion that the Expence you have incurrd is most unjustifiably Enormous.' Hove had not been sent to India with a bottomless purse. He was to go to Bombay immediately and get on the very next ship to London.

It may be that Hove never received Banks's letter. Certainly he didn't refer to it in his last letter from Bombay before leaving India: in it he simply stated that at the end of November 1788, having completed his observations of cotton growing as he had been instructed to do, he went to Bombay to get the next East India Company ship to London.[46] As it turned out, although the *Prince William Henry* was in Bombay making final preparations to leave, Hove and the ship's captain couldn't come to an agreement about Hove's plant collection, especially the living plants. The ship sailed from Bombay on 13 January 1789 without Hove and his collection.[47] Fortunately, a Danish East India Company ship was departing from a nearby port. The captain, though demanding payment up front for the passage, was much more accommodating about the plants. The *Norge* set sail on 2 February 1789 for the Cape of Good Hope, which it reached on 27 April. After a short stay, the ship resumed its journey towards England. At the Isles of Scilly Hove transferred to a passing pilot ship and arrived in Plymouth on 18 August 1789, more than two years after leaving for India.

By mid September Hove's collection was in London.[48] Those who saw, it or heard of it, were very impressed. Even before the cargo arrived in London, Banks knew its contents.[49] There were twenty-three kinds of cotton, which was the main object of the mission – they would soon be on their way to the West Indies to be distributed to interested planters;[50] but this was surpassed by the collection for Kew, which, was, of course, his cover – eighty varieties of living plants, including a nutmeg tree, a balm tree and a mangosteen, and more than 170 kinds of seeds.[51] When William Aiton at Kew saw the plants, he referred to them as Hove's 'treasure'.[52] Jonas Dryander, Banks's librarian, went to Kew to see the collection for himself. He was delighted to find that they had all arrived alive, an amazing accomplishment.[53]

Over the next few months, the duot settled on Hove's extravagance in western India and attention shifted to his actual accomplishments. On 16 January 1790, Banks was invited to a meeting of the Board of Trade to give his judgement on Hove's mission. The members present that day included Hawkesbury, William Grenville, the Home Secretary and the Board's vice-President, William Pitt, the Prime Minister, and other notables. If they were expecting Hove to be roasted, they were in for a big surprise.[54]

Banks began by emphasising that he had advised Hove to travel as cheaply as possible, told him to go on foot and live with the local people, as Pierre Marie François de Pagès had done previously in the area. Instead, as Hove recorded carefully in his journal, which Banks had in his possession, he had hired horses and a palanquin for his travels through Gujarat. He had strayed far from the original instructions, but he should not be blamed for this: he had been advised locally that this was 'the only practicable way of travelling'. As this extravagance would have attracted the attention of robbers, Hove tried to minimise the risk by hiring soldiers and buying protection from local rulers. Even so, he was robbed twice. No wonder that he ran out of money so quickly. Still, rather than give up, he had borrowed money from anyone who would lend it, in order that he could continue with his important mission.

Hove was, as Banks put it, 'imprudent [by] grossly departing from the conduct pointed out for him in his instructions', yet, in other respects, he deserved the admiration of the Board. In spite of personal danger and illness, he 'made himself master of the mode of cultivating cotton . . . and described the whole of it in his Journal very fully and distinctly'.

The Board members would by now be aware that Banks was praising his collector. He did so by listing his singular achievements, in addition to accomplishing his mission – to collect cotton seeds and observe production methods, which he had fulfilled admirably. He was the first European to describe how ebony and sandalwood were cultivated; the first European to witness the extraction of borax; in addition, his collection of plants for Kew was large and very valuable and included the mangosteen, 'which is allowed to be the finest Fruit in the East Indies and has not hitherto been transported to the West'.

In conclusion, and in defence of Hove, Banks argued: 'Though the Expence he has incurred is so enormously beyond what might reasonably have been expected, [I believe] that the whole Sum was fairly expended in the execution of his orders, that he did not embezzle or apply to his own use a single Halfpenny of it, and that the account of his Expenditure laid before Their Lordships is a just and fair one.'

After that praise from Banks, all the outstanding debts were paid and the case put to rest.

Hove was acclaimed a hero.

8

1787: Mr Nelson's Unfortunate *Bounty* Voyage

At the same time as he was instructing Anthony Pantaleon Hove on his delicate mission to western India, and the ships of the First Fleet were assembling in Portsmouth, Banks became involved in a much larger project, centred on the transfer of the breadfruit plant from Tahiti to the West Indies. Banks, as one of the few people in England who knew anything about this plant, and who had actually been to Tahiti, was the perfect person to advise on the project, even though he did not initiate it.

The focus on breadfruit must have taken Banks back to 13 April 1769, his first day in Tahiti. Banks recorded in his journal at the time that, even before the *Endeavour* anchored, canoes from the nearby shore surrounded the ship, looking to trade their breadfruit for the Englishmen's beads.[1] Banks had never seen a living breadfruit before but he knew what it was since both William Dampier and George Anson decades earlier had visited the islands of Guam and Tinian respectively, and had described it.[2] Banks also knew that the fruit, about the same size as a big grapefruit, grew on large trees, like very large apple trees, so, when, later that same day, he walked into the nearby woods, accompanied by many Tahitians, he knew exactly what he was seeing but was still surprised and thankful when it gave 'the most gratefull shade I have ever experienced'.[3]

During the following weeks when the *Endeavour* was at anchor, the Tahitians frequently supplied the company with breadfruit and Banks learned the different ways it could be prepared. When roasted, the fruit looked and tasted like a boiled potato; but when, out of season, it was allowed to ferment it was made into a sour paste, which Banks recorded was disliked by the whole ship's company.[4]

To Banks, the breadfruit symbolised the Tahitian way of life, which he saw as without cares or worries.[5] 'These happy people', he remarked in his journal, 'may almost be said to be exempt from the curse of our forefather; scarcely can it be said that they earn their bread with the sweat of their brow when their chiefest sustenance Bread fruit is procurd with no more trouble than that of climbing a tree and pulling it down.'[6]

Several years later, after he had returned from his circumnavigation, Banks responded to an invitation from the Dutch aristocrat William Bentinck, whom he was visiting in The Hague in February 1773, to recount his time in Tahiti in writing. This time Banks was more explicit, contrasting the ease with which Tahitians obtained their daily breadfruit and their sexual satisfaction, with the endless toil and sexual inhibitions of Europeans. In his own words: 'In the Island of Otaheite where Love is the Chief Occupation, the favourite, nay almost the Sole Luxury of the inhabitants . . . Idleness the father of Love reigns here in almost unmolested ease, while we Inhabitants of a changeable climate are Obligd to Plow, Sow, Harrow, reap, Thrash, Grind Knead and bake our daily bread and each revolving year again to Plough, sow &c &c subject to Famine if the sun should parch or the rain drench our superficial crop these happy people whose bread depends not on an annual but on a Perennial plant have but to climb up and gather it ready for the baking from a tree which deep rooted in the Earth scorning equaly the influence of summer heats or winter rains . . .'[7]

Banks only saw breadfruit as a plant food that provided Tahitians with easy nourishment. However, when Valentine Morris heard about it, he immediately began to envisage a new possibility for the plant. Morris had been born in Antigua where he inherited sugar plantations on that island and on St Vincent, as well as a country house in Monmouthshire near Chepstow.[8] Banks and Morris had known each other at Eton and it was probably shortly after Banks returned to London from his *Endeavour* voyage that he told Morris about the breadfruit.

It's unclear whether at this point, in mid April 1772, Morris already knew he was to be appointed Lieutenant-Governor of St Vincent,

which post he accepted in December 1772. He had, however, realised
already how useful breadfruit could be. Morris went to Banks's home
in New Burlington Street to discuss the matter with him but finding
him away wrote instead. He spelled out his idea briefly in his letter
of 17 April 1772. Morris wanted to get 'some information . . . whether
there was no possibility of procuring the bread tree, either in seed or
plant so as to introduce that most valuable tree into our American
Islands'. As the reason for this, Morris simply stated that he was
interested to 'procure to its habitants what must if once it could be
made [to] succeed be one of the greatest blessings they could possess.
As my motive for giving you this trouble is a humane benevolent
desire of benefitting so considerable a body of people . . .'⁹ Though
he avoided saying it, Morris was certainly referring to the slave popu-
lations he owned, like other British planters on the West Indies, as
those he thought would benefit from this transfer.

At about the same time that Banks was writing to William Bentinck,
John Hawkesworth in London was printing his three-volume work of
the Pacific travels of his countrymen, including, and especially, the
voyage of the *Endeavour*. The breadfruit featured prominently in the
section on Tahiti and though he lacked the rich prose of Banks,
Hawkesworth nevertheless extolled the fruit's virtues.

How Banks responded to Morris's letter isn't known. Once John
Hawkesworth's publication of the narrative of the voyage of the
Endeavour appeared in print in 1773, in which breadfruit was lauded
as a substitute for bread, with emphasis on how little labour was
required to grow it, the idea of transplanting the breadfruit tree from
Tahiti to the West Indies as food for the slave population quickly
gained interest and publicity, on both sides of the Atlantic.¹⁰ Edward
Long, a Jamaican planter who had returned to Britain in 1769 because
of poor health, was the first to propose the idea in print. In his three-
volume work on the history of Jamaica, which was published in 1774,
he pointed to breadfruit as one of the most important food plants
that should be introduced into Jamaica.¹¹ This work established Long
as the leading spokesman on the affairs of the British West Indies and
his influence was great.

The naturalist John Ellis was Royal Agent for the British possession
of West Florida – a colonial position, based in London, with responsibilities for the financial management of the colony – and also Colonial
Agent for the island of Dominica. He was the first to publish a
pamphlet on breadfruit, enlarging on some of the ideas he had

presented earlier in 1770 in an essay about the problems of transporting plants at sea over large distances.[12] Now, in 1775, he produced the first full-length description of the plant, referring his readers to various accounts of the plant in situ, including Cook's, and including important sections on its botany and methods of propagation. This wasn't just a natural history lesson: it was a call to sea captains and other interested parties to try and bring the breadfruit back for the West Indies; 'two or three trees suffice for the support of one man throughout the year.' 'They reckon', Ellis concluded, 'that an acre of land so occupied, affords more nourishment than any two acres of other produce.' Ellis repeated what Banks had noted, that the tree needed little attention and was pretty hardy. In short, 'it is, therefore an object of no small importance to our West-India planters.' To encourage action, Ellis recommended that the plants should be transported in a variety of specially made cases, including those made of wood and wire; and as a financial incentive, there was an advertisement at the end of the pamphlet, which reported that the Society of West India merchants, and the agents for these colonies, were already considering their next step and would, in time, be offering 'handsome premiums' to anyone who was successful in importing the breadfruit tree.[13]

Ellis was here referring to a resolution passed at the 7 February 1775 meeting of the West India Committee, a powerful interest and parliamentary lobbying group of merchants and planters in London, where the members agreed to underwrite the costs of introducing the breadfruit into the West Indies.[14]

The following month, the West India Merchants went public with their breadfruit interest and offered the substantial sum of £100 to 'any Commander of an East India Ship, or other Person' who brought the plant in a healthy state to England.[15] Two years later, the Society for the Encouragement of Arts, Manufactures and Commerce, England's second-oldest scientific society, of which Banks had been a member since 1761, joined in and offered a prize of £50 for each of three species of breadfruit brought to London in a 'growing state'.[16] Five hundred copies of the notice together with an extract from Ellis's pamphlet were published and distributed.[17]

Then it all went completely quiet. The American colonies rebelled against Britain. The French, in support of the colonists, deployed their warships in the Caribbean and the planters had more to worry about than breadfruit.[18]

As the conflict dragged on, the silence continued. The premium from the Society for the Encouragement of Arts was renewed every year and, every year, it was unclaimed. Perhaps this isn't surprising. Even without the upheaval of war, Tahiti was simply too far removed from the normal sea lanes used by the few British ships that sailed in the Pacific. Even if a ship ventured near Tahiti, the very practical problems of collecting the breadfruit plant, transplanting and trans-porting it nearly 20,000 miles, were daunting to say the least.

However, in 1784, following the ending of the war, Banks received a letter from Hinton East, the first of several from botanists and botanically minded officials living in Jamaica which had the effect of reviving the breadfruit venture.

East was a high-ranking official in Jamaica's government, having been, at one time, the colony's Receiver-General, a member of the House of Assembly and the Judge-Advocate-General of Militia.[19] The reason he was a correspondent of Banks's was that he had a substan-tial botanic garden, which he had begun in 1774 and where he grew many exotic plants imported from abroad.[20] East's letter, which appears to have been his first to Banks, described his garden and particularly the several examples of plants from the East Indies – mangoes, cinnamon and tea, for example – 'thriving amazingly'. The main thrust of the letter, however, concerned breadfruit. Here East spelled out why breadfruit would work so well in the West Indies. 'The Acquisition of the best kind of the Bread Fruit would be of infinite Importance to the West India Islands in affording exclusive of variety, a wholesome & pleasant Food to our Negroes, which [would] have this great Advantage over the Plantain Trees from whence our Slaves derive a great part of their Subsistence, that the former, wou'd be raisd with infinitely less labour and not be subject to be destroyd by evry smart Gale of Wind as the latter are,'[21] Banks was, of course, familiar with breadfruit's minor call on labour resources and its deep rooting system, but he may not have known about the 1781 hurricane, particularly how extensively it had uprooted plantain trees in several of the island's parishes.[22]

Banks was being drawn in. Fast on the heels of East's letter was a letter from Matthew Wallen, an Irish naval officer who had settled in Jamaica in 1747 and had started a coffee plantation. Wallen told Banks that the British forces had taken a French ship, which had come from Réunion in the Indian Ocean, on which they had many fine plants including breadfruit – though not from Tahiti, as Wallen swiftly added. Wallen was not so much pointing out that the French had taken the lead on transplanting a variety of breadfruit to their islands, though that was certainly true, but he was also contrasting their initiative with the lack of enterprise shown by the British who were doing little to bring East Indies plants to the British Caribbean.[23] Sometime later, Wallen provided Banks with a list he and Hinton East had drawn up of plants that should be introduced into Jamaica. 'When you see any of your Brother Planters', Wallen added, 'I think it would not be amiss to spur them on to get the Bread Fruit'.[24] A little over nine months later, on 6 May 1785, and clearly frustrated as what he saw as the inertia of the Jamaica Assembly in getting things done, Wallen returned again to the subject of breadfruit but this time was more forthright with his advice: 'The King', he wrote, 'ought to send a Man of War a Botanist & Gardener for the Plants we want.'[25]

The pressure to introduce breadfruit to the West Indies was growing and so too the idea that Banks might be influential in getting it done. The Society of West India Merchants added its voice in a letter to William Pitt, the Prime Minister, sometime in 1786. Meanwhile, Banks was asking Sir George Yonge, the Secretary of War, whose responsibilities included the botanic gardens in Jamaica and St Vincent, about what kind of plants from the East Indies were being grown successfully on the islands to get a clearer idea of which transfers had worked.[26] Hinton East, who had himself been receiving seeds and vital information about how to cultivate plants from the East Indies, visited Banks at his Spring Grove home in August of the same year to press his case for the introduction of the breadfruit plant.[27] Another hurricane had swept through Jamaica on 19 October 1786 and had damaged many trees: a supply of breadfruit was now urgent.[28]

By early 1787, these various pressures and interests had the desired effect. On 13 February, Pitt informed the Standing Committee of the West India Planters and Merchants that 'directions have been given for collecting as many of the Bread Fruit Trees as possible, and that every Opportunity be taken to have them conveyed to our West India Islands'.[29]

Though Pitt did not spell it out to the planters and merchants, he either knew or would soon learn that Banks was taking charge of the arrangements.

A few plants tucked away on board, out of harm's way in wire containers of the kind advocated by John Ellis and the French botanist, Henri-Louis Duhamel du Monceau, was not what Banks had in mind when he planned the breadfruit transfer. He knew enough about the perils of keeping plants alive at sea to know that mortality rates were very high and, for the breadfruit to have any chance of taking root in the West Indies, they would have to be planted there in huge quantities. At this point Banks's plan envisaged that one of the ships of the First Fleet, currently lying at anchor at Portsmouth and awaiting their final instructions to carry themselves and the convicts to Botany Bay, should be engaged for a breadfruit expedition.[30] The idea was that, after having unloaded its cargo in Botany Bay, the designated ship would proceed to Tahiti for the breadfruit plants and transport them to the West Indies before returning to England. The scale of the operation was immense and had never been attempted before. Banks immediately impressed on Pitt that the ship would need to be restructured as a plant transporter; and that both the commander and the crew would have to sacrifice space. Moving living plants by sea over long distances, Banks cautioned, was very risky and everything had to be done to reduce that risk. This meant that the great cabin, normally the finest and most private space on the ship, would have to be transformed into a greenhouse, and that the key to it was to be given to the gardener, who would be appointed to care for the plants on their long journey. Additionally, every bit of space on the quarterdeck would need to be appropriated for potting tubs, which would be lashed to the sides. The gardener would have to be provided with sufficient canvas to cover the plants when he felt it was necessary; and the crew would have to be prepared to help move the heavy tubs periodically from the nursery to the quarterdeck to benefit from sunshine. Fresh water had to be made available not only for watering but also, and more importantly, for washing sea water from the plants, that being their most dangerous enemy. No animals, apart from those that provided food, would be allowed on board, and everything possible would have to be done to eradicate rats and cockroaches.

Reading the instructions now, one might wonder: who would undertake such an expedition and for what gain? There is no mention of the latter in Banks's letter though the various premiums and prizes were still on offer. The West India Committee realised that some larger incentive would be necessary and in early March 1787 raised a subscription among its members for that purpose.[31] Captain Arthur Phillip, in his position as commander of the First Fleet, agreed to the idea of sending one of the ships to Tahiti and then to the West Indies.[32]

Banks now made his first and most important appointment of the expedition: the gardener. His choice was David Nelson. He and Banks had known each other since late April 1776 when James Lee, of the Vineyard Nursery in Hammersmith, recommended Nelson to be the botanist on the voyage of HMS *Discovery*, commanded by Charles Clerke, which was to accompany HMS *Resolution*, on Cook's third voyage to the Pacific.[33] Banks must have been impressed with Nelson when he came to his home in New Burlington Street, bringing his recommendation with him, for either next day or the day after, Banks advanced him twelve pounds and twelve shillings as his wages as botanist on the *Discovery*.[34]

Nelson was on the *Discovery* as Banks's employee, to collect and preserve seeds, plants and insects as the latter's exclusive property, as well as collecting seeds and bulbs for Kew.[35] On the first occasion he could, Captain Charles Clerke wrote to Banks from the Cape that Nelson, 'one of the quietest fellows in Nature', was performing well: he had just returned from botanising in the nearby countryside.[36] Nelson also impressed William Anderson, the surgeon on HMS *Resolution*, who was particularly taken by his botanical skills.[37] Cook was also impressed and told Banks that he would help in any way he could 'to add to your Collection of Plants & Animals'.[38] Nelson's plant collections, including the first plants from Hawaii, arrived in good shape in London in early October 1780.[39]

A few years later, feeling his confidence had been well rewarded, Banks employed him on a voyage to West Africa, but this trip failed, the ship returning to the Thames after getting no further than Plymouth.[40]

Now, in 1787, Nelson was employed for a third time by Banks and, at this point, it was up to Captain Phillip and the captain of whichever ship was sent to Tahiti from Botany Bay, to put the plan into action. There is nothing in the record to suggest that anything might change. On 22 March 1787, Banks learned from Evan Nepean, Under-Secretary of State at the Home Office, that Captain Phillip was on the point of deciding which of his lieutenants would command the ship that would be going to Tahiti.[41]

Then, suddenly, Banks changed his mind on how to proceed and alerted both Charles Jenkinson, Lord Hawkesbury, President of the Board of Trade, and his old friend Constantine Phipps, Lord Mulgrave, who was also a member of the Board of Trade, to a completely new plan.[42]

The breadfruit expedition, Banks insisted, was far too important to be a mere adjunct to the convict transport. It had to be a fully fledged voyage on its own, fitted out in London. Why the change? In his letter to Hawkesbury and Mulgrave, Banks explained the advantages of the new plan over the old by arguing that it would be faster to send a ship from London, and that the costs of outfitting it in London would probably be less than in Botany Bay. Added to that was the argument that in London it would probably be easier to find a suitable vessel and 'people may also readily be Found more capable of Conducting her through a voyage in which difficulties are likely to occur'. Banks thought a small two-masted vessel of around 200 tons with a small crew would do. Banks added a few more details as to the make-up of the ship but the only person he mentioned by name was David Nelson, whose virtues and talents he extolled, particularly emphasising that he had much experience in caring for plants at sea; that he had already been to Tahiti on HMS *Discovery*; and that he knew some Tahitian.

On 5 May 1787, Lord Sydney, the Home Secretary, put Banks's new, and more ambitious, plan into action. He wrote to the Admiralty saying that the King had agreed with the West India merchants and planters that introducing breadfruit into the islands would benefit both Britain and the Caribbean colonies. He requested that 'you do cause a Vessel of proper Class to be stored and victualed for this Service, and to be fitted with proper Conveniences for the Preservation of as many of the said Trees as from her Size can be taken on Board.'[43]

Events moved very quickly. There was hardly a single decision in which Banks wasn't involved. By 23 May, the Navy Board had completed the purchase of a ship called the *Bethia*, which had been approved by Banks.[44] A month later the ship was renamed Armed Vessel *Bounty*, and Banks and Nelson gave instructions as to how the ship's space should be transformed for its new use.[45] The great cabin was to be commandeered as a nursery; skylights were to be built and new openings were to be cut into the ship's structure to allow superior ventilation; and lead channels were to be made to collect surplus rain water for re-use.[46]

Though it was not Banks's idea to transfer breadfruit from Tahiti to the West Indies, this was his and Nelson's voyage.[47] It is remarkable how much responsibility Banks and Nelson, two civilians, were given in fitting out a naval vessel. The historian David Mackay wondered about this and came to the conclusion that the combination of Banks's

Pacific experience and his botanical knowledge made 'his direction of the enterprise invaluable and inevitable'.[48] With the death of the botanist John Ellis in 1776, James Cook in 1779 and Daniel Solander in 1782, Banks was really the only one left who had first-hand knowledge of the breadfruit tree and had actually spent time in Tahiti.

During almost the whole of the period that the *Bounty* was in dock, it did not have a designated commander. Nothing is known about what was going on behind the scenes during these months, what discussions were taking place, which names were being put forward, what negotiations ensued. All that is recorded is that on 6 August 1787, on the day after he returned from Jamaica, William Bligh wrote Banks a letter in which he said: 'I have heard the flattering news of your great goodness to me, intending to honor me with the Command of the Vessel which you propose to go to the South Seas.'[49]

Banks did not know Bligh personally but certainly knew of him.[50] Bligh was born in 1754. He first went to sea eight years later as 'captain's servant' and never looked back. After working his way up the career ladder and finding service on several naval vessels, Bligh got his first big break when, in March 1776, at the age of twenty-one, he was appointed master on Cook's HMS *Resolution* (the third voyage). Two months later, on 1 May 1776, he sat and passed the lieutenant's examination, and two months after that the *Resolution* left Plymouth on its marathon voyage in the North and South Pacific. The expedition took more than four years to complete and the ship did not return to England until 4 October 1780. For the next few years, Bligh served on line-of-battle ships but on 13 January 1783, once the American War of Independence was completely over, Bligh and many other junior officers were retired on half pay.

By the end of the year, Bligh was back at sea, this time as the commander of a merchant vessel owned by Duncan Campbell, a very wealthy merchant with a fleet of ships trading to the West Indies, and Bligh's uncle-in-law.[51] It was in this capacity, in which he had worked for Campbell for almost four years, that Bligh was returning to London from Jamaica.

Campbell and Banks knew each other through a number of contacts but especially because the former had been appointed 'Overseer of Convicts on the Thames' in 1776, when the problem of transportation first became acute, after the American colonies were closed as a destination. Campbell and Banks met when they had both been asked for their opinions on transportation at the hearing of the Beauchamp

Committee in 1785.[52] They knew each other well enough that when Bligh got back from Jamaica, it was Campbell who told him there was no point in going to Banks's house to thank him for the appointment as he was still in Lincolnshire.[53]

Campbell took credit for recommending Bligh to be the commander of the *Bounty* though he didn't reveal to whom he made this recommendation – he may have spoken directly to Banks or possibly to Evan Nepean at the Home Office, whom he also knew very well.[54] However it came about, Bligh jumped at the chance. He was the ideal choice: he had learned much about Tahiti when he was there for more than four months with the *Resolution* in 1777; he had seen breadfruit growing in Tahiti; he spoke some Tahitian; he knew his way around the West Indies; and he had served with Nelson, all of which experience would stand him in good stead. On 16 August 1787, after meeting Banks for the first time, Bligh received his official commission.[55]

Word had got out about the breadfruit expedition and Banks received at least two solicitations from men who wanted to be appointed as gardeners to the voyage, but he turned them down as he had already appointed William Brown to be Nelson's assistant.[56]

A few days after Bligh received his commission, Banks presented Nelson with his instructions.[57] Though he had already had a draft set given to him in May, this set was quite different in a number of respects, one of them highly significant. The primary purpose of the voyage, to collect and transplant the breadfruit tree from Tahiti to the West Indies, remained the same, but Banks added that other plants from Tahiti, especially a type of banana and the Tahitian apple, should also be collected. In addition, Nelson was to collect an array of tropical fruits and spices from wherever Bligh decided to call on the way to the Atlantic, whether Java or Mauritius, both places Banks thought likely stopovers.

The botanical cargo was to be far more diverse than just breadfruit. Banks told Nelson that he expected the *Bounty* to stop at St Helena before crossing the Atlantic towards the Caribbean. If this happened, Nelson was to leave some breadfruit and tropical plants for the island's Governor to be planted in the botanic garden, belonging to the East India Company who, effectively, owned the island. When he reached the West Indies, Nelson was to divide equally most of what was left of his botanical cargo between Alexander Anderson, the Superintendent of the St Vincent Botanic Garden, and Hinton East's garden in Jamaica.

Up to this point, Banks was instructing Nelson to fulfil the government's demands. But Banks had other plans for the voyage – he saw a new and unique opportunity to provide a vast range of tropical plants for the King's Garden at Kew. Banks told Nelson that he was to reserve examples of each specimen he had collected in Tahiti, and any other place the ship called at on the way back, for Kew; and that he was also to place on board whichever plants Anderson and East had collected and prepared for the King. Banks ended his instructions to Nelson with these words: 'I have now only to express my wishes for your Success in the management of this arduous Business, which is committed to your Charge, and to repeat to you that your diligence and attention will not pass unrewarded.'[58]

As the *Bounty* lay in dock, supplies were placed on board, including the specially made clay pots, around 800 of them, for the plants; and a large number of 'toeys' (iron objects in the shape of a traditional Tahitian adze), metal objects, glass beads, looking glasses, earrings, cloth and clothing with which Bligh was to barter with the Tahitians for breadfruit and other fruit trees.[59] Not everything went smoothly: there were delays caused by misunderstandings and unwanted interferences; and an inexplicable wait for sailing instructions, which took three months to get to Bligh. And then there was the threat of war with the Netherlands, some last-minute changes of personnel and very contrary winds.[60] But finally, on 23 December 1787, the *Bounty*, a relatively small ship of 215 tons with a company of 46 men, sailed out of Spithead for the South Seas on a voyage without precedent.[61]

Bligh's sailing directions instructed him to enter the Pacific by way of Cape Horn. He believed he was already too late in the season to take this route, but he nevertheless followed orders, though, through Banks's intervention, he was able to insist on having alternative instructions should he need them.[62]

To be able to make his way to Tahiti with the minimum of stops en route, Bligh had to call at Tenerife on the way south to resupply. During the few days that the ship was there, Bligh ordered the carpenters to repair everything that needed attention, filled the water casks to the brim and got as many fresh supplies – beef, pumpkins and potatoes – as he could manage.[63] On 10 January 1788, the *Bounty* was back in the Atlantic.

Bligh sailed in a southwesterly direction towards the coast of South

America. On 23 March 1788, Tierra del Fuego was sighted. Bligh decided to round Cape Horn to the south but the seas were heavy, with strong winds, sleet, snow and rain. The going was very hard. There was no let-up in the bad conditions but Bligh persevered. On 13 April, Bligh recorded in his log: 'Upon the whole I may be bold to say that few Ships could have gone through it as we have done, but I cannot expect my Men and Officers to bear it much longer . . .'[64] The weather deteriorated even further and the ship's pumps were constantly at work trying, mostly in vain, to keep the salt water out. The men and the ship were both complaining. On 17 April, Bligh had had enough. He decided to head for the Cape of Good Hope and enter the Pacific that way. 'The General Joy in the Ship was very great on this Account.'[65]

With the strong westerlies now in his favour Bligh made it across the South Atlantic in record time. On 24 May 1788, the ship anchored in Simon's Bay, Cape of Good Hope. The crew were fed and rested, the carpenters back at work. Bligh, with the help of Colonel Robert Gordon, who knew his botany, purchased seeds and plants which he thought might interest the Tahitians, and others which might be sowed at places where the *Bounty* called.[66]

The layover at the Cape lasted almost forty days. On 1 July 1788, the *Bounty* set sail into the Indian Ocean heading for Tasmania, the last scheduled stop before Tahiti.

On 21 August the ship anchored in Adventure Bay. Bligh had last been here in January 1777 on Cook's third voyage. The bay lies on the east side of Bruny Island, near the southeast corner of Tasmania. Adventure Bay had been a natural stop for European ships heading out into the Pacific from the southern Indian Ocean ever since Tobias Furneaux, who named it after his ship HMS *Adventure* in 1773 on Cook's second voyage to the Pacific, surveyed the bay and extolled its virtues as an excellent anchorage, as the source of a plentiful supply of fish and fresh water and the surrounding land as providing good wood.

Apart from watering and wooding, both Bligh and Nelson used the opportunity to plant some of the specimens they had collected at the Cape. In a less wooded part of the bay, where Bligh thought there would be less danger of fire spreading – the indigenous inhabitants used fire in their agriculture – he and Nelson planted a variety of trees: 'three fine young apple-trees, nine vines, six plantain-trees, a number of orange and lemon-seed, cherry-stones, plum, peach, and apricot-stones . . .'[67] Near where they found water, Nelson continued the plant exchange by sowing onions, cabbages and potatoes. Bligh explained in his log that he did this for the 'Natives, or those who may come after us'.[68] Perhaps it was also in imitation of his great hero – when he was last there in 1777, Captain Cook left behind a boar and a sow, presumably for the same reason.[69]

It was time to leave. On 4 September 1788, the *Bounty* began the last leg of the voyage to Tahiti.

Bligh knew these waters well. He set a course to the south of New Zealand and then northeasterly across the Pacific to Tahiti. On 26 October 1788, the *Bounty* moored in Matavai Bay, on the north coast of the island, which had been visited on several occasions by European ships since 1767 including the *Endeavour* in 1769, and the *Resolution* on three occasions, the last in August and September 1777 when Bligh was on board.

The plan that Bligh and Nelson were to follow was a simple one. The first thing was to find the breadfruit plantations closest to Matavai Bay. Then, Bligh, with the assistance of William Peckover, the gunner who knew Tahitian well, would begin to barter the sundry items they had brought with them; together with the plants they had collected along the way, which had been placed in the pots. The third step was to erect a tent, guarded by Nelson, Brown and other crew members, where the young breadfruit plants could be kept until the precise time, as determined by Nelson, that they could safely be moved to the ship. Only then would they leave for the voyage into the Atlantic and the West Indies.

The plan went well.[70] Nelson found nearby plantations and also the pomelo trees he had planted on his last visit. The exchange of European gifts for Tahitian ones was as smooth as could be. The only slight hitch was that, during the years in which they had been trading with Europeans, the Tahitians had become less enamoured of trinkets and more desirous of tools, particularly hatchets, of which Bligh had relatively few. Still, the variety of cloths and clothing he had brought with him seemed to make up this deficiency.

Within a week of arrival the tent on the beach had been put up and the carpenters began preparing the cabin for the breadfruit plants. The local people brought the plants to the beach, where Nelson and Brown potted them immediately – by the end of the first week of November there were already more than one hundred of them. A fortnight later, Bligh counted 775 plants potted already. The maximum capacity had been reached. The cabin was now ready and Bligh asked Nelson and Brown to collect other Tahitian species, particularly a variety of plantain and the Tahitian apple.

Bligh had expected that the whole operation in Tahiti would take three months and that he would be in the West Indies by the end of July 1789. This was overly optimistic. Though Nelson would have known that the breadfruit could only be propagated by root suckers or root cuttings, he did not know how long it would take for the roots to develop and for a viable plant to present itself.[71]

It was a very slow process. By mid December, though the plants generally looked healthy, they hadn't rooted. Then the weather turned. The rains were good for the plants but the accompanying heavy seas dangerous to the ship. On Christmas Day, Bligh decided to move it to another harbour protected by a reef. The pots all had to be put on board and the tent relocated. The next day, in a new and safer spot, the tent was re-erected and the plants moved back on land.

They resumed waiting but Bligh did not waste his time. He decided that the number of pots and the space in the cabin should not limit the number of plants he would take away with him. As Banks had suggested when he first planned the expedition, Bligh ordered the carpenters to dismantle the chicken coop on the quarterdeck and to prepare the area, protected from the elements by a roof, for the plants.[72] By the beginning of February 1789, the shipboard space was ready. More plants began arriving at the beach and were potted. Pleased with the progress, Bligh counted his booty: 1,015 breadfruit plants in total.

For the whole of February Nelson checked the collection daily, but it was only at the end of the month that he could tell Bligh confidently that there were signs of life. He predicted that in a month the plants would be ready to be placed on board. Bligh commented: 'Under these happy circumstances and the Plants easily cultivated. The success of the Voyage now only hinges on our passage home.'

Bligh now began the final phase of preparation – the wholesale cleaning of the ship and the eradication of any sign of animal life other than those invited on board. Using a combination of tobacco smoke and hot water, the crew got into all the crevices where cockroaches hid. The sailmakers completed the preparations by making coverings for the plants that would be placed on the quarterdeck.

Finally, on 29 March 1789, Nelson gave the signal that the plants were ready. Everyone pitched in and on 1 April the botanical collection was on board, all 1,015 breadfruit plants in 774 pots in the cabin and 63 tubs and boxes on the deck.

On 4 April, Bligh ordered the *Bounty* to set sail for the six-month-long final leg of the voyage.

The ship sailed west past various islands, stopping at a few but mostly sailing by in waters that were familiar to Bligh.

Then, quite suddenly and before dawn on 28 April 1789, about thirty miles from the Tongan island of Tofua, Fletcher Christian, the master's mate and several other men, cutlasses at the ready, burst into Bligh's cabin and tied him up; he was then forced on deck. The boat-swain lowered the launch – it was only 7 metres long, 2 metres wide, with 2 masts and 6 oars. Bligh and eighteen others were put on board. Some provisions were thrown in after them and the launch was pulled to the stern and cast adrift.[73]

From the launch, Nelson, who had remained loyal to Bligh – Brown joined the mutineers – watched as over one thousand breadfruit trees, in the cabin and on the deck, disappeared from sight with the *Bounty*. The plants, which as Bligh put it, 'I looked at with delight every day of my life' and which had flourished in his great cabin, were thrown to a watery grave in the Pacific a few days later.[74]

All their efforts were destroyed. The expedition was suddenly over, finished, in a remote part of the Pacific. The breadfruit would not be planted in the West Indies. The King would get nothing for his royal garden.

9

1790: The Second Circumnavigation
of Archibald Menzies

In late July 1789, Archibald Menzies arrived in Deptford after his three-year circumnavigation on the *Prince of Wales*. A month later, when he had yet to find his feet, Banks recommended him for another voyage, which was already in preparation.[1]

In the months before Menzies's arrival, William Grenville, the Home Secretary, was thinking about sending a ship into the South Atlantic to look for suitable bases for the expanding whaling industry.[2] The British whaling interests, headed by the firm of Samuel Enderby and Son, had been lobbying the government for several years. By 1789 they had managed to get some restrictions on their operations removed, establishing in particular their right to hunt whales in the Indian and Pacific Oceans, the domain of the East India Company and the South Sea Company.[3] Now they wanted to know what restrictions the Spanish might impose in the southwest Atlantic and the southeast Pacific on their fishing operations; and whether there were suitable places for their ships to replenish supplies within reasonable sailing distance of Cape Horn and the Cape of Good Hope.

Enderby had already asked Banks in August 1788 for his opinion as to suitable bases that were well clear of the Spanish.[4] His firm, Enderby explained, had a ship, the *Emilia*, fast, copper-bottomed and with a company of over twenty men, ready to sail into the southern Pacific via Cape Horn. Though the letter came out of the blue, Enderby reminded Banks that his firm had been in touch with Solander on a previous occasion and so had some connection to Soho Square already. Enderby had many questions for Banks: did he know where the sperm whales could be found; where good charts of the area could be had; whether the Spanish would be cooperative; and, most importantly, where the ships could take shelter and refresh – was Juan Fernandez Island settled? In return, Enderby suggested that as they were also sending ships into the Indian Ocean and were expecting to anchor in southwest Africa north of the Cape, if there was anything Banks wanted from there, Enderby would oblige.

No record exists of Banks's response but, by the time he wrote to Banks, Enderby would have had at least a strong suspicion that the British whalers were not particularly welcome, based on the experiences of the *Sappho*, a ship of a rival concern that had ventured into Spanish territory in April 1789 and had been ordered to leave its bounty behind.[5]

This was the context in which preparations for a new expedition were being made. Banks was involved from the beginning, recommending Menzies even before the ship or its commander had been chosen. By early October 1789, the instructions to the commander of the ship had been settled. The purpose of the voyage was to look for suitable whaling bases: to examine known islands and look for undiscovered ones in the South Atlantic; to examine that part of the coast of Africa which HMS *Nautilus* in 1786 had not; and to extend that coastal survey from the limit of Dutch territory at the Cape eastwards right up to the Portuguese settlements on Africa's eastern coast.[6]

While the details of the expedition, including its destinations, remained secret, some pieces of information did get out. On the same day that the Navy Board began looking for a suitable ship, several London papers were alerting their readers that another voyage to the South Seas was in the offing.[7] Rumours spread. George Dixon had heard about the expedition and seemed to know who was to have command of the ship, and to think that the northwest coast of North America would be one of its destinations.[8]

Dixon was wrong about where the ship was heading but he was right about who the commander was to be. Lieutenant Henry Roberts, aged thirty-six, was an experienced naval officer and had that special accolade of having served on Cook's second and third voyages, the latter in the post of master's mate. His special talent for drawing and making charts was highly prized and it was his work which appeared in the official published account of Cook's final voyage. The Navy Board eventually found the ship it was looking for. HMS *Discovery* (the same name as the *Resolution*'s companion ship) was delivered on 19 December. Another of Cook's men, George Vancouver, who had been on the second voyage and had served as midshipman on the final voyage, was taken on as first lieutenant, and Richard Hergest, who had also been with Cook on his last two voyages, joined as the second lieutenant. A second vessel, the *Chatham*, was found as companion ship.[9]

It was Banks's intention to make this voyage special as far as botany was concerned. Not only did he have an accomplished naturalist and gardener lined up, but, in response to the growing interest of George III and Queen Charlotte in their gardens at Kew, Banks hoped to use this voyage to bring new living plants back to England.[10] For this Banks got the Admiralty to agree that a glazed plant cabin should be erected on the *Discovery*'s quarterdeck, using a design similar to that which he had just installed on HMS *Guardian* for its voyage to New South Wales. The difference between the two cabins was that the one on the *Discovery* was only a third the volume of the one on the *Guardian* and it was not as tall. On 12 December 1789, Banks went to Rotherhithe to the private yard of the shipbuilders to supervise its construction.[11]

By the end of January 1790, the ships had received their nautical and astronomical instruments; the plant cabin had been erected; the crew assembled; and the all-important supply of antiscorbutics loaded. The ships were ready to sail.

And then, suddenly, it all stopped.

On 21 January the Duke of Leeds, the Foreign Secretary, received a letter from Anthony Merry, the British *chargé d'affaires* in Madrid, reporting that an English vessel had been seized by the Spanish in Nootka Sound. No further details were supplied. About a month later, however, the Marquis del Campo, the Spanish Ambassador in London, shed more light on the event.[12]

The ship in question was the *Argonaut*, commanded by Lieutenant James Colnett, with whom Menzies had sailed in the Pacific several years before. Colnett, it will be recalled, had remained behind in Canton with the *Princess Royal* while the *Prince of Wales*, carrying Menzies, returned to London. While in Canton, Colnett had been given command of a new expedition to the Pacific Northwest on another fur-trading venture, no longer headed by Richard Cadman Etches, but by a new consortium based in Canton. Three ships of the expedition had already left Macao for the other side of the Pacific.

Colnett followed in late April 1789 on the *Argonaut* and reached Nootka Sound in early July. There, the Marquis del Campo explained, Estéban Martínez, who had been sent by the Viceroy of Mexico to lay Spanish claim to Nootka Sound, promptly arrested Colnett, and sent him and his ship to San Blas, the new Spanish naval base on the Pacific coast of Mexico.[13]

The Pitt government reacted strongly to this act. Over the next few weeks, they secretly planned to send an expeditionary force to Nootka Sound for the purpose of establishing a settlement there. The idea was for the *Discovery*, which was ready to sail, to head for the Pacific Northwest via New South Wales, where men would be collected to build the new settlement at Nootka Sound. At least one heavily armed frigate would accompany the *Discovery*.[14]

Sometime in March, Evan Nepean at the Home Office, who had masterminded the expedition, asked Banks what kind of trade goods would be most suitable in order to secure cooperation from the local people in building a settlement at Nootka Sound. Banks, naturally, turned to Menzies for enlightenment and received a long list of trade goods which, in his experience, would work best. Menzies's answer was very interesting. Metal goods, in general, Menzies remarked, were in demand – not beads, he added, 'they were so over stocked with Beads as to ornament their Dogs with them.' But the actual demand for metal varied from place to place – iron in one, copper in another and brass elsewhere. The best plan, Menzies thought, was 'any Vessel going there ought to be supplied with Two Black Smiths & a Forge together with the necessary Utensils for working Iron, Copper & Brass into such forms as may best suit the fickle disposition of the Natives.'[15] When Banks costed out Menzies's list, the total exceeded £6800, a substantial sum.[16]

Preparations for the expedition proceeded over the next month, but not without some delays. Just at the point when the *Discovery* and the *Gorgon*, the forty-four-gun frigate which had been fitted out to accompany the smaller vessel, were ready to sail, the Nootka crisis, as it was being referred to, took an unexpected twist.

John Meares, a partner of the Canton fur-trading venture, arrived in London in early April with lurid tales of what had actually happened at Nootka Sound. He put his information together in the form of a memorial which he presented to the government. The story Meares told contradicted the one from or via Madrid: not just one ship but four ships had been boarded, the officers had been harassed and molested and even seized.[17]

Nepean's expedition to the Pacific Northwest was cancelled. Instead, Britain was placed on a war footing and all of the officers from the *Discovery* ordered to other stations.[18]

Although Britain and Spain were on the brink of war, a peaceful way out of the confrontation was found. On 28 October 1790, the Nootka Convention was signed between Britain and Spain to end the current crisis, but because of its many vague clauses it paved the way for problems in the future.

Peace raised the question of the *Discovery* and its future. Two possible voyages had already been cancelled but now, in November, orders were received to refit the *Discovery* and the *Chatham* and get them prepared for sea. Roberts was still in command of the expedition and, as far as anyone knew, the destination and the instructions were as they had been in October 1789 – that is to proceed to the South Atlantic to search for suitable whaling bases.[19]

And then, another twist. On 20 November, George Vancouver was ordered to attend a meeting of the Board of Admiralty.[20] Vancouver had been assigned to the *Discovery* as Roberts's first lieutenant but at the beginning of preparations for war against Spain, as the officers of the ship were sent to other stations, he was posted to HMS *Courageous* as third lieutenant on manoeuvres in the Channel. His commander was Sir Alan Gardner, under whom Vancouver had served in the Caribbean from 1786 to 1789. Vancouver looked upon Gardner as his patron. In September, Gardner had promoted Vancouver to first lieutenant.[21]

Gardner was now on the Board of Admiralty. Presumably, he used his new position to advance his protégé, for, on 11 December, the Admiralty suddenly ordered the company of the *Discovery* to be paid off. On 15 December 1790, Lieutenant George Vancouver replaced Roberts as commander. Nathaniel Portlock, who had commanded the *King George* on the first Etches fur-trading venture to the Pacific Northwest in 1785, was offered command of the *Chatham*, but near the end of December he resigned because of ill health.[22]

In yet another twist, it turned out that the *Discovery* and the *Chatham* were not going to the South Atlantic after all. Instead, they were going to the Pacific, in particular, the Pacific Northwest.[23]

Menzies could easily be forgiven for being confused. At least he had continued to be paid throughout all of the cancellations and changes of plans.[24] He busied himself part of the time in Soho Square surrounded by books and plant specimens, but became depressed, as he later admitted, of being in a 'long & tedious state of Suspense, more intolerable to me than the hardship and fatigue of traversing the wildest Desert'.[25] Now, in mid December with yet another alteration to the expedition, this time an actual change in destination, Menzies wrote to Banks in despair: he had just seen Henry Roberts who had told him about the new situation and that Vancouver had been put in command of the ships. Roberts, Menzies said, still intended to take a ship to the South Atlantic, maybe in the spring. Should Menzies go immediately with Vancouver on the *Discovery* or wait for Roberts on another ship? Roberts could not advise him.[26]

Banks acted quickly. When he received Menzies's pleading letter he wrote to Evan Nepean at the Home Office. He requested that Menzies should be appointed surgeon on the *Discovery*, and that he should have his own cabin, an assistant and an annual salary of £80 per annum. A week later, on 22 December 1790, Banks learned that Sir Alan Gardner had agreed to the terms.[27]

It seemed settled but then Banks received a letter from Portlock, telling him principally that he would not be commanding the *Chatham*. There were two things in Portlock's letter that must have disturbed Banks greatly. The first was that Vancouver had appointed his own man, Alexander Cranstoun, as surgeon. Menzies, therefore, would be travelling only as a naturalist and under instructions from Banks. Menzies reaction to this news, as he later recorded it in his journal, was to assume that Vancouver had objected to his appointment as surgeon, but he never learned if this was so.[28] Secondly, as Portlock put it: 'I deliverd your Message to Captain Vancouver respecting his waiting on you but think you may not Expect a call from him.' Possibly he was just too busy – Portlock thought the ships would be ready to sail in less than ten days, but there is no doubt that Banks would have taken this as a snub, especially when seen in conjunction with the appointment of Cranstoun.[29]

Banks hit back. Menzies salary would be increased to £150 per annum.[30] As for the snub, Banks would soon reveal his misgivings about Vancouver to Menzies. Meanwhile, Lieutenant William Broughton took command of the *Chatham*, in place of Portlock.

By early February 1791, Menzies and his servant had joined the *Discovery* and Vancouver's instructions had been agreed. His main focus was to survey those parts of the coast of North and South America which Cook had visited only cursorily or not at all, and to continue the search for a northwest passage, in and around Cook's River, and the Juan de Fuca Strait. He was to visit Nootka Sound to ensure that the stipulations of the Nootka Convention were being observed; and to explore the southern part of present-day Chile, below the latitude of 40 degrees south, paying particular attention to where the Spanish had settlements and whether there were any good harbours that might act as whaling bases.[31]

Soon after the instructions were agreed, William Grenville, the Home Secretary, asked Banks to specify the details of how the survey should be conducted. It may appear odd that Banks should be asked for advice about surveying, but it could be that Banks knew the experts in the field better than anyone in government. And indeed, it seems that on Banks's recommendation the surveying instructions were written out by James Rennell, Britain's leading surveyor, best known for his magisterial survey of Bengal.[32]

At the same time, Banks, probably feeling that Vancouver did not have his interests at heart, drafted a set of instructions setting out Vancouver's responsibilities to Menzies – what Banks called 'Instructions for Mr Vancouvers Conduct towards Mr Menzies'.[33] This was not as strange as it seems. As Banks was aware, while the naturalist or gardener knew how to look after the plants on a voyage, he needed the coop-eration of the captain to do so and there was no reason to expect a naval commander to know anything or care much about botany. So, when Vancouver received formal instructions from the Board of Admiralty on 8 March, he learned from them that he was to help Menzies bring earth to the ship; to take on board water for the plants; to furnish him with some amount of the trade goods to exchange for plants,[34] and, above all, to make sure that nothing should impede Menzies's care of the plant cabin.[35]

Having dealt with Vancouver, Banks now began drafting instructions for Menzies.[36] These fell into two main sections: a general one, couched in terms of promoting science and contributing to the stock of knowl-edge, in which Banks told Menzies to keep a careful and broad inventory of all he saw, to be recorded in a journal; and to collect specimens of whatever he came across in the natural and mineral world, and to observe and record customs, languages and religions of the people he encountered.

The second part was more specific and dealt exclusively with botany and supplying the royal gardens at Kew. The *Discovery* had a plant cabin erected on its quarterdeck by the end of December 1789 in readiness for the Roberts expedition to the South Atlantic.[37] How much of it survived or was altered during the intervening changes, cancellations and alterations, is unclear but the plant cabin that was built on the *Discovery* was twelve feet long, eight feet wide and three and a half feet high and was glazed.[38] It had shutters which could slide over the glass to protect it and had three arched covers on the top, all of which could be removed in order to allow maximum exposure to the sun's rays and rain. Inside the structure were square trays and pots.[39]

Just like its predecessor built on HMS *Guardian* in readiness for its voyage to and from New South Wales in September 1789, this plant cabin was designed by Banks. Though he knew from Riou that the plants had been doing well before the ship struck the iceberg and fell apart, Banks did not really know whether his design would work. This was a much more challenging project than that of the *Guardian* because Menzies would be collecting plants from many different habitats and trying to keep them alive in widely diverse climates. Even more challenging was that the *Discovery* would certainly be away for two or three years – the instructions to Vancouver were that the Admiralty did not expect him to return to England until after the survey season of summer 1793.[40] Could living plants survive such extreme shipboard experience?

With this is mind, Banks advised Menzies to collect ripe seeds, which could be dried and carefully pressed onto paper, and for which he should provide detailed information about their habitat, giving the gardeners at Kew the best chance of propagating them. Whenever an opportunity arose, Menzies was to send these packages to Banks who would then forward them on to Kew. The plant cabin would come into use only when Menzies thought that the seeds would not propagate. In that case he was to dig up the specimens and 'preserve them alive 'til your return'. Menzies was told to consider the plant specimens as the King's property, 'and on no account whatever to part with any of them, or any Cuttings, Slips, or parts of any of them for any purpose whatever but for His Majesty's use'.[41]

As it was part of the ship's architecture, Vancouver would encounter the plant cabin on most days. Riou had never complained about it but it became a serious bone of contention between Vancouver and Menzies.

* * *

The *Discovery* and the *Chatham* sailed from Falmouth on 1 April 1791. All was not well. Menzies couldn't, as he put it, 'help lamenting the Situation'.[42] About ten days before sailing and while at Falmouth, Menzies had received a copy of his instructions from the Admiralty. He was troubled by what had been omitted. There was, as he wrote to Banks, no mention of how he was to go about his business: no mention of his needing a boat, and assistance; and nothing about the trade goods which he was to exchange for plants. He wanted to know if Vancouver had received in his instructions anything about his obligations to Menzies.[43]

Unfortunately Banks did not know – the formal instructions to Vancouver were the responsibility of Evan Nepean and he was either unable or unwilling to let Banks have a copy of the documents.[44] Banks could only send Menzies the instructions that he, Banks, had drafted for Vancouver but he didn't know if these were the actual instructions that Vancouver had received from Nepan.[45]

Banks, however, did not leave it at that. He must have sensed possible friction between the two men and warned him. 'How Capt. Van will behave to you is more than I can guess unless I was to judge by his Conduct towards me which was not such as I am usd to receiving from Persons in his situation but as there was no imprudence in his not being civil to me & it would be highly imprudent in him to throw any Obstacles in the way of your necessary duty I trust he will have too much good sense to Obstruct it.' Banks insisted that Menzies should keep a record of any such instances in his journal so that, if necessary, they could be used against Vancouver.[46]

Menzies did not receive this letter until more than a year had passed when the supply ship *Daedalus* was sent to rendezvous with the other ships at Nootka Sound. In the meantime he was pretty much left in the dark.

These misgivings aside, Menzies was overjoyed to be back on a voyage of exploration. He promised Banks that he would throw himself into his tasks and that he would write to him, 'an Epitome of our proceedings', whenever he got a chance.[47] Not an easy promise to fulfil since they didn't expect many passing ships in the area they were going to survey.

Which way Vancouver should proceed to the Pacific was not specified, but soon after leaving Falmouth, he decided to go via the Cape of Good Hope, his preferred route.[48] Madeira was supposed to be the first stop – for wine and refreshments – but contrary weather ruled it out. Instead, the ships proceeded to Tenerife where they arrived on 28 April.

Menzies had never been on the island and his first impressions were not good – too hot and arid to collect plants: 'I cannot say that ever I traversed so much ground with so little pleasure,' he wrote to Banks in his first letter.[49] It was only when he and others went on an excursion into the interior of the island, to the town of Laguna, that Menzies was rewarded with a more fertile landscape. Though the plants were in greater profusion, few, he told Banks, were new to him. He predicted the Cape, their next stop, would be more interesting botanically.

The Cape, which was reached on 10 July, was new to Menzies and his first impressions were excellent: 'I am so charmed with the romantic appearance of the mountains,' he wrote to Banks in a short note on the day of arrival, promising more before departing for the Pacific – there were several East India Company ships at anchor in Simon's Bay on their homeward leg and Menzies was certain he could send the next letter with one of them.

Fortunately for him, there were two people at the Cape that knew the surrounding countryside and its botany very well. Menzies was in good company. Francis Masson was on his second visit to the country and still collecting for the royal gardens at Kew; and Colonel William Paterson, who had made four trips to the interior between 1777 and 1779, and who had recently published the narrative of his experiences, dedicated to Banks, was in town on his way to New South Wales. The two men accompanied Menzies to Table Mountain and in the countryside behind False Bay. Though Menzies described the botany of the Cape as being 'richer than any Garden . . . [wherever] I traversed . . . every situation ', yet, so thorough had Masson been in documenting it, that there was little new Menzies could collect.[50]

The next day the ships left. From this point on and for the next few years, Menzies would have to rely on his own instincts as a collector. Vancouver was taking the ships to their next stop, the southwestern corner of Australia, a space defined only as 'Lyon's Land' on contemporary charts: that is the area in and around Cape Leeuwin, the most southwesterly point of the continent. Almost two months later, on 29 September 1791, the ships entered a bay which Vancouver named King George III's Sound, took possession of the land for Britain and anchored in an adjoining harbour.[51] They were only a few kilometres from where the city of Albany would be settled more than thirty years later.

No one on the ships had ever been there before and Menzies took this as the start of a new botanical chapter. The very next day he scrambled his way up a nearby mountain to reach its summit by noon. His bird's-eye view did not disappoint. He could see the sound, the islands, harbours and inlets below him and fertile, verdant and woody countryside all around him. Like a surveyor giving his opinion on the suitability of the area for settlement, Menzies declared that it was 'capable with a little labor of sustaining thousands of inhabitants with the necessaries as well as the comforts of life'.[52] By the time he returned to the ship he had already made a substantial collection.

For the next two weeks Menzies explored the surrounding area and was well pleased with what he saw. There were new varieties of *Banksia*, of *Mimosa* and of eucalyptus but wherever he looked he saw 'such a variety of plants & shrubs in full bloom & entirely new to me, that I could not help leaving the spot with much reluctance'.[53] It took Menzies two days to arrange his collection.[54] The next day, 11 October, the ships left for Dusky Bay, on the southwest corner of New Zealand's south island.

The ships arrived at Dusky Bay on 2 November 1791. Cook had been here in March 1770 with Banks, and so had Vancouver in 1773. To Menzies, it was all entirely new and very different from southwestern Australia. Instead of the profusion of flowering plants, there were ferns and mosses. Menzies had a special liking for these – 'they are two tribes of plants', he noted in his journal, 'of which I am particularly fond, therefore no one can conceive the pleasure I enjoyed unless placed under similar circumstances.'[55]

Over the next fortnight Menzies continued his botanical excursions while the ships' boats were ordered out to survey those parts not covered by Cook. There were mosses galore but none that were new to him, though on one occasion he did manage to see an example of *Drymis winteri* (*Wintera aromatica* – Winter's bark), the very tree that Banks had asked him for when he was on the Colnett expedition. As it was believed to have antiscorbutic properties, Menzies decided to plant several in the cabin on the quarterdeck. The next day, 18 November, Menzies did the same with several New Zealand flax plants which, he hoped, would survive the voyage to end up in the royal gardens at Kew.[56]

Neither in King George III's Sound nor in Dusky Bay did anyone of the ship's company encounter local people, though there was ample evidence of their presence. The *Discovery* and the *Chatham* now set sail for Tahiti, well known to several of the ship's company, and where there was a highly developed and centralised society and economy.

The two ships became separated in bad weather on their way to Tahiti: the *Chatham* arrived first and anchored in Matavai Bay on 27 December and the *Discovery* followed on 30 December. This was Vancouver's fourth visit, the last having been in 1777. Menzies only knew it from tales and books.

Vancouver's original intention was to remain in Tahiti for only as long as it took to supply the ships with water and provisions. Perhaps Bligh's experience with the *Bounty* was in his thoughts – the fate of the *Bounty* was still unclear, though Edward Edwards on HMS *Pandora* had recently been on the island, spending the month of April 1791 rounding up the mutineers who had decided to remain there rather than sail on with Fletcher Christian. Whatever Vancouver may have felt, *Chatham*'s cutter needed serious repairs and, therefore, he decided to stay longer than he had intended.[57] Though he had been here fourteen years ago, few of the people he had befriended then were still alive – it was largely the younger generation that greeted him and the others.[58] An exception to this was one of the older chiefs who had known Cook and who asked after Banks, wondering whether he was still alive and when he was returning to the island.[59]

Partly because of the endless social obligations and partly because of the adverse weather conditions, Menzies had few opportunities to pursue what he called his 'researches'. Ten days went by after arriving in Matavai Bay before Menzies ventured into the countryside to look for plants. He made a collection of plants new to him but, on bringing them back to the ship, he was caught in a torrential downpour which damaged most of them. Before that happened, Menzies had been shown a grove of orange trees planted by Bligh. This horticultural distribution was a practice that went back at least to the days of the *Endeavour*. Menzies continued the tradition by contributing his own specimens: a number of orange seedlings, some lemon seedlings and a few vine cuttings, all of which he had brought from the Cape and which he had been raising in the plant cabin for the past five months.[60] In the garden of the chief of this district, Menzies also sowed English garden seeds which James Lee of the Vineyard Nursery in Hammersmith had provided 'to be distributed in the course of the voyage wherever they were most likely to be most useful and beneficial to mankind'.[61]

A few days later, local people showed Menzies a grove of shaddock trees which they said had been planted by Banks – judging from their size Menzies had no reason to think otherwise. While he botanised well into the interior of the island and came across new plants, the collections he made were, once again, nearly totally destroyed by rain.[62] There was little to add to the dried plant collection and nothing to the plant cabin. When he wrote to Banks from Nootka Sound nine months later and then from Monterey several months after that, Menzies had virtually nothing to say about Tahiti except to note that because of the rainy weather, 'my botanical collection from thence is but very indifferent'.[63]

Vancouver's instructions were to overwinter in Hawaii at the end of each surveying season on the Pacific coast, and this is where the ships headed next. They left Tahiti on 24 January. Despite having been away from England for nearly ten months already, Vancouver considered this the moment when the voyage really began. Hawaii was almost three thousand miles to the north. It took nearly six weeks to get there. After rounding the southernmost point of the island of Hawaii, canoes came out to meet the *Discovery* and the *Chatham* in order to trade. There were the usual vegetables from the area – taro, coconuts and a few yams – but also, and somewhat surprisingly, watermelons, which, the local traders said, had been propagated from seeds left by Cook (Menzies doubted this as he had not seen any watermelons when he was there on the *Prince of Wales* in 1788 and thought they had arrived more recently). In return, Menzies offered fifty of the 'several hundreds of young orange plants' which were in the plant cabin, and a number of English garden seeds. After being instructed in how to care for them, the Chief, who accepted these in trade, headed for the shore.[64]

The ships then set off for Kealakekua Bay where just over twelve years before Cook had been killed. Vancouver had no intention of staying there but did not leave before more trade was transacted. Menzies played his part by distributing 150 orange seedlings and some vines.[65]

The ships continued to work their way beyond Hawaii to the island of Oahu where they anchored, on 7 March 1792, in Waikiki Bay in the southeast. The stay was short. Two days later, they anchored in Waimea Bay on the southern side of Kauai, the most northerly main island of the Hawaiian Archipelago – where Cook had made his first landfall in 1778.

Menzies did not collect any plants in and around Waiema Bay. The ships remained there only as long as it took to get water and provisions for the next stage of the expedition. During their short stay, Menzies continued to distribute orange seedlings and English garden seeds, and, near the very end of the stay, cabbage seeds, which William Aiton had given him from Kew. All of these horticultural items Menzies entrusted along with strict instructions to the various sailors, stranded from an American fur-trading vessel, to plant the seeds throughout the island.[66] On 16 March 1792, the two ships left for the west coast of North America, to begin the hard work.

After a month at sea, landfall was made at Cape Cabrillo, just under 200 kilometres north of San Francisco on the Californian coast on 16 April 1792. Vancouver's destination was the Strait of Juan de Fuca. The ships sailed up the coast, making a running survey as they went. On 29 April, they met the Boston-based fur-trading ship, the *Columbia Rediviva*, and a few hours later they entered the Strait of Juan de Fuca.

As they began exploring the complex network of waterways that defined the area eastwards of the Pacific Ocean, Vancouver very soon realised that surveying was going to be tedious, and, as the ships were far too big for the intricate work, which needed to follow each inlet, river and rivulet threading out from the major waterways, small boats would have to be used.[67] In conjunction with the Admiralty, Vancouver had planned to use the summers to survey the continental coast from the bottom of Puget Sound – named by Vancouver after Peter Puget, second lieutenant on the *Discovery* – to as far north in present-day Alaska as they could manage. Intervening winters would be spent in Hawaii.

In order to explore the area's botany, Menzies had to join the surveying parties in the small boats that were often away for many days, so he also shared the difficult conditions under which the men worked. On the other hand, this was all new botanical territory to Menzies and he didn't have to wait long to sample it.

On 1 May, he made his first excursion since leaving Hawaii. It was on Protection Island, which lay off their first anchorage, in what Vancouver called New Dungeness – now in Washington State (it reminded him of the Dungeness headland in Kent). It was very exciting. Their eyes, Menzies, wrote, were at once 'dazzled' by the sight of the reddish-coloured flowers of wild Valerian growing behind the beach in large patches.[68] On climbing to the highest point of the island, Menzies could see the expanse laid out before him. 'A rich lawn beau-

Joseph Banks, oil portrait by Sir Thomas Lawrence, after 1795.

Archibald Menzies,
oil portrait by E.U. Eddis, 1836.

Francis Masson,
oil portrait by George Garrard, 1887.

John and Alexander Duncan.

Allan Cunningham,
oil portrait, *c.*1835.

Portrait of Sir John Barrow
(1764–1848).

Matthew Flinders,
*c.*1800, watercolour miniature portrait.

Captain William Bligh.

Robert Brown. Line engraving by
C. Fox, 1837, after H.W. Pickersgill.

George Macartney and George Staunton. Pencil, pen and ink, wash and watercolours. Images taken from Album of 372 Drawings of Landscapes, Coastlines, Costumes and Everyday Life Made during Lord Macartney's Embassy to the Emperor of China. Originally published/produced in 1792–94.

Daniel Charles Solander. Lithograph by Miss Turner after J. Zoffany.

Arthur Phillip, oil portrait, 1786, by Francis Wheatley.

Drawing of a flat peach made in Canton
by John Bradby Blake and Mok Sau *c*.1773.

Sketch by William Bligh of the plant nursery on HMS *Bounty*, 1787.

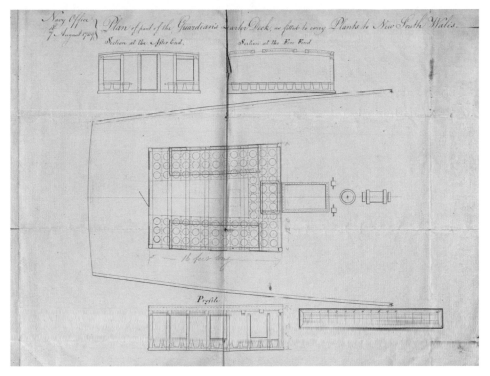

Sketch of plant cabin on HMS *Guardian*, 1789.

The mutineers turning Lieutenant Bligh and part of the officers and crew adrift from HMS *Bounty*, painted and engraved by Robert Dodd, 1796.

HMS *Providence* and HMS *Assistant* at the Lizard, 1791,
watercolour by Lieutenant George Tobin.

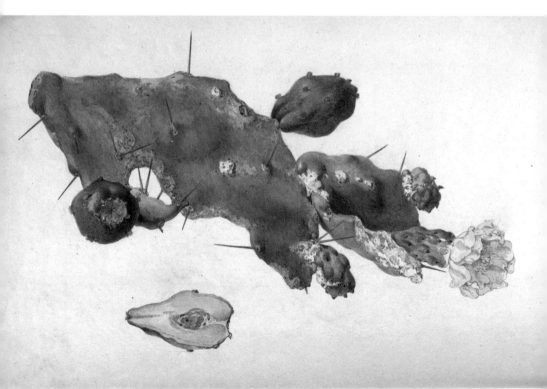

William Alexander's watercolour drawing of an
opuntia and cochineal from Rio de Janeiro, 1792.

Rio Janeiro, Dec. 11·1792. W. Alexander

tified with nature's luxuriant bounties burst at once on our view & impressed us with no less pleasure than novelty . . . It was abundantly croppd with a variety of grass clover & wild flowers, here & there adornd by aged pines with wide spreading horizontal boughs . . . the whole seeming as if it had been laid out from the premediated plan of a judicious designer.'[69]

For six weeks, the ships and the survey threaded their way into and out of each waterway connected to Puget Sound. Menzies collected when he got the chance and continued to be dazzled by plants he had never seen before. On at least two occasions, he collected live plants and placed them in the plant cabin on the quarterdeck.[70]

None of the waterways Vancouver examined held any possibility of being part of a navigable temperate crossing from the Atlantic to the Pacific Ocean. As he worked his way northward along what is now the coast of British Columbia, following the Strait of Georgia, he continued to examine more waterways but none of them were suitable. Menzies continued his botanising but whether he added anything to the plant cabin is unknown. One plant he did remark upon, though, was the *Menziesia ferruginea* (now known as *Rhododendron menziesii*) which he had seen growing in Alaska when he was there in 1788.

By late July, the ships were nearly at the very north end of what is now called Vancouver Island and they soon realised that it was indeed an island.[71] Continuing their way further north, on 11 August, the ships anchored in what they called Safety Cove between Calvert Island and the coast. From there the boats were sent out to explore the several waterways. Menzies went with them.

They had been gone only a few days when sails were sighted coming in from the south and heading towards the ships in Safety Cove. It turned out to be the *Venus*, a British vessel from Calcutta trading in sea-otter pelts on Vancouver Island and the Queen Charlotte Islands. Henry Shepherd, the ship's master, reported that the *Daedalus*, the storeship that had been sent from London in August of the previous year, had arrived at Nootka Sound – but the news was bad. Richard Hergest, the *Daedalus*'s commander (he had been with Cook on his second and third voyage) and William Gooch, who had been sent out to act as Vancouver's astronomer, had been killed when they were attacked at Waimea Bay on the Hawaiian island of Oahu in May by a group of local warriors.[72]

At this latitude, it was more like winter than summer. Menzies provided a rare description of what the surveyors were experiencing.

'The weather was now become so cold wet & uncomfortable', Menzies remarked, 'that the men were no longer able to endure the fatiguing hardships of distant excursions in open Boats exposd to the cold rigorous blasts of high northern situation with high dreary snowy mountains on every side, performing toilsome labor on their Oars in the day, & alternatively watching for their own safety at night, with no other Couch to repose upon than the Cold Stony Beach or the wet mossy Turf.'[73]

They must have been relieved to hear, when they rejoined the ships, that Vancouver had decided to close the surveying season and sail to Nootka Sound.

When, on 28 August 1792, the *Discovery* and the *Chatham* anchored in Friendly Cove, Nootka Sound, they found they were in the midst of a flurry of activity. There were the local villagers, the Spanish squadron, headed by Juan Francisco de la Bodega y Quadra, who was here to negotiate the terms of the Nootka Convention with Vancouver and ten other trading ships: British, American, French and Portuguese.[74] For the next month and more, most of the time was taken up by meetings between Bodega y Quadra and Vancouver. For Menzies, the arrival at Nootka Sound brought a big improvement in his lot. Alexander Cranstoun, the *Discovery*'s surgeon, who had been unwell since leaving the Cape, got worse and was invalided onto the *Daedalus*. Vancouver officially replaced him with Menzies on 9 September. The appointment did not alter Menzies's workload – there were several surgeon mates on the ships – but it did get him his own cabin where he could store his collections; his botanising was not disturbed either, though there was little new in the area to interest him.[75] But he did take advantage of the busy shipping activity in the cove. Not only did he write to Banks but he also prepared a packet of seeds for him which he entrusted to Zachary Mudge, the first lieutenant of the *Discovery*. Vancouver was sending Mudge back with despatches for the Admiralty via Macao on the *Fenis*, a brig flying Portuguese colours which was at anchor in the cove and soon to be on its way to its home port.[76] On 13 October, the two ships of the expedition plus the *Daedalus*, whose stores had been distributed between the other two ships, headed off for their winter rest – the *Discovery* and the *Chatham* initially heading south along the California coast for Monterey, while the *Daedalus*, whose destination was Port Jackson, was ordered to accompany the other ships to Monterey to collect stores for New South

Wales and to take them there via the Marquesas Islands and New Zealand.[77]

The voyage south went by San Francisco, which had been in existence as a Spanish settlement for only fifteen years, and which they reached on 14 November. Here they remained for ten days. Menzies saw a lot of plants new to him but, because of the time of the year, none were in flower and so were difficult to classify botanically. Two days after leaving San Francisco, the ships anchored in Monterey, the capital of Spain's province of Alta California, on 26 November.

Monterey was a much busier place than San Francisco. Much of the time was taken up by social engagements with the local authorities, Franciscan priests and Bodega y Quadra, who had told Vancouver that he would be there expecting to continue negotiations begun at Nootka Sound. As for Menzies, the botanical excursions were more successful than they had been in San Francisco. On one venture into the country to the east of the settlement's *presidio* (fortress) he came across a great number of plants he had never seen before, several of which were in flower.[78]

The Monterey excursions certainly swelled Menzies's dried-plant collection and he collected enough Californian seeds to make a second package for Banks to be delivered to the royal gardens at Kew. These he entrusted to William Broughton, the *Chatham*'s commander, whom Vancouver was sending back to London with a new set of despatches containing, among other items, copies of letters newly written by both Vancouver and Bodega y Quadra.[79]

On 14 January 1793, the day before departing Monterey for Hawaii, Menzies wrote for the first time of the problems he was having with the plant cabin. He was unable to get plants to survive. 'If it is uncovered in raining weather to admit air', he wrote in a letter he gave to Broughton, 'the dripping from the rigging impregnated with Tar & Turpentine hurts their foliage & soil – and if the Side Lights are opened Goats – Dogs – Cats – Pigeons – Poultry &c &c are ever creeping in & destroying the Plants.'[80] Banks received this letter six months later and it would not be the last time he heard about the plant cabin.

Before setting sail for Hawaii, Vancouver had purchased domestic animals which he intended as presents for notable Hawaiians. They included several young cows, two bulls and some breeding sheep.[81] The passage to Hawaii from Monterey took a month and though the

Discovery and the *Chatham* spent five weeks in the island archipelago, they never anchored in any one spot for more than a few days. Apart from one moment on 27 February 1793 when he collected seeds of the red-flowered *Rumex giganteus*, which he later learned had thrived at Kew; and an occasion when he discovered that some of the orange seeds he had distributed previously were now young trees, Menzies did little in the way of botanising during this break from surveying.[82]

On 30 March 1793, with the *Chatham* already on its way to the rendezvous point of Nootka Sound, the *Discovery* proceeded to cross the Pacific again for the second surveying season. The weather was poor. Cold winds and storms impeded progress and made it uncomfortable. On 2 May, the ship took refuge in Trinidad Bay, just north of present-day Eureka, California: they found fresh water and wood and were under way a few days later, but not before Menzies collected three new species of *Ribes*, the genus that includes the edible currants.[83] Nootka Sound was reached on 19 May and after a very short stay of three days, the *Discovery* made its way northwards to join the *Chatham* which was at anchor at the entrance to Burke Channel – in what is now northern British Columbia – the point at which the previous season's survey had ended. Before leaving, Menzies took the opportunity that Nootka Sound offered of writing to Banks, his last letter having been sent from Monterey. Aside from giving a few sparse details about what had happened since then, Menzies only added that he was continuing to work on his natural-history collection.[84]

The second surveying season began in earnest on 29 May 1793 in Burke Channel with the intention of moving up the coast to examine every waterway and island until the weather began closing in. Vancouver had hoped to have five clear months of surveying ahead of him but the difficult voyage from Hawaii had cut that down to four.

Menzies botanised whenever he could. His main strikes occurred in late June and early July and on one occasion on 4 August; but it wasn't until late August that he recorded that he was looking for live plants for the plant cabin.[85] On 28 August, when in the vicinity of Port Stewart, to the south of present-day Juneau, Menzies noted in his journal that he was setting aside the next three days to collect new live plants in the woods and put them in the plant cabin, 'with an intention of carrying them with us to the Southward to enure them to the Sea Air & tropical climates, & by this means I should be better able to ascertain the plants that were most likely to withstand our long Voyage to England & lay them in accordingly on our final departure from the Coast'.[86] On

3 September, Menzies hit lucky when he came across a small plant with blue berries: 'I took a drawing of this species & put some live plants of it in the frame on the quarter deck.'[87]

Less than three weeks later and at a place Vancouver called Port Protection, the northwest point of Prince of Wales Island, the winter weather hit hard. Vancouver decided the second surveying season was over and he moved his ships south to Nootka Sound, which they reached on 8 October. Menzies filled the plant cabin with more plants following the practice he had established further north.[88]

The California coast, once again, was the destination, but after being given a cool reception at both San Francisco and Monterey – the Spanish authorities, they were told, had withdrawn their permission to allow foreign ships to refresh – they sailed on to Santa Barbara, which they reached on 10 November and where they were more welcome. Menzies found several new species of *Mimulus*, whose dried specimens and seeds he preserved.[89] Continuing their way south, on 27 November, they put into the port of San Diego.

The Spanish had begun settling San Diego in 1769 when the *presidio* was built. It was the first European settlement in California. While Vancouver and the surveyors were busy finalising a copy of the charts they had made to be sent back to London, Menzies went out looking for plants but was generally disappointed. However just before they left, one of the Franciscan Fathers presented Menzies with a quantity of what he referred to as 'fruit in kernels . . . about the size of small kidney beans'.[90] This fruit, it turns out, was jojoba. To Menzies's delight, however, the Father presented him not only with the fruit but the plants that bore them. These were immediately planted in the plant cabin and they eventually made it to Kew. Possibly it was this experience that inspired Menzies to fill the plant cabin. Fortunately, he found plenty of new plants near the landing spot and employed two men to help dig them up and plant them. In the end, Menzies had managed to fill the entire space.[91] Before leaving, Vancouver bought two young bulls to take to Hawaii. The ones sent the previous year didn't make it but it was thought that younger bulls, whom the sailors would treat as pets, stood a better chance of surviving.[92] The ships left San Diego on 9 December 1793 to spend the winter in Hawaii with the new cargo of live plants and two bulls.

Menzies was clearly very happy with his bounty for the royal gardens. But the plant cabin had become, by then, an uncomfortable source of tension between Menzies and Vancouver. Menzies's frustration and anger boiled over on 18 November, the day they set sail from Santa Barbara south towards San Diego. So strained had Vancouver and Menzies relationship become over the plant cabin that Menzies resorted to expressing his grievances in writing. A copy of the letter was sent to Banks with an accompanying note.[93]

The shipboard animals, especially the poultry, had found their way into the plant cabin the previous night, Menzies stated, and they had done 'irreparable damages'. He wanted Vancouver to agree to make changes to the cabin to prevent such an occurrence, specifically to provide netting to cover the top and the sides when they were opened to allow in air and light. Also, he wanted Vancouver to appoint someone from the ship's company to look after the plant cabin on those occasions when Menzies was unable to do it himself – when he was attending to the sick or botanising on shore.

After a couple of months, Menzies told Banks, Vancouver gave in to the first of the demands about the netting but the second remained unmet. Menzies himself had to try and find someone to stand in for him in his absence. Menzies confirmed that the plant cabin had been a vexatious issue for some time – in his words Vancouver had harboured a 'disinclination for the success of the Garden [and it] has been pretty evident for some time back' – and that he had had only very few opportunities to botanise (four or five times, at most, he remarked).[94]

On 8 January 1794, the three ships of the expedition arrived in Hawaii (the *Daedalus* storeship had returned from Sydney and joined the other ships on the California coast). From Menzies's point of view, two events were the highlight of his stay. The first was his ascent of Mauna Hualalai, the island of Hawaii's third youngest volcano, at a height exceeding 8000 feet, the summit of which he reached on 19 January. For two days, Menzies and his party fanned out exploring nooks and crannies. Menzies devoted his time to botanising – 'In my rambles', he noted, 'I collected every plant I met with in flower or seed.'[95] Menzies was pleased because he had earlier promised himself that he would be sending Banks a collection of seeds on the *Daedalus*, which would be sailing back to Sydney and then on to London. Menzies returned to the anchorage at Kealakekua Bay on 25 January, where the *Daedalus* remained. About a fortnight later, on 8 February, the *Daedalus* left carrying Menzies's box of seeds, most of which had been collected on Mauna Hualalai for Banks, plus breadfruit plants, which Vancouver had put on board for Norfolk Island.[96] That done, Menzies set off for an even more challenging journey, the ascent of Hawaii's largest volcano, Mauna Loa, whose summit beyond the snow line and at an elevation of over 13,000 feet he reached on 16 February 1794.[97]

On 15 March, the *Discovery* and the *Chatham* left Hawaii for the third and final time. Vancouver had decided to make straight for Cook's River, as it was then called, from where the survey would begin and then move southward until it reached the point at which the previous summer's work had ended.

A little less than a month later, on 12 April 1794, the *Discovery* entered Cook's River: the ships had become separated during the voyage across the Pacific. This was to Vancouver a very important moment, for the Admiralty had specifically instructed him to survey this waterway beyond the point where Cook's survey ended. If there was a temperate northwest passage, this would be its northernmost point. For the next few weeks Vancouver made his way up the river in terrible conditions: Menzies remarked that the temperature fell to as little as seven degrees Fahrenheit and he had to treat several of the men for frostbite; Vancouver described the icy scene eloquently – 'the gale . . . was accompanied by so severe a frost, that the spray became instantly frozen and fell on the decks like sleet . . . and the water that was brought up with the lead-line, although in constant motion, cased it intirely with ice.'[98]

On 6 May, Vancouver had explored the waterway to its narrow end and could now declare with total confidence that it wasn't a river at all but an inlet of the sea. It now bears the name of Cook Inlet.[99] The next morning they found the *Chatham* and the ships were reunited for the southward survey.

Menzies had not done well. The snow and the ice prevented him botanising: he could only describe what he saw and that was mostly trees. But what depressed him most was what he confessed to Banks in these words: 'At the head of this great Inlet, we were much entangled & perplexed with drift Ice, and about the latter end of April, experienced heavy falls of snow, and intense frost . . . I am sorry to acquaint you, that this severe & quick transition of climate, killed the greatest part of the live plants I had in the Frame on the Quarter deck, & this is more to be regretted, as I am afraid it will not be in my power to replace many of them, particularly those from California

& the upper regions of Owhyee, which till this fatal period, were in a flowerishing state.'[100] One can only imagine Banks's disappointment when he read these words at home in Soho Square on 13 April 1795.

The drudgery of the survey continued through the next few months until 31 July 1794 when its southernmost point was reached. Vancouver found an anchorage at the southern tip of Baranof Island which he aptly named Port Conclusion.[101] Menzies went ashore as soon as he could and over the next few days made almost daily botanical excursions. The weather had improved greatly and plants were in flower. He busily collected live plants to replace those that had died.[102]

Only one degree of latitude remained to be surveyed. This and bad weather delayed the departure for Nootka Sound until 24 August 1794.

The ships were back at their usual anchorage in Friendly Cove on 2 September. There they found two Spanish ships that had arrived from San Blas. On one of them was Brigadier José Manuel de Alava, who had been appointed to continue the Nootka Convention negotiations with Vancouver; Bodega y Quadra had died earlier in the year. Both Vancouver and Alava had to wait for further instructions from their respective governments as to what to do next.

The wait, though anxious, suited Menzies. Free of surveying schedules, he made regular visits to the shore looking for new plants to restock the plant cabin, including finding an example of the Pacific madrone or strawberry tree.[103] Within the first week of being at Nootka Sound, Menzies could tell Banks that the plant cabin was nearly full again.[104]

As it turned out, no English or Spanish ships brought any news. Vancouver decided he had waited long enough and on 15 October the ships left for the voyage southward to Monterey. De Alava went too.

The voyage to Monterey proved to be very tough and longer than expected. The ships, yet again, became separated. The *Discovery* arrived on 6 November to find the *Chatham* had arrived several days earlier. Vancouver had no direct news from London but de Alava had learned that earlier in the year Spain and Britain had signed another Nootka Convention requiring another meeting between the respective parties at Nootka Sound. And that, Vancouver understood, did not include him. On 12 November, the day after hearing from de Alava, Vancouver decided it was time to go home.[105]

On 2 December 1794, the ships left Monterey.[106] Vancouver's instructions were to return directly without stopping on the South American coast, but there were no other strictures. Taking matters into his own hands, as they moved south towards Cape Horn, Vancouver continued exploring. Guadeloupe Island, off the coast from Baja California, was the first destination; followed by the tip of Baja California, then the Islas Marias; onward to Isla del Coco, where Menzies collected some herbarium specimens while the ships were at anchor between 23 and 27 January 1795;[107] then on to the Galápagos Islands where, on Albemarle Island, Menzies became the first known collector in the archipelago.[108]

The ships departed the islands on 10 February 1795 and took a course for Cape Horn but, a little over a month later, Menzies told Vancouver that some of the crew had scurvy. With winter storms expected at the Cape, Vancouver decided that he had to find a port of refuge. So, even though his instructions had forbidden him putting into a Spanish port, he decided he had no option but to make for Valparaiso, which was reached on 26 March 1795

When at Nootka Sound in September of the previous year, Menzies had remarked that he expected his cabin collection to survive the homeward voyage as long as they didn't stay long 'in any tropical port'. Unfortunately Vancouver's exploring junket in the warm Pacific – the equator passes through the Galápagos Islands – dealt yet another blow to the plants for Kew. In his letter to Banks from Valparaiso Menzies told the sad tale – during 'our long – ill chosen passage from Monterry . . . I am sorry to acquaint you [that the heat] proved fatal to many of my little favourites, the live plants from the North west coast, & California, notwithstanding my utmost attention, & endeavours to save them.'[109]

Not wishing to close the letter on a gloomy note, Menzies added that he expected to get another chance before entering the Atlantic to collect for Kew. True to his word, on 28 April, Menzies informed Banks that, once again, the plant cabin was nearly full. With another stop before the Cape he expected the collection in the cabin to be complete.[110]

On 7 May 1795, and with a favourable wind, the ships left Valparaiso. Though he had been instructed to survey the South American coast south of latitude 44 degrees, Vancouver decided he couldn't do it. So Menzies would not, after all, have another chance to collect and the collection he had made at Valparaiso would be the last. Vancouver also decided that rather than make the Cape of Good Hope the next port of call, as he had originally intended, he would instead make for

St Helena. As Vancouver predicted, the weather at Cape Horn was violent and didn't abate until they were actually in the Atlantic. The *Discovery* entered the bay at St Helena on 2 July 1795. The *Chatham* soon joined it.

Two weeks later, the *Discovery* left St Helena for home. At this point Menzies could look back with a mixture of satisfaction and disappointment: satisfaction that he had a full plant cabin and disappointment that its contents did not in any way reflect all the places he had been.

Kew would be the poorer for that. At least the plants that Menzies had on board now would not need to go through as many climatic transitions as had his previous cargo. But then, just as it seemed plain sailing, disaster struck.

It happened on 28 July 1795, just a few days after the ship crossed the equator. According to Menzies, who recounted it to Banks from Shannon where the *Discovery* first anchored in home waters, he had instructed his servant to cover the plant cabin but Vancouver had ordered him to go on watch instead, thereby leaving the cabin exposed. It was then that a 'very heavy and sudden deluge of rain crushed down the tender Shoots of many of the plants that never recovered it'.[111] Menzies, unaware of Vancouver's orders, complained to the commander about his servant's conduct but when he discovered that Vancouver was responsible, he asked for an explanation, 'coolly & without either Insolence or Contempt'. According to Menzies, Vancouver 'immediately flew into a rage . . . and I was put under arrest because I would not retract my expression, while my grievance still remained unredressed.'[112]

The tragedy was that so many plants were lost. As he remarked to Banks, 'I can now only show the dead stumps! Of many that were Alive and in a flourishing state when we crossed the Equator for the last time.'[113] Still at Shannon in early October, Menzies put a better gloss on the plant cabin saga by declaring that he had manged to save a few special plants. One of them, what Menzies called 'a most beautiful Pine from the Southern extremity of . . . Chili', was the monkey puzzle tree, the first to be introduced to Britain.[114] He had also brought back the *Drymis winteri* (*Wintera aromatica* – Winter's bark), a few evergreens from California and some other plants from Chile.[115] We can only imagine how many living plants Menzies had collected as the ships moved from islands to inlets, from Hawaii to Alaska, British

Columbia, California and Chile. Though Menzies told Banks that he had a list of all his collections, like the plants themselves it does not seem to have survived.[116]

The dispute with Vancouver – Menzies remained under arrest even when the *Discovery* arrived back home – rumbled on through most of October 1795 with letters and reports circulating between Menzies, Banks, Vancouver and Evan Nepean, who was now Secretary to the Admiralty.[117] By the end of the month, however, Nepean informed Banks that 'the dispute between Captain Vancouver and Mr. Menzies has been amicably adjusted.'[118]

The expedition had taken four and a half years from start to finish, a marathon trip. For Menzies the results were mixed: the seeds and dried plant collections were fairly successful; but the plant cabin suffered from events that were beyond Menzies's control.[119] Still, even in difficult times it seemed to work tolerably well.

For Vancouver the expedition took its toll. He had been unwell for long spells during the voyage and didn't live very long after its return, dying on 12 May 1798. Menzies returned to sea as a surgeon for several years but ill health forced him to resign from the navy in 1802. From then and for many years until his death on 15 February 1842, he ran his own medical practice in London.[120]

10

1791: The Gardeners
of the *Providence*

Aﬀter the mutiny, Bligh found himself adrift in the Pacﬁc in a small boat, with eighteen other men and few provisions. He had to decide what to do.[1]

Fortunately he knew the area. He made for the nearby island of Tofua to look for food and water. The bartering with the local people did not go well, and tensions rose to such a pitch that Bligh and his men had to flee. They just managed to get away, but not before John Norton, the quartermaster, was knocked down and killed by one of the islanders.

Bligh thought of returning to Tahiti, but, realising that that was where the mutineers had probably gone, he changed his mind. It was a critical moment as he later recounted in a letter to Banks. 'I was now solicited by every Person to take them towards home.'[2] Bligh explained to the men that in that direction the nearest land was the island of Timor, more than 4000 miles away. Bligh had never been to the island, which was partly under Portuguese and partly under Dutch control, but he did know that there was a settlement there. 'They all agreed to live on One Ounce of Bread per day and One Jill of Water,' Bligh continued. 'I bore away for New Holland and from thence to Timor . . . in a small boat . . . without a single Map, and nothing but my own recollection and general knowledge of the situation of Places, assisted by a Table in an Old Book of Latitude & Longitude to guide me.'

So, on 2 May 1789, Bligh and his men rowed and sailed west across the Pacific, towards the tip of Cape York, Australia and the Torres Strait, in the direction of Timor.

For the next few weeks they made their way from island group to island group hoping to find fresh food and water, and a less hostile reception than they'd had on Tofua. But the combination of heavy seas and storms and menacing islanders' canoes put paid to that plan, and instead Bligh had to rely on rain water and strict rations to get through. They were, on the other hand, making good progress and on 28 May, at one in the morning, the roar of surf signalled to Bligh that he was on the edge of the Great Barrier Reef and nearing the coast of New Holland. In the morning, Bligh found an opening and entered the inner passage, as it was called, where the water was calm at last.[3]

From there, the launch went northward to Cape York, stopping at various islands for clams and oysters before entering the Torres Strait on 4 June 1789. It was now a straight run to Timor, whose main settlement Kupang was reached on 14 June, after forty-eight days at sea in a small boat. It was an amazing achievement.

The stay in Timor, though welcome at first, proved to be longer than expected and fatal for some of the men who had survived the marathon voyage getting there. Several of them died from 'fever'.[4] Among them was the gardener David Nelson, 'whose good conduct in the Course of the whole Voyage, and Manly fortitude in our late disastrous Circumstances deserves this tribute to his Memory.'[5] He died on 20 July 1789.

For the journey home Bligh purchased a schooner on credit, to take him to Batavia, which he reached on 1 October. After a fortnight's stay, Bligh, who was very unwell, sold the vessel, and together with his servant and clerk took passage on the *Vlijt*, a packet belonging to the Dutch East India Company which was bound for Holland via Cape Town. After a short stay at the Cape, on 3 January 1790, the *Vlijt* continued its voyage to Europe.[6] The Dutch captain had orders from the Batavian authorities to drop Bligh and his companions off at the most convenient place in England before proceeding to Holland.

On 13 March 1790, a little over ten months after losing his ship, Bligh stepped ashore on the Isle of Wight. The story he told, which he delivered in a written account to George III and which also found its way into the London newspapers, was extraordinary to say the least.[7]

Because the ship had been lost, there had to be a court martial. It couldn't take place until all the survivors from the launch were back in England. So, it was more than seven months after Bligh's return that on Friday, 22 October 1790, the trial of William Bligh, John Fryer, the ship's master, three midshipmen and eight others – all that survived of the ship's company that had been on the launch – began.

Admiral Samuel Barrington presided with twelve naval officers, all but one admirals. Nothing could justify a mutiny in the eyes of the navy. The question was whether the accused had done all they could to prevent or thwart the loss of the ship. After evidence was taken and questions asked and answered, all the men were honourably acquitted.[8]

Now Bligh, who had been too anxious about the outcome to write to Banks before the trial, put pen to paper admitting that he had been worried about the outcome of the court martial and excusing himself for not writing to Banks earlier. With the trial behind him, he was keen to let Banks know that his name was cleared, and to solicit his help: 'I am concern'd at losing your kind assistance just at this time, but I hope and beg of you Dear Sir to do your endeavors to secure me the rank of Post, for if this opportunity passes off, it may be lost forever.'[9]

Bligh was concerned about his professional future; he wanted to be promoted to post-Captain, a rank which would ensure that he could rise to that of admiral in time. He had already been to see Lord Chatham, the First Lord of the Admiralty who'd assured him that the promotion would be forthcoming once he'd seen the King.[10] However, Bligh was sufficiently anxious to seek Banks's endorsement as well.

Banks provided it. He wrote to Chatham on Bligh's behalf. He admitted, interestingly, that he had had little to do with Bligh in the past other than hearing of his abilities, but now he held him in the highest esteem. A promotion, Banks maintained, was well deserved and overdue and he hoped that Chatham would see it that way.[11]

Five days later, Bligh was promoted and he thanked his new patron for his 'friendly & kind exertions' on his behalf.[12]

It was about this time that Banks received a letter from Hinton East, the Jamaican planter and politician who had contacted him earlier about breadfruit. He commiserated Banks on the loss of the *Bounty*, and all the breadfruit Bligh had collected so well. He expressed his anger at the mutineers who had ruined the project; or, as he put it, 'to have all these pleasing Prospects blasted by a Set of Miscreants raises such Resentment in my mind.'[13] The gallows, East recommended, was the best punishment for them. But what most enraged him was the fact that the island still did not have the breadfruit. Obliquely referring to Banks's influence with the King, East remarked that he was looking forward to a renewed attempt to get the plant before too long. East hoped that the 'Royal Personage to whose Benevolence, we are indebted for the late but unfortunate Attempt, will not let the Business rest, but make a second, send a Vessell of War to attend the one which is to convey the plants, that the Object may not be again lost from the same Cause'.[14]

Not long after receiving East's letter from Jamaica, Banks learned from Sir George Yonge, the Secretary of War, that the King had indeed given his blessing to a second breadfruit expedition.[15] More than that, he revealed that Banks would again be responsible for fitting out the voyage – 'I have spoke to Lord Chatham [First Lord of the Admiralty], who wishes to see you upon It . . . They have now Ships of all Sorts to spare them, and Lord Chatham said he only wish'd to know in what manner It would be necessary to Equip, & fit out the Ship destined for this Service – I told him you best could Explain this, & He desired I would send you, to Him – So go & Prosper.'[16]

Banks was delighted to get a second chance to transplant breadfruit. He was extremely disappointed that Kew hadn't received the plants he had planned for it; and that he had failed to find out whether plants did better on long sea voyages in their own space than if they were kept in specially designed containers and distributed around the ship as the captain saw fit – the currently accepted practice. Banks did not need to be convinced that Bligh was the obvious choice to command such an expedition: he had shown extraordinary navigational skills; he had more Pacific experience than anyone else who was available; and his promotion was a clear sign that the Admiralty had confidence in his abilities.

Bligh had already heard that his name was being associated with a renewed attempt at transferring breadfruit as early as 8 February 1791 when he wrote to his nephew Francis Godolphin Bond, who at the time was a first lieutenant on HMS *Inconstant*, referring to 'talk about my being sent out to Otaheite for the Bread fruit plant'.[17]

Bligh was right about the rumours but, as it happened, others had heard them before he did. Richard Salisbury, the botanist and horticulturalist, who was at the time living on an estate in Chapel Allerton, where he had a substantial garden, was certainly party to the rumours. Salisbury was so certain that the second breadfruit expedition was being prepared that he even recommended his gardener, James Wiles, to Banks as its head gardener.[18] Wiles was so delighted to hear this that, on 16 January, he wrote to Banks to tell him so and that he was looking forward to serving under Captain Bligh.[19] Banks, in reply several days later, seemed surprised to hear this, indirectly reproaching Salisbury for being 'rather . . . too sanguine in informing you that I have appointed you head Gardiner to the bread fruit Ship.'[20] That ship, Banks added, did not yet exist but he was confident it would soon, in which case he would be recommending Wiles as its head gardener.[21] Wiles admitted to Banks that his employer had probably jumped the gun, but felt confident enough to make his way to London in anticipation of the appointment.[22]

The rumours were true. On 23 February, Bligh was at Blackwall where he arranged the purchase of two ships from the shipbuilding company of Perry & Co.[23] He was moving with the utmost speed: he had made a plan of the ship's fitting to present to the Navy Board, and had recommended his nephew, Francis Godolphin Bond, to be his first lieutenant. There was a great sense of optimism in the air and Bligh was anxious to share this with Banks.[24] Two weeks later, Banks began to flex his own muscles to effect a speedy preparation. Apparently Chatham, the First Lord of the Admiralty, did not know about Bligh's visit to Blackwall – in fact he hadn't seen him for some time – and the fact that the ships had already been chosen.[25] Banks brought him up to date and reminded him that the expedition needed to arrive in Tahiti in the latter part of the dry season, so that it could be safely away before the storms arrived. This meant that the ships had to sail in May.[26] Chatham got the message.

Over the month of March, the machinery of government finally caught up with Bligh, Banks and other naval personnel who had been appointed to the ship. The first sign that everything was coming together was at the end of March when Bligh got his instructions from the Admiralty. Over the next two months other important decisions were made with respect to the voyage: the names that would be given to the ships – the *Providence* and the *Assistant*; the personnel of the ship; how many and which kind of guns it should carry; and the kinds and number of instruments. The result of all this was that the *Providence* would be well armed and carry fifteen marines to the rank of sergeant – significantly, the *Bounty* had had no armed men on board.[27] To reduce risks even further, the *Providence* would be sailing, with its tender, the *Assistant*, which would have its own smaller complement of marines. The *Bounty* had sailed on its own. Bligh, the Admiralty and Banks were taking no chances.

Though exonerated over the mutiny and promoted, Bligh was not allowed to forget the failure of the previous expedition. Indeed the Admiralty's instructions to him began with a reminder. 'The attempt', the Admiralty wrote, 'which was made in His Majesty's armed vessel the Bounty to convey Bread Fruit Trees and other useful Productions from the Society Islands & other places in that part of the World, to his Majesty's West India Islands, in pursuance of our Instructions to you dated the 20th of November 1787, having unfortunately failed . . .'.28

The instructions, sent from the Admiralty but written by Banks, confirmed the collection of breadfruit plants in Tahiti and their delivery to the West Indies as the central purpose of the voyage but there was much more to it than that. This was a voyage dedicated to an inter-oceanic transfer of plants on a scale never witnessed before. It involved not just Tahiti and the West Indies; but also, Timor, the Cape, St Helena and London.

The outward part of the voyage was all about transporting plants. It began with a purchase from the Brentford nursery of Henry and Hugh Ronalds of 'queen pines' (pineapples) and nectarines.29 The pineapples were to be a present for the Tahitians, whereas the nectarines were to be exchanged at the Cape for other plants.30 What else would be moved and where to depended entirely on where Bligh decided to buy the ship's provisions – Madagascar, for example, was mentioned in the instructions but Mauritius and even Norfolk Island and Port Jackson were also possibilities.31

On the homeward voyage, once Bligh had collected the breadfruit, he was to call at either Timor or Java where he was to replace the dead or dying breadfruit trees with fruit trees from the region – mangosteens, durians and jackfruits, to name the most important – and also to pick up a type of rice that grew in dry soil. At St Helena, Bligh was instructed to leave behind some of the breadfruit plants, some of the fruit trees and some of the rice in sufficient quantities for those on the island to try to cultivate them. In return he was to take plants from St Helena for the West Indies and for Kew. At the next stop, St Vincent in the Caribbean, Bligh was to leave half the bread-fruit trees (reserving for Kew some trees and seeds), in exchange for plants and seeds destined for Kew. At Jamaica, the next and final stop westward, Bligh was to leave with the Director of the Botanic Garden the rest of the breadfruit trees, other plants and seeds and return to England 'bringing with you such Plants, Seeds, etc as the abovementioned Director, or any other Person willing to promote Botany may put on board the ship for his Majesty's said Garden'.[32] Bligh was given an advance of £500 in order to purchase any plants he found interesting en route back from Tahiti.[33]

No doubt David Nelson would have been Banks's preferred choice to be the first gardener on the *Providence*. He had successfully collected and potted hundreds of breadfruit plants during the *Bounty*'s stay in Tahiti; Bligh had highly commended his work and his character; but he had died at Timor. William Brown, the other gardener on the *Bounty*, had stayed with the mutineers and was killed.[34]

As the two gardeners who had the most breadfruit experience were not available, Banks had to look elsewhere. James Wiles had already been recommended by Richard Salisbury, for whom he had worked for seven years as a gardener; and by Samuel Reynardson, the owner of Holywell Hall, Lincolnshire, and its famous gardens where Wiles's father had been the head gardener.[35] Banks was satisfied. On 11 April 1791, he appointed Wiles, who was now twenty-three years old, to the expedition.

The second gardener, whom Banks appointed a few weeks later, was Christopher Smith. About his early life little is known other than that he was Irish. He may have worked at Kew but there is no evidence.[36] What we do know is that William Pitcairn, the eminent physician, who had an extensive, four- or five-acre garden in Islington, and for whom Smith had worked, recommended him to Banks.[37] Pitcairn and Banks had had many opportunities to meet, certainly during the former's reign as President of the Royal College of Physicians (1775–84); and the two men, together with the Duke of Northumberland, James Lee of Vineyard Nursery, Hammersmith and Dr John Fothergill, had participated in a syndicate which, in 1780, sent William Brass, the Duke's head gardener, to West Africa to collect plants and seeds.[38]

Once the appointments were made officially – Banks had already inserted their names as the gardeners in the first draft of instructions, dated 29 March 1791, from William Grenville, the Home Secretary, to the First Lord of the Admiralty – Banks got on with the job of instructing them.[39] The success of the expedition lay entirely with them, Banks told them. It could not have been made any clearer – the plants came first. 'Guard yourself against all temptations of Idleness or Liquor,' Banks warned them. A moment's inattention 'may render the Great Expence incurrd by Government in this humane & liberal undertaking Wholly abortive'.[40]

Banks then got straight to the point. Tahiti was where the real work would begin. Bligh would obtain as many small breadfruit plants as possible from the Tahitians, and Wiles and Smith were to dig them up and pot them in the tubs that had been provided for their transfer to the West Indies. Then they were to care for them, 'watering & shading them with the most unremitting care & attention', until they were certain they had taken root, at which point they needed to inform Bligh. Meanwhile, they were also to collect specimens of other Tahitian plants, such as plantain, the Tahiti apple and a particularly large variety of yam, superior to anything then growing in the West Indies.

Once the plants had rooted, they could be transported and for the journey across the Pacific and Indian Oceans, there were strict and detailed instructions about how to avoid the plants getting into contact with salt, 'the Principal Enemy', as Banks put it. Furthermore, at each watering stop, Wiles and Smith were to give the plants an overall wash with clean fresh water.

Banks understood that even with the utmost care, some of the breadfruit plants would die, and when this happened they were to replace them with local plants as they went along. In Timor or Java, where they were most likely to stop, they were to purchase local fruit trees, such as mangosteen and jackfruit and also pepper, to put into the empty tubs.[41]

HMS *Providence* was a big ship, twice the size of the *Bounty*. In every other way, in its fittings and internal architecture, it was a copy of the *Bounty*. The captain's cabin, for example, was converted into a nursery for the plants. Whereas the *Bounty* had place for 600 tubs, the *Providence* had room for around 2000. As in the *Bounty*, gratings were to be placed in the deck and the sides of the ship to allow fresh air to circulate; watering cans and straw were provided; as were stoves, in which fires could be lit to keep the plants warm when the ship entered the South Atlantic.[42]

By early July 1791, the ship was ready to sail. Wiles and Smith joined the ship in Deptford, where the fitting took place and the stoves were brought onto the ship; at Galleons Reach, the ordnance and ammunition were placed on board.[43] A few days later, Bligh, in command of the *Providence*, and Lieutenant Nathaniel Portlock, in command of the *Assistant*, began to make their way downriver.[44] Two weeks later the ships were in Portsmouth awaiting their final orders, which came on 20 July. The ships were ready but adverse winds kept them at anchor for almost two weeks. Finally, on 2 August 1791, the ships headed off towards warmer climates.[45]

As on all ships, bonds of friendship were often made quickly and frequently endured well beyond the actual voyage. Such was the case between Wiles, Smith and one of the ship's midshipmen, the seventeen-year-old Matthew Flinders. This was Flinders's second appointment as a midshipman, the first being on HMS *Bellerophon*, under the command of Thomas Pasley, who had distinguished himself in both the Seven Years and the American Revolutionary War. Pasley had warmed to Flinders and recommended him to Bligh.[46]

The voyage southward through the Atlantic was uneventful but slow because of contrary winds. Bligh was not well and complained of fever and headaches. In his first letter to Banks, dated 30 August 1791, from Tenerife, their first stop, Bligh commented that he was 'writing with great torture'.[47] The heat, Bligh added, was extreme – which did not help his fragile state – and the 'Winds at Night are as blowing from a furnace'.[48] He hoped the sea air, especially south of the equator, would make him feel better.

At the Cape Verde Islands, where the ships stopped for only a day or two in early September, Bligh was still unwell though the ship's surgeon, Edward Harwood, had declared him out of danger. Still, he couldn't write the letter to Banks himself and had Harwood write it instead: 'my head is in a state, which will not allow me to keep my eyes fixed on the paper.'[49]

Arriving at the Cape of Good Hope on 7 November, Bligh declared to Banks that he was on the road to recovery though very shaky. 'I now however can bear some Noise and Bustle & give orders,' he wrote.[50] A stay of almost two weeks at Stellenbosch seemed to benefit him and by the time the ships were ready to leave, Bligh was flourishing. The same could not be said of the plants that Wiles and Smith had brought with them from England. Many of the nectarine trees had died, though surprisingly the pineapples were unaffected, as were the pine trees. Bligh recommended that fig trees should take the place of the dead nectarine trees: by caring for them, on their route to Tahiti, Bligh hoped that the gardeners would 'by their treatment gain some knowledge how to manage the Breadfruit'.[51] Wiles and Smith obliged and sought the help of Francis Masson, who was still at the Cape, in selecting the best varieties and procuring other plants to be potted into the garden's tubs.[52]

In his farewell message from the Cape, Bligh was very confident of the future: 'I hope that the Great Providence which has hitherto protected me will send your little Providence back to you with success in due time.'[53] On 23 December 1791, later than expected, the anchors were pulled up and the two ships headed into the Indian Ocean.[54]

Bligh was making for Adventure Bay, at the southeast corner of Tasmania, just as he had done on the *Bounty* in August 1788. This would be the third time he visited the place.

The *Providence* and the *Assistant* arrived in Adventure Bay on 9 February 1792, to take on water and wood. For Wiles and Smith, it was an opportunity for botanising. The haul was respectable and included examples of *Banksia*, 'Metrocedera' (*Metrosideros*, a tree of the myrtle family) and what was called mimosa. Bligh also set them to plant the botanical specimens they had brought from England and those they had picked up at the Cape. At a lake near the east end of the beach, Wiles and Smith planted '5 Figs, 9 Oaks, 3 Quinces, 1 Rosemary, 20 Strawberries and 3 Pomegranates'.[55] On a nearby island they sowed fir seed, and apricot and peach stones. Bligh also left behind a cock and two hens, hoping they would breed and become wild. How all these European imports would fare, he didn't know, but the outlook was not very good. Bligh didn't encounter any of the hogs that Cook had left behind; and all but one of the apple trees David Nelson had planted had disappeared. There was no sign of the potatoes.[56]

With the water casks full and the wood stores high, the ships sailed on Wednesday, 22 February 1792. Bligh kept following the *Bounty*'s route heading for the southern island of New Zealand, but giving it a wide berth, turning in a northeasterly direction straight for Tahiti's northern coast.

They anchored in Matavai Bay on 9 April 1792. It was Bligh's third visit to the island. For most of the ship's company it was their first, but they must have heard a lot about its charms. George Tobin, the *Providence*'s third lieutenant, in a retrospective journal entry, described the scene as the island revealed itself. 'At daylight we were gratified with a sight of the long wished for island, but at too remote a distance to distinguish, even with our glasses, more than its blue mountains. When about eight miles from point Venus . . . our expectations were more than realised in the many delightful views opening in successions as the vessels passed a short league from the shore. The heavy showers of the preceding night had given additional verdure to the lower grounds, while they served to form numberless white cataracts, serpentining amid the foliage on the distant mountains. The beach was tumultuously crowded with natives from their huts, scattered under the umbrage of the luxuriant breadfruit or towering cocoanut . . .'[57]

Bligh would have known that at least some of the crew would be hoping to enjoy the pleasures of Tahiti, the beautiful women and sexual freedom that they had heard about, but he was determined to maintain discipline, and to focus on their mission to obtain breadfruit. Everyone was to follow the rules and concentrate on work not pleasure.

Bligh was known to many people on the island and they welcomed him back. He could speak some Tahitian and there were several Tahitians who could speak English, two of them at least, very well. With that and the presents that he had assembled in the form of toeys, nails and other iron products; calico dresses, two hundred yards of chintz, ribbons, shirts and suits of clothing; knives, hatchets, files, rasps, looking glasses, beads, rings and combs; not to forget the many plants he had brought with him, from England, the Cape and Adventure Bay, Bligh felt confident that breadfruit plants would be delivered to him.[58] And so they were. The first consignment of plants, thirty-two in total, arrived on 17 April and were put in the containers they had brought from England.[59]

However, before Bligh could arrange for the breadfruit to be taken to the shore, various preparations and changes had to be made. Because of his friendship with Wiles, and because he himself was interested in the practicalities of the breadfruit transfer, Flinders described the process in some detail in his log.[60] Firstly, a tent was erected at Point Venus, the northernmost part of Matavai Bay, as a temporary shelter to store the plants that were still on the *Providence*, many of which had been on the ship since it had left England and others which had been either purchased or collected at the Cape and Adventure Bay. These included figs, pomegranates, pineapples, oranges and lemons – many of which Wiles and Smith had raised from seed on the voyage – aloes, banksia and maize.[61] Speed was essential as many of the plants were being attacked by flies and were becoming more damaged by this than they had been by the voyage.

Then the pots had to be put in the tent. Finally, Wiles and Smith had to choose a spot, somewhere near the tent, which provided shade and protection from sea breezes. They found a suitable place on the banks of a small river, 400 metres from Point Venus, under the shade of some enormous breadfruit trees. The ship's carpenters quickly erected a shed, or, in Flinders's words, a 'House, slightly constructed after the Oteheitean Manner without a roof, but which they covered at proper Times with Matts made of Cocoa nutt leaves called Pahwahs – the sides they covered with leaves of the Pandanus of Palm'. As soon as this structure was ready, the plants that were to be left on the island were moved there, as were the empty pots, from the tent.

This set a pattern for the days to come. Each day more pots were moved from ship to shore. Then it was time to get the breadfruit plants. Flinders remarked that the first method, that of sending Tahitians with empty pots to the breadfruit plantations and then returning to Point Venus with the plants potted, didn't work out. Instead, Wiles and Smith found a good supply of loam which they kept in the shed. The plants were brought in the morning and were potted in the evening. This process continued for several weeks. Bligh kept a careful count of the number of pots. By early May, the shed was full of 1000 pots planted with breadfruit. Wiles and Smith looked after them carefully, waiting anxiously for the plants to root so that they could be moved on to the *Providence*.

Meanwhile the ship's carpenters were busy building a greenhouse on the ship's quarterdeck to house some of the plants when they were ready to be moved on board.[62] It seems Bligh had the structure built on his own initiative, or, in his words, 'my own contrivance', and, after he had it further enlarged, it increased the ship's plant carrying capacity by one-third.[63]

By mid June the carpenters had completed the greenhouse on the deck. It was built of wood, with a skylight, over which netting could be placed.[64] A railing was fitted to protect the skylight and to enable coverings to be draped over the structure if necessary. The cabin was also prepared, including altering the ports so that fresh air could enter even when they were closed. Bligh was satisfied. 'The Plants are doing exceedingly well,' he wrote, 'which is a peculiar happiness to me.'[65]

With the plants on the shore and the ship altered for their reception all was ready apart from one thing. The *Providence* had to undergo a complete delousing. Bligh's preferred method of doing this was to use boiling water to kill the vermin. Bligh described the operation in his log as it began on 25 June 1792. 'The Coppers and all Vessels are got ready with the Water. A select number of People with Quart Pots are stationed and on my giving the word every hole and crevise is so deluged that few Vermine escape. The men are Clothed with Jackets to prevent being scalded, and the Ship for an Half Hour is a complete Vapour Bath.' This operation, and a parallel one which involved removing all the chests and boxes to the shore for delousing, continued for three days.

The ship was ready but the plants were not. A week went by. Bligh was keen to depart. He ordered fresh water on board. On Friday, 6 July threatening clouds rolled in. It rained heavily all that day. The following day, the weather returned to its normal state of light winds and mostly clear skies. 'My plants', Bligh noted in his log, 'have received vast benefit from the Rains, & I hope in ten days they will be fit to be received on board, as I am now anxious about my time.'

On 9 July Wiles and Smith told Bligh that the plants would be ready in a week. Five days later they gave the all clear and the plants began to be taken on board. Almost seven hundred pots, many with two plants in them, were moved on the first day, but on the second a heavy surf slowed the operation considerably; a day later, the weather improved, and the remaining pots were put on board. The ship dropped six inches into the water because of the weight.

For Bligh the planting and moving on board of the breadfruit was a repeat, albeit on a larger scale, of his experience with the *Bounty*, but for Wiles and Smith it was all new. On 19 July 1792, final farewells were made. Bligh was anxious to sail. The northwest monsoon, bringing heavy rains and difficult seas, was approaching. By noon, and with strong winds, the *Providence* and the *Assistant* weighed anchors and, for the first time during this voyage, set course to the west.

Timor was to be the next stop.

Bligh happily enumerated the size of the botanic haul from Tahiti. According to his reckoning there were 1151 vessels filled with breadfruit plants, which contained 2126 individual plants. In addition, there were 508 other plants making a grand total of 2634 plants on the ship. Wiles and Smith's reckoning was significantly lower: they counted 1686 breadfruit plants and 310 other botanical specimens.[66]

Whichever figure was correct, it was an impressive achievement, but now the question was how would they fare on the voyage to the West Indies. They would be passing through extremes of hot and cold weather both often fatal to plants.

Bligh followed his previous course on the *Bounty*'s launch: first Tonga and Fiji; then to the north of New Hebrides, into and through the Torres Strait; and finally arriving, on 2 October 1792, at their destination of Kupang, on the island of Timor.

Bligh wanted his stay to be as short as possible. The approaching monsoon, as he put it, 'alarms me much'.[67] The water cisterns were quickly filled and the breadfruit plants that had died in about 200 pots (a loss of around 20 per cent) were replaced by local plants as Banks had suggested. Banks had anticipated that there would be losses. This didn't seem to alarm Bligh or Wiles or Smith. Bligh didn't even try to explain it, but Wiles and Smith thought the lack of good, circulating air, in the cabin nursery was the cause.[68] The plants that were in the greenhouse on the deck were hardly affected. This was a lesson, one of several the voyage produced, that would come in useful in the future. Soon, Wiles and Smith collected enough local plants, chiefly fruit trees, to fill some 90 vessels.[69]

Only a week after arriving in port, the ships were ready to depart. Two buffalos were brought on board and at 1 p.m. the ships departed.

On 10 November, when the ships were about halfway across the Indian Ocean and sailing westward near the Tropic of Capricorn, Bligh reminded Wiles and Smith that his instructions allowed him to stop at Madagascar, should the ships require any provisions. He told them that he would prefer to go to St Helena, where they were expected, without stopping elsewhere, but if, for the sake of the plants, they wished to go to Madagascar he would do so.

Wiles and Smith agreed that the plants were in a healthy state and that they might actually suffer if they called at Madagascar. As they wrote to Bligh, 'the touching at Madagascar in the present circumstances, would not only be unnecessary delay, but in our opinions a very dangerous risk'.[70]

So they went straight on to St Helena in the Atlantic Ocean. The ships rounded Africa without calling at the Cape again.

Before he left Kupang, Bligh wrote to Banks for the first time since 18 December 1791, when the ships were preparing to leave the Cape for the Indian and Pacific Oceans. Almost a whole year had passed since the last letter and he knew that this time his expedition had been a success.

Bligh must have had a great sense of relief and accomplishment. It can be felt in his closing note to Banks. 'I trust after all that I have done I shall once see you to prove I have accomplished every thing that could be expected and to give Joy of the Plants being safely landed at Jamaica.'[71]

On 16 December 1792 the *Providence* reached St Helena. 'I give you joy of the success of your Plants,' Bligh wrote with delight to Banks.[72] It had been more than two months since the ships had left Timor. Bligh had been rightly anxious about his precious cargo and took stock as soon as the ships crossed the Tropic of Capricorn, just to the east of Mauritius. Almost 200 pots of breadfruit plants had perished since leaving Timor. Bligh explained that the plants could not bear the excessive heat during the earlier part of the voyage, especially through the Torres Strait. Those that were near the cabin wall suffered the most. Now he was in cooler conditions he did not expect losses at that rate again.[73] He was right because the next time he took a count, which was on 11 December, he reckoned that he had lost plants in only 73 containers.

Wiles and Smith were also delighted with the voyage so far though they, too, wrote about the losses. They were not alarmed. The deaths, they observed, were far greater in the interior cabin than they were in the greenhouse on the deck, something that both gardeners had noticed and reported on the voyage from Tahiti to Timor.[74]

The instructions to Bligh, Wiles and Smith had specified that when they got to St Helena they were to deliver some of the breadfruit plants and samples of other plants collected in Tahiti and en route to the Atlantic; and, in return, they were to take on board those plants that the Governor saw fit to present to Kew and those that he and the gardeners felt would be needed and would do well in the West Indies. Banks added that Wiles and Smith should look out for the island's indigenous fern tree (*Melanodendron integrifolium*, the black cabbage tree), several of which he wanted dug up with a ball of local soil, to bring back to England.

The process began on 20 December and one week later was completed. By then more than twenty breadfruit plants, half healthy and half in a delicate state, along with fifteen other varieties, had been moved into the botanic garden. Wiles and Smith were delighted to see that the breadfruit plants had rooted successfully and that the process of transplanting them into a new habitat went very smoothly.

In return for the breadfruit Wiles and Smith brought on board three fern trees, and a collection of plants, including coffee, green tea, almonds, several camellias and primula, which filled thirty containers and was fifty-eight plants in total.[75]

The *Providence*, lying at anchor in Jamestown's bay, was quite a sight. Not many ships arriving, certainly no armed naval vessels, would have looked anything like it. 'A floating Garden' was how William Doveton, the Secretary to the Council expressed it after he, the Governor and other representatives of the Council had visited the ship.[76]

With the plant exchange complete, the ships were ready to leave and on 26 December, they began the last leg of their monumental voyage to the Caribbean.

The 'floating garden' made its way swiftly on a northwesterly course helped by brisk southeasterly trade winds. Wiles and Smith busily cared for their plants. This was the crucial time. Most days were dry but on one occasion when light rain began to fall and continued to do so for two days, they brought as many plants as they could out of the cabin and onto the deck to benefit from this free supply of fresh water.[77]

Three weeks after leaving St Helena, they entered the warm waters of the Caribbean, seeing Barbados in the distance and then, closer, Saint Lucia, the tiny island of Bequia and, finally, St Vincent where, on 23 January, the ships moored in the harbour of the island's principal town, Kingstown.

Bligh knew how much this part of the voyage mattered. His anxiety showed in his frequent enumeration of the number of plants alive and well and those that had died. He had barely left St Helena when he took his first count in the Atlantic on 31 December when the ships were about 100 miles to the southeast of Ascension Island. He recorded the numbers of vessels and breadfruit plants in his log adding that he felt that Wiles and Smith were not very exact in their counts. On that day, Bligh noted that the *Providence* had 738 breadfruit plants, 52 fewer than the last time he took a count in the Indian Ocean on 11 December 1792. He was certain that the entire loss had happened en route to St Helena and that he had lost not one plant in the Atlantic, so far.[78]

Ten days later, and soon after crossing the equator, Bligh made another count and noted that he had lost a further 47 plants but, with over 700 doing well, he felt confident that he would be delivering a fine collection.[79]

His confidence was, as he himself admitted, overstated. He kept on losing breadfruit plants. Wiles and Smith concluded that the length of the voyage was the most important factor in accounting for plant deaths but not unrelated was their placement in the ship: those in the interior cabin always fared worse than those in the greenhouse on the deck.[80] A very valuable lesson for Banks to learn. By the time, on 23 January, the ships moored in Kingstown, Bligh reported to Banks that a further 53 had died since his last count. Still, the garden was flourishing. The total number of plants from Adventure Bay, Tahiti, Timor and St Helena topped 1200.[81]

The stay in St Vincent was short. A consignment of breadfruit and other tropical plants, about half of the total for the West Indies, was left in the island's botanic garden. Alexander Anderson, its superintendent, had already prepared a collection of plants from across the island and the botanic garden, which was to form part of the West Indian plant collection for Kew. These were put on the *Providence*, taking the place of the plants that had been left behind. Bligh was relieved. 'The greatest pleasure I have is thinking you will be very happy to hear of my success . . . I have done all that came in my way to do & all this is new,' he wrote to Banks just before the ships weighed anchor for Jamaica.[82]

The voyage westward across the Caribbean Sea to Jamaica was uneventful and quick. Just a week later the ships anchored in Port Royal, Jamaica, on 5 February 1793. In fact, the authorities were not expecting Bligh so soon and were not prepared. It seems incredible but they had not yet decided what to do with the breadfruit plants. There were anxious moments in the first few days as the Committee, at its opening meeting, came to no firm decision.[83] At its second meeting, on the following Sunday afternoon, the Committee finally came up with a plan: most of the plants would be held on deposit in the main botanic garden at Bath and some in a garden at Spring Garden, once the property of Hinton East, who had died in the previous year. The members also agreed with an idea that Banks had put into the instructions, that if Wiles wished to remain behind with the plants he was entitled to do so. The Committee agreed that Wiles was to take care of the breadfruit trees at the main botanic garden in Bath for an annual salary of £200, three times what he had received on the voyage. Bligh was asked to take the remaining plants directly to planters around the island, using the *Providence* and the *Assistant* to deliver them.[84]

This operation lasted until 21 February at which point the ships had landed over 600 plants in various parts of Jamaica. The second part of the operation, the boarding of the island's plants for Kew, should now have begun but that collection had hardly started; the authorities had once again shown a lamentable lack of preparation.

While waiting for the Kew plants to be collected, news came that Louis XVI had been beheaded (it happened on 21 January 1793) and that France was in political turmoil.[85] Bligh rushed to get his ship ready for its homeward voyage. Some of the plants for Kew were put on board on 30 March. Immediately afterwards they learned that revolutionary France, which in the previous year had declared war on Prussia and Austria, was now at war with Britain and that a naval engagement in the Caribbean was inevitable. Bligh received his orders to be prepared to join the squadron of Commodore John Ford, the Commander-in-Chief of the Jamaica Station. The plants for Kew were immediately returned to the shore.

For the next two months, Wiles, Smith and the island's botanical personnel prepared the plants for the moment when the *Providence* would be released from its battle orders and be ready to take the Kew collection back on board. This did not happen until the very end of May. Speedily, the pots, tubs and boxes full of hundreds of plants were put on board filling the room in the interior and in the greenhouse on the deck, which had previously housed the plants for St Vincent and Jamaica. Altogether, Bligh recorded that there were 796 containers holding 1283 plants.

Two days later, on 4 June 1793, Bligh weighed anchor and set sail for home. The 'floating garden' was at sea once more.

PART III

An Embassy, a Free Town and a Plant Exchange

Preface

B anks never met Carl Linnaeus though he very much wanted to. In 1767, having returned from his voyage to Newfoundland, Banks planned to go to Sweden, but the opportunity to go on the *Endeavour* took precedence. By the time Banks returned, Linnaeus was too ill to be visited.

Banks did, however, get closer to him than did most people in Britain. Banks came to know and, in a few cases, actually employed several of Linnaeus's students. Daniel Solander, who went with Banks on the *Endeavour* and subsequently helped him organise his natural-history collections, was the first of these. Andreas Berlin was the second, who also helped with the *Endeavour* collection; and third was Carl Peter Thunberg, who spent some time in the library and herbarium at Soho Square after he returned from Japan.

Adam Afzelius was the fourth of Linnaeus's students who had come to Britain. Banks had known him personally since 1787, and about him for several years before that. In late 1791, Afzelius, a follower of the ideas of Emmanuel Swedenborg, was selected by the Sierra Leone Company, of which the abolitionist William Wilberforce was a co-founder and director, to go to Sierra Leone as a botanist to assess the area's natural resources and to see whether they would support a settlement there of freed black people. Once Banks was alerted to Afzelius's appointment, he took responsibility for instructing him about what he should collect. Though Banks knew a lot about West Africa, in general, and Sierra Leone in particular, this was the first time that he had had a collector on the spot to supply Kew with plants that were entirely new to the gardens.

Although Banks was pleased to have Afzelius collecting in Africa, he would have liked him to be the botanist with Lord George Macartney's diplomatic mission to China, which began to be discussed at the same time. However, when the opportunity arose, Afzelius was already committed to Sierra Leone. In the event, although Banks had hoped Macartney's embassy would be scientific as well as diplomatic, he did not manage to get a naturalist included in it. The only collectors on the mission were a couple of gardeners, and Macartney and George Staunton, his secretary, both amateurs. It was disappointing for Banks, who had hoped to take advantage of the mission's unique access to China's interior.

The other project involving Banks around this time was an exchange of plants with India, particularly the Calcutta Botanic Gardens. This too used gardeners, not naturalists, but their expertise at caring for plants, particularly at sea, made up for their lack of botanical education. The use of new technology (building plant cabins on board ship) was very successful and promised well for the future. Plant exchanges would flourish.

11

1791: 'An Intertropical Abode': Afzelius in Sierra Leone

It was December 1791. For the previous four years, William Wilberforce had dedicated his life to the cause of abolishing the slave trade and was its most public proponent, leading the debate in Parliament as MP for Hull.[1] At the time, Wilberforce was staying with his cousin Henry Thornton in the City of London, where the two men, both known to Banks, were working on their new project to settle a part of Sierra Leone. This was one of the principal aims of their newly incorporated concern, the Sierra Leone Company. Thornton, a very successful banker and MP for Southwark, was its Chairman, and Wilberforce one of its founding directors.[2]

For a man who was known as an eloquent orator, Wilberforce's letter to Banks, written on 21 December and from Thornton's house, was remarkably simple and short. Wilberforce got straight to the point: could Banks recommend a 'good Botanist particularly conversant with Tropical plants, and who might probably be induced on moderate terms to go to Sierra Leone, in the service of the Company, to examine the hitherto unexplored forests of that neighbourhood'?[3] Wilberforce added that he was available to see Banks during the week and on two days in the following week.

Wilberforce and Thornton were members of the Association for Promoting the Discovery of the Interior Parts of Africa, more simply known as the African Association, an organisation that started on 9 June 1788. Banks was one of its founding members and its treasurer, so he would have known both men from its meetings.[4]

Wilberforce's request wouldn't have come as a surprise to Banks. He had been following the progress of the Sierra Leone Company since it was incorporated shortly after the passage of the parliamentary bill to that effect at the end of May 1791.[5] He had in his possession a copy of the Act that had brought the company into existence and a copy of the company's first report, published in October 1791, which also listed the names of its directors.[6]

This was not the first time that Banks had been involved in Sierra Leone. In 1771, just after his return to London from his famous voyage on the *Endeavour*, Banks met Henry Smeathman, a budding entomologist.[7] Smeathman, who was about the same age as Banks, was already in contact with the Aurelian Society, whose members included people Banks knew well, such as Daniel Solander, James Lee, the nurseryman, and Dru Drury, the silversmith and passionate entomologist.[8] Drury's close friend, Dr John Fothergill, a Quaker physician and keen naturalist with a substantial botanical garden to the east of London, had decided to sponsor a natural history expedition to Africa. Hardly anything was known in Europe about the fauna and flora of equatorial Africa and Fothergill's plans soon found favour with other naturalists, principally Drury and Banks.[9] Drury was mostly interested in insects, Banks in plants, and the English ornithologist, Marmaduke Tunstall, in birds.[10]

They all agreed to appoint Henry Smeathman as their collector and, on 26 October 1771, he left England on a trading ship and reached Sierra Leone on 12 December. He remained in the area, settling on the Banana Islands, just off the coast south of present-day Freetown as his base, for the next four years. At first Smeathman collected with the help of his assistant David Hill, but he soon realised that their lack of botanical knowledge was hampering the plant collecting. Drury promptly dealt with the problem by arranging for Andreas Berlin, one of Linnaeus's apostles, who had been working with Banks since October 1771 on the *Endeavour*'s collections, to go out to Sierra Leone and join Smeathman.[11] Berlin arrived in April 1773 but did not live long enough to make much difference: he died on 11 June 1773, most likely from 'fever'.[12]

Smeathman left Sierra Leone for the Caribbean in 1775 and returned to England in August 1779. During his time in West Africa he had sent Drury four large consignments of specimens. Banks was one of the main recipients, so he would have had a large collection of plants from West Africa in his herbarium.[13]

Smeathman appears to have been involved in various projects over the next few years including addressing the Royal Society on termites; opening a debating society; and spending time in Paris immersing himself in the new marvel of ballooning, communicating his enthusiasm to Banks.[14]

Smeathman did not, however, forget his experiences in Sierra Leone which were unique at the time and, in 1786, just before his death from 'fever', he published his thoughts about it in a small pamphlet, a copy of which he presented to Banks.[15]

This pamphlet outlined his proposal for the settlement in Sierra Leone of hundreds of mostly black people who had recently arrived in London and were homeless. Many of them were ex-slaves who had received their freedom as a reward for fighting on the British side in the American War of Independence. Once the British lost the colonies, those people loyal to them had to leave.

Smeathman's proposal was accepted by the government, and by the Committee for the Relief of the Black Poor, and the scheme was in operation by the end of 1786. After several delays, a small convoy of ships, headed by HMS *Nautilus*, left Plymouth with 411 settlers on 9 April 1787 and reached its Sierra Leone anchorage, in what is now Freetown, one month later.[16]

Banks, of course, knew the captain of the *Nautilus*, Thomas Boulden Thompson from the previous year when, aged only nineteen, following the sudden death of the ship's captain, he took command of the *Nautilus* during its survey of Das Voltas Bay in southwest Africa as a possible site for a penal colony.

On that voyage, Banks had placed Anthony Pantaleon Hove as the naturalist. Remembering Banks's botanical interests, Thompson, while waiting for the other ships to convene in Portsmouth in January 1787, asked Banks if there was anything he could get for him in Sierra Leone to add to his collections.[17]

Having discharged his duties and confident that the settlement could be safely left, even though around a quarter of those who had landed had died already, Thompson returned home in September 1787. Granville Town, named after its greatest advocate, Granville Sharp, continued to decline and in December 1789 it was burnt to the ground by the local people who objected to the incomers.[18]

Wilberforce and Thornton's Sierra Leone Company would be the second attempt to settle the area. This was to be a much larger project than the earlier failed attempt. They needed a skilled botanist to assess the area's natural resources, which was why Wilberforce had written to Banks.

Banks's reply was not very encouraging. 'I told him', Banks jotted down in a note on Wilberforce's letter, 'I knew of no regularly bred Botanist in England who was likely to be tempted': the best he could offer was a gardener who knew which African plants were growing in European gardens.[19]

To everyone's surprise, within a week of Wilberforce's writing to Banks, Carl Bernhard Wadström (about whom later) recommended Adam Afzelius to the Company's Court of Directors.[20] They accepted the recommendation without hesitation at their meeting on 28 December 1791.[21]

Adam Afzelius, born in Sweden in 1750, was one of Linnaeus's last students and the youngest of his apostles.[22] He was part of that exalted circle of Linnaeus's disciples which included Daniel Solander and Andreas Berlin. Carl Peter Thunberg wrote very glowingly of him.[23] Pehr Afzelius, his younger brother, had stayed with Banks in 1785.[24] So Banks knew of Afzelius long before he arrived in London in November 1789. By Christmas, he was installed at Soho Square.[25]

For the next two years Afzelius used the library and herbarium at Soho Square as his intellectual base, while immersing himself in the botanical circles in and around London, visiting the royal gardens at Kew, the Chelsea Physic Garden and other private gardens, including that of Dr William Pitcairn in Islington. Afzelius was consulted on Swedish plants, and visited John Sibthorp, who had recently retired as Sherardian Professor of Botany in Oxford, and was working on his *Flora Oxoniensis*; and he became a close acquaintance of James Edward Smith, President of the recently formed Linnean Society in London.[26] All of this activity endeared him to Banks: 'he is in truth an excellent Botanist', Banks remarked to Olof Swartz, the Swedish botanist and friend, one year into Afzelius's stay in Soho Square.[27]

In 1791, Afzelius had his first taste of botanical travel, something he was desperate to do.[28] During that summer, George Leonard Staunton, a diplomat and friend of Banks's, who was also a Fellow of the Royal Society and the Linnean Society, took his family on a tour of England and Scotland, designed particularly to impress on his young son, George Thomas, the range of 'the arts, manufactures, and natural curiosities, of the different parts of the country through which [they] travelled.'[29]

Staunton took with him his son's tutor, Johann Christian Hüttner, a classics scholar from the University of Leipzig and Adam Afzelius.[30]

The group travelled together for several months that summer, mostly in Scotland, until, in mid October, while in Edinburgh, Staunton received a letter telling him to hurry back to London on urgent business.[31] Afzelius continued young George Thomas's botanical education as the party made its way from Edinburgh south through England to London which they probably reached in mid November 1791.[32]

One wonders why Banks hadn't recommended Afzelius to Wilberfoce when he turned to Banks for help in December. Possibly he remembered the death of Andreas Berlin or he had other plans for him. Banks wanted Afzelius to inspect George Clifford's world famous Dutch herbarium, which he thought would be coming up at auction soon.[33]

In the event, Afzelius turned down this and other offers and accepted the Sierra Leone Company position.[34] As he told his brother, he saw the Sierra Leone Company offer as a turning point in his life. He would either die there or return a made man – 'where no Naturalist has been before, and where I thus expect a rich harvest of the works of unknown Nature'.[35]

As well as hoping to establish himself as a famous botanist, Afzelius had another reason for wanting to go to Africa. He was a committed Swedenborgian, a follower of Emmanuel Swedenborg, who died in 1772 in London, and was buried at the Swedish Church in Shadwell. Swedenborg, who was born into a wealthy Swedish mining family with strong religious connections, was a polymath, a mathematician, astronomer and engineer, who wrote profusely about the sciences associated with his far-reaching interests.[36] However, it was not for his science that a sect developed around his name after his death, but for a series of mystical revelations that he had had beginning in 1743–4, which put him, he stated, in direct communication with heavenly authority. The visions were so powerful that Swedenborg turned his back on the life he had been leading until then and turned instead to God, devoting himself to expounding his personal understanding of the spiritual world in many publications.[37]

Swedenborg professed to have travelled in his visions between the spiritual and natural world and was convinced that a new religious age was coming. Those who read and agreed with Swedenborg's writings formed themselves into a movement meeting in various places in East London until, in 1788, they settled into a routine centred on the New Jerusalem Church in Great Eastcheap.[38]

Swedenborg had had a major vision in which Africa was revealed to him in a completely new way. His spiritual authorities, he explained, had told him that Africa, especially its interior, held a special place in the divine cosmos. The people who lived there, he believed, possessed greater spiritual illumination than did Europeans: they were the 'human form divine', they 'had preserved a direct intuition of God' and that they already had a church of the New Jerusalem.[39] It was essential to join them.

According to his biographer, Afzelius had already become an adherent to Swedenborg's ideas before leaving Sweden, and within a short time of his arrival in London he contacted the industrialist Carl Bernhard Wadström and the chemist August Nordenskjöld, fellow leading Swedenborgians and abolitionists, both based in London and co-authors of a pamphlet expounding their own ideas of colonising Sierra Leone on Swedenborgian principles.[40]

Wadström had been to West Africa in 1787 and, upon arriving in London in 1788, had become closely acquainted with leading abolitionists, particularly Granville Sharp and William Wilberforce.[41] Based on his first-hand knowledge of conditions in West Africa, Wadström gave evidence in April 1790 to the House of Commons Select Committee, of which Wilberforce was a leading member, enquiring into the slave trade.[42] Wadström was one of a group of a dozen Swedes, which included Adam Afzelius, who had planned to go to Granville Town in 1788. Afzelius would have known about the first British attempt to settle Sierra Leone.[43]

It is not clear whether Banks knew Wadström personally but he certainly knew of him, for in 1790 he heard that Wadström had applied to go on a British voyage (the secret one that was planned with Henry Roberts and George Vancouver at the very end of December 1789 to search for whaling stations in the South Atlantic and the coast of Africa – Chapter 9), at his own expense.[44] Wadström was, initially, granted permission to go, but then the idea was abandoned, because, as Banks learned, '[they] will not permit a Foreigner . . . to go'.[45] Whether Banks knew of Wadström's Swedenborgian sympathies is unclear but given his interest in and knowledge of Sierra Leone, he may well have come across Wadström and Nordenskjöld's pamphlet outlining their ideas for settling West Africa in which they made explicit that Swedenborg's ideas of how to live were at the centre of their plans.[46]

At some point in early 1790 Carl Peter Thunberg, who was in Sweden, heard about the secret British expedition, and though he knew none of the details, nor for that matter that Wadström had applied to go, he wrote to Banks suggesting that Afzelius should be appointed as the ship's naturalist.[47] Banks explained that no foreigner would be allowed to join the ship though he agreed that Afzelius was an excellent choice. 'I feel a great Regard for Mr Afzelius's Character', he wrote, '& should be happy to do anything agreeable to him'.[48] Now Banks had his chance.

As soon as Banks knew that Afzelius wanted to go to Sierra Leone, he tried to do his best for him.[49] He wrote to Wilberforce on his behalf and with his consent. The Directors of the Company had already promised Afzelius, 'to make him a handsome Donation in case any Discovery shall be made by him of material Advantage to the Company', but Banks did not assume that they knew anything about botanists and how they worked in the field.[50] He began by pointing out that Afzelius was 'born a gentleman' and that he had been a demonstrator of botany at the University of Uppsala in Sweden, where, as he stated, '[it] is more diligently studied, and probably better understood than in any other part of Europe.'[51]

Banks then explained the details of a botanist's working practices.[52] Though the Company had agreed to provide Afzelius with a house and a table at which to work, the house was not yet built and much more than a table would be required for his work. Banks insisted that Afzelius should be treated in a manner commensurate with his education and professional attainments and that he should have a European servant who would not only provide the 'menial' services required but who would also be able assist him in his botanical work, particularly in drying plant specimens. In addition, Afzelius would need the services of several local people to help him in the field, guide him, dig up plants, and carry for him. Instead of being provided with a house and table, Afzelius should be put up at the Governor's house, where he should dine and have his own room. His servant could live nearby.

Banks then turned to Afzelius's botanical needs. He would require tools – axes, saws, spades and knives: phials and spirit to preserve delicate plants; paper and thread to prepare the dried specimens and tin boxes to keep them protected from ants and other vermin; sacks and bags to bring the plants from the surrounding forests; a microscope to examine the flowers; and a range of trade goods – beads, wire and rum – to present to the local people, 'to induce them to point out to him the vegetables in use among them, or whose qualities are known to them – by this means Medicines & dying Drugs etc may be discovered, whose properties would otherwise have remained unknown.' Reference books, Banks added, Afzelius would furnish himself.

All of this was to be paid for by the Company – not a penny of Afzelius's own money should leave his pocket.

Wilberforce, on behalf of the other directors, immediately agreed to every one of Banks's demands.[53]

Afzelius left London in the middle of March 1792 and arrived in Freetown on 6 May on the *Sierra Leone Packet*.[54] In his first letter to Banks, written on 7 July and taking advantage of the fact that the ship that brought him was about to return, Afzelius gave his first impressions of the new settlement, and he did not like what he saw. Afzelius estimated that about half of the settlement, some five hundred people, were ill and about two hundred had died already. The surgeon himself was among the sick. Fresh provisions and housing were in very short supply and it rained all the time. Afzelius recorded that in a fortnight in May it rained on ten days: in June it rained half the month. When it wasn't raining, the settlement was buffeted by tornadoes.[55]

Afzelius had arrived two months after the last of the ships bringing new settlers from Nova Scotia. On 15 January 1792, almost 1200 black Loyalists boarded a fleet of sixteen ships in Halifax which took them across the Atlantic to a new life. Many of them had joined the British forces during the revolutionary war and had taken refuge in British Nova Scotia and New Brunswick after the British lost in 1783. Buffeted by storms at sea, the ships arrived separately, the final one anchoring in what would soon be called Freetown on 9 May. The death toll was remarkably small – only sixty-five had died on the crossing.[56]

The death rate in the settlement, however, was alarmingly high and for the first few months its future was uncertain. Afzelius complained to Banks that the overcrowded conditions meant that he had hardly anything to send to him: '[I am not] yet able to make any collection or examination of Plants in a daily crowd, & where I scarcely have room to write this letter.'[57] On the other hand, he was certain that the area abounded with new and interesting plants and he was looking forward to collecting them.

The rains finally eased in November, fresh supplies from England arrived and houses began to replace the makeshift structures that had been erected at first.

By the time he wrote to Banks again on 30 December, Afzelius's situation was, like the weather, much better. He now had a house where he and his servant lived, a place for his collections, his instruments and his library. Governor John Clarkson, who had organised and brought the Nova Scotians to Sierra Leone, had provided Afzelius with a garden 'to make experiments on various seeds & plants'.[58]

After months of inactivity, Afzelius was able to work. To prove it, he sent Banks a box of living plants and seeds planted in soil which, he thought, would germinate by the time the *Felicity*, one of the Sierra Leone Company's ships, arrived in London. To give the botanical collection its best chance of surviving the voyage, Governor Clarkson, who was returning to London on the same ship, had agreed to care for it.[59]

Afzelius's letter to Banks was very short – probably he had very little time to write it and get the collection ready before the ship sailed. That may be why he didn't mention that August Nordenskjöld had died in Freetown, in a 'wet and delirious state' a little over two weeks earlier. Nordenskjöld had been hired by the Sierra Leone Company at the same time as Afzelius to report on the area's mineral potential, and, like Afzelius, he had been recommended by Carl Wadström.[60]

Afzelius himself began to feel unwell and rather than remain in the settlement under these circumstances and risk ending up dead like Nordenskjöld, he decided to return to England to recover.[61] He arrived back in August 1793 and though he thought he would be ready to return to Freetown by October, his confidence was misplaced.[62]

Governor John Clarkson returned to England on 10 February 1793 and the plants and seeds entrusted to him went first to Banks and then to William Aiton at Kew.[63] Banks was sorry to find that few living plants had survived because the ship had met frosty weather in the last stages of its voyage. It was a problem with which Banks was all too familiar – the disastrous effect of cold on tropical plants.[64]

Clarkson also brought back with him Afzelius's report to the Company. Wilberforce asked Banks to read it and while Banks found the letter 'sensible & perspicuous and gives a fair earnest [blank] that much may be expected from him, when he is put into a condition of exercising his talents to advantage', he was disturbed by what he learned from Wilberforce about the reaction to it of the Directors.[65]

The Directors were concerned that Afzelius had sent botanical specimens to Banks. Afzelius, Banks pointed out, did not receive a salary from the Company but was only maintained by them in return for his assessment of the area's natural-history resources. Afzelius, Banks continued, had made no attempt to hide the work he did for Banks: on the contrary, he had entrusted the collection to Clarkson and had even mentioned collecting 'for some particular friend of mine' in his report.[66] Banks added that the plants were not for him personally; he was just an agent for Kew: 'the moment the very bulbs and plants in question came to my hand, I sent them to His Majesty's Garden.'[67]

Instead of being critical of Afzelius for collecting for Kew, which Banks argued they had no right to be, the Directors should have concentrated their attention on another and more important part of Afzelius's report. Banks was referring here to a list of tropical plants that Afzelius thought could easily be grown in Sierra Leone – cinnamon, cassia, nutmeg, cocoa, clove, ginger and logwood – which, since they were not indigenous to the area, would need to 'be fetched from another quarter of the world'.[68] Echoing Afzelius, Banks pointed out to Wilberforce that some of the tropical plants mentioned were actually available from London nurserymen and especially from Kew, and 'a great addition to them is shortly expected from the arrival of the *Providence*, Capt. Bligh, who will probably return from the West Indies in the course of the next month.'[69] Having dropped a massive hint of a mutually beneficial *quid pro quo*, Banks filled in some details in case the Directors did not quite get the message. 'I am confident', Banks wrote, 'that his Majesty, if properly applied to, will order such as can safely be spared from Kew to the order of the Directors, if a proper convenience for carrying them out is erected on board a vessel intended for the settlement, which may certainly be done for less than the sum of 100£.'[70]

It didn't take long for the Directors to respond enthusiastically to Banks's suggestion. By mid June, James Rice Williams, Secretary to the Sierra Leone Company, informed Banks that the Directors wanted to thank Banks for his ideas and assistance, especially the plan of a 'Box . . . to be fitted to and sent out by the Company's ship *Harpy*.'[71] Williams was referring here to a plant cabin, of the same design that Banks had used on previous voyages where living plants were transported.

The Company moved into high gear. In late July, Henry Thornton, the Company's Chairman, wrote to Sir Evan Nepean, Under-Secretary of State for the Home Department, asking him to apply to the King for plants from Kew and those recently arrived on the *Providence*, which had just completed a successful second voyage to the Pacific and the West Indies. Banks's plant cabin, he added, was also being prepared.[72] Nepean, who had already worked with Banks on several projects, naturally turned to him for help, explaining that his boss, Henry Dundas, the Home Secretary, supported the whole idea. After all, Nepean added, Dundas knew that 'you have . . . frequent opportunities of conversing with his Majesty on subjects of this sort.'[73]

Banks moved quickly and by the last week of August, he and Aiton at Kew were discussing which plants from Kew and which from the *Providence* should be assembled, and how they would be sent in exchange for plants from Sierra Leone which they didn't already have at Kew.[74] Then everything ground to a halt. Banks had learned that Afzelius was back in London.[75] He advised the Company to put the project on hold until Afzelius returned to Freetown so he could receive the plants and care for them in their new habitat. The plant cabin, Banks added, could easily be taken down and re-erected on another ship at little expense.[76]

Towards the end of October, Afzelius contacted Banks telling him that he had recovered and was expecting to sail to Freetown in the middle of November. The Court of Directors, he added, had asked him to restart the plan of sending plants from Kew and Afzelius wanted his advice.[77] As it happened, Afzelius did not manage to leave England until 20 March and didn't arrive in Freetown until 22 April 1794.[78]

As soon as Afzelius had left, and anticipating that he would be in Freetown before long, Banks reopened the project of exchanging plants from Kew for plants from Sierra Leone. On 24 April 1794, he informed James Rice Williams at the Sierra Leone Company that the right time of the year for transporting tropical plants was approaching and that the plant cabin should be re-erected on whichever ship the Company selected for the task.[79] Williams responded positively and quickly that the Company's ship, the *Harpy*, which was the largest in their fleet, and big enough to take a plant cabin on its quarterback, was due back from Sierra Leone and would be fitted out shortly for its return voyage.[80]

Over the next few months, final arrangements were made: the cargo was chosen – Tahitian, Timor and West Indies plants brought by Captain Bligh, and others, unspecified from Kew – and so was the gardener, who would accompany the plants on both legs of the voyage. The King insisted that the gardener should be from Kew and, so, on 24 July 1794, David Barnet was put in charge of the collections.[81]

The plants and the plant cabin were sent downriver by barge from Kew. With the plant cabin fitted, the *Harpy* departed from Gravesend on 31 August 1794 and arrived off Cape Sierra Leone on 9 October 1794. Apart from the plant cabin, there were over one hundred passengers, and goods amounting to the value of £10,000 on board. Instead of the quiet anchorage they were expecting, what they saw made the captain reverse course immediately and head for open waters.[82]

Britain had been at war with France for eighteen months and on 28 September, a flotilla of seven well-armed ships approached Freetown and began firing at the town. Several people were killed and many more injured. The firing came from a French squadron though some of the ships were English (ships that had been captured as prizes).[83] There was no possibility of resistance and the Governor surrendered. The French landed, houses were broken into and plundered, animals were killed. On 2 October, several of the main buildings in Freetown were burnt, as were a number of the Company's smaller vessels and domestic homes. A few days later, the burning of the town resumed and the church, several shops and houses were set alight.

It was in the aftermath of this very distressing and still visible event that the *Harpy* arrived and tried to flee. Instead, the ship was pursued, boarded and impounded.

Afzelius's carefully tended 'garden of experiments' had been ransacked and almost wiped out – he reported that he had had over 500 Sierra Leone plants growing there at the time.[84] As for the Kew plants on the *Harpy*, Afzelius recounted how he went to the shore to speak to someone in charge to try to rescue them. 'But the French', he wrote, 'were so ill-natured, that they would not let me come on board, and in fact it was already too late, for the most of the plants had been thrown overboard, and some I saw myself share the same fate.'[85] The plant cabin was destroyed.[86]

A few days later, on 13 October, leaving mayhem behind, the French squadron departed, taking the *Harpy* as one of its prizes together with most of the supplies, including food, intended for the colony. David Barnet, the Kew gardener in charge, was, according to Afzelius, deeply distressed by the events and quickly became ill. He died on 1 November 1794.[87]

For Afzelius, however, the drama was not yet over. When he returned to his house, he could hardly believe his eyes. 'I saw my garden destroyed, my quadrupeds, birds, 7 lizards which I kept alive in my piazza, all killed, my bottles containing quadrupeds, birds, amphibia, fishes, vermes, flowers and fruits broke to pieces, my boxes of insects partly taken away and cut open, my dry plants, fruits & seeds, shells, books & manuscripts scattered over the whole floor. All in a huddle, continually tramped upon and covered with dirt, grease, molasses, rum, porter, bread, meat & bones etc.'[88] On top of all that his instruments were destroyed and so was the journal he had kept since arriving.

Afzelius apologised for his scrawl and for the bad paper – 'in these distressing and penurious times I am very glad to have any writing – apparatus at all.'[89]

Banks reacted with horror and 'unfeigned sincerity' at the enormity of Afzelius's losses and told him that he would be sending him supplies by the next available vessel and that he would make everything good at his own expense.[90]

Conditions in the colony following the French invasion were very difficult and continued to be so for some time. Afzelius frequently expressed his desire to return to London but, at the same time, felt that he had to stay long enough to recover his natural-history losses at least. It was not until 13 July 1795 that he thought he had collected enough material, much of it from excursions in the surrounding countryside, to share with his friends in London.[91] They were placed on the Company's ship *Amy* in a large wooden box which contained dried specimens for Banks, James Edward Smith at the Linnean Society, the directors of the Sierra Leone Company and Carl Peter Thunberg in Uppsala.[92]

Afzelius's collecting activities continued unabated, even though his desire to leave was growing.[93] Finally, near the end of April 1796, he decided to leave the colony on the *Eliza*.[94]

The date of sailing was a little delayed but the *Eliza* left on 17 May 1796. Afzelius carried with him a substantial collection for his own researches: more than 2000 dried plants, boxes and bottles of fruits and seeds and many examples of animals, fishes and other marine creatures.[95] During the voyage, he carefully tended a small garden on the quarterdeck where he watched seedlings sprouting, including examples of coffee and tamarind. Though Afzelius rarely demonstrated any emotions in his journals, on the homeward voyage his writing reveals a sense of quiet confidence and accomplishment as he watched the shoots appear.

It didn't last long. When the ship was in the middle of the North Atlantic, at about the same latitude as Halifax, they ran into a storm. The seas were so high that the quarterdeck was frequently washed over and Afzelius's precious seedlings were destroyed by salt water, the greatest enemy of living plants at sea.[96]

The *Eliza* arrived in Torbay on 5 July 1796. Soon Afzelius was busy working on his collection in Banks's house in Soho Square. As a mark of his contributions to botany, Afzelius was made a Fellow of the Royal Society on 19 April 1798.[97] He returned to Sweden in 1799 and stayed working at Uppsala University for most of the rest of his life.[98] Though he had planned to publish a major flora and fauna of West Africa, it never happened.[99]

12

1792: Macartney, Staunton and the China Embassy

No European country (apart from Russia) had managed to obtain any diplomatic concessions from the Chinese Emperor, let alone establish a permanent diplomatic mission in the Imperial City of Peking, in spite of trying for a century and a half.[1] Some European Catholics had been allowed to stay in China to provide mathematical and astronomic services to the imperial court, but other foreigners were unwelcome, and in 1757, the Qianlong Emperor set up the Canton System whereby no foreign merchant was allowed any further into China than the port of Canton, and they could only stay there for the trading season.[2]

Both the British government and the East India Company were eager to negotiate a more liberal trade arrangement with China, especially to improve their access to the interior of the country. The East India Company had been extraordinarily successful in becoming the most powerful trading enterprise in Asia, and it was desperate to extend its operations into China. The British wanted to persuade the Emperor that allowing them more freedom to trade in China would benefit both China and Britain. A diplomatic mission seemed the best way to go about this.

In 1787, Henry Dundas, MP for Midlothian, and effectively running the Board of Control for India (an institution set up by Pitt's 1784 India Act to oversee the East India Company), began making plans for such a mission to persuade the Chinese Emperor to relax the trading restrictions for the British. Dundas was both part of the government and well known to the East India Company.[3] He decided that Charles Cathcart was the ideal choice to lead the embassy.[4]

Cathcart was an ex-military man, the son of a diplomat (Lord Cathcart), and had experience of successful negotiating in Mauritius. As MP for Clackmannan he had impressed the Prime Minister, William Pitt, as well as Dundas.[5]

Cathcart, aged just twenty-seven, jumped at Dundas's offer to lead the embassy to China.[6] On 21 December 1787, HMS *Vestal*, a 28-gun frigate, left Spithead for China. There had been many delays and difficulties in getting the expedition together. The commander, Erasmus Gower, and the naturalist, Johann Christian Fabricius, both opted out after Cathcart selected them for the mission. This was bad enough, but even worse was the fact that Cathcart was a sick man, ill with consumption. This was known, and it was actually written into the commander's instructions that the ship should return to England if Cathcart died.[7] It was hoped that the voyage and change of climate would improve Cathcart's health, but instead he got worse. Richard Strachan, the ship's commander, wrote from the Cape, expressing his fear that Cathcart wouldn't survive.[8] At the end of May 1788, in the seas near Java, Cathcart was so bad that the physician, John Ewart, advised they should make for Macao, instead of the northeast coast of China.[9] Less than a fortnight later, as the ship passed the eastern coast of Sumatra, Cathcart died.

Strachan headed back to England, as instructed – the mission was abandoned. Cathcart was not the only casualty: others died on the voyage home, including Jean-Charles-François Galbert, the interpreter.[10] 'Mr Galbert having been the only man in Europe Master of the Court and Common Language of the Chinese – It is impossible to replace him'.[11]

The embassy hadn't even reached China. The government and the East India Company were at a loss as to what to do next. New initiatives were planned and then abandoned – by the end of 1789, the whole idea of an embassy to China seemed to have been scrapped.[12] However, only a couple of years later, it was revived. Henry Dundas was now Home Secretary, and he was still keen to establish direct diplomatic relations with the Emperor in Peking. The government and the East India Company began to think again. A new initiative, with a new ambassador, was being considered.

Dundas was planning to have Lord George Macartney lead the

diplomatic mission. As early as 11 January 1791, William Cabell, a high-ranking clerk in the Board of Control for India, had sent Macartney 'seven volumes of Manuscript Papers' as background information on the current state of Chinese–British relations. Cabell continued sending such papers, under instruction from Dundas, throughout the year.[13]

Macartney was born in Ireland in 1737, educated at Trinity College, Dublin, Lincoln's Inn and Middle Temple, but never practised law.[14] Instead, he went on a grand tour, lasting four years. Though he had hoped for, and had been promised a seat in Parliament, his would-be benefactor fell from power, and Macartney's hopes fell too. However, through the patronage of Henry Fox, Lord Holland, a very influential government minister, Macartney was offered and accepted the diplomatic post of Envoy-Extraordinary to the Court of St Petersburg, in 1764, at which point he was knighted and painted by Reynolds. After that, in 1769, he was made Chief Secretary in Ireland. In 1775, he became Governor of Grenada. From 1781 to 1785 he was President of Fort St George, commonly known as Governor of Madras, the first major British settlement in India. In spite of his wide experience and many successes, Macartney was not offered another post, although he was not yet fifty. Instead he settled in a new house in Mayfair and immersed himself in London society.

Macartney's sister-in-law, Lady Louise Stuart, described him thus: 'He is very fond of his fine house and his great room, partly for its convenience and partly for its grandeur, and that makes him fonder of home than he used to be.'[15] He joined Samuel Johnson's Literary Club in 1786, and bought a country house in Surrey.[16] He also nurtured his Irish connections, attending to his ancestral demesne of Lisanoure, Co. Antrim, becoming a trustee of linen manufacture in Ulster, and taking his seat in the Irish House of Lords (he was made a Baron of the Irish Peerage in 1776).

It seems that Macartney first met Banks in 1786 at a meeting of the Literary Club. Banks was one of its senior members, having joined when it was a smaller, more intimate gathering of men like Joshua Reynolds, James Boswell, Edmund Burke and Oliver Goldsmith, all founder members. However, when Banks contacted Macartney in 1792, it was not about literature; Banks had heard that Macartney was to head the new attempt to send an embassy to China. If it succeeded Macartney would become the British Ambassador to the Court of

Qianlong. After almost six years, Macartney was back in demand as a diplomat.

At the end of 1791, Dundas invited Macartney to meet him in Whitehall to talk about the mission to China, and a week or so later, Macartney responded with a long report giving 'my sentiments on the most likely means of rendering such a measure effectual'.[17] Macartney's vision was of something altogether grander than the Cathcart effort which had ended so miserably. His years of diplomacy had taught him that the scale and scope of the enterprise would be the key elements in its success. It was to be a regal affair; King George III of Great Britain approaching the Qianlong Emperor of China as one great ruler to another.

Macartney emphasised that this was not the time to complain about past treatment. Though he completely understood that the East India Company's objections to the Chinese had been increasing over the years because of the tight restrictions imposed by the Canton System, and that it was important to address this, Macartney insisted that 'merely obtaining a redress of grievances . . . might create suspicion of disgust'. Instead, their complaints should be subsumed under 'a more dignified form, as extending to more elevated Views'.[18] Macartney reminded Dundas that the mission was about the meeting of equals and the advance of knowledge. 'For as our own Sovereign is so justly celebrated in Foreign Countries', he began, 'on account of the Voyages projected under his immediate auspices and direction for the acquisition and diffusion of knowledge, it might be made to appear as a natural consequence of the same system and disposition that among other purposes he should have been desirous of sending an embassy to the most civilized as well as the most ancient and populous Nation on the Globe, in order to observe its celebrated institutions, and to communicate and to receive the benefits which must result from an unreserved and friendly intercourse between that Country and his own.'[19] Instead of arousing suspicion or even alarm, which Macartney thought would happen if the embassy took grievances as its driving force, his suggested approach would be flattering to the court and would 'naturally lead to a preferable acceptance of a Treaty of Friendship and alliance with this Country'.

The first impression was most important, Macartney continued – if you get that right you have a chance. 'A certain degree of dignity should therefore be observed in the embassy as well as in the person of the Embassador [sic], whose demeanor must, besides, partake of that temper and decorum which seem to be the qualities chiefly valued by the Chinese.' Therefore, they should not send a frigate, but 'a King's Ship – one of fifty Guns'. A part of the squadron in India should be enlisted and the retinue should 'consist of Persons . . . who, by being verse in such Sciences or Arts as are admired in China, might tend to increase the respect for the Country of which such Men were Natives.' The embassy should bring with it the means of performing experiments and the latest technologies – steam engines and cotton machinery – to impress their hosts and in return, Macartney was certain, the Chinese would allow the visitors to inspect the 'abundant specimens of all their useful production'.[20]

Banks hadn't been involved in the first diplomatic mission, in spite of knowing Charles Cathcart and Johann Fabricius, its intended naturalist, but he was determined not to let this one escape him.[21] In mid January 1792, Banks arranged to meet Macartney to discuss what contribution he could make. A few days later, on 22 January, Banks wrote to him to outline how he envisioned the diplomatic mission could be broadened to include science.[22] By this time significant progress had been made and all of the principal players in preparing for the embassy were in place: on the government side was Henry Dundas, Secretary of State for the Home Department and effective President of the Board of Control for India; and on the East India Company side was Sir Francis Baring, its Chairman and head of the banking concern that would, in 1804, be called Baring Brothers & Co. In less than a month, therefore, many of the basic blocks of the embassy were put in place which included the choice of who should command the King's ship and a rough idea of the make-up of the personnel.[23]

It was at this point that Banks wrote to Macartney to advise him about how and why the diplomatic mission could be used to extract vital scientific and technological information for Britain's benefit. The letter, in which he outlined his ideas, is a blueprint for how diplomacy could act in the interest of science.[24]

Banks began by offering Macartney the loan of seventy-eight chapters of a Chinese encyclopaedia, which was in his library and which he had selected in order to describe the contemporary state of Chinese agricultural technology. The work in question (its title is not mentioned in the letter) is the *Shoushi tongkao*, which first appeared in China in 1742, in the sixth year of the Qianlong Emperor's reign.[25] This may have been the only copy outside of China. Banks invited Macartney to study this rare document. Even though it was written in Chinese, which neither could read, the illustrations, he told Macartney, 'afford an universal language your Lordship will be able as well to understand the Point at which their Science is now Fixd nearly as well as if you was acquainted with the Chinese Language'. Banks also sent Macartney the first volume of Benedict Hermann's *Beyträge*, a work published in 1786 and which contained a statistical abstract of China.

Banks stated that the embassy should aim to give Britain first-hand knowledge about Chinese progress in what he called 'the usefull as well as the ornamental Sciences'. Banks mentioned the great Chinese inventions of gunpowder and paper, both ancient, and known in Europe for some time; he then singled out porcelain and tea as two of the most mysterious products of China – 'to Learn these arts alone would be to give to Europe an invaluable blessing'.

Banks then told Macartney how he should go about getting such knowledge and the enormous benefit it would be to Britain: 'a few Learned men admitted among their workmen might in a Few weeks acquire Knowledge for which the Whole Revenue of the immense Empire would not be thought a sufficient Equivalent.'

And finally, Banks insisted that no expense should be spared – 'I sincerely hope that no stint whatever in Point of Expense will be suffered to interfere with the Proper preparations for Commanding Respect & Procuring information . . . [the] nation ought not to be Economical but Profuse in the Extreme.'

Less than a week later Macartney wrote to Dundas that he and Banks were in touch on the subject of 'such new and ingenious discoveries as might be most useful and acceptable to the Chinese'. He added that he had asked Banks to handle the orders of models and machines because 'as they are made under the Eye of so accomplished a Master they are likely to be executed in great perfection'.[26]

Banks next wrote to Sir George Leonard Staunton. Just as in politics Dundas had become Pitt's right-hand man, Staunton and Macartney had formed a two-man team.

Like Macartney, Staunton was born in Ireland, and in the same year, but their backgrounds and early career paths were quite different. It was only by chance that they got together. Staunton had trained in medicine in Montpellier, where he went after completing his studies at the Jesuit College in Toulouse. In 1759 he arrived in London where he wrote on medical subjects and seemed to be set upon a medical

career. Three years later, however, Staunton took a chance when he
decided to seek his fortune in Dominica, which Britain had just
captured from the French during the Seven Years War. Here, for eight
years he practised medicine but also began to study law while he led
the life of a plantation owner. Politics now became an interest but
rather than pursuing it on Dominica he decided on Grenada as a
better bet.[27]

After a short visit to London, where he married, he returned to
Grenada in 1772 looking to make money and gain political influence.
He never returned to medical practice.

It was in Grenada, this latest addition to the expanding British
Empire, in 1774, that he met Macartney who had been appointed
the island's Governor. Staunton's progress within Grenadian politics
was rapid, adept as he was at settling disputes between the British
and the French settlers, where his perfect French came in very useful.
Macartney was so impressed by Staunton's diplomatic skills that he
had him installed as the island's Attorney General. The two men
became politically inseparable. When the French invaded the island
in 1779, both men fled, leaving their fortunes behind. Soon after, in
1781, Macartney was appointed Governor of Madras, taking Staunton
with him as his private secretary. Once again, and in a totally different
political environment, Staunton proved his diplomacy, receiving as
a reward both an annual pension from the East India Company and
a knighthood for the Kingdom of Ireland from George III. Writing
in 1784 to his friend Charles James Fox, a prominent politician,
Macartney summarised Staunton's chief qualities: 'His sagacity, and
singular talents for public business, his extensive knowledge of most
parts of the world, – his spirit, integrity, and fidelity . . . give me a
right to speak of him in high terms and . . . that his assistance will
be of infinite use to the public, in almost any department he could
be employed in.'[28]

From 1786, while Macartney was honing his new persona of land-
owner – 'the taste for it grows upon me wonderfully', as he put it to
Staunton – and enjoying London high society, Staunton was becoming
more involved with London's scientific community, moving inexorably
towards Banks's social and intellectual world.[29] In 1787 Staunton was
elected Fellow of the Royal Society. Two years later he joined the
newly-founded Linnean Society.[30] An Honorary Doctor of Laws from
the University of Oxford followed in 1790. Banks congratulated him
on it and, typically, asked him to find out whether the quantity of

wheat annually brought to market in Oxfordshire had increased or decreased in the past twenty years.[31]

Macartney began to receive information about China from Henry Dundas during 1791. He lost no time in sharing it, with Staunton, whom he intended to appoint as his private secretary, and possibly Ambassador to China, if he himself became incapacitated (presumably thinking of Cathcart's demise en route on the previous mission). In mid October, Macartney contacted Staunton, who was in Edinburgh on a family field trip, to hurry back to London for urgent discussions about China.[32]

Staunton replied to Banks's letter on 26 January 1792 from Paris. He was delighted to hear of Banks's interest: 'Nothing can be more advantageous or more gratifying to the Persons embarking in the present Undertaking than the favorable light in which you consider it'.[33] Staunton knew that Banks would influence both the government and the East India Company to take a broader view of the embassy's remit. It would seek to obtain scientific and technical knowledge from the Chinese as well as collecting plants and other items of interest.

Though he had brought George Thomas, his ten-year-old son, and his tutor, Staunton was not on holiday. He was in Paris on very serious business in which both Banks and Macartney were involved.[34] He was there to find an interpreter for the embassy.

Macartney had already warned Dundas that finding the vital Chinese interpreter would not be easy: there was no one in Britain who knew Chinese, and relying on the Catholic missionaries at the court in Peking, who had other allegiances, would be dangerous.[35] Canton was no help as no British subject there knew Chinese; and he was disinclined to call on any of the professional interpreters who were used to help trade, the 'linguists', as they were called, who 'might have local views and Connections, or feel themselves under too much awe to be able faithfully and completely to render the sense of the most decent Representations'.[36]

Macartney could not guarantee that any interpreter they might find in Europe would not have other allegiances, but he was confident that such a person 'might contract in the necessary intimacy of a long voyage, an attachment which would ensure fidelity and zeal in the Service'.[37] His character would doubtless reveal itself over time and Macartney and Staunton would get a sense of how much he could be trusted.

'Not a moment is to be lost,' Macartney told Dundas.[38] The very next day, 8 January 1792, Dundas agreed about the interpreter and told Macartney to put his plan into action.[39]

What Macartney had in mind is what we would call a covert operation: someone sent to Europe, confidentially, appearing as no more than a 'traveller solely curious of knowledge and enquiry'; he would 'soon find out a Proper Man . . . who would answer the purpose without his being made acquainted with the main object until it was quite certain that it might be entrusted to him'.[40]

That 'traveller' was to be Staunton. Macartney had already discussed the problem of the interpreter with him and shared his anxiety that they might not be able find the perfect person. Macartney had little good to say about Galbert, Cathcart's interpreter – 'a worthless fellow' – but he couldn't fault his superior linguistic skills. He told Staunton not to be too fussy.[41]

Staunton was in Paris from 20 to 28 January 1792. During that time he made two attempts to find an interpreter: one at the headquarters of the Lazarists, who had taken over from the Jesuits as the official French missionaries, sanctioned by both Louis XVI and Rome, but found no one there who knew Chinese; and then at the Foreign Missions Society, which also sent missionaries to China. Staunton had more success there, in the sense that there was someone who had once known Chinese, but the elderly man had forgotten much of the language, and didn't want to return to China.[42]

However, there was a contingency plan and this required careful preparation, key contacts and intermediaries. Now Banks came into the picture.

On 28 January 1792, Staunton, Hüttner and young George Thomas set off for Naples. Macartney was confident of success. His informant, Louis Dutens, whom he had known for over thirty years and who had been chaplain in several British Embassies in Italy, had assured him that there were always some Chinese students studying to be missionaries in Naples.[43]

Dutens was referring to the Collegium Sinicum, founded by Matteo Ripa, a secular Catholic priest (he did not belong to an order), as a place where young Chinese men, brought over from China, were trained as missionaries and then sent back to their homeland to spread the Catholic message and convert the Chinese to Christianity.[44] By the time Staunton was on his way to Naples, the college had already prepared some fifty Chinese men for their missionary work. How many there were in residence at the present time Staunton did not know; first there was the problem of how to contact them and this is where Banks could help.

Banks had two close contacts in Naples, both of them significant enough to open the doors of the college. The first was Sir William Hamilton, who was the British Ambassador to the court in Naples, a position that he had held for about thirty years.[45] He knew everyone of any importance in Naples. His own interests were in collecting vases and volcanology and it was through the latter and the Society of Dilettanti in London that he came to know Banks.[46] Hamilton was already a Fellow of the Royal Society before Banks became its president, and he had published a large number of essays in the *Philosophical Transactions* (the official journal of the Royal Society), Banks and Hamilton had corresponded frequently and regularly since 1775. Louis Dutens was also a friend of Hamilton's and had been a witness to his marriage to Emma on 6 September 1791.[47]

The other Neapolitan contact of Banks was the natural philosopher and antiquary Gaetano d'Ancora who had dedicated his most recent book to Banks. He knew Hamilton well too and had also met Staunton in Naples the year before.[48] Banks wrote to both men and they quickly began to use their influence on his behalf.[49] Hamilton was very confident – he had the ear of the King of Naples and the superiors of the college were, he told Staunton, 'my friends having rendered them once a signal service'.[50] Four Chinese priests were at the College, all of them 'well conversant in their own and Latin language' and he was certain that at least one or two of them would take up Staunton's offer. The Chinese students were given places to study in Naples on the understanding that, once their education was completed, they would return home as missionaries. Hamilton added that he was certain that the superiors 'to flatter him and Don Gaetano' would produce one or two Chinese students by the time Staunton arrived in Naples. The fact that Staunton would be paying their passage back to China would clinch the deal.

Because of bad weather and an accident crossing the Alps, the Staunton party did not arrive in Naples until the second week of March.[51] Fortunately the delay didn't matter. Banks's intervention with Hamilton and D'Ancora had paid off handsomely. Two Chinese students at the college who had already finished their studies agreed to return to China with Staunton and to act as the embassy's interpreters.

Jacob Li and Paul Ke, both forty, had been at the College since

1773 and on 20 March they left for London en route for Peking. Hamilton wrote to Macartney to tell him that they were on their way.[52]

Li and Ke's journey to England with Staunton took them through Germany and the Low Countries. Staunton paid for all their travel expenses. In Brussels, Staunton also had to buy new clothes for both of them, 'they having travelled until then in ecclesiastical Habits, in which it would not be convenient for them to appear in England.'[53] They arrived in London at the end of May where they stayed in Staunton's home.[54] There Li and Ke waited for over four months while the ships were prepared for their voyage to China.

Not long after Macartney learned that the two interpreters were heading to London, Banks also heard directly from Staunton of his good fortune and wrote back congratulating him on his success in 'so important an article as that of interpreters'.[55]

In a tone characteristic of the time and the person, Banks noted that he would feel himself flattered if he 'thought [his] letter had realy produced the Effect' that Staunton attributed to it. As he added, his purpose in writing to Hamilton was 'to Provide a Spur for my friend . . . who we know is not always so active in Serving his Countrymen as he might be & I could contrive no more Effectual one than making him understand that if he was neglectfull the business might be done by some one Else.'[56]

Macartney must have been mightily relieved when he learned of Staunton's success in Naples. He had confessed, when Staunton was just about to leave Paris, that Naples was their only chance: 'I doubt', he told Staunton, 'we shall be able to find the proper subjects in this hemisphere; there is no one at Gottenburg, Copenhagen, or Lisbon, that we can hear of, fit for the purpose.'[57]

While Macartney was waiting for Staunton's return, he and Banks were working together to finalise the rest of the embassy's personnel. Macartney was leaving nothing to chance. Already by the end of the first week of January, he had devised what he called a 'Tableau or Sketch of an Embassy', in which he listed the important personnel. For example, Macartney envisioned a suite which included, in addition to him, Staunton and the two interpreters, two under-secretaries, a comptroller, a surgeon, a 'Gentleman of Science capable of making Philosophical experiments' and a 'Machinist under him', persons from Birmingham, Manchester and Josiah Wedgwood's pottery, who knew and could show off the latest innovations in hardware, cotton and porcelain to the Chinese; an artist, six musicians and a number of servants; and finally a number of military personnel to show off the latest in British firearms technology. As for presents – and presents were absolutely essential he insisted – Macartney had in mind such technically advanced items as 'the Steam Engine, the Cotton Machines, Chain Pump, Balloons, Telescopes, etc., etc., together with Pictures and Engravings . . . curious Fire Arms for the exhibition of late discoveries in Gunnery and Projectiles and a couple of elegant Post Chaises on Springs painted from Chinese designs'. All of this to be conveyed to China in a fifty-gun ship commanded by Captain Erasmus Gower, whom Macartney knew from India (Gower should have been the commander on Cathcart's embassy).[58]

Since early January and with Banks's help, Macartney had been negotiating about the presents with the East India Company, who had agreed to foot the bill; the process of selection went on for months. Meanwhile, Macartney also wanted to expand the scope of the commercial side of the presentation. Here he sought the President of the Privy Council Committee for Trade, Lord Hawkesbury's advice: he hoped to take 'an assortment of such articles as may be thought best calculated to strike the eyes of the People of Peking and the neighbouring Provinces'. As so little was known about the taste of the Chinese, Macartney felt that the Chinese reaction to a selection of these items might give the embassy an insight into their preferences. 'The idea', which as Macartney explained it to Hawkesbury has a decidedly modern ring, 'is not to sell or dispose of these things for money in the first instance but to make presents of them to the Chief Merchants of the Emperor and to other Persons at Peking who may be most likely to establish a taste for them among the People, to encourage a demand for them and promote their future sale.'[59]

Over the following months, Macartney and Banks were busy contacting their friends and colleagues in Manchester, Stoke-on-Trent and Birmingham, to ensure not only that the products of Britain's innovative industry found their way as presents for the court but also that the right people were sent along with the embassy to find out what the Chinese were doing that the British could profit from.[60]

While most items had to be fought for and numerous compromises made along the way to a final selection, many original ideas had to be abandoned altogether. Macartney wanted to show off a steam engine to the Qianlong Emperor but after visiting Matthew Boulton's factory in Birmingham he felt it was just too difficult to transport – the engine's weight and size were simply too big to handle.[61] The idea of a balloon – all the rage in Britain, France and the United States – was also abandoned primarily because James Sadler, who had been considered as one of the 'scientists' to accompany the embassy, didn't want to go.[62]

In the end, manufactured goods, which Macartney felt certain were Britain's strongest suit when it came to impressing the Emperor, gave way to scientific instruments, including globes, telescopes and the famous Planetarium.[63] No one from Birmingham, Manchester or the Potteries accompanied the embassy. This was probably a mistake.

While Macartney and Banks did not always get their way in choosing the presents and the personnel, when it came to the size of the embassy, they were satisfied. Macartney's wish for a fifty-gun Admiralty ship to carry the entourage, was upgraded to the sixty-four-gun ship HMS *Lion*; and the East India Company contributed the *Hindostan*, one of its largest and fastest ships, to carry the bulk of the militia and some of the passengers, and added the *Jackall*, a tender.

After months of delay, the embassy, consisting of three ships and upwards of eight hundred men, assembled in Portsmouth on 21 September 1792 for its epic voyage to China.

Eleven-year-old George Thomas Staunton was there, too. His father had insisted that he was included. It was the educational opportunity of a lifetime, and there was a bonus to it. As George Thomas recounted many years later, it was in Naples that he first saw a Chinese person and that, as he put it, 'my ears were first familiarised to the sounds of a language in which, during the next five-and-twenty years of my life, I had so much exercise.'[64]

Beginning in Naples and then continuing as he travelled through Europe to London with his father, his tutor and the two Chinese interpreters; and in the Staunton household for four months, George Thomas learned, and became extremely proficient in Chinese. Then followed ten months of even more intensive study as the ships made their way to the coast of northeast China.

George Thomas was preparing for the experience of a lifetime, and he had his journal ready. He made his first entry on Saturday, 15 September in London: 'This morning my papa, Mr Huttner and myself in one charge and the two Chinese in the other and a servant riding behind set out on our Journey to China.'[65] The next day he saw and went on board HMS *Lion*, which was to be his home for almost two years. 'Here we had a comodious staircase, to enter the ship by, we first saw Lord Macartneys cabin which was large comodius and well furnished. We then saw my Papa's and Sir Erasmus Gower's which are both of the same size, well enough but small, my cabin is a part of the cuddy and separate from it by a curtain.'[66]

After several days at anchor in Spithead, in the Solent between Portsmouth and the Isle of Wight, waiting for an improvement in the weather – George Thomas was often sick – the big day finally arrived. Wednesday, 26 September 1792: 'This morning we unfurled our Sails and Set Sail.'

13

1793: The Accidental Naturalist in Qianlong's Empire

Erasmus Gower first went to sea in 1755 when he was thirteen years old. By the time he took command of HMS *Lion*, Captain Gower, aged fifty, had circumnavigated the globe twice. He had seen service on several naval stations, especially in the East and West Indies, achieving success and subsequent promotion on all of them. As a ship's captain, he took his orders from the Admiralty. On this trip, conveying the Macartney embassy to China, Gower got his orders instead from Henry Dundas, the Home Secretary and thus indirectly from the King. When, on 8 September 1792, he received his instructions for the voyage, Gower would have seen their unusualness immediately.[1] Normally the commander of the ship was the ultimate authority. But Dundas told him that he had to take instructions from Macartney and, in the event of his death, from Staunton, his secretary. He was to take HMS *Lion* directly to a port on China's eastern or northeastern coast as near as possible to Peking without calling at Canton. He and the ship's company should avoid any interaction with the Chinese that might result in 'pernicious consequences and indeed destructive to the objects of the embassy'.[2] It went without saying that Gower was responsible for everything else – sailing the ship and ensuring the health and safety of its company.

Gower wanted to get to the Chinese coast as quickly as possible. Though the winds were favourable at this time of the year, nothing was predictable. Bad weather could easily delay the voyage: ships often found themselves in the doldrums – that part of the Atlantic Ocean near the equator where the two prevailing trade winds meet, sometimes becalming ships for weeks on end. Illness among the crew could also cause delay. Perhaps the most unpredictable factor of all was the response of the Chinese authorities to the arrival of a fleet of three ships from Britain. Furthermore, Gower had a schedule to keep. The embassy had to arrive in time for the celebration of the Qianlong Emperor's eighty-second birthday on 17 September 1793. If they were late, the whole enterprise might fail.

In spite of Banks's close relationship with all the people in charge of the embassy – Henry Dundas, Francis Baring, George Macartney and George Staunton – he did not always get what he wanted. Although he was convinced that James Lind should be the embassy's physician, Lind wasn't appointed.

Banks had known him for at least twenty years and Lind had been on Banks's last expedition to Iceland in 1772. Moreover, Lind had been to China on two occasions: the first in 1766 before graduating as a physician from the University of Edinburgh; and the second in late 1779 or early 1780.[3] He knew enough Chinese to help Banks make sense of his Chinese books and manuscripts and provided him, as we have seen, with a clever and very practical way of ensuring that collectors of Chinese plants for Banks only collected new varieties that Kew didn't have already.[4]

Lind was the very first person that Banks proposed to Macartney to join the embassy.[5] Banks extolled his medical experience and his wide-ranging scientific interests: 'he is . . . well acquainted with Experimental Philosophy Chemistry & Mechanics [and] he is also a tolerable astronomer & has paid particular attention to Amusing Experiments.'[6] Above all, he knew more about China, its people, its scientific achievements and its language than anyone else, in Britain. Even more important, and in this Lind was absolutely unequalled, he knew, as Banks put it, 'what they stand in need of'.[7] His advice on which British manufactures and technology the embassy should take to the Emperor would be invaluable.

Macartney agreed Lind was the perfect choice, and when Banks told him this, Lind immediately rushed to London from Windsor, where he had a lucrative private practice (which included being physician to the royal household).

Macartney invited Lind to his Curzon Street home to discuss the appointment.[8] Banks, Macartney and Lind himself all wanted it, but there was a big hurdle – money.

Everyone began calculating. If Lind accepted the post he would be away from home and his practice for two years, and possibly three. Macartney thought that he could get Lind an annual salary of £500 plus a £1000 present from the East India Company at the embassy's conclusion. If the Company were unwilling to make this gesture, Macartney offered to give Lind £500 from his own pocket.[9]

Banks knew that these sums were inadequate; he explained to Lind that this offer was the best Macartney could manage – 'the Economy of the Company is so Frugaly administered that he will not have it in his Power to make such an offer as you for the sake of your family & in justice to your business . . . will be under the necessity of requiring'.[10] Time was of the essence. 'In order to Cut the matter short', Banks wrote hurriedly to Lind, 'I shall be Obligd to you if you will Fairly State what Sum of Money you would think a fair Compensation.'[11]

Banks had reminded Macartney when Lind's name was first mentioned that Parliament had voted Lind the sum of £4000 for joining James Cook on his second voyage in 1772. Macartney's offer was not even half of that amount. So, when Lind told Banks that he would need £6000 to be paid in advance, that was the end of the matter, although Lind really did want to go. He even contacted Henry Dundas directly, but was told that the East India Company, who were paying the bills, would never pay the sum he needed, nor would the government.[12]

Two months later, Banks recommended Hugh Gillan, another Scottish-educated physician, to join the embassy instead of Lind. Gillan had no preconceived ideas about his salary and, indeed, as he told Banks, the chance to go to China was more than enough to satisfy him. The East India Company offered Gillan only £200 per annum, and he took it.[13]

Staunton had joined the embassy as Macartney's secretary but he had another role as well; he was to be Banks's botanical agent. It was a role that he gladly accepted but which he knew was probably beyond him – '[my] own very imperfect Qualifications, make me fear not a little . . . for being of much use to knowledge.' What he lacked in ability, he emphasised, he would make up for in enthusiasm: 'we will not desist', he told Banks, 'Unless we find the Impracticability of the Attempt.'[14]

As far as Banks was concerned, every port of call was an opportunity to collect. The closer they got to China the more exciting the potential harvest. Just before Staunton left for Portsmouth, Banks formally entrusted the mission's botanical investigations to him. Why Staunton instead of a trained botanist? It is true that China's botanical riches were not uppermost in the minds of the embassy's planners. Perhaps they thought that such matters would be taken care of later once the Chinese authorities acceded to British demands to open their country to direct trade with Britain. Certainly money, as became clear in Lind's case, was an issue. It seems a foolish economy in retrospect.

Staunton certainly had expected to have a botanist appointed to the embassy. He said as much to Banks in the early stages of the mission's preparations and even had someone in mind. 'I hope they will determine on sending . . . a good Naturalist' – Staunton had been confident that with Banks's involvement this would be the case.[15] Staunton naturally suggested Adam Afzelius, as the two men had already spent time botanising together during the summer of 1791. Banks would have agreed wholeheartedly with his choice. Not only was he a trained botanist and a friend of Thunberg (who was an expert on Japanese plants), but he had also studied oriental languages, which included Chinese, and was proficient enough in them to be made a senior lecturer in 1777.[16] But by the time Staunton recommended him, Afzelius had already agreed to go to Sierra Leone.[17]

Banks shared Staunton's dismay at the planners' disregard for natural history. 'Everyone Complains', Banks wrote, 'that the arrangement is made without any regard to natural history which Some think should have been the first thing attended to as it Certainly is the Foundation of the whole.' The mechanic, the dyer and the potter, he explained, could not possibly know whether the ore, the plant and the clay were new to European science simply by looking at the finished product. 'The naturalist alone', Banks continued, 'can by considering these investigate their advantage to the produce of the rest of the World & take advantage of the affinities he may find to things already Known.'[18] Staunton fully agreed with Banks's sense of 'uneasiness', as he put it, and added that Banks might 'exert your powerful influence with the first Person in the Nation'.[19]

It may be that there was no one available of whom Banks approved. He said as much to his good friend Olof Swartz, the Swedish botanist and student of Carl Linnaeus the Younger, when the embassy was on the point of departing – Banks, it seems, would have liked Swartz himself to go but he couldn't or wouldn't.[20] Another possibility is that there was a problem of space, something that was always at a premium on a ship. Banks no doubt remembered that it was an argument over space that led to his leaving HMS *Resolution* in 1772. It was not the living space that a naturalist needed that was difficult; it was the tools of his trade. The library of books, paper, preserving liquids, bottles, collecting boxes, dried plants, etc. took up so much room, room that was required for the presents for the Emperor. Staunton, of course, already had a cabin on board.

In the absence of a naturalist, Banks did manage to get two gardeners to join the embassy, David Stronach and John Haxton, whose experience working with live plants would come in very useful.[21]

Banks had admitted to Olof Swartz that the whole botanical exercise, even if the mission had included a naturalist, might amount to very little – 'possibly they will be so watched during their stay in China', he wrote, 'as to have few opportunities of Exercising their Talents for Botany'.[22] Still, Banks had to do what he could to make the botanical side of the expedition, whatever form it might take, as successful as possible. He began by drawing up a plan, what he referred to as 'Hints on the Subject of Gardening', which he shared with Staunton and Macartney as well as the two gardeners.

The first part of the document was an essay on Chinese horticultural methods and details of a few highlighted Chinese plants, such as the magnolia and the peony, some of which were already growing at Kew and others about which Banks knew only from second-hand sources, especially the encyclopaedic and multi-volume *Mémoires concernant l'histoire*.[23] From a practical point of view, it was the second part, which had details of the Chinese plants Banks most wanted, that really mattered. Banks advised against discrimination; they should gather and dry and preserve as many plants as possible and leave it for Banks, with the resources in his library and herbarium, to decide what was worthwhile collecting – 'to be able to distinguish' as Banks put it, 'with absolute Certainty whether a plant is realy new or not is scarce possible to the most sagacious Observer'.[24] The rule-of-thumb, Banks counselled, was not to be tempted to ignore a plant because it didn't 'look' especially interesting. In his own words: 'the most Obscure & minute & those that have the Least pretension to Elegance & beauty are the Lest likely to have been before taken notice of.'[25] Banks insisted: 'To leave behind one Scarce & Curious plant under the mistaken Idea of its being a Common one will be a Source of Vexation for Ever afterwards.' Banks emphasised that the collectors should not ignore common varieties, nor should they restrict themselves to only one specimen of each plant. Science, Banks believed, was to be shared. As he put it: 'many individual specimens of Each Species [should] be brought home in order that different Botanists may be consulted.'[26]

In addition to the list he included in the document, Banks also provided the embassy with pictorial representations of the plants he most wanted for Kew. These pictures he took from three main sources: Englebert Kaempfer's *Icones*, a picture book of Japanese plants, which remained in manuscript during Kaempfer's lifetime, but which Banks had published in 1791 at his own expense – he presented Staunton with a copy of the book before he left for China;[27] to that he added six plants described in Kaempfer's 1712 magnum opus of Japanese plants, the *Amœnitatum exoticarum*; and finally, four plants taken from Carl Peter Thunberg's *Flora Japonica*, published in 1784, after the Swedish author's sojourn at the Dutch East India Company's trading post in Nagasaki, where he was the head surgeon for fifteen months.[28]

There were over twenty plants depicted in these sources, plants 'that are likely to be more acceptable to his Majesties Botanic Garden than the rest'. Banks did not specify whether he wanted the living plant or seeds, presumably leaving it up to the collectors and the circumstances in which they found themselves.[29]

To assist his collectors even further in getting plants for the King's Garden, Banks told them to contact Alexander Duncan when they got to Canton. Duncan was the English factory's surgeon and Banks's collector. Duncan had with him the copy of Banks's 'Book of Chinese Plants', which they should borrow to use during the trip through China. Banks explained his system of crosses that ranked Chinese plants in terms of their desirability in England.

Though the squadron of three ships, HMS *Lion*, the *Hindostan* and the *Jackall*, left Spithead on 26 September 1792, it was not until 1 October that they were able to leave England behind, as contrary winds forced them to await better conditions in Torbay. This was the last time that bad weather affected the mission, for the voyage to the Indies was done in record time.

After ten days the ships arrived in Funchal Bay, Madeira, their first stop, on 11 October. Macartney wrote to Banks with the good news that they were finally on their way to China and that everyone on board was performing their duties as he had hoped and Banks had planned: 'There is the fairest prospect of a pleasant progress & happy issue to our expedition.'[30]

Macartney also informed Banks that both Staunton and Hugh Gillan, the ship's physician, who had been chosen by Banks for the voyage, had begun what Macartney called 'their philosophical adventures'. Specimens of plants had already been collected, Macartney assured Banks, and the two accidental naturalists, Staunton and Gillan, had already been to see the crater of one of the island's most dramatic extinct volcanoes.[31]

On the same day, 21 January 1793, that he received Macartney's letter, Banks also had a letter from Staunton written two days earlier.[32] Both men had hurried to write because they knew that a ship at anchor near them was planning to sail back to England. The three-month period delay between writing the letters in Funchal and Banks opening them in London was an example of the vicissitudes of communicating by ship.

In his letter to Banks, Staunton simply expanded a little on what Macartney had told him: he and Gillan had been to the volcano and they had tried to collect some botanical specimens. Staunton feared that he couldn't have collected anything new since Banks had been there himself on the *Endeavour*. Still, as Macartney put it, 'It will give them a habit of botanical exercise & facilitate more curious researches.'[33]

Their next port of call was Santa Cruz, on Tenerife, which they reached on 21 October. After a stay of several days, Gower steered south towards the equator and on 2 November he put into Porto Praya (now Praia) on the island of St Jago (now Santiago), the largest of the Cape Verde Islands.

Gower had hoped to replenish his supplies of water and food before leaving the Atlantic for the long run to Batavia. Water supplies were critical; by the time HMS *Lion* anchored in Porto Praya, almost one-third of the water supply had been used up already.[34]

Gower needed to top-up the water tanks and get as much fresh meat and vegetables as he could. Unfortunately the Cape Verde Islands were now in the third year of a long drought. There was very little fresh water to be had; the crops had failed; and people and cattle had died.[35] Gower cut their stay short and decided to cross the Atlantic to seek supplies in Rio de Janeiro, something he had hoped not to do.[36]

After a record run of twenty-three days across the Atlantic, during which time, on 17 November 1792, the usual rituals took place on board as the ship crossed the equator, HMS *Lion* anchored in the harbour of Rio de Janeiro on 30 November.

With its population approaching 50,000 people, Rio de Janeiro was, after Lisbon, the second largest city in the Portuguese Empire, but its primary importance to British ships was as a way station. It had abundant fresh water and food (especially fruits and vegetables), excellent repair and medical facilities, and because of Portugal's close ties with Britain, Rio de Janeiro was particularly welcoming to British shipping, both naval and mercantile. Hardly a day passed without a British ship anchoring in the harbour.

Though Gower had not intended to stop in Rio de Janeiro, Staunton and Banks had hoped they would, and Banks had special instructions for both Staunton and Gillan in case they did.[37]

Banks was after two items in particular. Ipecacuanha was still top of his list. Banks had tried to get specimens of ipecacuanha from Rio de Janeiro five years earlier in August 1787, when he had asked Arthur Phillip, on his way to Botany Bay with the First Fleet, to look out for the plant. Phillip, who knew Brazil well, managed to get several varieties for Banks and had them sent on to London.[38] However, when Banks received them, he was disappointed. Though Phillip had sent him two different plants – one black and one white – neither of them came with their flowers or fruits and without these, as he told Staunton, 'the distinguishing Characters cannot be investigated'.[39]

While ipecacuanha's emetic qualities were widely known, hardly anyone in Europe knew anything else about the plant, and without flowers and fruits it was impossible to classify it. Banks explained the problem to Staunton before he left England. 'All I Know on the Subject of it', Banks wrote, 'is that as a valuable part of the materia medica it is in some degree a disgrace to the national Character of our medical men that we are still wholly ignorant to what Family of Plants it belongs.'[40]

So, when he got to Rio de Janeiro, Staunton tried to find out more about the plant but he failed; the physicians whom he consulted had no new information for him. Still, he had a letter of introduction to a former Governor of the island of Santa Catarina, to the south of Rio de Janeiro, who promised to supply him with the plant itself. When Staunton finally saw it, it was clear that Banks would again be disappointed: for it had neither flower nor fruit.[41] Nevertheless, he sent it with other material on a whaling ship bound for London.

While obtaining the ipecacuanha plant was Staunton's responsibility, the task for Hugh Gillan, the ship's physician, was tracking down cochineal. Banks knew from Arthur Phillip that Brazil was producing cochineal from a native variety of cactus, which grew on the island of

Santa Catarina. Banks wanted a specimen of the cactus complete with
the small white insects, which produced the red dye when crushed.
Gillan moved quickly – within a week of arriving in Rio de Janeiro,
he had managed to get several cactus specimens complete with the
insects. Banks wanted live plants and insects and Gillan, probably with
the help of the gardeners, Haxton and Stronach, carefully put three
plants, with their soil attached, into a large wooden box, which they
sent on the same ship, the *Hero*, that Staunton was using.[42] Gillan
gave very careful instructions to the captain of the ship to make sure
that the plants and their insects survived the voyage, which was expected
to take two months.

On 16 February 1793, Samuel Enderby Jr., the son of the owner of
the *Hero* whaling ship, told Banks that the boxes sent by Staunton
and Gillan had arrived in London.[43] As Staunton feared, Banks was
disappointed in the ipecacuanha specimen because it lacked flower
and fruit. Subsequently, Banks told Staunton that he was beginning
to think that the 'Rio naturalists do not like to Let us Know What
it is' – the fact that Phillip's plants had also lacked the critical parts,
Banks concluded, was not a coincidence.[44] By contrast, Banks told
Staunton that among the few plants he had collected in St Jago, and
which he was afraid were worthless, were some real gems, including
three new species of plants and a 'remarkable Grass'. Banks hadn't
quite finished going through the collection but he assured Staunton
that he would be sending the duplicate plants to Staunton's wife who
would add them to their son's burgeoning herbarium, as agreed.[45]

Banks took longer in replying to Gillan than he did to Staunton.
He was thrilled by Gillan's work. The cacti and insects were alive and
doing well and the accompanying coloured drawings of both, done
by William Alexander, the embassy's artist, were excellent.[46] Banks
moved the specimens to the hothouse he had built in his Spring Grove
home, about ten miles to the west of Soho Square. He had been using
it since 1779 as a kind of field station for botany and horticulture.[47]

It was there, as Banks told Gillan, that things started going badly
wrong. Banks had moved cacti from Kew into the hothouse in order
to provide the insects with a succession of plants on which they could
feed. Although the insects did move from the Rio de Janeiro cacti to
the Kew cacti, the result was very disappointing. They hardly grew at
all – no larger than a turnip seed, was how Banks described their size
– while the Rio de Janeiro cacti withered and died. The most likely
reason for the failure, Banks thought, was the fumes from the mech-

anism used to heat the hothouse: not that the cacti were the wrong kind. On the plus side, Banks had learned that the insects and their hosts could be transported by sea, even in cold climates. So, all was not lost, he reassured Gillan, while thanking him profusely.[48]

After a two-week stay in Rio de Janeiro, during which time Gower replenished all of the ship's stocks, especially water, but also meat, fresh fruits and vegetables, it was time to leave. In company with the *Hindostan*, HMS *Lion* left the harbour on 18 December 1792, sailing through the South Atlantic and straight across the Indian Ocean to the Dutch enclave of Batavia on the island of Java.

Staunton, Gillan, Haxton and Stronach had ample time to study Banks's document as the ships made their way towards the Sunda Strait separating the islands of Java and Sumatra. About ten weeks after leaving Rio de Janeiro, the ships were within sight of the coast of Java, and on 25 February 1793 they entered the strait. Both ships were hoping to meet other ships returning to Europe from China in order to send letters home and to hear news from Canton. Everyone who could took the opportunity of writing. As it happened, one of the East India Company ships on its way back to London from Canton encountered the *Hindostan* but not HMS *Lion*. So Macartney and Staunton, both on board HMS *Lion*, missed their chance to send letters.

The East India Company ship was carrying good news from Canton but it would not be until the ships reached Batavia, where a packet of despatches was waiting for Macartney, that those on board would know for sure exactly where they could go. Gower had instructions to make for the port closest to Peking, but they could only do this if the Chinese authorities allowed it. Normally British ships could only go as far as Canton.

Previously, in March 1792, those planning the embassy (Dundas, Macartney and Baring) had instructed a negotiating committee – the Commissioners of the Secret and Superintending Committee – to go to Canton to get permission for the embassy to bypass Canton and go directly to Peking by sea. They were to negotiate with the Viceroy (the Emperor's civil representative) and the Hoppo (the Imperial Customs Officer). They were to emphasise that the embassy was from the King of Britain who was trying to make amends for not having sent a deputation to congratulate the Emperor on his eightieth birthday. The King's high-ranking representative was bringing presents for the

Emperor which would be damaged by the long overland route from Canton to Peking.[49]

The commissioners had arrived in Canton on 20 September 1792, several days before the embassy's ships even left Portsmouth. They began work immediately and though, at times, it looked as though the Chinese would not give way, the commissioners were confident enough of success that on 25 November they sent a note to the effect that everything was 'proceeding in a fair train' to a post box at North Island, just off the coast of Sumatra in the Strait of Sunda, for Macartney to collect.[50] The commissioners' confidence was justified for on 3 December, an imperial edict was issued which gave the embassy permission to arrive by sea, and to make its way from a port on the Yellow Sea to the city of Tientsin from where the mandarins would accompany them to an audience with the Emperor. This good news, including a copy of the edict in Chinese, was put into a packet of enclosures, which was waiting for Macartney in Batavia.[51]

The embassy's ships arrived in Batavia on 6 March 1793. The Dutch welcomed the ambassadorial suite and Macartney picked up the despatches that had been awaiting him.[52] Macartney had the interpreters translate it into Latin on board HMS *Lion*, to be sent back to London on a passing ship.[53] It was in Batavia that many of the British men first encountered Chinese people en masse, as the Chinese had long been trading and living in this part of Asia.[54]

Ten days after arriving in Batavia, on 17 March, the ships left for Tientsin (Tianjin), Peking's nearest port, up the Hai river from the small town of Dagu on Bohai Gulf, part of the Yellow Sea. Their route took them through the Strait of Bangka, into the South China Sea, and along the coast of Cochin China (today's Vietnam). Sickness among the crew forced the ships to halt on 26 May in Tourane Bay (Da Nang, Vietnam), where plentiful supplies of fresh provisions, some of them presents from the local ruler, were brought on board.[55] Macartney had decided that though all the signs for a welcome reception in Peking appeared to be in place, he wanted to be certain that nothing had changed since the packet of enclosures was sent. There was not only a possibility that the attitude had changed, but that the Emperor himself might have died (it was his eighty-second birthday they were hoping to mark). So, Macartney needed to make for Macao where the East India Company officials could give him their up-to-date news.[56]

Nearly a month later, against unfavourable winds, the ships arrived in Macao on 20 June. Almost immediately, Paul Ke, one of the two Chinese interpreters, pleaded to be allowed to stay in Macao as he was afraid of how he would be received in Peking.[57] This was a setback but the news from Canton was that the embassy was still welcome, and the Emperor (still alive) had even made it easier for the ships to get to their destination by alerting pilots along the Chinese coast to assist the British ships who lacked accurate charts of the area.[58] The British left Macao on 23 June favoured by monsoon winds.[59]

Western ships hadn't been seen along the Chinese coast north of Canton for generations and their sighting caused quite a stir. In Chusan (Zhoushan), to the south of Shanghai, where the ships anchored on 3 July, local people clambered on board HMS *Lion* to get a close-up view of such a large and completely different-looking ship from theirs. 'The decks were so crowded with visitors', Staunton remarked, 'and others were waiting with such eagerness to come on board, that it became necessary to dismiss, after a short visit, the first comers, in order to gratify the curiosity of others.' Some of the most curious visitors even made it into Macartney's cabin (how we are not told). Once there they were both surprised and gratified to discover that Macartney had hung a portrait of the Qianlong Emperor. Staunton continued: 'Some of them . . . immediately recognised it, and prostrating themselves before it, kissed the ground several times with great devotion; on rising, they appeared to feel a sort of gratitude towards the foreigner who had the attention to place the portrait of their sovereign in his apartment.'[60]

It was in Chusan that the first Chinese pilot was brought on board to navigate the way to Tientsin. No sooner was he on HMS *Lion*, than he told Gower that, because they drew far more water than their Chinese counterparts, the British ships would not be able to make it upriver to Tientsin, but would need to anchor in the Bohai Gulf. Word was sent back to Chusan so that preparations could be made to transfer the presents and the ambassadorial suite from the British ships to the junks and smaller ships, which would take them upriver at Dagu. The British ships left Chusan on 8 July and a fortnight later sailed into Bohai Gulf. On 25 July, and at dawn, the voyage to China came to an end as the ships dropped anchor at the mouth of the Hai River.

The next few days saw little activity but the Chinese knew that the embassy's ships had arrived. On 28 July, as Macartney recorded it, 'several inferior mandarins came on board, and informed us that everything was preparing for our landing.'[61] This was Macartney's first encounter with members of the Chinese bureaucracy but his journal entry for this day makes no further mention of the mandarins.

It was very different, however, three days later, on 31 July when two high-ranking mandarins boarded HMS *Lion* at noon. Seven large junks came alongside the British ship. They were loaded with provisions. Macartney expressed delight at the overwhelming generosity, but also some trepidation. 'The profusion of these was so great', he wrote, 'and so much above our wants that we were obliged to decline accepting the larger part of them.'[62] The list of provisions was staggering: it included '20 bullocks, 120 sheep, 120 hogs, 100 fowls and 100 ducks', not to mention boxes upon boxes of fresh and dried fruit, vegetables, chests of tea, flour, rice, bread and baskets of earthenware.

Wang, a military mandarin, and Qiao, a civilian mandarin, also made a great impression on Macartney.[63] They seemed 'to be intelligent men, frank and easy in their address, and communicative in their discourse'.[64] Once the business talk was over, the two mandarins sat down to dinner with Macartney and his suite. 'Though at first a little embarrassed by our knives and forks', Macartney noted, '[they] soon got over the difficulty, and handled them with notable dexterity and execution.' The mandarins tasted the wines and spirits, listened to the music provided by the musicians that formed part of the embassy's suite, and generally seemed to enjoy themselves – 'they shook hands with us like Englishmen at their going away.'[65]

Over the next few days, the huge task of moving the personnel, their belongings and the presents from HMS *Lion* and the *Hindostan* onto the junks that would be carrying everyone and everything upriver to Tientsin progressed. It took more than a week and by the end of the preparations the ambassadorial flotilla consisted of almost forty junks. The plan was to follow the river upstream to Tientsin, a very large city of 700,000 people (Macartney reported) with thousands of ships, many of which were used to trade in the Yellow Sea and the East China Sea, even as far as Canton. There the flotilla would rest a few days and then continue to the city of Tungchow (Tongzhou) where the river ended and from where it was necessary to follow the major road to Peking. The mandarins suggested that the suite might prefer to stay in Yuanmingyuan, just to the north of Peking in the country-side, where the Emperor had extensive gardens, palaces and guest accommodations; but staying in Peking was also a possibility.[66] The actual audience with the Emperor would take place at his summer residence of Jehol (Chengde), to the northwest of Peking, beyond the Great Wall.

The plans for getting the embassy to their base were followed precisely. The suite, belongings and presents arrived in Tientsin on 11 August; everyone and everything else arrived in Tungchow on the 16th; they passed through Peking on the 22nd and had their first view of Yuanmingyuan, just to the northwest of the capital, on the 23rd.[67]

Macartney was at a loss for words to describe the scene. 'The various beauties of the spot, its lakes and rivers, together with its superb edifices, which I saw . . . , so strongly impressed my mind at this moment that I feel incapable of describing them,' Macartney confessed in his journal. He did manage to note, however, that the whole place was one imperial residence; with a park that was eighteen miles in circumference and 'laid out in all the taste, variety, and magnificence which distinguish the rural scenery of Chinese gardening'.[68]

Yuanmingyuan was not unknown to Europeans. On the contrary it had been well described before by Father Jean-Denis Attiret, a French Jesuit missionary and personal painter to the Qianlong Emperor, in a letter he wrote to Paris from Peking in 1743. It was subsequently published in the widely read *Lettres édifiantes* in 1749 and published separately in London in 1752.[69] Macartney, in common with any well-educated Englishman, would have read Attiret's description: Banks had a copy of it in his library.[70]

While he was overwhelmed by the sheer magnificence of the surroundings, Macartney was less impressed by the quality of the accommodation and persisted in requiring a residence in Peking. So, on 24 August, the decision was taken for Macartney and some of his suite to remove to Peking while those most closely associated with the mathematical and astronomical instruments, intended as presents for the Emperor, would remain in Yuanmingyuan.

The construction of Jehol (Chengde), located about 250 kilometres to the northeast of Peking, about halfway between there and Mulan, the Manchus' traditional hunting grounds, began in 1703 when the Qianlong Emperor's father decided to build a summer residence in the Yanshan mountains to escape Peking's searing heat.[71] By the time that the present Emperor had contributed his own additions, Jehol had become a major Qing site, covering an area of over 5.6 square kilometres. At its heart was a summer palace – Bishu shanzhuang (literally Mountain Resort for Escaping the Summer's Heat) – which lay surrounded by temples and other palaces. Also surrounding the summer palace was a large area whose topography was intentionally designed to replicate the empire's own topography: hills symbolised the empire's northwest region; the lake stood for southern China's cultivated region; and the flat plain, the Mongolian grassland. Jehol was a microcosm of China, an expression of the all-embracing power of the Emperor.[72]

The Macartney embassy set out for Jehol from Peking on 2 September in order to be at the Emperor's summer residence in time for the prolonged celebrations for his birthday, which would begin before the actual date, 17 September, and continue for several days afterwards. At the very front of the train leading out of Peking was a carriage, which Macartney had brought with him from London as a present for the Emperor, in which he and young George Thomas Staunton were seated. Four Mongolian horses, described by Macartney as small in size, drew the carriage. Behind the main carriage, George Leonard Staunton was carried in a palanquin; horses and carriages carried the rest of the ambassadorial train, about seventy people in all: about two hundred porters conveyed the presents, bedding and baggage.[73]

The terrain became increasingly steep and narrow as the embassy headed north. On 5 September, at about halfway to Jehol, they saw one of the most amazing feats of engineering ever built and hardly ever seen by a Westerner – the Great Wall. It was so huge, Macartney remarked, that if 'all the masonry of all the forts and fortified places in the whole world besides were to be calculated, it would fall considerably short of the Great Wall of China'.[74] Macartney and his entourage were so taken by the Wall and examined it in such detail, that the mandarins who were accompanying them began to wonder at their behaviour. Wang and Qiao, the first mandarins whom Macartney had met and who were now leading the delegation, admitted that though they had been on this route more than twenty times, they had never given the Wall a moment's thought.[75]

Three days later, on 8 September 1793, and after assembling his train in its finest livery, Macartney arrived in Jehol. Word had come down from Wang and Qiao and been translated by Jacob Li, the embassy's interpreter, that the Emperor had seen the procession approaching the summer residence and that he was pleased by the sight.[76]

*   *   *

Saturday, 14 September 1793. 'This morning at four o'clock a.m. we set out for the Court under the convoy of Wang and Chou (Qiao), and reached it in little more than an hour, the distance being about three miles from our hotel. I proceeded in great state with all my train of music, guards, etc. Sir George Staunton and I went in palanquins and the officers and gentlemen of the embassy on horseback. Over a rich embroidered velvet I wore the mantle of the Order of the Bath, with the collar, a diamond badge and a diamond star. Sir George Staunton was dressed in a rich embroidered velvet also, and, being a Doctor of Laws in the University of Oxford, wore the habit of his degree, which is of scarlet silk, full and flowing.'[77]

In the middle of the garden of the imperial palace was a round 'spacious and magnificent tent [twenty-five yards in diameter], supported by gilded, or painted and varnished pillars . . . within the tent was placed a throne . . . with windows in the sides of the tent, to throw light particularly upon that part of it.'[78] Behind it was a smaller, oblong tent with a sofa in it where the Emperor could rest should he wish. Small tents had been erected on either side of the Emperor's main tent for the visiting dignitaries, members of the imperial household and high-ranking court officials. One of these tents was reserved for the Macartney suite and there they waited for an hour until music and drums announced the arrival of the Qianlong Emperor. At this point Macartney and his men emerged from the tent and walked out on a long green carpet. From what looked like a 'high and perpendicular mountain, skirted with trees, as if from some sacred grove' the Emperor, seated in an open palanquin and carried by sixteen men, and accompanied by a vast retinue of mandarins of all ranks and courtiers, came into view. They all wore buttons on their hats, showing their rank, ranging from the highest, transparent red, through the eight ranks to the lowest, brass, a distinguishing mark introduced by Qianlong. 'Clad in plain dark silk, with a velvet bonnet, on the front [of which] was placed a large jewel',[79] the Emperor passed by the waiting dignitaries on his way to his own tent. All of his subjects, residents and visitors alike, greeted their ruler by kowtowing, both knees and foreheads touching the ground; the members of Macartney's suite 'paid him . . . compliments by kneeling on one knee.'[80]

Once the Emperor had ascended the throne and the members of his suite had arranged themselves in their correct positions within the tent, Macartney was ushered in. Holding 'the large and magnificent square box of gold, adorned with jewels, in which was inclosed his Majesty's letter to the Emperor, between both hands lifted above his head' he ascended a few steps towards the throne and, bending on one knee, presented the box to the Emperor.[81]

Presents and pleasantries between the two men were then exchanged. As soon as Macartney stepped down from the platform, Sir George Staunton, George Thomas Staunton and Li ascended and paid their respects in the same way as Macartney. For the young and impressionable George Thomas, it was a moment that would stay with him for the rest of his life. He described what happened next in his personal journal. 'The Emperor gave my Papa such a stone as he gave the Embassador and took one of the little yellow purses hanging by his side and gave it to me.'[82] The Emperor had been told that this was no ordinary child but a linguistic genius. Staunton continued: 'He wished I should speak some Chinese words to him which I did thanking him for the present.'[83] Qianlong also asked him to draw a sketch of the purse, which young Staunton did a couple of days later.[84]

Presents were then distributed to everyone in Macartney's suite and once the ceremonies were over, the guests, attendants, mandarins and the Emperor were treated to a sumptuous banquet with tea as the accompanying beverage, all of which was conducted in silence: the 'calm dignity and sober pomp of Asiatic grandeur'. The whole experience lasted five hours during which time the Emperor also received the compliments of a number of ambassadors from present-day Burma and western Mongolia.

On the day after this ceremony, Macartney and some of his suite were taken on an extensive tour of the imperial gardens at Jehol, the Wanshu yuan, the Garden of the Ten Thousand Trees, while the Emperor excused himself in private prayer at one of his shrines.[85] Macartney and young Staunton took to horseback for the first part of their visit to these spacious gardens, and, seeing how the park was landscaped with contrasting arboreal growths, quickly understood its name. However the real beauty of the place was unveiled when they boarded a 'large magnificent yacht, ready to receive us' on 'an extensive lake . . . the extremities of which seemed to lose themselves in distance and obscurity.'[86] As Macartney looked around him from the gently moving boat, he did find the words to describe what he saw unlike at Yuanmingyuan. It is one of the most evocative descriptions in his journal. 'The shores of the lake have all the varieties of shape which the fancy of a painter can delineate, and are so indented with bays or broken with projections, that almost every stroke of the oar brought a new and unexpected object to our view; nor are islands wanting, but they are situated where they should be, each in its proper place and having its proper character. One marked by a pagoda, or other building, one quite destitute of ornament, some smooth and level, some steep and uneven, and others frowning with wood, or smiling with culture.'[87] Whenever they saw an interesting structure, Macartney and Staunton got out to look closer and what they saw astounded them. 'They are all furnished in the richest manner, with pictures of the Emperor's huntings and progresses; with stupendous vases of jasper and agate; with the finest porcelain and japan, with every kind of European toys and sing-songs.'[88] Macartney was over-whelmed with excitement but then his eye caught a sight that dismayed and troubled him deeply. He was looking directly at 'spheres, orreries, clocks, and musical automotans of such exquisite workmanship, and in such profusion, that our presents must shrink from the comparison and hide their diminished heads'.[89]

By contrast, the actual birthday celebrations on 17 September were a bit of an anticlimax. There were entertainments galore but the Emperor himself was never seen, preferring instead to remain behind a screen the whole time. However Macartney and Staunton were taken to visit the side of the imperial gardens they hadn't seen before. The contrast between the two sides could not have been greater, Macartney remarked: '[it] exhibits all the sublime beauties of nature in as high a degree as the part which we saw before possesses the attractions of softness and amenity.' It was an extended forest, 'one of the finest . . . in the world, wild, woody, mountainous and rocky, abound-ing with stags and deer of different species . . . immense woods . . . oaks, pines and chestnuts grow upon perpendicular steeps . . . at proper distances you find palaces, banqueting houses and monasteries adapted to the situation and peculiar circumstances of the place.' Wherever Macartney turned he wrote: 'I saw everything before me as on an illuminated map, palaces, pagodas, towns, villages, farm houses, plains and valleys watered by innumerable streams, hills waving with woods and meadows covered with cattle of the most beautiful marks and colours. All seemed to be nearly at my feet and that a step would convey me within reach of them' – the voice of a committed Romantic.[90] From leaving their accommodation at 3 a.m. until their return, the visit lasted fourteen hours.

Though the festivities continued through the next day, the message

Macartney received from Wang, the military mandarin, was that it was time to go. The Emperor had decided to leave for Yuanmingyuan and the Macartney suite had to be on their way before he left. The date of 21 September was set for Macartney's departure from Jehol.

The presents from the Emperor to the embassy were wrapped ready for their shipment back to England and other preparations for the journey back to Peking made. This was all as expected but Macartney now sensed a cooling off, 'a decided disinclination towards the embassy'. And he was right.

Macartney's entourage reached Peking on 26 September, where he was told to prepare himself to pay his compliments to the Emperor at a spot outside Peking that the Emperor would pass through on his way to Yuanmingyuan. Macartney obeyed and when he returned to Peking, on 30 September, he found the conversation with the Emperor's representatives turned not to the issues Macartney wished to discuss with them, the very *raison d'être* of the embassy, but instead focused on the arrangements that needed to be made for him to leave Peking. Macartney understood the reason for this (why no substantive talks were taking place) when, on 3 October, he received a copy of the Qianlong Emperor's Edict to King George III.[91] The news could not have been worse. There would be no special trading arrangements with the British; they would not be allowed to gain a foothold in another port and would remain confined, as they already were, to Canton; and there could never be a permanent diplomatic mission in Peking, not from Britain or any other country; and, finally, though the Emperor thanked the King for his presents, they were nothing special. What the British had to offer the Chinese they found to be of no use whatsoever: 'We have never valued ingenious articles, nor do we have the slightest need of your country's manufactures.'[92]

It was all over. The mission had completely failed to impress the Emperor. The date of 7 October was agreed as the embassy's departure date. Perhaps it was as well that they were leaving. Macartney had learned that war between France and Britain was impending: this news had come, astonishingly, in a letter from Canton carried by Paul Ke, the Chinese interpreter from Naples who had left the embassy at Macao, and who had eventually travelled to Peking to visit his family.[93] Macartney's services and those of HMS *Lion* would be needed at home.

Macartney wanted to leave China finally from Canton, and the

Chinese authorities agreed to this; the only question was how he was going to get there. HMS *Lion*, Macartney learned, had had to leave its anchorage in the Bohai Gulf because it was not safe – the northeast monsoon would soon be blowing – and the ship's men were very sick. The Chinese authorities had agreed to let the ship shelter in one of the harbours of Chusan, where they could attend to the sick.[94] The *Hindostan* was also at anchor there.

So, Chusan was the next destination. Macartney's route took him back to Tientsin where the Grand Canal begins its southerly course to end at the city of Hangchow (Hangzhou), with easy access to Chusan, a distance of around 1000 miles. All the while Wang and Qiao, the two mandarins who had been with Macartney since he first arrived in China, were with him still. So was Sung-yun, a Mongol Prince and Grand Councillor, a high-ranking member of the inner imperial policy-making body who had considerable experience dealing diplomatically with the Russians. Over the course of the journey southward Sung-yun and Macartney found they had a lot in common resulting in a relationship that benefitted Macartney.[95]

All seemed to be going well, though slowly, until three weeks into the journey, on 29 October. Sung-yun told Macartney that he had just learned that HMS *Lion* had sailed from Chusan for Macao almost two weeks previously and that only the *Hindostan* remained there at anchor. When Macartney argued that he now had to take the inland route to Canton from Hangchow because there was no room on the East India Company ship for so many passengers, Sung-yun agreed to take the request higher up the chain of command. Near the mouth of the Yangtze River, on 8 November, Sung-yun received word that Macartney's request had been granted and that Changlin, the new Viceroy of Canton, would be replacing Sung-yun at Hangchow, where the Grand Canal ended, and would convey the embassy to its final destination.[96]

They reached Hangchow on 9 November and Changlin replaced Sung-yun as planned. As disembarking proceeded and as preparations for the onward journey took place, Macartney received a letter from Gower telling him what he already knew – that Gower had to depart for Macao because he was in desperate need of medical supplies, particularly bark (cinchona, an antimalarial) and opium; and that he would return to Chusan on a favourable southwesterly monsoon.[97] As Macartney would now be taking the interior route to Canton, he needed to send a letter to Gower in Macao, to reach him before he

left, so that Gower would know not to return to Chusan but to make
for Canton instead. Changlin and Sung-yun acted immediately and
assured Macartney that a special messenger would be sent at once
from Hangchow to intercept Gower in Macao.[98]

It was under a cloud of uncertainty, on 14 November, that the
embassy took to the road, Macartney being carried on a palanquin
until they reached a river where they boarded shallow-draught barges
for the next stage of the journey. So it continued, partly on rivers and
partly overland for more than a month until, finally, on 18 December,
on the outskirts of Canton, in one of the Hong's houses, the East
India Company Canton Commissioners greeted Macartney. They gave
him packets of letters and lots of news, especially and most welcome,
that his letter to Gower had arrived in time and that the ships were
waiting in Canton to take the embassy home.

After a period of almost three months, Macartney's fleet, now in
company with the East India Company ships that were going back to
England at the end of the trading season, left Macao on 14 March
1794. Almost six months later, on 4 September, they sailed into
Portsmouth, a little less than two years after their departure.

Macartney put a brave face on it, even remarking optimistically
that he interpreted some of the signals he had received from Changlin,
the new Viceroy of Canton, as holding out the possibility of future
diplomatic exchanges;[99] but nothing he could do or say changed the
fact that from a diplomatic and commercial point of view, the embassy
was a failure. What happened to shatter the British hopes of a change
in the Chinese *modus operandi* has been the subject of a substantial
body of scholarly studies, using both western and Chinese sources to
penetrate the inner workings of the Chinese imperial attitudes and
policies regarding the British.[100] It seems that the reasons why the
embassy did not meet its objectives are complicated – the Chinese
attitude to foreigners; the operations of tribute and gift-giving; the
subversive role of European court missionaries; and the not unimpor-
tant factor of the increasing territorial expansion of the British in India
and other foreign policy issues. All of these go well beyond the simplistic
view that the embassy failed because Macartney did not kowtow to
the Qianlong Emperor but accorded him the same respect he would
have offered to his own sovereign – kneeling before him on one knee
with head held high.[101]

Macartney came away with nothing, but Banks did a little better botanically, though it was still far short of his hopes. Throughout the months-long preparations for the embassy, Banks had stuck to the view that language was the only barrier to learning about Chinese science and technology. He believed that the recruitment of the two Chinese priests from Naples as interpreters would open the way to acquiring that knowledge. These hopeful plans were based on the premise that the embassy would be permanently based in Peking in the first instance for two or even three years.[102] Even after the embassy returned to Peking from the ceremonies at Jehol, members of the embassy, Macartney included, assumed that they would be spending the winter in the capital city. Banks knew perfectly well that the questions he posed in his 'Hints' document could not be answered quickly but he had enormous faith in the social bonds that science engendered: he expected that the embassy's plant collectors would be meeting Chinese gardeners who 'will be Able & no doubt willing to Answer all Questions, which are stated to them with clearness & precision'.[103] What no one could have foreseen was that the embassy's stay in China, when it was not on the move, and when Staunton, Gillan, Stronach and Haxton would have the opportunity of talking to Chinese gardeners, would be as short as it was – two weeks at most. Back in January 1792, when he first put his ideas to Macartney, Banks wrote optimistically that 'a few Learnd men admitted among their workmen might in a Few weeks acquire Knowledge for which the Whole Revenue of the immense Empire would not be thought a sufficient Equivalent.'[104]

Besides not allowing them much time, the embassy was not given much freedom of movement either. This was what Banks had feared the most, as he had remarked to Olof Swartz. Visits were supervised and travel accompanied. Both Macartney and young Staunton, when they were on the lake in the Wanshu yuan, the Emperor's palace grounds in Jehol, noticed that the lotus flower was everywhere but there was no one they could question on how its growth was managed – although Banks had singled it out as of the greatest interest.[105] Heshen, the Qianlong Emperor's Prime Minister, who accompanied Macartney and Staunton in the tour of Wanshu yuan, whisked the two visitors from one jaw-dropping vista to the next. Even on the odd occasion when one of the collectors was given some free time, other unforeseen impediments spoiled their chances. Haxton, who recorded his experiences in China in a journal, noted on just such a difficulty as the embassy was on the approach to Tientsin in its southerly course to Canton. 'Having on 14 Oct 1793 obtained Leave to go on shore for the purpose of Collecting I was surrounded by a crowd of People who when they saw that my Employment was Collecting Plants & Catching Insects began hooting & Running after me & as the soldier who protected me rather encouraged them they began to pelt me till I Returned to the boats.'[106] Haxton believed that the close watch kept on the embassy's members had as much to do with ensuring their protection as it had to do with control – 'the Lower orders of Chinese are so prone to maltreat Strangers', he commented.

No wonder that Banks's list of desiderata was largely unfulfilled. Still, in spite of the many problems, the embassy's 'Gentlemen', Macartney included, did not come away as empty-handed botanically as they did diplomatically.

In his published narrative, Staunton recorded that members of the embassy collected almost 400 species of plants while they were in China.[107] He did not state who collected which plants and the printed lists often give no more than the genus name. On 12 November 1793, the first occasion that he wrote to Banks from China, Staunton was very apologetic, pointing out that their stay in China had been very short and that 'my voyage in point of Curiosity has been almost totally frustrated.' Reading between the lines, it would seem that the embassy's plant collecting was haphazard, akin to grabbing whatever was available as quickly as possible – 'picked up along the road', in Staunton's own words.[108]

It was almost a year and a half between the time that the plants were brought to Soho Square in September 1794 and the time that Banks and Jonas Dryander, his librarian, finished cataloguing them in late January 1796. Though Banks explained to Staunton that this kind of work was what excited him most, and that some new species would be added to his herbarium, he also admitted that there were many problems with the collection: the quality of the specimens often varied; no care had been taken to distinguish who actually collected each specimen (both Macartney and Staunton's names were jumbled); most seriously of all, Hugh Gillan, the ship's physician, though he had agreed to do it, did not bother to make notes describing the live plant of which a particular specimen was an example. Without such notes, an exact botanical identification was impossible. One of the gardeners, Banks pointed out, had also been careless in not providing examples of the flower and the fruit for many new species.[109]

In the end, Banks could not really vent his frustration on the collectors. They were not trained naturalists and they were not responsible for the absence of such a person on the embassy. Staunton gently reminded him of this when the collections had been catalogued: 'If ever there should be another expedition of the same sort', Staunton wrote, 'it is to be hoped that more careful and able Persons may be found for the department of natural History, as well as that better opportunities and more leisure will be allowed in the Country for finding out and preserving every object worthy of curiosity.'[110]

At some point while they were in China, Macartney, Staunton, or one of the gardeners, had managed to collect seeds and specimens of living plants. Banks had provided the collectors with instructions on how to package seeds and keep plants alive on the return voyage to England.[111] The plants, in six pots, had been put aboard the *Bombay Castle*, one of the East India Company ships travelling in convoy with the embassy's ships and had arrived in London in late September 1794.[112] When in 1813, the second edition of Aiton's *Hortus Kewensis* was published, five plants introduced by Staunton were already growing at Kew, including the *Rosa bracteata*, sometimes referred to as the 'Macartney Rose'.[113]

Banks was pleased with the success of the living plants at Kew. It would be fitting, Banks felt, if they had been collected at Jehol, though without the accompanying notes, no one could be sure. As he put it to Staunton: 'If [this] is the case it would be pleasant . . . to learn that our King at Kew & the Emperor of China at Jehol Solace themselves under the Shade of many of the Same Trees & admire the Elegance of many of the same flowers in their respective gardens.'[114]

14

1794: To Calcutta and Back

In early September 1792, the East India Company officials in Bengal informed the Court of Directors in London that Thomas Hughes, who had been working as a gardener in the Calcutta Botanic Garden since October 1789, had died and that he should be replaced as soon as possible.[1] The Court of Directors turned to Banks for a recommendation. Somehow, Robert Murray, a gardener who had worked for William North's nursery in Lambeth, heard of the vacancy and, on at least two occasions, wrote to Banks asking for his support. Murray also provided a reference attesting to him being 'a very industrious scientific man'. Despite or perhaps because of Murray's insistence, Banks did not recommend him but suggested Christopher Smith instead.[2]

Smith was the perfect candidate for the Calcutta position. As Banks told William Ramsay, Secretary to the Court of Directors of the East India Company, Smith had worked for William Pitcairn, a renowned physician at St Bartholomew's Hospital, who had an extensive garden in Islington where he cultivated 'scarce and foreign plants'.[3] These included, Banks emphasised, plants from India. That was qualification enough but Smith had accomplished something even more impressive. He had been the assistant to James Wiles, the main gardener on HMS *Providence*. Under the command of William Bligh, the ship, which had left England in August 1791, had achieved a complex and highly successful multilateral exchange of plants between London, the Cape, Tasmania, Tahiti, Timor, St Helena, St Vincent and Jamaica. It was now returning home from Port Royal, Jamaica.

As Banks explained it to Ramsay, Smith alone had cared for the plants on board the ship, all of which were destined for Kew, as Wiles had decided to remain behind in Jamaica, at the botanic garden in Bath. Smith's experience of transferring live plants from one part of the world to the other was now substantial.

When the *Providence* arrived in Deptford on 7 August 1793, the pots, tubs and boxes of plants were transferred onto lighter craft for their trip upstream to Kew. According to Smith, the collection consisted of 703 containers of plants, which he could name, together with a further 354, which he couldn't.[4] Altogether there were almost two thousand plants, the majority of which had come from Jamaica and St Vincent, making it the single largest shipment of plants from the West Indies to London to date.[5] Though dwarfed in sheer numbers by the West Indian varieties, it was the arrival of live plants from Tahiti and other stops on the return journey – Possession Island, Timor and St Helena – that was most remarkable. There were four breadfruit plants and twenty-eight other plants representing sixteen specimens, which had travelled more than halfway across the world and which had survived aboard ship for more than a year.[6] As if the figures did not speak for themselves, Banks added that Bligh would vouch for the success of Smith's experience in handling tropical plants, and for his character.[7]

For now, Banks arranged for Smith to go to work at Kew to care for the plants he had brought back.[8] It was several months later that Banks arranged a meeting between Smith and Ramsay at India House. This took place sometime after 10 December, probably within a day or two of Smith's election, on 17 December, as a Fellow of the Linnean Society, an honour that recommended him to Banks and undoubtedly impressed Ramsay.[9] Soon Smith's appointment was officially ratified.[10] Everyone was pleased.

So an excellent replacement for the gardener, Thomas Hughes, had been found, but meanwhile, the first superintendent of the Calcutta Botanic Garden, Colonel Robert Kyd, died on 27 May 1793 when he was only in his late forties. Though the Calcutta Botanic Garden was part of the East India Company, it was Kyd who had first conceived of it and who, as its first superintendent, had nurtured it through seven years into success (Banks had fully supported Kyd's work during this period of time). His death was an immense blow.[11] William Roxburgh, an East India Company botanist who had been in charge of its botanic garden in Samulcottah, to the north of Madras, accepted the appointment to replace Kyd. In November, Roxburgh became the new superintendent of the Calcutta Botanic Garden. A month later he wrote to Banks, with whom he had been corresponding for at least fifteen years, to give him this news, as well as sending him a box of plants by one of the East India Company ships returning to London.[12]

Although Banks saw Kyd's death as a great loss, in fact Kyd hadn't provided Banks with many plants from India. For the first few years when he was the superintendent in Calcutta, Kyd had concentrated on establishing the garden under his care rather than satisfying the needs of Kew. During this period, Banks relied almost exclusively on the good offices of John Murray, a prominent civil servant in the employ of the East India Company in Calcutta, who sent Banks boxes of seeds annually as presents for the King. Though generous and interested in plants, Murray knew little about them. The best he could manage was to give the local names of the trees and flowers whose seeds he had had local people collect for him.[13] The gardeners at Kew had no alternative but to sow them and wait to see what might emerge.

When, in 1790, Kyd finally did send Banks live plants, the shipment was a disaster. Only one out of every five plants arrived in a satisfactory state and there was some doubt as to whether even they would survive.[14]

With Smith's appointment, and with his knowledge of Indian plants and how to keep plants alive at sea over long distances, Banks saw a solution at last for the relative absence of plants from Bengal and its environs at Kew. The problem was that Smith was going to Calcutta for some time, possibly permanently. While he might be able to choose the perfect plants in Calcutta, Banks feared that they would fare no better on the return trip to England than those that Kyd had sent.

What Banks needed was someone else, in addition to Smith, who would travel out to Calcutta with him and return alone with the chosen collection; preferably someone who was also familiar with Kew.

In May 1794, Banks contacted William Aiton at Kew and asked for his help in finding such a person. On 29 May, Aiton wrote back that he had just the man for Banks. Peter Good was his name and he had been at Kew for two years. Aiton thought very highly of him and Good was keen to go and work abroad.[15] He was sending Good to Banks with a letter of introduction.

Good must have made the right impression for only five days later, Banks wrote to William Devaynes, Chairman of the East India

Company, outlining his scheme, one which would not only bring Indian plants to Kew but would also directly benefit the Company and specifically its botanic garden in Calcutta.

This was Banks's plan as he explained it to Devaynes. Peter Good would go out with Smith and return with the plants for Kew. To give him the practical experience he would need to take best care of the plants on the return journey, Banks suggested that the ship should take out to Calcutta specimens of fruit trees and other useful plants and that these should be entrusted to Smith's care. Banks knew what was growing in the Calcutta Botanic Garden and he had also learned from Kyd of his wish to grow certain varieties of European fruit trees that he was sure would succeed.[16] Kyd had wanted this for some time but he knew the difficulties only too well.[17] In the first year that he was in charge in Calcutta, for example, two gifts of an assortment of fruit trees arrived on an East India Company ship from London, but hardly any of the plants had survived the voyage.[18]

The fruit trees and other plants, Banks advised, should be housed in a plant cabin, which could also be used for the plants on the return voyage. Banks had already had several good experiences with these glazed structures, but this would be the first time one would be built on an East India Company ship.[19]

Devaynes was convinced. The very same day, he informed Banks that the Court of Directors had agreed to accommodate Good on the ship, that the Company would be footing the bill and that Good should begin to prepare for the voyage to India.[20]

As Good had not previously worked as a gardener on a long-distance voyage, Banks was careful to inform him of certain basic shipboard rules: specifically, that he should always obey orders from the ship's officers, and though he would not normally be called to assist in the ship's daily business, there might be occasional emergencies when his help would be required. Then, 'you must do it readily cheerfully and willingly', Banks warned.[21]

Good's main responsibility, however, was to Smith and to the plants. On the way to Calcutta, he was to follow all of Smith's directions and to learn from him about how to care for plants at sea, and particularly how a plant cabin worked. While the *Providence* hadn't had a specially built plant cabin on its deck, the arrangement of the ship's garden and the alterations made to the deck by Bligh acted very much like such a cabin. So Smith's experience was extremely valuable; Good should learn everything he needed from him because on the homeward journey, the plants for the King would be kept in the plant cabin, and it would be entirely Good's responsibility.

To ensure the greatest possible success, Banks insisted that Good should learn as much as he could while in Calcutta from the superintendent of the botanic garden about the kinds of plants and seeds which were not yet growing in Kew; and how to plant and pack these specimens for their homeward journey.

As soon as the ship was ready to return to London, Good was to load the plants and place them in the plant cabin personally, and no one else was to be allowed access to it. The cabin was strictly for the protection of the King's plants. On no account should Good allow anyone into the cabin to help themselves to a bulb, some seeds or even a cutting. If somehow this did happen, Good was to report the event in writing to William Aiton.

Finally, Banks told Good that his care of the plants and the plant cabin was an absolutely full-time job and that he should never 'sleep out of the ship'.[22]

In the past, Kyd had told Banks how he wanted to have peaches, nectarines, apricots, cherries, plums and grapes growing in the botanic garden – he knew English fruit trees such as apples, pears and currants wouldn't survive.[23] Banks also had a copy of the plant inventory of the Calcutta Botanic Garden of early 1790 and, so, had a pretty good idea of what specifically it didn't have.[24] Banks obtained as much as he could from Kew but there was a shortfall, which Banks made up by purchasing from London nurserymen. In addition, from Lee and Kennedy, the Vineyard Nursery in Hammersmith, Banks bought a number of West Indian plants. Garden tools, which were always in short supply in Calcutta, he also added to the list. The East India Company immediately reimbursed Banks for these expenses.[25]

Banks was leaving nothing to chance. As soon as he had given Good his instructions, Banks wrote to Devaynes to emphasise that the ship's captain would also have to play a role in protecting the plants and that it might be a good idea to add a few words to this effect to his usual sailing instructions.

Banks, who, at this point did not know the name of the captain, simply asked Devaynes to pass onto the commander several instructions about the plant cabin, all of which would help ensure a successful plant transfer. They seem obvious, but Banks must have felt that they needed to be spelled out. The captain was to help the gardeners bring the plants on board and to disembark them; he was to make sure that the shipboard animals, invited, or not, should not be allowed near the cabin; and that it should also be off limits to the crew.

Several further points were not so obvious and had to do specifically with the management of the plant cabin. When the ship rounded the Cape into cold weather, the captain should provide old canvases to cover the cabin, and possibly offer 'artificial heat, of which 3 or 4 Tallow Candles will in so small a space – furnish a great deal', should it get very cold.' As for watering the plants, the captain should make water available so that Smith would be able to ensure proper hydration. Finally, Banks insisted that the cabin on the return voyage was for the exclusive use of the King's plants. If, Banks added, the captain should bring the King's plants home in a healthy state, he would not only do himself great credit, but he could be assured that 'Plants from thence [Kew] for his Friends will never be refused him.'[26]

While Banks did not yet know the name of the captain, he did know the name of the ship. At over 900 tons, the *Royal Admiral* was one of the bigger ships in the East India Company's fleet and had recently completed its sixth voyage to and from India and China. Banks knew about the ship because a day or so after 20 August 1793, he received a letter from Hugh Parkins, an East India Company officer, who happened to have been a passenger on the *Royal Admiral*, as it sailed to London from Canton.[27] Parkins told Banks that his ship and HMS *Lion*, carrying the Macartney embassy to China, had met on the coast of Java in April. Staunton had taken the opportunity of writing to Banks and entrusting a letter, as well as a set of books and plants he had collected in Java to Parkins. When the *Royal Admiral* arrived, Parkins delivered the letter and books to Banks but, as he explained, the severe weather and constant gales, as the ship headed into the South Atlantic from the Cape, had so traumatised the plants that he had thought it best to leave them at St Helena to recover. Before he left the island three of Staunton's plants had already died.[28] A point that underlined Banks's advice to Devaynes about the need for protection from the elements when transporting plants on the high seas.

By the end of July 1794, almost everything was in place to bring Banks's grand scheme for the transfer of plants to and from Calcutta to fruition. The gardeners had been chosen and given their orders; the plants assembled for the outward voyage; and the ship and its plant cabin readied for departure.

Banks had still not met the captain though he had been to India House in July on several occasions hoping to, but the introduction promised by his friends in the Court of Directors hadn't been made. So with time running out, Banks wrote directly to Captain Essex Henry Bond 'personally recommending to your protection the outward bound Cargo of Fruit Trees & usefull Vegetables . . . & bringing home in the Cabbin which carries them out, a homeward bound Cargo of curious Plants for the use of His Majesties Botanic Garden at Kew'.[29]

Bond was in Portsmouth, on board the *Royal Admiral,* when he received Banks's letter. The youngest child of a wealthy City merchant, Bond had entered the East India Company's service in 1777, at aged fourteen, as the servant of Captain Joseph Huddart, commander of the *Royal Admiral* on its first four voyages, and had worked his way up the ranks until he replaced Huddart as the ship's commander in 1790. This would now be Bond's third voyage to the East Indies as the ship's captain.[30]

It is fortunate for us that Banks did not meet Bond in person because in the letter he revealed something that he hardly ever talked about in the rest of his voluminous correspondence: he explained how he saw himself. Banks described himself as a 'Projector', someone full of projects and that this one, finding plants for the the King's Garden at Kew, was his 'Favorite Project'.

Banks did not add that he had been engaging in this project whenever he got the opportunity since the voyage of HMS *Bounty* in 1787, but he did explain to Bond how he saw his role. 'In truth', Banks wrote, 'tho I call it my Project . . . the Plants now on Board are wholly the property of the Company & those which are brought home will be carried straight to Kew Garden – without passing in any shape through my hands.'[31] In short, Banks's efforts were entirely disinterested, he didn't stand to gain personally in any way, except in the King's eyes.

Banks also reminded Bond that the success of this plant transfer depended greatly on him. If he succeeded, Banks promised him, 'you will lay me under obligations which Sir I hope I shall not readily forget.' And, Banks continued, 'You will have the satisfaction of adding to the private pleasure of an illustrious Family, whom all Britons adore & you will have a claim for all kinds of Civilities which the Superintendant of the Royal Botanic Gardens can at any time hereafter show to those who have essentially promoted the superiority they now hold over all similar collections now in Europe.'[32]

The very next day, Bond, who was in the last stages of preparing the voyage, replied to Banks. The plants could not be in better hands on both legs of the voyage, Bond assured him. The weather, which he felt would be kind to the voyage at this time of the year, was sure to play its part in ensuring success; but in addition, Bond told Banks that he had already had a successful experience transporting plants long distances. Together with the convicts whom he had conveyed to Port Jackson, Bond had taken on board several dozen fruit trees at the Cape and these, he had learned, were doing exceedingly well in the colony.'[33] Banks could not have wished for more.

The *Royal Admiral* sailed out of Portsmouth harbour on 14 August 1794. On 4 September, the ship arrived in Madeira and the next day Christopher Smith wrote his first letter to Banks. It must have put a smile on Banks's face. 'It is with pleasure I tell you', Smith wrote, 'that my little Vegetable family continues in Luxuriant health partly owing to their own good Stamina and somewhat to my Paternal attention.' Despite the fact that it had hardly rained during almost three weeks, the plants had required little watering, Smith added. This he put down to how deeply the trees had been planted in their boxes. All was well, including Peter Good, who was also 'very attentive'.[34]

Smith wrote again a week later but this letter does not appear to have survived. Though the ship stopped at the Cape, Smith did not write from there but waited instead until he reached Calcutta on 20 February 1795, just a little over six months after leaving Portsmouth.

The plants, Smith wrote to Banks, were in excellent condition. The shipment of three hundred and nine separate plants represented, as Smith put it, 'the largest Collection of plants that ever has been introduc'd from Europe into India'.[35] Captain Bond was as good as his word having 'paid great attention to the plants . . . and always ready to supply . . . everything that was necessary for their preservation'.[36] Many of the plants were entirely new to Bengal – Banks had chosen wisely.[37]

Smith had apparently received instructions from Banks that when he was at Madeira and the Cape he should augment the shipboard collection of fruit trees by purchasing local varieties which, in his estimation, would succeed in Calcutta. For this Captain Bond advanced him with the necessary funds – a little over £30.[38]

In Madeira Smith bought several varieties of grapes, including those used to make Madeira wine. At the Cape, where Smith benefitted from the generous advice of both Francis Masson and Colonel Robert

Jacob Gordon, who knew the Cape's botany better than any other European, he was able to buy more vines, some vegetable plants and many flowers.[39]

In Calcutta Smith lost no time in moving the plants from the ship to the garden. And as soon as that was done, he began collecting the plants for the homeward voyage on 4 March. Six weeks later the collection was on board the *Royal Admiral*.

The ship only stayed in Calcutta as long as it took to unload one set of plants and replace them with the Indian specimens. On 26 April, Christopher Smith drew up his final list of 375 plants, some singles, others duplicates, all of which, Smith noted 'I have potted with my own hands.'[40] He was clearly pleased with his collection. He had annotated an earlier version of his list; one plant, which had been named after Sir William Jones, the jurist, orientalist and founder of the Asiatic Society of Bengal, Smith described: "There is no plant exceeds this for beauty when in flower'; another Indian plant, Smith simply called 'glorious'.[41]

The *Royal Admiral* now began its homeward voyage. The ship sailed first to Madras, where more plants were put on board, most of them coming from Dr Andrew Berry's garden.[42] Sir Paul Jodrell, physician to the Nawab of Arcot, donated a large cinnamon tree as a present to the King.[43]

Trincomalee, on Sri Lanka's eastern coast, was the ship's next stop and there it remained for the second half of August. Almost twenty plants were collected though no further details are available.[44]

After a short layover in St Helena, the *Royal Admiral* sailed straight for home. But, instead of heading into the Channel, Captain Bond, acting on intelligence, false as it turned out, that a French fleet had taken position there, sailed into Bristol in late January 1796. Customs officers in the city would not allow the plants to be taken from the ship until they had orders to do so from London.[45]

Banks tried to help move things along because the plants were suffering from a lack of warmth even though the weather at this point was unseasonably mild.[46] The East India Company Court of Directors ordered their agent in Bristol to provide a wagon to bring the plants to London and Banks arranged for a Kew gardener to travel to Bristol to help Good unload them from the ship. Eventually the customs officers let the plants go.

When the plants finally arrived at Kew, Banks reported that although they had arrived in Bristol 'in the finest order possible' some had been lost by the delay in Bristol and more had died on the road trip to London than had died at sea. Nevertheless, he was able to state with pride that 'next to the Collection which Capt Bligh brought home it is the largest addition Kew Gardens have ever received at one time'.[47]

From start to finish, Banks's 'Favorite Project' took less than two years to complete. Everyone had played their part well and the plant cabin had worked. How this combination could be repeated on other projects was something that would occupy Banks's thoughts in the future.

Christopher Smith remained in the East Indies, as expected. In 1805, and no doubt recommended by Banks, he became the superintendent of the East India Company's first botanic garden in Penang but unfortunately didn't live long enough to do anything other than get it started – he died in 1807.[48] Meanwhile Peter Good's standing with Banks could not have been better. Soon, he would be on yet another voyage of exploration taking him even further afield.

PART IV

Fifth Quarter of the World

Preface

B anks had been to the eastern coast of what Cook named New South Wales in 1770 and stopped briefly at Botany Bay in the south and for longer at Endeavour River in the north. Eighteen years later, Arthur Phillip established the British penal colony at Port Jackson, a little to the north of where Banks had first stepped ashore.

In 1800, although Banks had been kept informed about progress in the settlement, apart from gifts of seeds sent to him from contacts and well-wishers, his knowledge of the area's natural history remained limited. That was about to change dramatically when George Caley, a self-trained botanist, disembarked from the ship that had brought him to Port Jackson. He was to be Banks's first dedicated plant collector in New South Wales. Caley was no ordinary collector: he was knowledgeable, independent, adventurous and unpredictable, but he always took advice from the local Aboriginal people, one of whom he came to rely on and brought with him back to London.

Caley sent back seeds and dried plants from a large geographical area which included entirely new specimens from Van Diemen's Land and Norfolk Island but his movements were limited by how far away from the colony he could get by land and by whether he could join a local surveying voyage to take him further.

From Banks's point of view, Caley's expert botanising was but a tiny sampling of what might lie well beyond the confines of the convict colony. There was no doubt that the land mass which was called New Holland was vast but other than the fact it had coasts on all sides, little was known about it. In particular, it was not clear whether the land mass was a whole or whether it was split into two by a river or inland sea running somewhere from north to south.

Lieutenant Matthew Flinders, whom Banks had known since he served as a midshipman under William Bligh on the breadfruit voyage of HMS *Providence*, also wondered about New Holland. Only a circumnavigation, he proposed, could settle the issue and only Banks, to whom he turned, could get the right people in the Admiralty to put his plan into action.

Banks needed no convincing. He immediately saw that this was a unique opportunity for an extensive scientific voyage, far beyond what Flinders himself envisaged. Memories of the *Endeavour* and the *Resolution* must have flooded back, as Banks outfitted the ship with an entourage to collect, record and draw the natural history and astronomy of a land largely unknown to Europeans. In the interests of botany, he had appointed a naturalist, a Kew gardener, who had extensive knowledge of caring for plants at sea, and a highly accomplished botanical artist. Banks wanted not just herbarium samples but living plants that could be conveyed and stored along the route in a plant cabin brought from England before being transported to Kew. No British exploration expedition before this had paid such attention to the pursuit of botanical knowledge.

15

1795: The Farrier's Son Finds Banks

Arthur Phillip, the Governor of the convict colony in Port Jackson, New South Wales often lamented the absence of a botanist. On several occasions, including in his first letter to Banks, Phillip remarked that 'a good Botanist in the Country would find full employment'.[1] If not a botanist, a good gardener, Phillip added, would be a great asset.

Phillip did all he could to get seeds to Banks by the first available returning convict transport, and, considering how busy he was with other things, he accomplished a great deal. However, he was not the only one in the colony looking for seeds to send back to London; Kew was not the only garden in Britain. Phillip alerted Banks to the 'private' trade that was going on in New South Wales. The *Alexander*, one of the ships of the First Fleet, Phillip wrote, was on its way back to England but he had had little to put on it. Not that seeds were lacking: rather, as Phillip put it, 'the Man I employ'd for three months to collect Seeds, sold them to the people belonging to the Transport & which I did not know 'till the Ships were ready to Sail.'[2] No doubt the seeds went to London, where they were sold in the commercial horticultural market.[3]

The government showed little interest in appointing a botanist to the colony though they were not against adding a collecting role to that of Superintendent of Convicts, as they had in the case of the gardeners Smith and Austin. However Banks clearly wanted his own botanist, someone whose commission was only to collect for Kew, and it seems that he had difficulty finding the right man for the job.

Both Banks and Phillip knew of a man who would fit the bill

perfectly. Francis Masson was based in Cape Town, and, as Kew's first plant collector, his botanical knowledge was unrivalled. Phillip took advice from him when the First Fleet anchored in False Bay, as did William Bligh when he was there in June 1788 with the *Bounty*, en route for Tahiti.[4] In March 1789, Banks was about to ask the King to order Masson to New South Wales when he learned from William Aiton at Kew that Masson had 'an aversion to the place'.[5] Banks backed off. Masson would soon be ready to leave the Cape but New South Wales was not where he wanted to go.[6]

Banks told Masson that he was confident that he would find someone else, but he was wrong. Instead he had to be content with men who were not botanists and who could only collect for him occasionally. David Burton was a case in point. With a background in gardening and surveying to qualify him for the post, Burton secured Banks's recommendation for him to go to New South Wales as Superintendent of Convicts for a period of three years at an annual salary of forty pounds.[7] At some point, Burton believed he was also employed to collect for Kew, but Banks insisted that this was not so: rather, Banks offered him a supplemental income of £20 per annum to collect seeds and plants on his behalf and to take care of the collections Phillip had been making. The sole condition of his employment was that he did not collect for anyone else. Burton accepted the terms and left England on 15 March 1791, on HMS *Gorgon*, which was carrying thirty convicts and vital provisions for the colony.

Burton arrived on 21 September 1791 and was soon busy surveying and collecting seeds and plants. A large collection was ready by the end of the year to be put on board the *Gorgon*, which was scheduled to return to London. Phillip thought highly of Burton but his work was soon cut short. While out shooting ducks in early April 1792, Burton's rifle went off accidentally injuring his arm. He died a week later from the infection.[8]

For Banks there was yet more bad news. On 15 October 1792, Phillip wrote him his last letter from New South Wales. He was tired and his health was failing. It was time to go home. 'I find it more & more necessary,' he remarked to Banks.[9] Almost two months later, Phillip left on the *Atlantic*, never to return to Australia.[10] On 19 May 1793, the ship arrived in England. Phillip had brought with him fifty tubs of plants, seeds, four live kangaroos and two Aborigine men, Bennelong and Yemmerawanne, who had expressed an interest in visiting England.[11] Two of the kangaroos were given to Banks who then presented them to the King – they were alive and doing well at Kew soon afterwards.[12] The two Aborigines stayed in London at various addresses, eventually taking up residence in Eltham. There, after an illness that lasted for several months, Yemmerawanne died on 18 May 1794, aged only nineteen. He was buried at Eltham Parish Church. Bennelong, left alone and homesick, wanted to go back to Port Jackson.[13] HMS *Reliance,* the ship that he was to take home, was already being fitted out for the return voyage. It was carrying live plants chosen by, and housed in a glazed cabin designed by, Banks.[14] On board, as well as Bennelong, was George Bass, the surgeon, and Matthew Flinders, the ship's senior master's mate. The latter two men were intending to explore the coast of Australia to the south of Port Jackson.

John Hunter, who had been appointed as Phillip's successor as Governor, was also on board the ship; he had already been to New South Wales as second-in-command to Phillip on the First Fleet. When the *Reliance* arrived in Sydney on 7 September 1795, Hunter took up his post and Bennelong went to stay with him in his house. Bass and Flinders set out on their first voyage of exploration in a small boat, the *Tom Thumb*, which Bass had brought from England. Like Phillip, Hunter frequently communicated with Banks. His letters were long and rich in details of the colony's progress but unlike those of Phillip, there was only the occasional mention of seeds or plants.

Banks had tried to entice Masson to reconsider his decision by telling him that the bounty and the beauty of New South Wales plants far surpassed those of the Cape.[15] As only Britain had access to this botanical paradise, Masson, by filling the King's Garden with these delights, would bask in its reflected glory, but he didn't take the bait and remained, for the time being, at the Cape.

Six years passed and then, unexpectedly, a botanist of unusual background and personality walked straight into Banks's life. There was nothing, initially at least, to suggest that he would become Banks's first botanist in New South Wales.

George Caley was born in Yorkshire on 10 June 1770 but spent his early years in Manchester. He received a modicum of classical education at the Free Grammar School in Manchester, but for most of his youth he worked as a stable boy for his father, who was a farrier. It was not long, however, before George's interest in horses led him into the field in search of herbs from which to prepare recipes for alleviating equine diseases. From this, as Caley recalled later, it was a short step to becoming interested in the plants themselves. The botanical societies, whose meetings were held in local pubs, were a strong feature of the artisan life in Manchester and the nearby Lancashire mill towns, and Caley sought out their company.[16] The societies held regular meetings at which the plants brought by the members were discussed; botanical books were purchased to stock a modest library; and field trips were organised in the surrounding countryside. These botanical discussions, and encouragement from William Withering, the noted botanist, chemist and physician and member of the Lunar Society, inspired Caley to become a naturalist.[17]

On 7 March 1795, not yet twenty-five years old and without any introduction, Caley wrote to Banks directly. The letter impressed Banks who wrote back immediately.[18] Caley's letter appealed to Banks; by writing of Banks's botanical explorations and 'the noble examples you have set forth to encourage the study of nature'; by referring to learned botanical treatises, including Erasmus Darwin's translation of Linnaeus's *Systema Vegetabilium*, which was dedicated to Banks;[19] by emphasising how obsessed he had been by studying botany over a period of eight years – 'and now my mind being so fired with it'; and by sending him examples of plants growing which he thought were new to science, particularly a species of moss he came across while botanising near Altrincham, Cheshire and which, as Caley emphasised, was not mentioned in Withering's magnum opus, *A Botanical Arrangement of All the Vegetables Naturally Growing in Great Britain*. Though his upbringing, Caley admitted, was not of the best kind, he was appealing directly to Banks for help: 'hoping that you would give me some encouragement to assist some Botanist or employment of that like, which you would think proper, which would raise my spirits to the highest pitch.'[20]

In his reply, Banks pointed out that if Caley were intent on becoming a botanist he would have to be content with never having much money: 'I do not know', Banks wrote, 'there is any trade by which less money has been got than by that of Botany.' If he could master the subject by learning the names of the plants in the botanic gardens by working as a gardener, and if he was willing to work hard and study not only 'the culture of plants, but also Exotic Botany', then Banks was willing to recommend him and in time, as he put it, 'give you such assistance as will make your station less disagreeable than otherwise it would be'. [21]

So, in the early part of April 1795, Caley accepted Banks's advice and went to London. His first job was in the Chelsea Physic Garden but soon afterwards he moved to William Curtis's Brompton Botanic Garden, where he stayed for more than two years. At some point, probably near the end of 1797, Caley went to work at Kew but it did not suit him at all. He found, specifically, that the schedule did not allow him the leisure time he needed in order to satisfy his desire for botanising as opposed to gardening. Also the salary was less than he had been receiving at Brompton. As he told Banks in early January 1798, when he explained his disappointment with Kew, 'what plants I have become acquainted with since I have been in this country [by which he meant London and the southeast], have been chiefly obtain'd at leisure hours' – a time devoted to 'cultivating my mind'.[22] Caley was unused to working set hours – from 6 a.m. to 7 p.m. – and much preferred the task-oriented schedule he had followed in Manchester.

However, he wasn't just writing about the problem with the work schedule; Caley had heard that Banks was planning to send a botanist to New South Wales, and he wanted the job.

Caley was right in that Banks had been thinking about sending a botanist to New South Wales for some time. In a letter to Governor John Hunter on 30 March 1797, Banks explained that he was waiting for the day when he could 'solicit the King to establish a Botanist with you'. What was holding him back was the uncertainty caused by the war with France. He was hoping that 'better times' would return. For the present, though, Banks could only remind Hunter that 'Kew Garden is the first in Europe' and that he should look out for plants for the King and Queen whenever he or others ventured into new parts of the country.[23]

In early December, Banks responded to Caley by inviting him to discuss the matter in person. Caley recalled later that Banks talked about a ship being prepared to go to New South Wales under the command of Philip Gidley King and was expecting that the ship would sail in about four months, around April 1798.[24]

Caley recalled that he and Banks seemed to have agreed about New South Wales and its prospects. 'You said . . . that the climate would be very healthful, and there would be no danger to run. I immediately told you, that I was afraid of no clime, "you said you did not believe that I was, nor even afraid to face the Devil himself".'

Caley clearly thought that the job of collecting for Banks in New South Wales was his, and when it seemed that the sailing date of

King's ship to New South Wales would be delayed, he decided not to stay on at Kew waiting for the call. Caley explained, '[had] I remained . . . it would have] deprived my intellects . . . of that culture that nature had intended for them.'[25] He had already acquainted himself thoroughly with the plant specimens from New South Wales that were growing in Kew, and there was no reason, therefore, for him to stay any longer.

When Banks learned that Caley had decided to quit Kew he wasn't pleased but his response was restrained; he simply reminded Caley that working in gardens was the way the great British practical botanists had laid the foundations of their botanical knowledge. On the other hand, Banks added that 'I always wish those with whom I have any connection to speak their minds at large.'[26] Caley did and decided to leave London. By the end of February 1798, he was back in Manchester having, in a way, broken his understanding with Banks.[27]

By July 1798, Caley was feeling anxious and fearing that Banks had dropped the idea of sending him as the botanist for New South Wales because he had quit Kew. He wrote to Banks to argue again that his reasons for leaving Kew made sense. Gardening was not what he wanted to do. He preferred 'Botany instilled by Nature'; or, as he put it later in the letter, 'I have a great love for plants in their wild state, I shall never swerve from it.' Near the end of the letter Caley made a point with which Banks must have agreed. Caley told him that he had heard it said that 'the chief of the plants were known and sent from Botany Bay'. By this he meant that the most beautiful and most colourful plants would have been collected first because they would have been noticed first, and most of them would have been collected from, or near, Botany Bay; while more mundane specimens, which might prove much more useful, had doubtless been overlooked. Caley would find them. In his final sentences Caley pleaded that he himself should not be overlooked but given the chance to go to New South Wales.[28]

Banks replied at once. He denied he was angry with Caley but he stuck rigidly to his advice that working as a gardener was the training necessary for being a 'botanical traveller'. Others, Francis Masson included, Banks pointed out, had done this and 'without complaining'. Banks reassured Caley that 'no person is appointed to go to Botany bay in your stead'.[29]

Caley wanted desperately to go botanising in the wider wilder world. He was clinging on to the New South Wales opportunity because that's what he understood Banks to have offered him. If that proved impossible, he would find somewhere else to go.[30] But Caley did not want to let go of his relationship with Banks. On 22 July 1798, he wrote Banks another long letter going over many of the arguments he had previously made but now emphasising his zeal and his abilities in the wild. 'Let me be tried upon the lofty mountain, the dark and intricate wood, the wide-extended plain, the marsh and peaty bog; but for all that I have said take me into a field or where the earth is covered with verdure, I do not know that vegetable production within the circle of one yard around me.'[31] Comparing himself now to the finest gardeners in Kew, Caley pulled out all the stops: 'I do not hesitate to say', he declared, 'that I could find more new plants, and send over the finest ones . . . I make no doubt but you will think I am boasting to a degree of insanity.'[32]

On 27 August 1798, Banks replied. Again, his tone was very measured, with not a hint of anger. By then Caley was convinced that he had 'lost [Banks's] favour' and that he would not regain it unless he returned to Kew. Banks agreed with Caley that he was 'by nature Eminently Qualified for a Collector'; though he added that Caley's estimation of himself was 'far higher' than his own.[33]

As for having lost his favour, Banks did not address this directly but agreed with Caley's understanding that he would not be sent out to New South Wales 'as a Collector for the King's Garden' because he was not fully acquainted with the plants that were already growing there.[34]

'Collector for the King's Garden'? This was the first Caley had heard of it. He had been under the impression that he would be going to New South Wales as a plant collector for Banks and that he would be looking out for all manner of plants: ornamental, industrial and medical. There had been a misunderstanding but Caley was still keen to go. He was sure he would need only two weeks in the King's Garden to get to know the plants there.[35]

Banks repeated that at no time during their long correspondence had he felt anger, but he did feel regret that Caley 'indulged a temper of a kind Which no Person in my opinion can justify'. Then Banks added that the ship currently being prepared to go to New South Wales would not be taking a botanist, but only a gardener who wanted to settle there. Caley was still going to have the chance to collect for Kew, and he might not have to wait longer.[36] It was now 12 September 1798.

Caley was still hoping to set himself up as a freelance collector, selling 'showey plants' to make some money, while collecting on a less remunerative basis for the government for plants that would have public utility. New South Wales was, to him, just one possibility; he was also thinking of the Cape and the West African coast.[37]

On 12 November everything changed. Philip Gidley King told Banks that he had an unexpected vacancy on board and could now welcome Caley as 'one of my Family'.[38] Caley could go to New South Wales after all.

What had happened was this.

Back in May 1798, Banks had written to John King, an official at the Home Office thus: Britain had now been in possession of New South Wales for ten years, yet, over that period of time, Banks pointed out, nothing had been discovered there that would be of material benefit to the mother country. The reason for this was that the interior of the country had been neglected in favour of the coastal settlement. It is inconceivable, Banks argued, that a land mass the size of Europe would not have navigable rivers that would allow exploration 'into the heart of the interior'. 'If properly investigated', Banks concluded, 'such a Country, situate in a most Fruitfull Climate, should . . . produce some Native raw material of importance to a Manufacturing Country as England is.'[39]

This was not just a passing idea but a practical proposal because Banks had just the right person for the job and he had already volunteered his services at a very reasonable rate. Mungo Park had recently returned from his widely celebrated and heroic journey into the interior of Africa and was apparently willing to explore the interior of New South Wales next. Banks, who had sponsored the African adventure, spoke extremely highly of his friend, assuring King that there was no one more qualified as an explorer than Park.

Banks wanted the government to requisition a vessel in New South Wales for this undertaking, one which would have ten men on board, six of whom would accompany Park on the journey to the interior. As for who would command such a ship, Banks once again could recommend just the man, Lieutenant Matthew Flinders, who was already in New South Wales.[40]

By early September, an understanding had been reached between Mungo Park, John King and Philip Gidley King, and only Park's salary seemed not yet agreed.[41] However, less than a week later, Park unexpectedly pulled out. Yes, he still wanted to go to New South Wales

– after all, as he told Banks, the whole plan was his idea; and although the salary was too low, that was not the reason why he was now turning down the opportunity. It was because he wouldn't be ready in time; he had been told that the ship would be departing in only ten days.[42]

James Dickson, the Covent Garden nurseryman, Park's brother-in-law, a friend of Banks and Caley, thought that the real reason for his withdrawal was that Park was involved romantically with a woman in Scotland and he didn't want to leave her.[43]

Over the next few weeks, Banks and Robert Moss, an under-secretary at the Home Office, tried to persuade Park to change his mind by offering him a better salary and carte blanche as far as the instruments, arms and presents for the locals were concerned, but Park stuck to his position.[44] He wasn't going to New South Wales. Banks was disgusted. In place of his former high praise, Banks now referred to Park as 'a fickle Scotsman'; that he wasn't 'worth a farthing' and that he was now 'totally useless to us'.[45] It seems that Dickson was right; in August of the following year, Park married Allison Anderson and settled down in Peebles as a doctor.[46]

Once it was certain that Park wasn't going to New South Wales, Philip Gidley King offered his berth to Banks for Caley. Banks wrote immediately to Caley with the good news that he was would be going to New South Wales after all. The government, Banks explained, was not willing to pay for a botanist at this time, but Banks would employ him to collect 'Specimens of such curious Plants as you may find' for Banks himself, and seeds for Kew. As for duplicates, Caley was at liberty to sell or dispose of them in any way he wished; and the same went for any plants he might find that would be of value to manu-facturing or commerce.[47]

Caley agreed by return of post and soon afterward joined HMS *Porpoise*, King's ship.[48]

The voyage of the *Porpoise* had an additional purpose. It was Banks's next attempt, following the disaster of HMS *Guardian*, to transplant European commercial plants to the colony. In early 1798, Colonel William Paterson, a captain in the New South Wales Corps, a naturalist and explorer himself, and a frequent correspondent of Banks, was back in London on sick leave. He told Banks that only a handful of the almost one hundred varieties of fruit plants, herbs and vegetables originally sent to the colony were still growing.[49] This fact Banks relayed to John King at the Home Office. Philip Gidley King, who had been back in London for more than a year recovering his health after his five-year period in command of the settlement on Norfolk Island, was planning to return to New South Wales to replace John Hunter as the colony's Governor. As he had taken care of the plants on HMS *Gorgon* during its voyage to New South Wales in 1791, it seemed appropriate that the ship taking him back should carry the plant cabin, for which the *Gorgon* did not have room.[50] As for caring for the plants, Banks had the perfect man in mind – a young market gardener, who came highly recommended by Banks's uncle and who wanted to settle in New South Wales with his new wife.[51] Banks had known about George Suttor's wish to emigrate to New South Wales as early as December 1797, and had suspected that 'his motive for going is the love of a very young woman'.[52]

By mid May 1798, it had been decided that HMS *Porpoise* was to be the ship tasked with replenishing the colony's European plants.[53] Banks's plans for the plant cabin were solicited and very soon the cabin was built.[54] Not long after, Suttor was invited to go to Kew and collect the plants that were to go in it.[55]

By early September Suttor had filled the boxes with a large assortment of different kinds of plants: culinary, medicinal, garden and fruits. Hop trees, which Banks insisted should be tried in the colony in order to wean the colonists off spirits and onto beer, were added.[56] These cuttings were to be supplied by Reverend Peter Ashleigh, of Kent and Essex, an Eton contemporary of Banks's.[57] The previous attempt to transport hops had ended in failure when the trees died aboard the *Gorgon* in Cape Town in July 1791 en route to New South Wales.[58]

By the end of October 1798, it looked as though everything was ready but then it all went wrong. First the weather delayed the *Porpoise* leaving the Thames Estuary; then when finally it entered the North Sea violent winds caused such damage that it was forced to make an unscheduled stop at Sheerness on 11 December.[59] More delays followed and it was not until early February 1799 that the ship got to Portsmouth. The ship's master was so alarmed at 'the Cranknes of the Ship' – it tended to keel over in the wind – and believing that the reason for its instability was that the plant cabin was too heavy and too high up, insisted that it should be dismantled and rebuilt between decks.[60] Arguments went back and forth between the ship and the Navy Board but in the end they agreed with Philip Gidley King that the cabin could remain on the deck but that the plants would be moved between decks as the weather en route changed.[61] The ship was once again ready to sail at the end of April but now it seemed an armed convoy was needed to protect it from a possible attack by a French fleet.[62]

This threat, as well as yet more repairs on the *Porpoise*, and inclement weather, kept them in port. Other ships came and went but the *Porpoise* remained at anchor. Philip Gidley King was disappointed and anxious. He had his family with him living onshore in Portsmouth. His debts were increasing. He also feared for the plants – if the ship left at the wrong time of the year, the plants were likely to meet weather they couldn't survive.[63]

Finally, on 5 September, after a seven-month stay in Portsmouth, Philip Gidley King learned that the *Porpoise* was to sail the next day.[64] Four days out, disaster struck yet again in the shape of powerful winds in the Bay of Biscay. First the tiller broke, and then the rudder's casing gave way, and the ship began to take on water. The order was given to return to Portsmouth again, where it arrived on 17 September.[65] Less than a month later, King was in London advising the Admiralty that the ship was not worth repairing again, and that it should be replaced by a Spanish-built prize ship, about the same size as the *Porpoise*, and that the plant cabin should be transferred to the new ship. King could not wait for that to be done but would return instead to New South Wales on a whaler, the *Speedy*, scheduled to sail shortly. Caley went with him. The ship actually sailed on 20 November 1799, carrying, in addition to King and Caley, a young ensign, Francis Louis Barrillier, who was going out to serve in the New South Wales Corps, and a number of passengers and fifty-three female convicts.[66]

The refit of the new *Porpoise*, as the Spanish ship was now called, took several months and it wasn't until 17 March 1800 that the plants, the plant cabin and George Suttor and his wife departed for New South Wales. The long delays, not surprisingly, took their toll on the plants. Many died and had to be replaced even before the ship sailed, and the voyage to New South Wales was very unkind, especially to the hop cuttings, none of which survived.[67] The difficulties of transporting plants by sea (and there was no other way for many destinations) were almost overwhelming, even worse than getting the right collector to the right place.

16

1800: Caley and Moowattin

'His disposition is Singular & Whimsical & May at first seem to require more indulgence than you may at first be inclined to grant to him it is said to be a Proverb in New South Wales that Caley & the common hangman are the only two people who do as they please.'[1]

This was how Banks described Caley. They were the most unlikely of friends: one, the president of the Royal Society; the other, a farrier's son from Manchester. Yet, there was a clear bond between them from when Caley sent his first letter to Banks. Maybe Caley's energy and insistence on being free to roam the natural world – observing, collecting, preserving, packing and shipping it – reminded Banks of himself on the three voyages that he had made when he was a younger man. Maybe Banks was reliving his own experiences vicariously through Caley. Maybe Caley's unrefined, determined and irascible ways endeared him to Banks. Or more likely it was because Caley was a good botanist, a rare commodity as Banks knew only too well. Whatever it was, the two men had a relationship that lasted almost a quarter of a century. For ten of those years Caley was in New South Wales.

Caley went to New South Wales ill equipped for preserving plants, and through no fault of his own. From Cape Town Caley complained to Banks that he was without paper for pressing his specimens. Philip Gidley King, who was in command of the *Speedy*, the ship on which he had sailed, had told him that there was no room on board for the paper. 'I told him that it was an article I could not do without,' Caley wrote: 'I would as soon leave one chest of cloaths behind as it'.[2]

That was one problem. The other was that on the voyage to Cape

Town and in the town itself, King and Caley were at odds: King had wanted Caley to take advice from John Barrow as to where to go for Banks's botanical requests, but Caley had his own ideas. Barrow had been at the Cape since 1797 when he had become private secretary to Lord George Macartney, who had been appointed Governor of the Cape Colony (Barrow had been on Macartney's embassy to China in the role of comptroller to the East India Company). Barrow had planned a botanical itinerary for Caley, even arranging for his food and lodgings en route. Caley, however, did not even bother to meet Barrow. Instead, he took up residence in Cape Town and then set off into the countryside on his own. Only on his return did Caley contact Barrow when he asked him to send the collection he had made to Banks on the next available ship. King was furious with Caley. 'I hope he will give *you* satisfaction', he wrote angrily to Banks, 'but really he appears to me very wild.'[3]

Caley thought that what he had collected so far was poor. He had not had enough time to do a proper exploration of the area and he was certain he had found nothing new. He was right. When the seeds and dried specimens arrived at 32 Soho Square a little over four months later, Banks wrote at the bottom of Caley's accompanying letter: 'Dried plants mostly English & all very common nothing of any value.'[4] Not a great start.

The *Speedy* had arrived in Port Jackson on 15 April 1800. Caley, perhaps unsurprisingly, decided to make his base in Parramatta, a little over twenty kilometres to the west of the main settlement where he would be undisturbed in his botanical pursuits. Most of the 5000-strong New South Wales population lived in and around Port Jackson but Parramatta, built on the banks of a river, had better soil than the port, and soon the colonists were growing wheat there. Though Caley lived in a very modest dwelling, a small timber house with a thatch roof, Parramatta itself had some fine houses and decent streets. Most of the 1400 or so people who lived in the village were employed on the land. Caley was allowed to dry his specimens in the village's finest building, Government House, built by Arthur Phillip.[5]

A little over five months after the arrival of the new convicts and settlers on the *Speedy*, King gave Banks some shocking news. He told him in his letter of 28 September that Caley was in love with and intended to marry the widow of Edward Wise, a master weaver who had been travelling with his family on the *Speedy*.[6] Wise had fallen overboard somewhere between Cape Town and Port Jackson.

When Banks heard the news he was incensed. 'This marrying has been often in my way,' he wrote to King, alluding to other instances, such as that of Mungo Park and Matthew Flinders, when women impeded Banks's plans. Banks preferred to hire or recommend single men. Good botanist or not, Caley's plans to marry were, to Banks, thoroughly incompatible with collecting plants. 'I did not hire him to beget a Family in N.S.W.', Banks wrote irritably to King. If Caley persisted in getting wed, Banks had no doubt what to do. 'I must Certainly get Rid of him.'[7]

As it happened, Caley did not marry Mrs Wise at this point but continued to see her and to support her family, a small boy and girl.[8] For the time being, Caley's job was safe.

Caley was itching to get on, but he lacked supplies. He was still waiting for the paper: it had been put on HMS *Porpoise*, which had set sail from Portsmouth on 17 March 1800. He had to make do with 'a few old newspapers, and some other waste papers'; but, as he put it, '[these] are but mere trifles to what I want'. His enthusiasm and self-image were not dampened in the least. 'In the course of one year', he reminded Banks, 'provided no obstacles fell in my way, I could collect half of the specimens in the Colony.'[9] For the time being he was happy to explore the immediate vicinity where he was already making new discoveries: 'Every inch of ground', he wrote, 'I consider as sacred, and not to be trampled over without being noticed.'[10]

That was music to Banks's ears. But Caley wanted more. He wanted to go where no white man had been before, especially to the mountains: 'I have always been partial to mountainous places,' he confessed. Though botany was to be his principal pursuit, he was also intent on exploring the area's zoology and mineralogy.'[11] To do all that, however, required equipment he did not have, so he proposed that Banks should send him the following – pocket knives, blow pipe, fish hooks, pins for insects, a pocket barometer, small bottles with wide necks, strong thread of various colours, and most important of all a gun and a double-barrelled pistol.[12]

Near the end of December, Caley told Banks of his great relief that the paper had arrived at last and that he was now free to travel. 'The absence of the paper threw a gloom on me of late that I cannot describe to you in words.'[13]

While he was waiting for the supplies from England, Caley made his first collecting trip away from the colony, on a survey of the southwestern mainland, on board the *Lady Nelson*, commanded by

Lieutenant James Grant, between March and May 1801. It was, Caley admitted, very disappointing. There was nothing new to be found and life on board ship did not suit him as a collector. 'When I am at home', Caley wrote, 'I can collect without being noticed by others, but when I belong to a ship, some one or other is sure to be with me and partaking of what I collect.'[14] Caley had a very clear idea of how and what he wanted to collect and ships were not for him. Over the next few years, he turned down maritime invitations as soon as they were offered. 'I expect to be again in the woods, and hope hereafter so to continue,' he wrote to Banks.[15]

Between October 1801 and March 1802, Caley made four trips to the Nepean River, to the west of Port Jackson and beyond the boundary of the colony. These trips and other local excursions brought a harvest of botanical material, plants and seeds. By the time these were ready to be shipped on the *Speedy*, which was returning to England, Caley had three boxes for Banks, one of which contained two hundred and forty papers of seeds.[16] This was his second shipment from the colony to Banks, and much larger than the first, sent some nine months before.[17]

Banks was very pleased with Caley's work but was unable to tell him so; for most of 1801, he had been ill and apologised to Caley for not writing to him for more than a year. Now, on 14 August 1802, he was able to put his thoughts down on paper. Banks conceded that there was little new in Caley's collections, but he was appreciative of his 'diligence & industry'. Even better, Banks was coming around to Caley's own estimation of himself as a very fine botanist. 'Your descriptions of Plants do you Credit,' Banks wrote. 'Tho Roughly drawn they Show an attention to the Structure of the Plants you have described from which a good Botanist of Perfect Education may derive advantage.'[18] As for Caley's insistence that doing things his way would work out, here too, Banks was beginning to agree. 'I sent you out to act as a free man in New South wales', Banks acknowledged, '. . . and as a free man I have no doubt you will do better both for me & for yourself than if you follow the opinions of other people.'[19] The supplies he had asked Banks to send him were already on their way and, to top it all, Banks was raising his salary by almost one half 'as a Testimony of my approbation of your Conduct'.[20]

It was not just in London that Caley was being appreciated at last but also in New South Wales. Governor Philip Gidley King, who had earlier been one of Caley's main critics, also began to come round.

Caley, King noted to Banks, had been busy exploring the interior and always had Banks's requests as his principal object. 'With all his faults, which he cannot help,' King wrote about Caley, 'I believe his Clever & faithful.'[21]

With King's support, Caley could act more freely. Botany, as Caley understood it, was his principal concern but his interests, his observing and collecting, went well beyond plants. He was particularly intrigued by the kangaroo. He felt that the further into the interior he could penetrate, the more he could discover about the animal. He was well aware that Banks and Cook had been the first Europeans to describe it.[22]

Although during Arthur Phillip's time as Governor of the colony, many kangaroo skins and a few live kangaroos had been shipped back to London, there was no understanding at the time that there were different species. This was partly because the only kangaroos that were caught or shot were from the local area. There was also a lack of interaction between the colonists, who knew nothing about the animals, and the Aborigines, who knew them well.

As early as 25 August 1801, Caley assured Banks that he would be sending him new examples of kangaroos. He was just waiting to be a little more settled financially so that he could afford dogs to hunt them. He knew of at least half a dozen species of kangaroo and would send him specimens of each of them in time.[23]

After a while, Caley produced some startling information. He had been learning more about kangaroos from the indigenous inhabitants, and he was probably the first white man to do this. He was certain, for example, that the kangaroos that were already in England belonged to three species, what he termed the 'great Forest Kangaroo, the Kangaroo Rat and the Brush Kangaroo'. He even gave the three species their native names: Patagaran'g, Cuniman'g and Wal'aby, respectively. This was all new to Banks. Caley added that he had also learned that there were three more species that were unknown outside Australia: he didn't know what to call them; he only knew them by their native names and by description. One was the 'Pattymablon', reddish in colour, and smaller than the forest kangaroo. Caley remarked that this kangaroo lived in thickets where even dogs couldn't reach them. The 'Werin'e' had a hairy tail and lived on rocky ground in the Blue Mountains. Finally the 'Beton'y' had a slender tail and lived in such impenetrable thickets that the indigenous people could only flush them out by setting fire to the area.[24] It would be two more years before Caley managed to get the skins of the Werin'e and the Beton'y kangaroos.[25]

Caley had got to know his Aboriginal neighbours in Parramatta and they had shared their knowledge with him. Caley told Banks that his friendship with the local people was a great advantage to him. 'I shall always strive', he commented, 'to be on good terms with them, which may be done by giving them small axes, old cloaths & filling their bellys – By doing so I can make myself popular among them, & gain information of many things that could not be acquired but by such means.'[26]

Although he never explicitly said so, the indications are that Caley knew enough of the local languages to rely on the Aborigines for vital geographical information. For example, on one of his trips to the Nepean River in December 1802, they told Caley that his plans were not feasible and that he would be endangering both his and his horse's lives if he continued in the direction he was taking.[27]

Caley was planning a big trip, though that was all Philip Gidley King knew about it. Caley kept his plans to himself. King just hoped that Caley kept Banks informed about his movements: and he did.[28]

Caley's destination was the Blue Mountains, a ring of rugged mountains, many of them exceeding 1,000 metres in height, which encircled the New South Wales colony beyond the Cumberland Plain, the foothills beginning about fifty kilometres to the west. By the time Caley planned his trip at least ten other white colonists had already been there, not only to explore the area but also to get a glimpse of what lay beyond.[29]

Caley first went into the Blue Mountains in October 1802, for a short time. He saw several plants that were new to him, but he was waiting to get his supplies from Banks and for the use of a good horse before he ventured further. Caley told Banks that the general opinion in the colony was that the mountains were impassable, but he didn't think so. The government had already sent Francis Barrillier, with whom Caley had sailed on the *Speedy*, but even though the expedition was well supplied, including the use of two Aboriginal guides, Caley didn't think they had got very far, certainly not as far as they claimed. Caley was confident as usual: 'I am so vain as to think', he declared, 'that with another man besides myself & a horse, that I can go farther than what this party will.'[30]

The arrival of the East India Company ship, *Alexander*, in the latter part of 1802 with everything he had asked for, especially the gun and pistol – 'just what I wanted' – ensured that Caley would soon be on his way. So he was, though initially for not as long as he had hoped. Before heading off to the west, Caley sent Banks his first consignment of dried plants from the Blue Mountains.[31] Unknown to Caley, back in London, Banks had just taken delivery of Caley's seeds and plants from the Hawkesbury and Nepean River and was thrilled with them. He had already written his thanks to Caley and praised him to Governor King.[32] Those from the Blue Mountains were sure to please Banks as much if not more.

On 13 May 1803, returning to Parramatta after another short trip, Caley told Banks that in the following year he was going back to the Blue Mountains but, this time, he would be penetrating even further to the 'ruggedest part'. Repeating his claim that he could go further than anyone else, Caley declared that he would 'spare neither expence nor labor'. He knew already that the terrain was impassable for a horse and he was planning to go on foot with 'some good travellers' men whom he knew and who could endure.[33]

After a short trip to the Hunter River, in November 1804, he found himself with enough free time to tackle the Blue Mountains again. He was taking along three other men as well as his dog. 'Botany', Caley exclaimed, 'is not the primary object but that is on an enthusiastic pride of going farther than any person has yet been.'[34]

Caley had hoped that Banks would find the results of the expedition to the Blue Mountains interesting and of value to the colony. Certainly Caley did his best but what he saw and reported was more disappointing than interesting. Caley and his companions set out on 3 November and were in the foothills of the mountains soon after. That was the easy part, but for the next ten days, the party descended into deep canyons and climbed even higher mountains, making their way very slowly in a westerly direction. So rough was the terrain that the four men on one occasion covered barely two kilometres in a day. By 15 November, less than two weeks into the trip, Caley's companions were becoming tired of the expedition and were refusing to go any further. Caley, alone, climbed a thousand-metre mountain, which he named, in honour of his patron, Mount Banks. (It is about ten kilometres from the present town of Blackheath – about 80 kilometres to the west from Parramatta as the crow flies.) From the summit, in every direction, Caley saw yet more canyons and mountains. The party's provisions were low and his companions exhausted. Caley decided he had done enough and that it was time to return. The next day, 16 November, he recorded his thoughts in his diary. 'Now being at the furthest of my journey', he wrote, 'and just upon the point of setting . . . upon our return, I found myself elevated in a manner I knew not why, no further than the sake of return. The journey being a dangerous one, I knew well before I set off, but as my adage is "nothing ventured nothing won" being imprinted in my mind, surmounted all the obstacles that fell in my way.' Eight days of walking later, Caley and his dog were back in Parramatta.[35]

As soon as he could, Caley wrote to Banks about his experiences. He explained that the length of time the party could be away was governed by the amount of provisions they could carry. The furthest point he had reached, he named Mount Banks. The going was very hard. 'The roughness of the country', Caley admitted, 'I found beyond description I can not give you a more expressive idea than travelling over the tops of the houses in a town.' Though as he had warned Banks earlier, the trip was not really about botany, nevertheless, Caley claimed to have found about thirty new plants and thought that there were more in and around Mount Banks. However, he wasn't planning any more trips – it was too far from the colony, and too difficult and dangerous for more such excursions.[36] Caley's Blue Mountain plants were shipped to London in early January 1805.[37]

In April 1805, Caley told Banks that he wanted to return to England – he had, in his estimation, been away too long and not accomplished as much as he had hoped.[38] Governor King was under the impression that Caley had already decided on which ship he would sail.[39]

However, he changed his mind – he had rediscovered his zeal for exploration. He had a new companion. Moowattin, or Dan, as Caley often referred to him, was about fourteen years old and had been living in the colony for some time, having been adopted by Richard Partridge, the hangman, as a small boy.[40] He spoke English as well as his native language, Dharug or Eora, as spoken by the indigenous people of the Sydney region.[41] In October, Caley and Moowattin travelled to Norfolk Island, which they explored for a few weeks.[42]

Writing later to Banks, Caley had almost nothing to say about Norfolk Island, dismissing it with the words: 'At Norfolk Island I was as long as I wished to be.'[43]

Still, while there for only two weeks, Caley and Moowattin did find enough to keep them busy botanically. One specimen that caught

Caley's eye was a *Eucalyptus* (or *Corymbia*) which Caley referred to in his notes by its Aboriginal name, *Coroy'ba*. 'It grows to an immense height', Caley wrote, 'and may be said to be the tallest in the forest, though not the thickest. Dan says it is taller than the pines at Norfolk Island and I myself am of the same opinion.'[44]

Where Caley really wanted to go was Van Diemen's Land (now Tasmania). Banks had long wanted Caley to explore the island for plants, arguing that because its climate was more like England, they would have a better chance of surviving than those from Sydney – '[they] die in every frosty winter if exposed to the Air'.[45]

In 1803, some fifty men from Sydney had established a small settlement on the eastern banks of the Derwent River in the south of Van Diemen's Land. This was in order to claim the land for Britain in the wake of the visit, a year before, of a French expedition to the area commanded by Nicolas Baudin.[46] By 1805 the settlers had moved to a better site on the opposite bank of the Derwent River which they named Hobart.[47]

This is where Caley and Moowattin arrived on 29 November. They started by exploring the area around Hobart. After a few days, Caley was ready for something more challenging. The mountain had several names already – Table Hill, Snowy Mountain and Skiddaw, after a similar peak in the Lake District, which name Caley preferred.[48] It is now known as Mount Wellington and its height of almost 1300 metres dominates present-day Hobart as it then dominated the Derwent River. The mountain is frequently covered by snow.

Caley was not the first white man to climb the mountain, but, as usual, he did it his way: 'I gained the summit by a route which had not before been attempted, by being considered impracticable,' he told Banks.[49] When, eventually, Caley and Moowattin reached the mountain's summit, the weather had turned against them: the thermometer fell to near freezing point, the wind blew fiercely and it was sleeting. 'Never did I feel such piercing cold . . . my fingers were so benumbed', he wrote, 'that I could not hold a pen'. It was only the warmth of the conversation with Moowattin, who had also never experienced such cold before, that kept his mind on things other than the weather.[50]

Though theoretically Van Diemen's Land seemed promising, in practice it was disappointing. There were hardly any alpine plants growing on the mountain and certainly none that Caley could collect in such difficult conditions. More to the point, Caley concluded that the climate was not really that different from that of Port Jackson and so the plants were no hardier. He saw, however, that the island was interesting from the point of view of exploration, and he proposed three inland expeditions to David Collins, the settlement's commander.[51]

These never materialised. Not long after the ascent of Skiddaw, Caley and Moowattin left Van Diemen's Land, where they had been for almost a month, arriving back in Port Jackson on 23 January 1806.[52]

Besides the stimulus of a new and knowledgeable companion, Caley was probably hoping that the new Governor, William Bligh, who was expected to arrive in the colony very soon, would be more supportive of his plans than Philip Gidley King had been.[53]

In the event, Caley did make at least four more trips to the interior of New South Wales, on which Moowattin accompanied him. On one of these, an expedition in July 1807, Caley and Moowattin struck out in a southwestern direction to an area that Caley had been to before. This time, however, the two followed the Georges River, which flowed into Botany Bay. As they followed the river upstream, they came across a magnificent waterfall, 'the greatest natural curiosity yet known in the colony', which measured over fifty feet in height. Caley named it Carrung-Gurrung. The river that fed it, Caley named Moowattin, in honour of his friend and guide.[54] Here Caley came across several new species: two specimens of what was called *Embothrium*, a red-flowering bush; a *Correa* also with red flowers; and a *Pultenaea*, commonly known as the bush pea, with purple flowers. As it was accessible by horse, Caley intended to return in the following summer.[55]

Caley was brimming with ideas for more exploration and more botanical collecting. He knew, however, that putting the ideas into practice was going to be difficult. The projects he was planning were going to be expensive and would depend on the generosity of the Governor. Caley wasn't hopeful. As he put it to Banks: 'being on too saving a plan, which I considered to be as wearisome, as undergoing the journey; for if the latter fatigues the body, it enlivens the mind, whereas the former causes not only fatigue, but a depression of spirits.'[56]

Two days after expressing his pessimism about future explorations, Caley wrote again to Banks to repeat his desire to go home: he wanted Banks to recall him – 'the place is miserable without any signs of improving'.[57] He was not getting the support he had hoped for from Bligh.

As it turned out, Caley did only two more trips, one in October and another in December 1807, about which he confessed, 'I was so overcome with fatigue that I was obliged to return the same way as I went out.'[58] He pleaded again with Banks to recall him.

In a reflective frame of mind, Caley summed up his experiences of living in and exploring New South Wales. 'I oftentimes occupy my thoughts, what new plants there are in some parts of the Blue Mountains undiscovered, and lament that I am deprived of the means of going thither. However great may be my natural failings, and however incapable I may be of following Some pursuits, I have the vanity to think that Nature has given me preeminence as a traveller in this country, in exploring his works. Though I flatter myself to be possessed of this innate principle, yet I have never had a fair trial There has always been something to disconcert my views. Now the time is past. What grieves me most, is always being busied in doing Something, which on reviewing appears as if I had done nothing – I mean that I can transmit to you.'[59]

On the other side of the world, Banks had also decided that Caley had done enough. Banks told Caley that he was letting Caley go; Banks was chronically unwell and no longer had 'the pleasure I usd to have in the pursuits of Natural History'. To reward Caley for his past services 'in the extensive manner in which you have employed yourself of Collecting great Quantities of Articles', Banks was offering Caley a lifetime annuity of £50 per year regardless of whether he chose to remain in New South Wales or to return to England.[60]

Caley had already made his decision. On 12 May 1810, he left New South Wales for good on HMS *Hindostan*.

He took Moowattin with him.

Caley valued indigenous knowledge and wrote eloquently to Banks about Moowattin's talents. 'He would point out the tracts of different animals, which would be unnoticed by a European,' Caley explained. 'Many new birds would be readily obtained; as the natives in general are excellent marksmen and quicker sighted than our people. The specimens of different trees might be procured which would otherwise had been missed or left for the want of a climber.'[61] As to the often-quoted remark that the indigenous people of New South Wales were 'the most idle, wretched, and miserable beings in the world', Caley argued strongly against that. 'I could single out several', Caley commented, 'that surpass numbers of Englishmen in mental qualifications.'[62]

Caley reserved his highest praise for Moowattin. It is worth repeating his words in full. 'The native I have been speaking of is the most civilized of any one that I know who may still be called a savage. And the best interpreter of the more inland natives language of any that I have met with – I can place that confidence in him which I cannot in any other – All except him are afraid to go beyond the limits of the space which they inhabit, with me . . . And I know this one would stand by me until I fell, if attacked by any strangers. Moowattin is in great measure bred with me.'[63]

It is in this letter that Caley told Banks that he wanted to bring Moowattin back to England with him. Caley had a very good reason for bringing Moowattin to London. 'By shewing him the different Museums', Caley wrote, 'we should gain a better knowledge of the Animals of this part; for I have heard of several that are not in England.'[64]

The *Hindostan* dropped anchor in Portsmouth on 25 October 1810, a little over five months after leaving Port Jackson. Everyone got off the ship apart from Moowattin. He was confined to the *Hindostan* until someone in authority could be found who would vouch for his orderly conduct. In early January 1811, after Moowattin had been left on the ship for two months, Banks provided that guarantee. Banks thought that Caley had made a mistake in bringing Moowattin to London but agreed to maintain him during his stay.[65]

On reading Caley's account of Moowattin's stay in London, which he committed to a daily diary, it is clear that Moowattin's time was fully occupied.[66] Life revolved around social occasions, with Caley's old friends; cultural activities, the theatre and public events, such as the military parade on Wimbledon Common celebrating Wellington's key victory over the French at the battle of Fuentes de Oñoro; meetings with other naturalists, including several appointments with Banks himself, sorting through specimens and discussing the koala; and going to Kew to see the Australian plants growing in the gardens.

However, there were also problems, especially drinking bouts that ended in Caley having to extricate Moowattin from various difficult situations. Whether it was because of these incidents; or whether it was because Caley was on the point of returning permanently to Manchester to join the widow Mrs Wise, whom he had married on his return to London; Banks insisted that Moowattin should return to New South Wales. On 11 August 1811, Banks wrote to Robert Peel, then Under-Secretary for War and the Colonies, to find a passage for

Moowattin on any of the ships that were being fitted out for a voyage to New South Wales.[67]

Two months later, on 11 October, Moowattin, accompanied by George Suttor, the gardener who had taken Kew's plants out to New South Wales on the much-delayed HMS *Porpoise*, boarded the *Mary*. This was a convict ship, at anchor in Portsmouth. It arrived in Port Jackson after a long voyage on 12 May 1812.

Suttor wrote to Banks with news. 'The Native which your humanity fed and clothed in England returned in health to his native country; the polished manner and comfortable living made but a Slight impression on his mind a few days after landing he left my house which I wished him to have made his home, made away with his clothes & ca fowling piece which Mr Brown had given him for some Peach Cyder and returned to his countrymen and Native life.'[68]

Moowattin's end, at the age of twenty-five, was tragic. On 28 September 1816, he was tried for the alleged rape of a settler's daughter, living near Parramatta. On 7 October, the court passed the death sentence on Daniel Moowattin and on 1 November, he was hanged. Moowattin was the first Aborigine to be legally executed in Australia.[69]

Even before Moowattin left London, Caley had moved back to the Manchester area to the then rural village of Chadderton, where he occupied himself with classifying and mounting his collection from New South Wales, and returned to the provincial natural history world that had nurtured him when he was younger. As he worked through his collection, he kept in touch with Robert Brown, who was now Banks's librarian and whom he had known in New South Wales. When not focused on taxonomic matters, his mind drifted back to New South Wales and his longing for what the landscape and its contents gave him: as he told Banks, 'if it was not for the thoughts of seeing my collection put in a state of preservation, all my thoughts would be engrossed with New South Wales, for in the mountains of that country my desiderata lie – and believe I shall never rest till I have visited some places there I have years ago marked out.'[70] It wasn't to be. He never saw New South Wales again.

In 1816, on Banks's recommendation, Caley was appointed as the third superintendent of the botanic garden on St Vincent in the West Indies. Caley arrived on the island, his first taste of the tropics, with his wife and stepdaughter on 1 August, but it was clear from the very beginning that it was the wrong place for him – his first letter to

Banks, written on 27 August, catalogued a string of complaints, which he continued to add to for the whole of the time that he stayed in the West Indies.

Caley's Caribbean ordeal lasted six years. He left St Vincent in 1822. Banks, his patron and friend, had died two years earlier. Caley returned to London and settled into a quiet life, adding to his library and sometimes visiting Kew. He died in 1829.

17

1800: Not Since the *Endeavour*

By the latter part of the eighteenth century, Banks knew more about Australia, or New Holland, as it was still called, than almost anyone else. Politics, trade, relations with Aborigines, exploration and natural history, were all topics in which Banks kept up-to-date. The first and second governors of New South Wales, Arthur Phillip (1788–92) and John Hunter (1795–1800), both wrote diligently and at length to Banks. Other senior figures in the colony such as David Collins, the colony's judge-advocate and William Paterson, responsible for the New South Wales Military Corps, were also regular correspondents.[1] So was David Burton, the surveyor, gardener and botanist, whom Banks had recommended to go to New South Wales as Superintendent of Convicts and who managed to send Banks some plants and seeds before fatally shooting himself by accident.

All of these contacts and their energetic communications ensured that Banks knew as much as anyone about life in the colony. The problem was that the colony of New South Wales, concentrated around Port Jackson, was only a small part of the British claim of New South Wales, which, itself, was a part of a larger land mass, then called New Holland. While the western part of New Holland had been visited and charted by Europeans since the seventeenth century, the northern and southern parts had yet to be properly surveyed. The interior was unexplored.

So, when, in 1798, Banks first suggested an expedition into the interior led by the explorer Mungo Park, he intended that Matthew Flinders, who was already in New South Wales, should command a ship to take the expedition to a convenient place on the coast from where they could start their journey. When Park withdrew from the expedition, Banks did not give up on the 'Scheme of Discovery', as he called it. He told Robert Moss, Under-Secretary of State at the Home Office, 'I will readily give my assistance against the next outfit, whenever that time may come.'[2]

Banks didn't have to wait very long. On 6 September 1800, Matthew Flinders wrote to Banks from Spithead on board HMS *Reliance*, which had recently arrived from Sydney.[3] Flinders had been away from England for more than five years during which time, especially on the sloop *Norfolk*, he had surveyed the east coast of New South Wales, north to Hervey Bay, and south to Cape Howe and, eventually, circum-navigated Van Diemen's Land thus proving that it was an island, separated from the mainland by a narrow body of water which was given the name Bass Strait after George Bass, the naval surgeon on the expedition.[4]

Flinders was writing to Banks to report on his exploring and charting activities, but this was only one of his reasons for contacting him so soon after his return. Flinders wanted Banks's support for an ambitious new expedition. He was convinced, as were others at the time, including Governor John Hunter, that the western part of New Holland might be separated from New South Wales by a strait or inner sea leading from the Gulf of Carpentaria in the north, to somewhere on the, as yet unexplored, southern coast.[5] On his coastal voyage, Flinders had not, as he told Banks, found any great rivers emptying into the Pacific, and this suggested that if there were rivers, they were actually flowing in another direction. Only a coastal survey, by implication covering both the northern and southern coasts, could settle the matter.[6] The kind of survey he had in mind involved a 'circumnavigation of New Holland', an ambitious project. Flinders volunteered himself to lead the expedition and ended by saying that he would welcome the opportunity of calling at Soho Square, when he would show Banks the charts he had made of the east coast of New South Wales and of Van Diemen's Land, and deliver the letters entrusted to him.[7]

Banks's ill health prevented him from answering Flinders's letter immediately. Two months passed but eventually on 16 November, Flinders had his invitation. Conveniently, he was staying at an address in Soho, no more than four streets away from Banks's home.

* * *

Flinders had first met Banks in 1793 when he delivered a letter from James Wiles, the gardener on HMS *Providence*, who had been his messmate on the ship and with whom he had made a close friendship. Wiles had decided to remain behind in Jamaica with the breadfruit plants and Flinders visited Banks at Revesby Abbey to claim £30 that Wiles had borrowed from him and which Banks was repaying from Wiles's wages.[8]

At the time, Flinders was only twenty-one years old, less than half Banks's age. Born in Donington, Lincolnshire, twenty-five miles to the south, across the fens, from Revesby Abbey – he was a 'Countryman of mine', as Banks referred to Flinders later.[9]

Flinders had gone to sea at fifteen and served on two ships before joining HMS *Bellerophon* in 1790 under the command of Captain Thomas Pasley. A year later, on 8 May 1791, Flinders joined HMS *Providence* under William Bligh, to whom Pasley had recommended his young protégé.[10] Not long after his return from the marathon second breadfruit expedition in August 1793, Flinders returned to the *Bellerophon* where, in less than a month, he was promoted from able seaman to midshipman. After seeing some military action against the French, Flinders returned home but not for long; in early September he joined HMS *Reliance* as senior master's mate. The ship left England on 2 February 1795 bound for Sydney. There he was based for the next five years carrying out a number of very important surveying expeditions. The Admiralty promoted him to lieutenant in January 1798.

Banks and Flinders must have discussed their ambition for a circumnavigation of New Holland at a meeting at Soho Square in November 1800, and after that Banks began actively planning for it. The circumnavigation was now seen as a matter of some urgency. Several months previously, in June, Banks had received a request from Louis Guillaume Otto, a distinguished French diplomat who resided in London, where he was acting as the French government's commissioner for the exchange of prisoners. (Two years after this contact with Banks, Otto was praised widely for his part in negotiating the Treaty of Amiens, which, for one year, ensured peace between France and Britain.)[11] Otto also acted as a conduit between British and French scientists, especially between the Royal Society and the Institut National des Sciences et Arts, whose communications had been interrupted by war.

On this occasion, Otto was asking Banks, in his role as President of the Royal Society, to help him secure passports for French ships that were preparing to leave France for a voyage of exploration 'to continue the useful discoveries that your navigators made in their voyages around the world'.[12]

Passports were essential during periods of war in order for ships to identify themselves as non-combatant and so be allowed to sail freely. Though Otto did not specify which or even the number of ships nor their destination, Banks answered promptly and positively, reassuring him that the appropriate government body would issue the necessary papers since they, like himself, were keen to promote the advancement of knowledge.[13] Two months later, Banks learned from Antoine Laurent de Jussieu, the director of the newly created Muséum d'Histoire Naturelle in Paris, that two ships under the command of Captain Nicolas Thomas Baudin were on the point of departing for the southern oceans, now that they had received the passports.[14] This was the second time that Banks had been involved in arranging a passport for Baudin: the first was in 1796 when Baudin led an expedition on his ship the *Belle-Angélique* to Trinidad to recover a French natural-history collection that had been seized by the British.[15]

The news of the French expedition's departure in October 1800, in addition to Flinders's enthusiasm for a British one, spurred Banks to act.

On 12 December 1800, a meeting was convened at the Admiralty. Banks was there, as was Earl Spencer, First Lord of the Admiralty, Sir Andrew Snape Hamond, Comptroller of the Navy, and Matthew Flinders. All of the essential decisions about the expedition were taken at this meeting: the ship, the *Xenophon*, was chosen (later named the *Investigator* for the voyage);[16] its establishment of eighty-one men decided; and it was agreed that a naturalist, a botanic artist, a landscape and figure painter, a gardener and an astronomer would be appointed and paid for by the Admiralty.[17]

As might be expected, Banks was put in charge of selecting what he often referred to subsequently as the 'Scientific Men', which included the two artists.

Even before the meeting at the Admiralty, Banks had already begun the process of tracking down a suitable candidate to be the naturalist. This was new: in the past he had had to content himself with gardeners. Being allowed a naturalist on a voyage was an exciting development.

Banks had heard of Robert Brown but had not yet met him, even though he had been studying in Banks's herbarium and library at Soho Square during the summer of 1798.[18] The Abbé José Francisco Correia da Serra, the Portuguese naturalist who had first met Banks in 1795 and who then became a regular visitor at Soho Square and Revesby Abbey, first alerted Banks to the presence of Brown. In a letter to Banks in mid October 1798, Correia da Serra mentioned that as soon as Brown had heard that Mungo Park had declined to explore the interior of New Holland he had volunteered to take his place. Correia da Serra sang Brown's praises pointing out that he was a 'professed naturalist', a Scot by birth, serving as a surgeon's mate in the Fifeshire Fencibles, an army regiment currently on duty in Ireland to help prevent a pro-French insurrection on the island.[19]

Brown was born in 1773 and after grammar school and a three-year stint studying the arts in Aberdeen, he went to Edinburgh to study medicine, but found his medical studies far less interesting than botany – like Banks himself.[20] Any spare moment he had he would be off botanising and visiting the city's nurseries. It was during his Edinburgh years that he met Mungo Park and, through him, the nurseryman James Dickson, who had married Parks's sister. Dickson, who owned a nursery in Covent Garden, was also a close friend of Banks. Brown, who joined the Fencibles in 1794, found himself in Ireland with the regiment a year later.

When the Flinders project was in its early stages of preparation, even before the Admiralty meeting, Banks remembered Correia da Serra's praise of Brown and contacted Dickson to find him. Dickson replied that he had last heard from him two months earlier and that he was in Ireland with his regiment.[21] Banks wrote to Brown at once after the Admiralty meeting with an offer; he needed an immediate answer. He told Brown to expect a salary of £400 per annum, that the expedition would last at least three years and that the ship might sail within weeks.[22]

Brown jumped at the opportunity. He couldn't leave immediately because he was the only serving surgeon in the regiment at the time but, nevertheless, he thought he could get out of his duties and be in London by 26 December.[23] As it happened he arrived a day earlier and finally got to meet Banks over dinner at Soho Square. Once he had found lodgings nearby in Soho, Brown went every afternoon to Soho Square where he and Jonas Dryander, Banks's librarian, whom he already knew, began to prepare a comprehensive herbarium of New Holland plants, including those from the *Endeavour* as well as those which had been arriving from New South Wales over the past decade.

By the end of January Brown had met everyone else, including Flinders, who was associated with the expedition so far. He had reacquainted himself with Correia de Serra and James Dickson and spent most Sunday evenings with Charles Francis Greville, a close friend of Banks's, a member of the Dilettanti Club, and nephew of Sir William Hamilton (Greville's mistress for four years was Emma Lyon, before she married Sir William); Greville was also a highly respected mineralogist and horticulturalist and had a famous garden at Paddington Green.[24]

Brown had done all he could to prepare himself for the voyage. His commanding officer, Colonel James Durham, however, was not happy either with Brown being on leave in London or with his desire to retain his commission while serving on the *Investigator*. Though Banks did his best to convince Durham to relent, it was only when Banks took it up with his superior, the Marquess Cornwallis, who had seen action in North America and India, and was now Lord Lieutenant and Commander-in-Chief of Ireland, that the way was smoothed for Brown to get what he wanted.[25]

It wasn't just Brown whom Banks contacted and appointed without delay. A few days after writing to Brown to offer him the appointment of naturalist, Banks informed Earl Spencer, the First Lord of the Admiralty, that he had appointed Ferdinand Bauer as the expedition's botanic painter.[26]

Like Brown, Bauer came very highly recommended. Born in 1760 in Feldsberg, Austria (now Valtice in the Czech Republic), Ferdinand Bauer, together with his older brother Franz, moved to Vienna around 1780 where he began to work for Baron Nikolaus von Jacquin, Professor of Botany and Chemistry at the University of Vienna, and another close correspondent of Banks's.[27] Bauer was occupied in illustrating Jacquin's fundamental book on rare plants. It was in Vienna that Bauer's life took a diversion when in late 1785 he was introduced to John Sibthorp, Sherardian Professor of Botany at the University of Oxford, who was passing through Vienna on his way to the eastern Mediterranean, for a botanical tour, together with the geologist John Hawkins. Sibthorp invited Bauer to join him and Hawkins as the tour's botanical artist. The party left Vienna in early March 1786 and by early December of the following year, Sibthorp and Bauer, bringing over 1500 sketches with him, were in Oxford.[28] For the next seven years, until 1794, Bauer remained in Oxford illustrating Sibthorp's magnum opus, the *Flora Graeca*.

By the time Ferdinand had completed his work for Sibthorp, Franz Bauer had already been working as a plant illustrator at the King's Garden, Kew for six years. Now, in 1794, both Bauers were in London, the elder Franz at Kew and the younger Ferdinand in London where, over the next few years, he received several commissions for botanical art, including an important one from Aylmer Bourke Lambert, a wealthy botanical collector, who asked Bauer to illustrate his planned book on the botany of the pine family – Lambert also had strong connections with New South Wales corresponding with the same people as Banks.[29]

Whether Franz Bauer introduced his younger brother to Banks or whether Lambert did – they moved in the same scientific circles and Lambert showed Bauer's work to Banks – by the time that the Flinders expedition was being planned, Banks would have known quite a bit about Ferdinand Bauer and, there is evidence that he had already considered Bauer for a voyage to the West Indies for when his work for Sibthorp was completed.[30] Banks was very impressed by Bauer's talents; they may have reminded him of Sydney Parkinson, Banks's artist on the *Endeavour*. Certainly, when it came to telling Earl Spencer that he was about to appoint Bauer, Banks did not stint on his words of praise: '[he is] in my opinion as good as Can be found in all Europe'.[31]

In the same letter to Earl Spencer, Banks revealed that he was looking for a landscape and figure painter and that the search was not going well. He had tried to get William Alexander, who had gone with the Macartney embassy to China between 1792 and 1794, but Alexander had turned down the offer because he couldn't leave his wife, who was unwell.[32] Alexander was ideal; not only had he demonstrated his talents for naturalistically depicting landscape and figures – a selection of these had been included in Sir George Staunton's *An Authentic Account of an Embassy . . .* , published in 1797 – but he also drew plants and animals with a perceptive eye.[33] Not many artists combined these skills or wanted to venture on a three-year voyage to a relatively unknown part of the globe.

This was a setback. Banks then tried to get Julius Caesar Ibbetson to go. Ibbetson, who was recommended by Charles Greville, had accompanied Charles Cathcart on the disastrous first attempt to establish direct diplomatic relations with the Chinese Emperor in 1787. When asked to join the Macartney embassy, Ibbetson declined and recommended his student William Alexander to replace him.[34] Ibbetson declined the New Holland expedition too, probably because his wife's death a few years earlier had left him in charge of three young children.[35]

Banks asked Earl Spencer's wife Lavinia, an accomplished artist in her own right with connections in the London art world, to help him in his search. Whether it was Lady Spencer or Charles Greville or someone else who made the recommendation is unclear, but by 12 January 1801 it seemed settled that William Daniell would be going to New Holland.[36]

In 1785, at age sixteen, William Daniell had accompanied his uncle Thomas Daniell on a trip to Canton and on an extensive tour of India, taking in Calcutta, Madras, Mysore and Bombay. They returned to London in September 1794 and worked together to turn their sketches into an exhibition of oil paintings and a series of views from their trip was published under the title, *Oriental Scenery*. At the same time as his uncle was elected an associate of the Royal Academy, William Daniell joined the Royal Academy Schools.[37]

Daniell was sufficiently well travelled and accomplished to appeal to Banks, and it seems he was to be the expedition's landscape artist. Then, at the end of April 1801, Daniell started to make conditions for his engagement. Banks was not pleased and demanded that Daniell make up his mind one way or the other: time was running out.[38] Daniell answered by return of post that he would only go if the ship had a passport from the French authorities. Then, only a few days later, he withdrew completely. Banks asked Robert Brown to speak to him, to find out whether the passport issue was the real reason for his change of heart or whether there was something else.[39]

There was indeed, for on 11 July 1801, Daniell married Mary Westall, the sister of Richard Westall and the half-sister of William Westall.[40]

Richard Westall was older and more famous as an artist than his brother, William Westall, who was just twenty years old and had only recently been made a probationer at the Royal Academy. It seems that at some point when Daniell was dithering about the appointment Banks contacted William Westall and got from him an assurance that he would be willing to go instead. It seems that Banks even put his name on an Admiralty contract before Daniell had officially withdrawn from the expedition.[41] Westall was the last member of the scientific team to be recruited and it was only shortly before the *Investigator* set sail.

* * *

Choosing the gardener was much more straightforward. Peter Good, who had so successfully managed the plant exchange between Kew and Calcutta, was the obvious choice.[42] When the *Royal Admiral's* bounty of Asian plants arrived at Kew in early February 1796, Good was with them and remained at Kew to ensure that they were well cared for. He was still there in mid October 1800.[43] But shortly after he was enticed away to work for General William Wemyss who had a magnificent garden at his home, Wemyss Castle, in Fife. Banks contacted William Aiton, head gardener at Kew, to find Good for him and to act as his go-between.[44] Aiton wrote to Wemyss telling him what Banks wanted and also offering a replacement for Good, a gardener hand-picked by himself from Kew.[45] Wemyss passed Banks's offer on to Good and on 14 January 1801, he wrote to Aiton that Good was delighted to accept it.[46]

Though he didn't mention it in the letter to General Wemyss, what Banks was planning was that Good would be looking after a plant cabin, built on the quarterdeck, which would be taking fruit trees out to New South Wales – Flinders had advised sending gooseberries and currants, which the colony lacked, plus some other plants such as cherries, nectarines and raspberries which had rarely survived the journey because the heat of the tropics destroyed them en route.[47]

The Admiralty instructions to Flinders in May 1801 revealed an elaborate and clearly thought-out plan. A plant cabin was to be provided for the collection of New Holland living plants.[48] On the voyage to Sydney it would be packed away in its frame. Not until they arrived was the ship's carpenter to erect it on the quaterdeck following Flinders's instructions. The arrangement of it and the provisioning of boxes with soil were to copy that of the plant cabin on HMS *Porpoise*, which was already in port. The *Investigator's* newly erected plant cabin was to house all the living specimens that Brown and Good would collect during the ship's survey of New Holland. These were destined for the King's Garden in Kew. Every time the *Investigator* returned to Sydney, the plants would be removed to the Governor's garden and cared for there until the ship was ready to return to England. At that point the plant cabin would be dismantled and the plant cabin from HMS *Porpoise*, which was larger, would take its place on the quarterdeck ready for the return journey. Banks had been impressed that Good had brought the plants home from India during a particularly unfavourable time of the year and 'in better order than any had ever arrived from that Climate', and he was confident that he could do the same with New Holland plants.[49]

Banks made two more appointments. He had already been in touch with William Milnes, the estate steward at Overton Hall, Derbyshire, one of Banks's properties, to find him someone with an interest in minerals who would be willing to go on the expedition. Banks was not asking for a mineralogist, just someone like a miner, who knew about minerals and who would collect specimens when directed to by the naturalist.[50] Though Brown was no geologist, Banks had solicited from the geologist John Hawkins, who had been on the trip to the eastern Mediterranean with John Sibthorp and Ferdinand Bauer, a kind of geological crib sheet about where to look for interesting minerals and how to take specimens.[51] Milnes's first choice wasn't allowed to go by his parents;[52] but he found a replacement called John Allen, a 26-year-old lead miner, whose family lived on the estate.[53]

When arranging the appointment of the expedition's astronomer in December 1800, Banks naturally turned to Nevil Maskelyne, the astronomer royal. By return of post Maskelyne told Banks he had the ideal candidate but wasn't certain he would accept the post because of ill health.[54] In fact, when Maskelyne saw John Crosley the next day, he was perfectly well and keen to go.[55] By 23 January 1801, after some bureaucratic hurdles had been cleared, Crosley was appointed, and the best instruments to take with him decided upon. Crosley, who was in his late thirties, had been Maskelyne's assistant and had some travelling experience, having been to the West Indies to test timekeepers, and also to the northwest coast of North America.

Certainly Banks had been given a free hand in choosing the 'Scientific Men' and it is quite clear from other interventions he made with the Admiralty that his remit was far greater than these appointments and included drawing up Flinders's instructions and sailing directions.[56] Banks made no secret of the fact that the Admiralty, because of other and more pressing commitments (that is, the war with France), had 'from time to time intrusted me with the execution under their orders of almost every detail of the lesser articles of the outfit of the Ship'.[57]

Not since the *Endeavour* had Banks so influenced the preparation of a voyage of exploration and collecting.[58] One can imagine with what excitement and anticipation he viewed the departure of the *Investigator* when it finally left Spithead on 18 July 1801.

Armed with its proper passport, the ship set sail for Madeira, its first stop, then called at the Cape before crossing the Indian Ocean to the far southwestern point of New Holland.

18

1801: Australia Circumnavigated and Beyond

On the day before he left Portsmouth, Flinders received his sailing instructions. These stipulated that he should make for King George III Sound (now the site of the city of Albany – Vancouver had named the sound in honour of the King in late September 1791) in the very southwest of New Holland and continue along the south coast towards Sydney.[1] He was 'to use [his] best endeavours to discover such harbours as may be in those parts; and in case [he] should discover any creek or opening likely to lead to an inland sea or strait', he should either examine it or not, as the circumstances allowed. The idea was that this part of his voyage was not the real survey. That wouldn't begin until after Flinders reached Sydney and joined the *Lady Nelson*, the ship that had been chosen to accompany the *Investigator*.[2] Then the two vessels would begin a series of detailed surveys of New Holland beginning with the south coast and going from east to west.[3]

Just when Flinders decided to disobey his orders is unclear, but he did.[4] Although he had been told to put into King George III Sound only for provisions, Flinders resolved instead to start his survey immediately. It was a stroke of luck for Brown, Good and Bauer that he did this, for it gave them the opportunity to collect a wealth of new flora and fauna.

For nearly a month, while Flinders surveyed the main bay and the adjoining bodies of water, the 'Scientific Men' collected on shore. Brown, in particular, was impressed by the variety of *Banksia* and found seventeen species not known in England.[5] Some plants were taken to be dried as herbarium specimens, but around seventy, including the *Banksia*, were collected in their living state and planted in boxes for the voyage to Sydney.[6] Brown and Good knew they would make a valuable addition to the King's Garden.

Because the botanising was unplanned the plant cabin that the ship was taking to be erected in Sydney remained below decks. Brown and Good had to use boxes, made on the spot, to transport the living plants. The sailors and the marines helped fill the boxes with earth. When the boxes, with their specimens, were put on board, after a delay due to bad weather, they were placed in the stern section on the quarterdeck – the safest place for what might be a rough voyage.[7]

Brown estimated that he had found almost five hundred species of plants, many of them unknown in Europe. This was his best botanical experience on Australia's south coast.[8] As the *Investigator* made its way eastward, despite examining many inlets, coves, sounds and islands, and having anchored in almost twenty places, the opportunities for collecting were minimal – most stops were for only a few hours and never longer than a week.[9]

While the botanising was disappointing, partly because of the lack of time and partly because many of the plants they saw were not in flower, Flinders assiduously carried out his detailed survey. For most of the ship's company, the voyage was tedious but on 8 April 1802 there was some excitement when the *Investigator*, after having departed Kangaroo Island (in what is now South Australia and offshore to the southwest of Adelaide) met up with Nicolas Baudin's ship, *Le Géographe*, in a place which Flinders named Encounter Bay.[10] Baudin's expeditionary ships, the *Géographe* and the *Naturaliste* (temporarily separated from Baudin's ship at the time of the encounter), had been at sea since October 1800. After having surveyed part of the western coast of New Holland, Baudin was now continuing with a detailed survey of the south coast, but, in contrast to Flinders, from east to west. The British and French ships remained in Encounter Bay for only two days after which they resumed their separate ways. Flinders and Baudin met again in Sydney a few months later when Flinders also met Emmanuel Hamelin, the commander of the *Naturaliste*, which was already in port when the *Investigator* arrived.[11]

Flinders wanted to get to Sydney before winter set in but he felt that the bay he next encountered had to be explored before they left, so the *Investigator*, which had arrived on 26 April, remained there six days.[12] This gave Brown and Good a chance to collect, and they found several plant species they had not encountered before.[13]

The survey of the south coast of New Holland was now complete and Flinders now made his way straight to Sydney, where the ship anchored on 9 May 1802, a little under ten months since leaving Portsmouth.

Though Flinders had written a short note to Banks from the Cape, Brown had declined to write on the grounds that he had, as yet, little to report. Now, in Sydney, and with more time on his hands, Brown wrote his first letter to Banks.[14] He reported that he had found over seven hundred plant species, most of which were from the southwestern corner of New Holland. Bauer had worked hard making more than four hundred drawings of plants and animals – he had also devised a system of colour coding his sketches so that they could be worked up into colour drawings when he returned to London.[15]

Good was busy preparing the seeds that had been collected along the way for the voyage to London.[16] The *Speedy*, the ship that had brought Caley to New South Wales, was on the verge of returning to England and Good wanted to make sure he didn't miss its departure. On 26 May, he put the seeds of 253 species of plants in a small box – to help preserve the seeds, the small box was put into a larger box which Governor King was sending to Banks.[17] The ship left on 6 June.

Not everything went as smoothly as the preparations. The living plants that had been potted into boxes in King George III Sound did not do well. Of the seventy plants that had been placed on the *Investigator*'s quarterdeck (without the use of a plant cabin to protect them), only ten – 'unfortunately the least interesting' – survived the voyage.[18]

For the next twelve weeks Flinders prepared the *Investigator* to continue the survey, this time in company with the tender, the *Lady Nelson*, under the command of Lieutenant John Murray. Flinders and Governor King decided to continue the survey up the east coast, into the Torres Strait (which separated the northeastern part of New Holland from the south of New Guinea); and from there, westwards to the Gulf of Carpentaria and beyond, until finally continuing south down the western coast.[19]

Flinders replaced some of the men he had lost through death or desertion with a number of convicts whose good conduct on the voyage would ensure a pardon on their return. Flinders also asked

King to allow him to take two Aboriginal men with him as guides: one of them, Bungaree, had been with him on the *Norfolk* on the coastal survey to the north of Sydney in 1799.[20] The other man, Nanbaree, came recommended by David Collins, the colony's judge advocate.[21]

As for the *Investigator* it was repaired and refitted for what might prove a long voyage – he was expected to be away for a year.[22] As far as its infrastructure was concerned there was only one new addition: the ship's plant cabin. It had been brought from England in pieces and now had to be assembled. When Flinders realised what it would look like and what it would weigh when the boxes were filled with earth, he decided it was too big and would, in bad weather, make the ship unstable. In consultation with Brown, they decided that a smaller plant cabin – roughly one-third the size of the original – would be better for the ship: and it would probably do for the plants especially if the ship returned to Sydney occasionally.[23]

Brown, Good, Bauer and Allen, meanwhile, passed the time collecting in the vicinity of Sydney and making drawings. At first they went on their own but, shortly after arriving in Sydney, they met George Caley, who willingly offered his services as a guide: he was pleased, as he later told Banks, to have a real naturalist to talk to.[24] Brown found the area rich in flora and Bauer increased his stock of drawings to seven hundred.[25] Seeds were placed in paper and given to Governor King for safe-keeping; new species of living plants were potted and joined the few that had survived the previous voyage. They, and Bauer's drawings, were left in the care of Governor King.

On 21 July 1802, in company with the *Lady Nelson*, the *Investigator* left Sydney to head north along a coast that Banks knew well. For the next three months the two ships followed the coastline, and explored the small coves and islands that dotted the route. Brown and his company of scientific men investigated each stopping place and tried to collect seeds, plants for drying and living plants for the plant cabin. Many more plants were seen than taken.[26] The first collection of living plants, fourteen in total, included a very striking red flowering plant which Brown first labelled *Embothrium banksianum* and which was subsequently published as *Grevillea banksii*. This was at Port Curtis where they stayed for a few days from 5 August.[27]

Unfortunately, as Brown later wrote, it was the wrong season for collecting: most plants weren't in flower or fruit and there were few

which were new to him.[28] Nevertheless, as the ship sailed north, all
the scientific men kept busy. By mid October the plant cabin had
almost thirty plants, from fifteen species, growing well.

The ships had been away for less than three months and were barely
halfway on their voyage north, when, on 18 October, Flinders told
Lieutenant Murray that he was sending the *Lady Nelson* back to Sydney.
Flinders explained that the ship was not in good shape – it had already
suffered several accidents including damage to the keel – and it would
be better for everyone if it returned to Sydney.[29]

For those on the *Investigator*, the news that the *Lady Nelson* was
going back to Sydney was welcome. It was a chance to send letters.
Flinders wrote to King, to Banks and others as well. Brown and Bauer
also took advantage of the opportunity.[30] There were some changes in
personnel. A few sailors were exchanged between the two ships and
Nanbaree decided to return to Sydney with the *Lady Nelson*. Bungaree,
who had proved to be invaluable in the various encounters with local
people, remained on the *Investigator* to continue his work.[31]

Now on his own, Flinders decided to get to the top of New Holland
by sailing outside the Great Barrier Reef. It took two days of very
careful and patient navigating to find a way out of the reef, but on
20 October, he found a narrow channel, now called Flinders Passage,
and with a favourable wind the *Investigator* sped towards the Torres
Strait.

They reached Murray Island on 30 October but the stay was very
short. For the next few days, Flinders steered the *Investigator* very
carefully through the reefs and islands that dotted the shallow stretch
of Torres Strait. A few days later, after passing through the Prince of
Wales Channel to the north of Hammond Island, to the northwest
of Cape York, at the very top end of New Holland, Flinders changed
direction and, for the first time since leaving Sydney, sailed south into
the Gulf of Carpentaria.[32]

A crucial part of the survey and the botanical exploration was about
to begin. From now on this was new territory.

Named after a Governor-General of the Dutch East Indies in the
seventeenth century, the Gulf of Carpentaria had only been visited by
Europeans, all of them Dutch, three times, most recently in 1756.[33]
Only the east coast had been charted. Flinders only had a map produced
in 1663 and augmented on a few occasions thereafter.[34]

They entered the Gulf on 3 November and spent the next four

months surveying the coastline. Flinders was not only trying to produce the first detailed and accurate survey of the region, but he was also trying to establish whether New Holland was separated from New South Wales by a channel of water.

The survey was extremely painstaking. Every cove and every river on the Gulf's western coast was carefully examined. Brown, Good and Bauer had little opportunity to see much of the eastern side of the gulf as Flinders hardly touched that coast, relying on the Dutch surveys. They did better on the southern and western side of the gulf.

For some time, while they worked away at surveying and botanising, the ship had been leaking, not much at first but it was increasing. Now at anchor in the Wellesley Islands near to the southernmost point of the gulf, the carpenters, who were caulking the ship, reported to Flinders that they had discovered that one of the timbers was rotten. Soon more rotten parts were discovered, not just in the timbers but elsewhere: 'almost every hour, report after report of rotten parts found in different parts of the ship, timbers, planking, bends, tree-nails &c until it is become quite alarming'.[35]

The time had come for a complete overhaul of the ship to ascertain its precise condition. On 24 November, Flinders ordered John Aken, the master, and Russel Mart, the chief carpenter, to make a thorough examination and to report the results to him. Two days later, the report, in writing, was handed to Flinders. It was daunting – the ship was rotting. Nothing could be done about it at present because laying the ship on its side to make the necessary repairs might compromise its structure and make things worse. They concluded that though in a year's time there wouldn't be a sound piece of timber in the ship, 'if she remains in fine weather and happens no accident, she may run six months without much risk'.[36]

When surveying Flinders followed a strict principle that his work would be so accurate and so thorough that 'there shall be no necessity for any further navigator to come after me'.[37] The state of the *Investigator* was going to make that difficult. After agonising as to what to do, he decided to continue with the survey as best he could.

On March 6, the *Investigator*, having completed the survey of the gulf, and having sailed through many of the islands scattered to the north of Arnhem Land, Flinders made a momentous decision. He would make for Sydney counter-clockwise sailing westward along the coast of northern Australia. Many of the crew were ill and his own health was deteriorating. He thought he could make it to

Sydney within Aken and Mart's estimate of how long the ship could survive.[38]

So Flinders set sail, but three weeks later he was forced to call at Timor because of the crew's poor health.

On 31 March 1802, the *Investigator* anchored in Kupang, Timor's main port – another unexpected opportunity for sending letters. Three days earlier, Flinders had written to Banks. 'As yet we are barred out of the interior part of New Holland', meaning that the Gulf of Carpentaria, and the rivers that emptied into it, did not lead anywhere. 'Would I could make a river', Flinders went on, 'then should not the very center of this great country escape our examination: but if no river or strait exists, I fear our utmost exertions will not find any.'[39] Flinders also told Banks about the poor state of the *Investigator* and that he had decided to put into Timor because of scurvy among the crew. As soon as the ship was provisioned with fresh produce he would go straight to Sydney where, he hoped, to find another ship, either HMS *Porpoise* or HMS *Buffalo*, which were both in the port, or any other ship he could hire or buy to continue the survey.[40] Flinders wrote to Evan Nepean at the Admiralty advising them of his proceedings since leaving Sydney the previous July.[41]

Robert Brown also wrote to Banks. The land around the Gulf of Carpentaria, Brown told him, was not very interesting from a botanical point of view. The plant species were few in number, and most were identical to plants from the Endeavour River area (and so in Banks's herbarium already) and others were identical to plants from India; 'the number of absolutely new species hardly amounting to 200.'[42]

On 8 April, having loaded the ship with fresh water, molasses and hogs, the *Investigator* left the port and headed into the Timor Sea. Just north of the Tropic of Capricorn, on 24 April, Brown checked on the plants in the plant cabin. The news was not good. Only three of the twenty-seven plants collected on the east coast were still alive. Those more recently collected and potted had fared better, but, even so, mortality stood at around 50 per cent.[43] So much for transporting living plants by sea. By contrast the seeds that were sent to Banks on the *Speedy* had arrived in excellent condition and they were all already growing in the gardens at Kew. In Banks's estimation: 'much hopes built on the success of them which we expect will create a new Epoch in the prosperity of that magnificent Establishment by the introduction of so large a number of New Plants.'[44]

On 9 June 1803, two months and a day after weighing anchor in Kupang, and having only made a single stop at the Recherche Archipelago in southwestern New Holland, the ship was back in New South Wales.

Despite Flinders's best efforts, several of the *Investigator*'s crew lost their lives during the run to Sydney. Although Peter Good was not one of them, he too died, like them, of dysentery, only three days after they arrived.

Now Flinders had some more difficult decisions to make.

The survey was 'scarcely half completed' and was expected to take a further two and a half years. The *Investigator* was too far gone to be repaired.[45] Flinders now had two options: HMS *Porpoise* or the *Rolla*, a ship that had recently arrived with convicts and supplies from Cork. Both ships were at anchor in Port Jackson.

The *Porpoise* needed repairs and a refit, which would take at least six months; and the owners of the *Rolla* wanted £11,550 from the government to make it available. Its refit would also take six months.

Governor Philip Gidley King, whom Flinders consulted at every step, suggested sending the *Porpoise* back to England. This seemed to offer Flinders a way out of the impasse. Flinders and the company from the *Investigator* could travel as passengers on the *Porpoise*, and once he was in London Flinders could negotiate with the Admiralty for a new ship to continue the survey.

Once the return of the *Porpoise* to England was decided, Brown put a large collection of seeds in boxes that were themselves placed in strong cases into the ship.[46] The plant cabin from the *Investigator* was erected on the quarterdeck and the living plants from the circumnavigation were placed in it: Brown chose a man from Sydney to take care of them on the voyage.[47] The plant cabin, which had been brought to Sydney on HMS *Porpoise* in 1800 and which had been in store, was handed over to Brown to use for the living plants that he would collect in New South Wales.[48]

On 10 August, the *Porpoise*, in company with the *Bridgewater*, an East India Company ship, and the *Cato*, a merchant ship, left Sydney, without Brown and Bauer who had jointly petitioned Flinders and Governor King for permission to stay behind. They reckoned that the time that Flinders would be away – about eighteen months – would be better spent in New South Wales, where they could organise and augment their collections. John Allen, the mineralogist, and William Westall, the landscape artist, joined Flinders for the homeward voyage.[49]

Flinders had decided that the ships should sail outside the Great Barrier Reef until they reached the Torres Strait.[50] It all began well; the voyage benefitted substantially from a steady southwestern wind. As the ships progressed the wind direction shifted but always in their favour. However, only a week after leaving Sydney, disaster struck. In the middle of the night, first the *Porpoise* and then the *Cato* hit an uncharted reef, well to the east of the Great Barrier Reef. The *Porpoise* was a wreck; the *Cato* not much better. The *Bridgewater* was nowhere to be seen. At daylight Flinders surveyed the scene. The *Porpoise* was lying in shallow water. Everything on deck was destroyed: the plant cabin smashed and the living plants killed by sea water. The hatches, fortunately, remained shut so what was below decks, such as the stores, provisions and Brown's seed collection survived. All the men, apart from three who had drowned, were assembled on the sandbank and, with the provisions saved from the ships, they set up camp as best they could.

Flinders waited for a while for the *Bridgewater* to appear and rescue them, but with no sighting after five days, he decided that their only hope of survival was for him and a few of the men to take one of the ship's cutters and row back to Sydney.[51]

After some seven hundred miles and nearly two weeks of rowing south in the Pacific, on 8 September 1803, Flinders, unshaven and unrecognisable, staggered ashore to tell Governor King what had happened. King put three vessels at Flinders's disposal to collect the survivors and get Flinders to England. HMS *Cumberland*, a small schooner of 30 tons, which Flinders and a hand-picked crew would sail home; the ex-convict ship, *Rolla*, of 450 tons, to take most of the survivors to Canton from where they would find passage to England on the returning ships of the East India Company; and the *Francis*, another small schooner which would return to Sydney with whoever wanted to go there.

With Flinders in charge of the flotilla, the three ships left Sydney on 21 September to sail northwards hoping to find the castaways. On Friday, 7 October, after making a small correction in their bearings, they spotted the flag flying above the sandbank, the highest point on the reef.[52] Everyone was overjoyed and relieved. As Flinders later recalled: 'A salute of eleven guns from [the bank] was immediately fired . . . on landing, I was greeted with three hearty cheers, and the utmost joy of my officers and the people; and the pleasure of rejoining my companions so amply provided with the means of relieving their distress, made this one of the happiest moments of my life.'[53]

Then Flinders chose his men for the voyage to England. A few of the survivors boarded the *Francis* to go back to Sydney. Most of them went on the *Rolla* to Canton. This included William Westall, who remained in Canton for two months before travelling to Bombay where he stayed for almost four months before returning to England. John Allen, who also went to Canton with the rest of the *Investigator's* men, would prove invaluable to Banks in the Chinese port. Even though Brown thought him 'of little use', he had actually learned enough about plants to do Banks a great favour.[54]

The *Rolla* had an uneventful voyage to Canton as did the *Francis* to Sydney. Flinders continued to be dogged by misfortune. He was forced because of the worsening condition of the ship to put into French-held Mauritius on 15 December 1803. France and Britain were again at war and Flinders and the *Cumberland's* company were ordered to sail to Port Louis, the island's main harbour where they were immediately detained because the ship didn't have the right papers – Flinders himself was suspected of being a spy. Though all his companions were freed at some time on prisoner exchanges, Flinders remained confined for six and a half years until 28 March 1810 when he finally received his liberty from Charles Decaen, the island's Governor and Flinders's jailer.[55]

Meanwhile in New South Wales, Brown and Bauer continued botanising.[56] For most of the time that they were based in the settlement, they pursued separate interests in different places. Sometimes alone and sometimes in the company of Caley, Brown collected widely in and around Sydney, but his collecting took him much further afield than that. During 1804, he spent almost seven months in Van Diemen's Land, on two separate visits; and a week in Port Phillip. Bauer, for his part, explored the area to the north of Sydney and spent several months on Norfolk Island. It was a very productive time for both.

The agreement made with Governor King was that Brown and Bauer would stay in and around the colony for eighteen months before going home. In mid March 1805 Governor King found he needed a ship to sail back to England as soon as possible. Fortunately, the *Investigator*, which had been lying at anchor and used as a store ship since it was decommissioned in June 1803, had already undergone a radical transformation.[57] When examined carefully in May 1804, the ship's hull proved to be sound, so King ordered the ship to be cut back – the upper deck removed and some of the masts shortened significantly. With these changes and other repairs it was deemed seaworthy. A test voyage to Norfolk Island in early 1805 and back to Sydney on 11 March (Bauer returned from his visit to Norfolk Island on the ship) confirmed that the *Investigator* could make it back to England.[58]

King told Brown of the decision to send the *Investigator* home with Brown and Bauer on it. After a number of arguments between King and Brown that lasted almost two months and concerned where in the ship Brown could store his plant collection, an agreement was struck.[59] Brown boxed up all his dried plants and seeds and zoological specimens; Bauer took his sketches and drawings. The living plants were left behind with Caley in Parramatta, together with the plant cabin from the *Porpoise*, awaiting a convenient opportunity to send them back later.[60]

On 23 May 1805, the newly revamped *Investigator* set sail for England. Captain William Kent, the commander of the ship, was instructed to take it to England via Cape Horn.

Nearly five months later, on 13 October, and without any misfortune, the *Investigator* arrived in Liverpool.

In January 1806, by which time the boxes and cases from the *Investigator* had been opened and their contents organised and evaluated, Banks was able to tell William Marsden, the first secretary at the Admiralty, of the preliminary results of the Australian venture.[61]

Brown, Banks reported, had amassed a unique collection: the seeds that had already arrived had sprouted in Kew – 'at this Time [it] Constitutes a large Portion of the newest ornaments of that Extensive & Possibly unparaleled Collection.'[62] Now he had brought back his collection of dried plant specimens, a collection that amounted to 3600 examples – 1700 collected during the circumnavigation, 1000 from Sydney and its environs and 700 from Van Diemen's Land (the remaining 200 were from Timor) – many of them were new to science.

As for Bauer, he, too, had been extremely productive. He had brought back more than 2000 sketches. 'The Quantity of Scetches', Banks remarked with admiration, 'he has made during the Voyage & prepared in such a manner by References to a Table of Colors as to

Enable him to Finish Them at his Leisure with perfect accuracy is beyond what I confess I thought it Possible to Perform.'

Banks predicted it would take several years for the collections to be arranged and before Bauer could produce the coloured versions of his drawings. As for publishing the results, an outcome that both Banks and the Admiralty insisted upon, the wait would be a little longer. Brown's *Prodromus Florae Novae Hollandae et Insulae Van-Diemen* came out first, in 1810, followed, in 1813, by Bauer's *Illustrationes Florae Novae Hollandae.*[63]

Finally, in 1814, the two volumes of Flinders's, *A Voyage to Terra Australis*, came out.[64] It united the work of the 'Scientific Men': there was the narrative of the voyage, a selection of Westall's landscape drawings, ten of Bauer's drawings of Australian flora, Brown's botanical glossary, coastal views and the charts. Flinders's map showed a single land mass with New Holland on the west and New South Wales on the east. Though this was not the first time that the land mass had been depicted in this single continuous way, it was the first time that it was given its new name – Australia.[65]

PART V

Botanical Diplomacy and the Tropics

Preface

Because of the unpredictability of ocean voyages and because Banks often found himself responding to opportunities for botanical projects that came his way rather than initiating them himself, he became involved in many diverse expeditions at the same time. This had been the case for almost twenty years and continued to be so.

In the opening years of the new century, many of Banks's projects, which had in some cases been initiated many years before, were still ongoing in many widely different parts of the world: Francis Masson, who had first gone to the Cape in 1772, was now collecting plants in Upper Canada; the cochineal story, which began in 1786 in Madras, had now moved to Central America where a Scottish physician had excitedly informed Banks that the insect and the cactus were growing and accessible in nearby Guatemala; the *Investigator*, under the command of Matthew Flinders, had just returned to Port Jackson from its circumnavigation of Australia, and plans were under way for yet another voyage; and George Caley, the fearless self-taught botanist, was at his home in Parramatta preparing a new and much larger expedition to return to the Blue Mountains.

In 1803, Banks enlarged his botanical projects portfolio even more when he sent William Kerr, a Kew-trained gardener, to Canton to collect plants for the royal garden. What was important about this appointment is that Kerr was not, like the brothers John and Alexander Duncan before him, employed by the East India Company, but was paid by Banks to collect for Kew. As it turned out, Kerr's presence and work in Canton led to some very unexpected and valuable exchanges of plants, a version of diplomacy unimagined before and much more successful than Macartney's attempted embassy a decade earlier.

Kerr remained in Canton for seven years before being promoted to a new post in Ceylon. Though he was more successful than most of Banks's contacts in Canton, Kerr's collection had still been largely confined to local plant varieties: the longed-for plants from China's interior were still out of direct reach. With the launch of the third attempt to establish diplomatic relations between Britain and China in 1816, Banks's hopes for obtaining these rare plants were rekindled but dashed. The Amherst embassy, as it was called, was even less successful than that of Macartney's.

For the time being, Canton continued to be the only part of China available to the West and the knowledge of its natural history remained as it had been, second-hand. Though travel to and within Brazil was not as restricted as that within China, nevertheless, when Banks was in Rio de Janeiro on the *Endeavour* in 1768, his attempts to collect plants were largely thwarted. Although James Bowie and Allan Cunningham, two Kew-trained gardeners, had only meant to go to Brazil to change ships for their intended destinations of the Cape and Port Jackson respectively, they ended up staying for almost two years and provided Kew with many new plants from the area around Rio de Janeiro and São Paulo.

Though Banks was not enthusiastic about getting tropical plants for Kew – they were more difficult and expensive to grow than temperate varieties – he couldn't resist the chance of getting plants from a part of the world that was entirely unrepresented in Europe's gardens. When John Barrow, the powerful Second Secretary of the Admiralty, suggested an expedition to Africa to solve the riddle of the Niger River, Banks, now in his seventies, jumped at the chance to expand its remit to include natural history and a substantial scientific team. It was to be his last venture into global botanical projects.

19

1803: William Kerr in Canton

According to the second edition of Aiton's *Hortus Kewensis*, only nine Chinese plants had been introduced into the King's Garden during the thirteen years in which John and Alexander Duncan were in Canton – less than one plant in each trading season. Not very impressive, but that was about to change dramatically as Banks would soon be appointing someone with extensive botanical experience to go to Canton and supply Kew with a wide variety of Chinese plants.

On 9 February 1796, New Year's Day in the Chinese calendar, the 85-year-old Qianlong Emperor abdicated in favour of Yongyan, his heir and fifteenth son. Taking the name of the Jiaqing Emperor, the 35-year-old Yongyan was an emperor in name only. Qianlong still held on to the most important reigns of power and made the decisions on issues of policy and appointments.[1] The Jiaqing Emperor had no choice but to accept the terms of the abdication. Then, on 7 February 1799, almost three years to the day of the abdication, the Qianlong Emperor died.

The internal politics of the Chinese Empire were thrown into disarray as the new Emperor dealt with appointments his father had made which were not to Yongyan's liking. But as far as external relations were concerned, it was pretty much business as usual. A few details changed but the Canton System, enforced under the Qianlong Emperor, remained in place.

While the system hadn't changed, Canton's trading pattern had. The principal alteration was that many of the European trading companies, except the British, had left. During the 1770s, there had been ships from all over Europe, from France, Holland and Spain, in addition to those from Britain and the Nordic countries. In the year of the Qianlong Emperor's death, by contrast, there were in addition to the thirty British ships, only four other European ships (one from Sweden and three from Denmark) trading in Canton.[2] Into this vacuum, came the Americans. Beginning in 1784, their presence grew substantially during the years of the French Wars (from 1793). In 1799, there were eighteen American ships trading out of Canton, making them the second largest foreign presence after the British, a situation that continued for decades to come.[3]

Whether it was the toll taken by the French Wars on merchant shipping returning to Britain from the East – privateers and French frigates operating out of Mauritius and Réunion often attacked vulnerable British ships – or for other reasons, the flow of Chinese plants to Britain fell to a new low in the years following the Duncans' departure from Canton.[4] Banks could claim to have introduced only four Chinese plants to Kew between 1796 and 1803.[5]

Towards the end of 1801, however, Britain and France took the first hesitant steps towards making peace. Both sides signed the Treaty of Amiens. As a result of these negotiations, on 25 March 1802 and, for the time being, British merchant shipping in the Atlantic, Indian and Pacific Oceans could proceed unmolested. Reflecting the new maritime conditions, the number of ships sent from Europe and the United States in the first trading season following the peace jumped by almost 20 per cent and included one French ship, something that hadn't been seen in Canton for several years.[6]

Even though Banks didn't believe the peace treaty to be very peaceful – 'no more than a turbulent & quarrelsome Truce', as he put it to Robert Brown, the botanist on Flinders's HMS *Investigator* – he nevertheless did not want to be left behind in advancing botanical science and so renewed his involvement in the pursuit of Chinese horticultural knowledge.[7] Banks began planning a collecting programme with a specific gardener in mind around the time that the Treaty of Amiens was concluded.[8] The methods employed and the results of the collecting activities were dramatically different from previous efforts. Though the Canton System still set the framework, it became less rigid, and with good relations between the East India Company supercargoes and the Hong merchants, Banks and the supercargoes were able to push the system to its limits.

*   *   *

Banks's ideas for introducing new Chinese plants to Kew coincided
with the King's renewal of interest in his gardens after a period of
neglect. The King told Banks that he wanted plants from South
America, specifically from the area around the River Plate. So Banks
contacted Lord Hawkesbury, President of the Committee for Trade,
and an old political friend, to use his influence to convince the Spanish
authorities to allow a Kew gardener to stay in Buenos Aires, entirely
at the King's expense, and to collect seeds as appropriate.⁹ Hawkesbury
replied that he would write to John Hookham Frere, who was going
to Madrid to renew diplomatic relations as the King's ambassador after
an interval of six years.¹⁰

The man Banks had in mind was a young gardener, 'who had no
education beyond his line of life' and whose character, therefore,
'cannot give any kind of political umbrage'.¹¹ However, events in Madrid
were not progressing smoothly, and by late December, Frere had not
yet been installed and the issue of sending a gardener to Buenos Aires
remained unresolved.

Banks wanted to use the time while waiting for an answer from
Madrid productively. He told Hawkesbury that he would like to send
the gardener to China in the meantime. Hawkesbury insisted that this
was not a good idea as he expected an answer any day.¹² Eventually
Frere took up his ambassadorial position but the gardening job never
materialised.

By late 1802 then, Banks had already planned a new China plant
collection and identified the man he wanted to carry it out. This was
to be the first time that Banks had initiated the collection process by
sending his own hand-picked gardener to China, instead of, as in the
case of the Duncans, relying on the good will of amateurs. The young
man that Banks had in mind was 23-year-old William Kerr, born in
the Borders, and currently employed at Kew.¹³ William's father John
worked for the well-established and highly regarded nursery of
Archibald Dickson's at Rassendeanburn, near Hawick, where William
was born.¹⁴ Banks had known William Kerr for some time and thought
him 'a considerate & a well behav'd man' – not someone who would
annoy the Chinese, any more than he would have annoyed the Spanish
had he gone to Buenos Aires.¹⁵

To activate the plan, Banks had to enlist the help and support of
the East India Company. On a visit to their headquarters on 5 April
1803, Banks put his idea to Jacob Bosanquet, a director and Deputy-
Chairman of the Company.¹⁶

Bosanquet reacted warmly and positively. He told Banks that he had passed on the plan to the Chairman and several directors and that their reaction gave him the confidence to say that it was as good as agreed.[17]

In his reply to Bosanquet, Banks made it clear that there was more to his idea than simply that of 'adding something to the laudable amusement of our beloved sovereign'. The plan, as he pointed out, 'carries with it a fair Prospect of procuring vegetable articles of food for Commerce & for ornament to our west Indian colonies & to the mother country of advancing materialy the Science of Botany'. Banks also intended that the plants from China would travel to Europe housed in a plant cabin like the one he had built for the East India Company ship, the *Royal Admiral*, which, several years earlier, had successfully carried plants from Kew to the Calcutta Botanic Garden. Finally, Banks expected that the gardener he would appoint would remain in China for three years and that another gardener would accompany the plants back to England.[18]

Several days later, Banks received official confirmation from the Court of Directors, in the form of a letter to the supercargoes in Canton, that he could go ahead as planned. The letter explained that a Mr Kerr would be coming out to Canton as the King's gardener, and that he would be there under the supervision of David Lance, who was returning to Canton as a supercargo after an absence of almost fifteen years. Both men would travel together from London on the next available ship. The Court was certain, the letter remarked, that between Mr Kerr and Lance, 'the greatest care would be taken to prevent any Umbrage being given to the Chinese'.[19]

Banks now drew up his instructions for Kerr.[20] This was Kerr's first big opportunity and Banks began by reminding him of this: that the assignment, if carried out satisfactorily, would elevate him in a way impossible otherwise; and that it was a learning experience, an apprenticeship as a botanical 'Traveller' – to 'Learn to conduct yourself with propriety under the various Circumstances & in the different Kinds of Society, in which a Traveller must of necessity be Continualy Engagd'. David Lance would be his teacher. There followed instructions on the kind of contextual information Kerr needed to send to William Aiton at Kew. In addition to the plants themselves, notes on the necessary climate, soil, and manure for each living plant should be sent to Aiton. Banks asked Kerr to provide, at the same time, a dried specimen, in flower and in fruit, so that a correct botanical description could be made.

Banks set many questions for Kerr to answer and the kind of details he required were no different from those which he had asked of the gentlemen of the Macartney embassy. Specific plants were highlighted, especially those that grew in China's more northerly provinces, plants that would have the best chance of surviving in England's climate. The weeping thuja, a striking evergreen whose branches commonly sprawl along the ground, would, Banks thought, 'be the Chiefest ornament of our groves' if Kerr could get a sample of it.

Banks also wanted specific information about Chinese horticultural methods, especially as they applied to fruit growing. He wanted to know, for example, how they cultivated their fruits; which methods of husbandry they used; and how they processed manure and applied it. Banks grew many fruits in his estate at Spring Grove, to the west of London, where he also experimented with breeding and cultivation. He expected that knowledge of Chinese methods would give him valuable new insights.

Banks conceived of Kerr as a new kind of collector, one who would get practical experience of Chinese horticulture, instead of just buying plants from the Canton nursery as others had done before him. In particular, he hoped and expected that Kerr would tend a garden in Canton provided for him by Lance. Here, with the help of a Chinese assistant, Kerr was to grow and care for those plants he intended to ship back to Kew. Moreover, he was always to have on hand examples of those plants growing in boxes or in Chinese pots of the correct dimension to fit an on-board plant cabin, ready, at a moment's notice, to be put on a ship returning home.

Banks gave Kerr two main lists he was to use to guide him to the desired Chinese plants. One of these, Banks culled from the English translation, published in 1788, of the *General Description of China*, by Abbé Jean-Baptiste Grosier, a Jesuit priest, who, though never having been to China himself, devoted over forty years to studying the country. Grosier's work was considered authoritative and this English translation was, in fact, the thirteenth volume of his massive history of China. In many cases, Banks could only provide Kerr with a transliteration of the Chinese name for the plant in question.

The other list Banks provided for Kerr was contained in his anno-tated Book of Chinese botanical illustrations, which had last been in Canton in early 1794, and which Dr Hugh Gillan, the physician to the Macartney embassy, had brought back with him. Banks gave it to Lance for safe-keeping on the voyage out. It still had the system of crosses corresponding to the degree of desirability of a plant that Alexander Duncan had put to such good use when he collected for Banks in Canton in the early 1790s.

Banks knew the difficulty in getting precisely those Chinese plants that were not yet in Kew. His annotated book would, without doubt, help, but even that wasn't enough. He told Kerr to look out for Chinese botanical drawings in Canton wherever they were depicted – on paper and even on furniture because, as he put it, 'the Plants Painted by the Chinese . . . are so exact & so little exaggerated as to be intelligible to a Botanist.' Not only would these depictions provide Kerr with accurate representations of Chinese plants but they could advance his knowledge: 'In this manner', Banks noted, 'you may Procure the names of valuable Plants Produced in distant Provinces & after perhaps be Enabled to procure by Mr Lance's Correspondence the Plants themselves.'[21] As for the Chinese language, Banks mentioned nothing about it in his instructions. His only contribution in this respect was to provide Kerr with a list of Chinese characters that George Thomas Staunton, who had returned to Canton in 1798 as a Company employee, had drawn up from Banks's books and manuscripts.

Kerr was now prepared to go. The only thing that remained to be resolved was the issue of money. A little over a week after receiving the go-ahead, and just a few days after settling details with Kerr, Banks went to the headquarters of the East India Company to see Bosanquet. He and Banks hammered out a deal by which the King would pay for Kerr's salary and the Company would meet all other expenses – his maintenance and all the costs associated with preparing the Chinese plants for their return to England and for their carriage and care.[22]

Banks realised full well that Kerr's success in Canton depended on Lance and his relationship with the principal Hong merchants. Lance, who was forty-six years old at the time, was what might be called 'an old China hand'.[23] He had started working for the East India Company in 1773 when he was just sixteen and became a supercargo six years later. For the next sixteen years he remained in Canton but, like other supercargoes, he engaged in private trade in addition to his responsibilities to the Company. He made a fortune and when he returned to England in 1789, he married the niece of Thomas Fitzhugh, a senior Canton supercargo, and settled down near Southampton.

Lance was now returning to Canton but not before being asked to execute a secret diplomatic mission to Cochin China (present-day

Vietnam) to convince the King to remain allied to the British rather than falling in with the French. On 3 April 1803, after having received reports that the French were sending six ships and 1300 troops to the East Indies, Lord Castlereagh, the President of the Board of Control, impressed the Chairman and Deputy-Chairman of the East India Company of the need for immediate action, to communicate directly with the Emperor of Cochin China.[24] A week later Lance was selected for the mission. Banks was probably told about this conversation for soon after Castlereagh's meeting with the East India Company, Banks knew not only that Lance was on his way to China but that he was going to Cochin China on the way.[25]

It is not clear whether Lance and Banks knew each other before this but they certainly had friends in common, especially Alexander Dalrymple, a Fellow of the Royal Society, a long-time associate and employee of the East India Company, and, since 1795, hydrographer to the Admiralty. It was Dalrymple's memorandum to the East India Company's Court of Directors, which led to Lord Castlereagh's intervention and Lance's appointment.[26]

Whatever their relationship, Lance asked little of Banks apart from requesting a number of fruit plants, which he wished to take with him to Canton. These included apples, pears and plums; soft fruits such as raspberries, strawberries and gooseberries; and peaches, nectarines and apricots. At least one rose was also included. These were not intended for the Chinese, as Banks understood it. All these European fruits and more, Banks was certain, were already growing in China.[27] Their actual destination remains unknown. Perhaps they went to Lance's Chinese acquaintances; or, perhaps, Lance was going to distribute them among his fellow supercargoes, to be grown in the factory's garden, located between the river front and the main building; or possibly they were for Macao, where some of the supercargoes had their own private gardens.[28]

The *Coutts* was the ship chosen to take Kerr and Lance to Canton. It had already made two round trips to China since 1796, the year in which it was built. Banks's selection of fruit trees was carefully loaded onto the ship and stored 'in a place of security', as Robert Torin, the ship's captain, had promised Lance.[29] By late April all was ready and, on 6 May 1803, the *Coutts* left for China.

This was a little late in the year for a voyage to China. The Canton trading season had ended a couple of months earlier and the Company ships that had traded there were already on their way back to England. The *Coutts* was to sail with several other ships, some bound for China and others for India. Though the small convoy left during an interval of peace, less than two weeks after the ships departed the English shore, Britain declared war on France and the conflict between the two countries resumed. Fortunately for the *Coutts*, Charles-Alexandre Linois, who commanded the fleet that Napoleon had ordered to sail to the Indian Ocean and South China Sea, was at least two months ahead: the French squadron, which had left Brest on 6 March 1803, was already at the Cape and would soon be sailing into the Indian Ocean.[30]

The *Coutts* had an uneventful voyage to the Cape. The only sign of trouble was when, in the warmer waters of the Indian Ocean, the surgeon, John Livingstone, noted signs of scurvy among the crew. Liberal amounts of lime juice averted more serious consequences. Not long after, however, while nearing the coast of Cochin China, the captain of a British brig, the *Eleanora*, came on board with the news that the country's ruler, the Emperor Gia Long (also known as Nguyen Anh) had left his capital Saigon and was somewhere in the north. There seemed no point in delaying the ship's progress to Canton by trying to find the Emperor or by speaking to officials who probably had little or no power, so the decision was taken to abandon their diplomatic mission and sail north to Canton.[31]

It was then, on 23 September, that the ship was struck violently by a typhoon. First the foresail was split, then the topmast was blown away and finally the main mast broke. The ship rolled badly and took on so much water that the pumps had to be run constantly. Had the ship been further away from Canton when the typhoon struck, the result would have been catastrophic, but the combination of the captain's skill and the proximity of the port saved them. Less than a week after the *Coutts* survived the typhoon, it limped into the anchorage at Whampoa, alongside the other East India Company ships there for the 1803–4 trading season.[32]

Most of the fruit plants that the *Coutts* was carrying did not survive the typhoon though they were, Kerr noted, in good shape until this point. The plant boxes were smashed and their contents washed into the sea. Those that Kerr could save were not in a good state and he was left with only the rose to propagate.[33]

Banks's hope that Kerr would have a garden to tend was fulfilled. Lance arranged with Mauqua, one of the Hong merchants, to let Kerr use a piece of land about a mile further upriver from the site of the thirteen factories. As for getting plants, Kerr soon understood that, in common with the other collectors who had preceded him, he would have to rely on Canton's 'Nursery Gardens', a further two miles upriver from Kerr's garden.[34] It seemed that when the nursery merchant did not have a particular plant in stock, he would substitute another for it without saying what he'd done. Kerr realised that he would have to learn some Chinese, enough, at least, to be able to recognise the Chinese names for the plants he wished to buy.[35] As he remarked: 'at first [it] appeared to me an arduous task, but repeated impositions resulting from my deficiency in this respect at least roused my attention so far that a great number of the names are now become familiar as well as a number of other words in the Language.'[36]

That he needed to learn Chinese was one important insight that Kerr had soon after his arrival in Canton. The other, and of equal importance, was his realisation that, as far as plants were concerned, the Chinese were intrigued by novelty. Moreover, Kerr saw that novelty had a bargaining value. He had brought some seeds from Kew with him including those of the mignonette, a highly fragrant plant, and when it was sowed in his patch of ground, it flowered easily. As he told Aiton: 'its novelty & agreeable smell soon got a number of Admirers & with difficulty I have been able to keep one Pot for Seed. The others are likewise esteemed as valuable curiosities. These things however trifling they appear have a very good tendency to promote the interest of our object.'[37]

Kerr was very active during the first few months he was in Canton even though he was so short of time to assemble his collection as the trading season was almost over. Though Kerr spent time at the nursery gardens getting to know them, his success at getting plants depended to a greater extent on the relationships that existed between the super-cargoes and the Hong merchants. The grant of the land to Kerr by Mauqua was one example of this, but there was more to come.

Given his past, often frustrating relationships with the East India Company, Banks must have been delighted at the generous manner in which the 'Select Committee of Supra Cargoes' in Canton responded to the Court of Directors' instructions concerning Kerr, Lance and the plans for collecting plants for the King's Garden. Since the super-cargoes' response could not have preceded Lance's arrival – he probably delivered the instructions himself – he personally must have played an important role in their response. The Committee said that not only did they not expect any difficulties from the Chinese in fulfilling Kerr's requests: on the contrary, the Chinese merchants 'have assisted us considerably in this pursuit'.[38] As proof they cited Mauqua's gesture but they added that Puankhequa II, president of the Hong merchants and, therefore, one of the most powerful men in Canton, had contributed substantially to the effort by making presents of 'many of the choicest plants'.[39]

Then the Committee told the Court that Puankhequa thought that the mandarins would enjoy receiving European plants as gifts in return and that Kerr should list the most desirable ones for Banks to send. This was a new and exciting possibility for an exchange of plants that could produce untold riches, especially for Kew.

The Committee also had an idea of how this could be done. Banks provided part, or perhaps all, of the inspiration for this. Drawing on the success he had had on previous occasions – with plant cabins as a means of transporting live plants – those on HMS *Bounty*, HMS *Guardian*, HMS *Providence* and the East India Company Ship *Royal Admiral* would have sprung to mind – Banks had decided to use a plant cabin to transport the Chinese plants to Kew. The idea was that the cabin would be built in Canton in preparation for Kerr's first collection, according to Banks's designs conveyed by Lance.[40]

The Committee took Banks's idea one step further. Not only should the plant cabin be used in transporting plants from Canton but, the Committee added, it should also be used to send European plants, flowers and fruits 'that will be objects either of Utility or Curiosity in this Country' in the other direction. The Committee was certain that in return the Chinese merchants would wholeheartedly involve themselves in this project 'to procure for us such Plants and Flowers from the various Provinces of the Empire'.

Where direct diplomatic attempts such as the Macartney embassy had failed, the Committee hoped that plants and flowers would pave the way for a new commercial relationship: 'We hope by this Exchange', the Committee noted, 'that the Commerce or Comforts of both Countries may be promoted and the least benefit that can be derived from it will be the extension of a liberal and elegant pursuit.'[41]

Through his purchases at the nursery gardens and Puankhequa's presents, which were arranged by James Drummond, the president of the supercargoes, Kerr's collection exceeded ninety plants.[42] Many of them went into the plant cabin that was built on the deck of the

Henry Addington whose captain, John Kirkpatrick, had a botanical bent and had introduced Chinese plants to Britain on earlier occasions.[43] He volunteered to carry the collection and the cabin.[44] Other plants collected by Kerr were placed on other ships, notably the *Warley* and the *Hope*, both commanded by botanically minded captains. Other ships in the sixteen-strong fleet leaving Canton at the end of the 1803–4 trading season also carried plants for private collectors. Seeing the fleet leave, Kerr wrote enthusiastically to Aiton: 'You will not be surprised to hear that more Plants have gone from Canton this Season than ever did at one time before, being scarcely a Ship in the Fleet without some; without doubt my Mission has excited their attention to this as well as a kind of emulation.'[45]

With that Kerr left Canton, as foreigners were required to do at the end of the trading season. He retired to Macao to await the arrival of the next trading fleet

The sixteen ships of the China fleet passed by Macau on 6 February 1804 on their way home. One of the last passengers to join the *Earl Camden*, commanded by Nathaniel Dance, was David Lance. Ever since his arrival in Canton he had been unwell, and he felt that to stay in the area during an even hotter season would be further detrimental to his health. James Crichton, the surgeon to the factory, agreed with his patient and he was given permission to leave.[46]

The fleet was also joined by some of the passengers from HMS *Porpoise*, including John Allen, who had chosen to proceed home via China. Allen, the Derbyshire miner, had gained valuable experience about plants and how to care for them during the circumnavigation of Australia by HMS *Investigator*. Now, in Canton, he had a chance to put those skills to the test and, if successful, endear himself even more to Banks, who had selected him for the *Investigator*'s voyage. Allen boarded the *Henry Addington* as a passenger but his real purpose was to tend the plant cabin and its valuable contents.

The voyage home was not going to be easy. Intelligence had reached Canton that the truce ushered in by the Treaty of Amiens had ended and Britain and France were back at war.[47] Admiral Charles-Alexandre Linois and his ships were somewhere in the area, but no one knew precisely where. On 7 February, a day after passing Macao, Commodore Nathaniel Dance was clear of land. His contingent of sixteen East India Company ships had now swelled to a total of twenty-eight vessels, thanks largely to the addition of eleven ships involved in Asian

private trade. Dance was heading for the Straits of Malacca, favoured
by the northeast monsoon.

Linois, who knew exactly when Dance and the fleet had sailed, was
waiting for them.[48] On 14 February, two of Linois's complement of
five ships, three of which were heavily armed, spotted sails approaching
from near the island of Pulau Aur. As they came into view, the number
of ships the French could see was far more than their intelligence had
told them to expect. Linois was cautious about his next move assuming,
incorrectly as it happened, that the extra vessels were warships from
the Royal Navy's eastern squadron.

The night passed. In the morning, with the French and the British
showing their colours and, thus revealing the kinds of ships in each
squadron, the two sides approached each other but it was not until
the early afternoon that Linois opened fire. The British ships at the
front of the formation returned fire immediately. For less than an hour
both sides fired at each other, without sustaining any damage. Linois,
convinced that he was attacking a fleet with naval firepower, withdrew.
The British ships gave chase more as a gesture than as a determined
effort. At nightfall the China fleet anchored at the entrance to the
Straits of Malacca as though nothing had happened. Soon the fleet,
now joined by the two naval vessels they were expecting, continued
their homeward voyage. They saw no more French ships and arrived
back in England in early August 1804. It was a close call.

Allen, who had tended the plant cabin for almost seven months,
wrote to Banks as soon as he could.[49] This would have been the first
Banks would have heard of the fate of the plants on board. As it
turned out, the plants on the *Henry Addington* fared best. There were
sixty-two cultivated plants and thirty wild plants housed in the plant
cabin. Kerr provided a detailed description of each plant. One example,
that of plant twenty-four, will give a sense of how Kerr described his
specimens. 'Hum sew', he explained to Aiton '[is] a beautiful ferrug-
ineous Shrub the Leaves in Shape resembling those of Camellia. I have
not seen the flower but some Chinese drawings suppose it to be a
species of Magnolia, it is highly valued & cultivated in great Numbers
chiefly in Pots as are most of the Chinese plants for Ornaments, those
planted in the ground attain the Height of 12 or 14 feet.'[50] This plant
survived the voyage home, but twenty-four of the total, almost 30 per
cent did not. Most of the plants on other ships died on the voyage
or were thrown overboard. Banks told Kerr this, complimenting him
on the care he had taken with the plants but also cautioning him for

the future. 'I should advise you', Banks wrote, 'to send the more Curious ones always in the Cabins that are sent & trust as little as Possible to other opportunities.'[51] On 17 August, Banks and the King inspected Kerr's 'rich' collection.[52] Aiton's *Hortus Kewensis* listed fourteen plants growing in Kew that Kerr had sent on the *Henry Addington*: they were 'flourishing exceedingly in His Majestys Garden'.[53] Begonias, gardenias, asters and a juniper tree were some of the most notable introductions.

It was a singular achievement.

20

1812: And Still Not First-Hand

The *Hope*, an East India Company ship under the command of Captain James Pendergrass, left Portsmouth for Canton on 25 April 1805. This was its fifth voyage to the Chinese port city. Pendergrass knew the ship very well as he had been part of its company since it first went to sea in 1797. He began his maritime career as a midshipman aged twelve, and he was now on his second voyage in command of the *Hope*. Not only was he a committed mariner, but Pendergrass knew about plants, as he had been collecting Chinese varieties for Sir Abraham Hume, the *Hope*'s owner. Hume, whose father and uncle had made a fortune with the East India Company, was an avid collector of art and precious stones. He and his wife Amelia also shared a passion for horticulture and their garden at the family estate of Wormleybury, near Broxbourne in Hertfordshire, was known for its exotic collection of plants, particularly those from China.[1]

Magnolias, chrysanthemums and peonies, just to mention the most important introductions, found a comfortable home in Wormleybury. Beginning in 1798 and continuing for some time, it was Pendergrass who brought the plants from China for Abraham Hume.[2]

Now in 1805, and no doubt because of his Chinese botanical interests, Pendergrass was undertaking a task that he, and possibly no other ship's captain, had ever accomplished before. The *Hope*, equipped with a plant cabin built at Kew, was now on its way to China with a payload of over three hundred species of garden plants and fruits; as well as seeds of annual flowers and edible plants, all from Kew.[3] There were grapes, plums, cherries, pears, apricots, figs, rhododendrons, azaleas, roses, irises and much more besides. Banks and Aiton were taking no chances about what might spark an interest in Canton. Though as Banks himself had pointed out earlier, the Chinese probably had many of these fruits already, he could not be sure, since he was relying on second-hand information. David Lance, who had been back in England for just over half a year, knew more and probably guided Banks and Aiton's choice. He had also put together two boxes of plants, as perhaps the most important part of the shipment, as a present for Puankhequa II, president of the Hong merchants.[4] The rest of the boxes in the ship's plant cabin were for other mandarins and merchants who had supplied plants for Kerr's first shipment home.[5] Kerr was told to let James Drummond, the president of the supercargoes, decide how these presents should be distributed, in order that the recipients would 'make returns in Kind by Procuring from the northern parts of the empire Plants likely to suit the Climate of England'.[6]

Not long after writing to Kerr alerting him to the nature of *Hope*'s collection of Kew plants, Banks wrote directly to Puankhequa II. This may have been the first time that he had written to one of the Hong merchants. It was certainly his first letter to Puankhequa II. In it, Banks, who referred to himself as presiding over the 'Department of the Arts and Sciences in this Country', thanked the Hong merchant for the assistance and the 'kind protection' he had extended to Kerr, and, in return, offered his 'Services in any manner that may either tend to the Improvement or Advantage of your Country or your own personal gratification'.[7] As a mark of gratitude, Banks informed Puankhequa II that the *Hope* was on its way to Canton bringing plants for him from Lance as a gift. Banks had also made certain that the ship was carrying gifts for Puankhequa II from him – two carpets with borders, 'a Pair of Girandoles with Shades & Some Spare ornaments in case of accident 36 Glass Cups marked with your Initials & a Set of Views of London'. This was, Banks hoped, the beginning of a 'National Exchange equally advantageous to both Countries'.[8]

Despite the best of intentions, a plant cabin and a captain who cared about plants, the voyage did not go well and almost all of the collection was destroyed. Nearing the Cape, the *Hope* was engulfed by a violent storm in the middle of the night. The plant cabin broke loose and the boxes of plants scattered over the deck and many over-turned. During the time it took for the cabin to be repaired and for the boxes to be returned to their places, the damage was done. When the *Hope* arrived at its mooring in Whampoa, Kerr inspected the collection, such as it was, and discovered that only four fig trees, three rhododendrons, one pear and a few bulbs had survived: the pear died soon after Kerr took custody of the plants.[9]

In his effusive letter to Banks, in which he remarked that his name was well known to him, Puankhequa II thanked Banks for the presents that did arrive and for which he was very grateful: 'My Apartments', Puankhequa II wrote, 'will be adorned, and my Table will be graced by the several articles, and they shall be so disposed as may appear most worthy and honorable to the munificent donor, as well as recall him oftenest to my remembrance.'[10] In return, Puankhequa II sent presents for Banks: lanterns of rosewood and glass, porcelain cups, lacquered stands and dishes, eight pots of the choicest peonies, and 'A Peculiarly Curious and ancient Dwarf Tree'.[11]

The *Hope*'s plant cabin had now been repaired and Kerr's collection was placed on board, as Banks had planned. Puankhequa II's present of peonies was placed in the plant cabin. One of them, given the number 69 on its tin tag and which Kerr only knew by its Chinese name, was 'in every probability . . . a new and undescribed species', It had never been seen before in Canton, Kerr added.[12]

For this shipment Kerr carried out another of Banks's ideas, sending a Chinese gardener with the plants to take care of them during the homeward voyage. Once he got to London, Banks would arrange for him to go to Kew and to work there as a gardener learning about European plants, which he would then care for on the return voyage to Canton.[13] By the time the *Hope* was ready to leave its anchorage at Whampoa, Kerr had found the right person. He was 'A Chinese boy whose friends are all Gardeners, & who himself has been brought up in the same business.'[14] On the ship's muster, his name was put down as 'A hie'.[15]

Kerr did not arrange for Puankhequa II's dwarf tree to be sent on the *Hope*, presumably to spread the risk. The other presents for Banks probably went along with the tree, placed on board the *Henry Addington*. Kerr commented that the tree had been growing in Puankhequa II's garden for almost thirty years but that he thought it was at least a century old.[16]

When, in late August 1806, the *Hope* returned to England, Captain Pendergrass reported to Banks that the collection was in good shape (although some of the plants had not survived), and that he should arrange for someone to collect the plants soon. Once he had left the ship, Pendergrass warned, there would be no one around to care for them and anything could happen.[17]

As it turned out, one of the casualties of the voyage was Puankhequa II's dwarf tree. The dead tree was sent to Frogmore, to Queen Charlotte's country residence, where she could at least admire 'the Art peculiar to the Chinese, of dwarfing into picturesque the most lofty Tree of the Forest'.[18]

A Hie was not the key to the success of the English–Chinese plant exchange as Banks had hoped, although he remained at Kew, at the King's expense, until the plant cabin had been erected for the return voyage. A Hie was not a popular guest and no more Chinese gardeners were invited.[19] Before he left England, however, he was given a silver watch, worth £5, 10 shillings, and an engraved seal as presents.[20]

While the *Hope* was on its way back to England, the China fleet for the 1806/7 trading season was being prepared and with it another attempt was made to send plants from Kew to Canton. The ship chosen for this task was the *Thames* under the command of Captain Matthew Riches. Just as on the *Hope*, the Kew plants would be carried on the *Thames* in a plant cabin erected on its deck.[21] Towards the end of March 1806, Banks wrote to Charles Grant, the Company's Director, to seek the Court's approval for this alteration to the ship. This would now be the second plant cabin at sea since the first one was still on board the *Hope* for its homeward voyage.

William Ramsay, the Secretary to the East India Company, dealt with Banks's request. The two men had known each other for at least fifteen years. A week after writing to Grant, Ramsay was able to inform Banks that the Court had agreed to have a plant cabin built on the *Thames*. It was the same cabin that had been used to transport Kerr's first shipment of living plants on the *Henry Addington*.[22]

As Matthew Riches, the ship's captain, had never transported plants before, Ramsay asked Banks to brief him, through the Company, on what steps he should take to safeguard them.[23]

The document that Banks produced provides a good insight into his ideas about transporting live plants across thousands of miles of ocean. As he pointed out, he had been closely involved in this difficult logistical exercise since before Bligh's highly successful *Providence* expedition. Bligh had brought thousands of breadfruit plants and other vegetables from Tahiti to the West Indies, and from there Caribbean plants to Kew 'without Experiencing any material Loss'.[24]

Captain Riches, Banks wrote, should understand the advantages that Britain and the East India Company hoped to gain by helping to promote 'this very usefull Barter between the West & the East'.[25] The plant cabin was the key to its success and it would give the plants the protection they needed. Riches would not have to do more than be moderately attentive to them: he only had to ensure that sea water never got near them; to give them exposure to the air at the appropriate times; and to water them frequently. Timing and testing were essential to success, and Banks provided Riches with some useful pointers, mostly involving licking his hand or the leaves of the plants, to know how salty the air was; and how to spot signs of over- or under-watering.[26]

Aiton prepared the shipment and provided a Kew gardener to house the plants properly and securely in the plant cabin.[27] As it turned out, the ship's steward had some gardening experience and was happy to care for the plants. For this favour, he was to receive a small reward and a certificate attesting to his good work.[28]

Thomas Manning, who, through Banks, had obtained the East India Company's permission to go to Canton to improve his knowledge of Chinese, was on board the *Thames*. Nearing the Cape, he wrote to Banks. He was fond of plants and while at sea he had taken a great interest in the collection, so much so that many on board thought him a botanist. He was able to tell Banks that most of the plants were thriving, though a few had perished, and that the steward was tending them carefully. Manning also told Banks about the general attitude on board to the presence of the cabin and its contents. Banks wouldn't have been surprised to learn that most of the ship's company had little regard for this cargo: the officers often plucked fragrant leaves to put in their button holes; the first mate looked upon the cabin with suspicion – 'He says it wracks the ship to pieces'; the captain thought much the same; and 'when the beams creak in the Cuddy they turn to me sometimes & damn the Flower-pots'.[29]

Although the *Thames* reached Canton and returned to England safely by the end of December 1807, the Kew records are silent on what became of the plant collection, but as Banks's practice of sending Kew plants to China in plant cabins continued unabated, one assumes it was reasonably successful.[30]

As for shipments to Kew, they, too, continued on an annual basis and grew in volume. Four ships, for example, returning to England in February and March 1809 after the trading season, came back with a substantial collection of plants spread over four plant cabins, one per ship. The following year, Kerr sent back plants on two ships, both of them equipped with plant cabins.[31]

Now it was time for Kerr's final shipment from Canton. Apart from a short trip to Manila, he had remained in Canton and Macao for seven years collecting plants and seeds for Kew. His collections were extensive, which shows that, as far as the movement of plants was concerned, the Canton System was not quite as impregnable as it had been or seemed to be. Kerr used every resource available to him, relying on the favours of the Hong merchants, especially Puankhequa II; the city's nurseries; Chinese gardeners and seedsmen; missionaries in Peking; and the supercargoes who tended their gardens in Macao during the off-season. Banks was very pleased with his work: 'Kew Gardens bears Testimony to your Success with many fine Plants,' he wrote.[32]

On 30 June 1810 Banks wrote to Kerr again to say that the King had been so impressed by his diligence, ability and success in Canton, he was giving him permission to resign his 'charge of Botanic Collector' and to take up a new post as 'Superintendant & Chief Gardener of the Royal Botanic Garden of Ceylon', with immediate effect.[33]

Kerr had earned his promotion but now Banks was left without a gardener in Canton. What next was anyone's guess. European politics were uncertain: not only were Britain and France still at war but Napoleon had just enlarged his continental empire by taking Rome.

Still, business between East and West continued and Banks found himself welcoming a potential new collector for China in his Soho Square home in March 1812. John Reeves was thirty-eight years old and had been working for the East India Company in London for four years as a tea taster and chief inspector of teas, though he had been in the private London tea trade for almost twenty years. In 1811 Reeves was promoted to be the assistant inspector of teas in Canton and would soon be leaving London for China.[34]

An older cousin of Reeves's, also known as John Reeves, had introduced him to Banks.[35] The elder Reeves and Banks had known each other for some time.[36] They not only wrote to each other but they had met on numerous occasions so that his recommendation of the younger Reeves was reliable.[37] Coincidentally, the elder Reeves also knew George Leonard Staunton and that the two Chinese interpreters were lodged with him.[38]

But it was Banks who introduced the younger Reeves to George Thomas Staunton, who, during the two decades between 1792 and 1812, remained the only Briton proficient in Mandarin.

George Thomas Staunton had been deeply affected by his Chinese experiences: it was not only that extraordinary moment when he and the Qianlong Emperor spoke together during the Macartney embassy. As he explained it many years later, even his first sight of the Chinese in Naples and the sound of their language had made an indelible impression on him that lasted his whole life.[39]

So, when he returned to London from China in early September 1794, aged thirteen, George Thomas continued his study of Chinese with Ahui, one of the two Chinese servants that his father had brought back with him.[40] For the next six years his proficiency increased. Though George Thomas enrolled in Trinity College, Cambridge, in 1797, he only stayed there a year. With his father's help, he was appointed writer for the East India Company's factory in Canton, and, in early 1800, he was back on Chinese soil. It was the first time that anyone employed in the British factory was fluent in Mandarin.

From then on, Staunton's star rose spectacularly. He was promoted to supercargo in 1804 and became chief translator and Secretary to the Select Committee of Supercargoes in 1808. His command of Mandarin was substantial enough for him to undertake, in 1808, when on leave in London, a translation of the Qing penal code, which was published in 1810.[41]

Staunton returned to Canton in December 1810 to resume his responsibilities there. He was, however, not happy with his situation and, in January 1812, he left for London on sick leave.

In March 1812, Banks, not knowing that Staunton was on his way home, wrote a letter to him in Canton introducing Reeves as his new collector.[42]

Reeves, Banks told Staunton, was 'a man of more than ordinary Talents'. He had learned some Chinese – frustratingly Banks does not disclose how. Banks had taken to him instantly: 'I have from the traits of his Character that I have observed conceived a Friendship for him.'[43]

With a copy of this introductory letter in hand, Reeves left Portsmouth a fortnight later on 25 March and arrived in Canton on 20 September 1812. Staunton, of course, was not there and they only met two years later when Staunton returned to Canton. Nevertheless, Reeves seemed to have entered the Anglo-Chinese world easily enough.

By the time he wrote his first letter to Banks, Reeves was able to tell him quite a bit about what he had learned so far about China and the people he had met.

Banks had asked Reeves to do a number of things for him in addition to looking out for plants, and these included finding out about the mythological or religious figures – he didn't know which – depicted on Chinese porcelain. This 'commission' as Reeves referred to it was actually on behalf of Dorothea, Banks's wife, who had an extensive China ware collection.[44] Banks had asked similar questions of Staunton when he was in Canton.[45] Reeves had already made enquiries for Banks from Sin Chong, one of the Chinese merchants who traded in Chinese ware in Canton, but he had not answered any of Reeves's questions clearly. Further enquiries led nowhere and Reeves was frustrated. On the other hand, he had met and dined on several occasions with an extremely wealthy Chinese gentleman who, because of his 'mode of living & partiality to Englishmen, is known generally by the name of 'the Squire', and who was Puankhequa II's brother.[46] His garden was exquisite: according to Reeves it contained between 2000 and 3000 pots of chrysanthemums (double-flowered) and many camellias.[47]

It sounded very promising. Reeves had made the right social connections but, as far as collecting plants went, it wasn't good. 'It was my intention, when I left England', Reeves wrote apologetically, 'to forward some plants, if I found any worth notice – but, Tyro as I am in botany, how can I presume to send any to you – without the danger of incurring the blame of sending those of little value as being already known in England.' Though Reeves had been to one of the city's nurseries he had not seen anything he thought was new.

Still, in spite of his ignorance of botany in general, Reeves reminded Banks that he was a tea expert and intended to find out all about the tea plant. Tea plants are what he promised in the future.[48]

Over the next few years, Reeves sent Banks a number of living plants, including the promised tea plants; several other camellias; *Aglaia odorata* (the Chinese perfume tree); and a variety of small sweet orange 'which Puankhequa had taken off one of his Trees for you'.[49] These, as was customary, were entrusted to the care of sympathetic captains, some of whom were taking plants back for themselves.[50] At one point Reeves asked Banks to send out from England a 'conveyance' to take the Chinese plants home.[51] This could only refer to one of Banks's plant cabins which had been so widely used when Kerr was collecting

白牡丹花

Breadfruit, 1792, watercolour
by Lieutenant George Tobin.

White peony by an unknown
Chinese artist, c.1794–6.

A view of Funchal, Madeira, oil painting by William Hodges (1744–97).

Transplanting of the bread fruit trees from Otaheite,
painting by Thomas Gosse, 1796.

The library at the home of Joseph Banks, Soho Square,
undated drawing attributed to Ralph Lucas (1796–1874).

Kew Gardens: the Pagoda and Bridge, 1762,
oil painting by Richard Wilson.

Cape Town, 1791, watercolour by Lieutenant George Tobin.

Adventure Bay, 1792,
watercolour by Lieutenant George Tobin.

Point Venus, Matavai Bay, 1792,
watercolour by Lieutenant George Tobin.

Presidio at Monterey, *c*.1792,
watercolour by William Alexander (after John Sykes).

Friendly Cove, Nootka Sound, *c*.1792,
watercolour by William Alexander.

Qianlong's court (showing George Thomas Staunton on one knee in front of the Emperor), 1793, watercolour by William Alexander.

A scene at Tientsin, 1793, engraving by William Alexander, *The Costume of China*, 1805.

View of Sir Edward Pellew's Group, Gulf of Carpentaria, *c.*1802,
oil painting by William Westall.

The European factories, Canton, 1806,
oil painting by William Daniell.

The road to the home of Baron von Langsdorff on the outskirts of Rio de Janeiro, 1817–18, watercolour by Thomas Ender.

A panorama of São Paulo, 1817–18, watercolour by Thomas Ender.

for him. Whether the plant cabin was in fact used is unclear from the surviving correspondence, but it seems likely that Banks would have used the method again.[52] Reeves also recognised the value of exchanging gifts with the leading Hong merchants, including both Puankhequa II and his brother 'the Squire'. He thought that they would both benefit from receiving examples of lemon trees – which, Reeves noted, did not grow in China – and tulips and hyacinths.[53] Banks tried to pass the message on about lemon trees to William Kerr, who was now the superintendent of the Ceylon Botanic Garden in Colombo. Unfortunately, by the time Banks's letter got to Ceylon, Kerr was dead. In his place, Sir Alexander Johnston, a senior colonial and legal officer in Ceylon, made sure that the desired shipment would be sent to Reeves in Canton.[54]

Just before Reeves returned to London in 1816 he sent Banks a large number of living plants in pots. They went on four East India Company ships and included camellias, azaleas, gardenias and a fruit from Puankhequa II's tree.[55] Reeves was back in London in May 1816 and during his stay he remarried, saw his children, bought a house south of the river and went to East India House to see something very special that was part of the Company's museum.[56]

What Reeves went to see was something that William Kerr had arranged during his stay in Canton.[57] Not long after he arrived, Kerr had learned that the Court of Directors in London, with Banks's assistance and support, wanted him to commission and assemble a set of drawings of Chinese plants to be displayed in their new India Museum, adjoining the Company's headquarters in Leadenhall.[58]

In Canton, Kerr had quickly found a Chinese artist to produce the drawings. He was probably assisted by John Bradby Blake's former go-between, Whang at Tong, who was still active in Canton and working for one of the leading Hong merchants.[59] Banks's Book of Chinese Drawings, which Kerr had with him, would have enabled him to show the artist what was required. The first set of drawings was much admired in London, so the Court of Directors asked Kerr to arrange for a second set to be made. They arrived two years later in 1806: there were nearly four hundred drawings in total.[60]

These were the first Chinese drawings the Company had commissioned. They seemed to be intended as an adjunct to the 'Cabinet of Natural Productions' in the India Museum, where specimens of Asian plants, fruits and seeds were displayed.[61] The drawings were to help the viewer imagine the splendour of the living plant. For Kerr, the drawings also had a scientific purpose. Each time he sent a shipment of living plants to Kew – which he did on at least a dozen occasions between 1803 and 1810 – he cross-referenced their entry in the garden's records of incoming plants to the drawings, each of which were numbered in sequence at the top right-hand corner, so William Aiton at Kew could use the drawings to help him identify the plants correctly.[62]

Reeves wanted to continue this practice so he borrowed some examples of the plant drawings that William Kerr had commissioned for the East India Company.[63] By April of 1817, Reeves was ready to return to his work in Canton, but with a new commission: he was to collect plants for the Horticultural Society, which had been founded in March 1804 – Banks was one of the seven founding members; and to commission Chinese artists in Canton to produce drawings of Chinese plants, presumably using the copies he had made or had had made in London as an example of what he wanted.[64]

It is unclear whether Reeves continued to collect for Banks for the first few years of his return to Canton. There is no record of any collections from Reeves arriving at Kew during this time, and there is no correspondence either.[65] Then, in 1819, Banks wrote to Reeves asking him to find a particular variety of red rose. Reeves couldn't find that particular rose but discovered, instead, two rose varieties new to him – one, very dark red and another a dwarf pink one. He promised to send them but, sadly, by the time he replied, Banks had been dead for five months.[66]

In many ways Reeves operated under the same handicap as all of Banks's previous collectors in Canton: they could only send what they could find locally and, more unusually, what was given to them as a gift. That was the reality of plant collecting in China and though Banks had pushed the system to its limits, he simply could not get access to the interior.

The last opportunity for collecting in the interior had been in 1793 when the Macartney embassy travelled through China from Peking to Canton; but from a botanical point of view little came of it, both because of a lack of scientific knowledge and access. Now, in early 1815, during a period of peace, the idea of sending another embassy was being given serious consideration by the main participants, the Admiralty, the Board of Control, the East India Company and Joseph Banks.

John Barrow, Second Secretary to the Admiralty, instigated the process. Barrow had long had an interest in China. He had been part of the Macartney embassy, his first political post, acting as its comptroller – in charge of the presents to the Emperor and court mandarins – and private mathematics tutor to George Thomas Staunton. Barrow then went on to become Lord Macartney's private secretary when he was appointed, in 1797, Governor of the new British possession at the Cape, where he remained until 1804. Barrow published his version of the embassy's experiences in China in the same year as he became the Second Secretary. A few years later this was followed by his biography of Macartney.[67]

Barrow and Staunton, in spite of an age difference of seventeen years, had become close friends and stayed in touch. As Staunton recalled years later, it was when he had returned to London from Canton in June 1808 that he began urging the government and the East India Company to renew the attempt to make diplomatic contact with the court in Peking. In this he had Barrow's support. Indeed, in November 1809, Barrow was fully convinced that a new embassy would be attempted and that Staunton, who could read and speak Mandarin, would be the designated ambassador.[68] Staunton was called to a meeting at the East India Company expecting this appointment to be conferred; instead he was told that as an employee of the East India Company he could not be considered for a diplomatic post.[69]

Shortly after, the whole idea of an embassy was dropped, and there it rested until February 1815 when Barrow wrote to the Earl of Buckinghamshire, the President of the Board of Control, outlining a new plan.[70] Though Buckinghamshire was enthusiastic, the Court of Directors of the Company was less so. Three months passed after Barrow's first communication. Communiqués flowed back and forth. Finally, in May, the Company decided, after a meeting with Buckinghamshire and Lord Liverpool, the Prime Minister, that the time was not right to send an embassy to China.[71]

Then, a couple of months later, at the end of July, after having received advice from Canton, the Court changed its mind again and Barrow's plan was back in favour. There were no further U-turns. By the end of August, the search had begun for the ambassador; and, after a shaky start, when more than one person declined the offer, in late September, William Amherst, the 1st Earl Amherst, who had had only two years' experience as a diplomat, accepted the post.[72]

Having actually been to the court in Peking, Barrow effectively took over the preparations for the embassy, in particular the choosing of the presents. The general principle he adopted was 'that the Chinese court prefer Articles of intrinsic value, to those of mere shew or curiosity, though the latter are not to be wholly omitted'. Following this principle, Barrow suggested: 'A Golden Cup with Cover and Salver of the same beautiful workmanship', silver vessels, blue, purple, and yellow fine merino woollens, velvets, perfumes, liqueurs, mirrors, and a portrait of the Prince Regent, in a gold box set with diamonds and 'another Gold Box, set with Stones to carry the Princes' Letter to the Emperor'. They shouldn't send animals, carriages, maps of India and China – '[they] would create jealousy in that people' – military weapons and books of botanical drawings – 'they excel us in this branch of Drawing'.[73]

With a great deal accomplished already, on 3 October 1815, Barrow wrote to Banks. They had been corresponding for some time and had known each other, at least, since 1795 when George Leonard Staunton was preparing his official account of the Macartney embassy for publication. While he was at the Cape, Barrow had arranged for the shipment of plant specimens collected by George Caley, Philip Gidley King and Francis Masson and so knew something about Banks's efforts to secure plants for Kew. A year after his return from the Cape and his appointment to the Admiralty, Barrow was elected a Fellow of the Royal Society and, from then on, he and Banks became close. Barrow helped at the Royal Society, at its dining club and spent most Sunday evenings at Banks's Soho Square home.[74] More recently Barrow had been involved in preparing Flinders's account of the voyage of HMS *Investigator* for publication, in arranging for an exploration of the Congo River and in offering Banks naval support for further collecting expeditions to the Cape and Australia.[75]

Barrow's October letter to Banks told him that Lord Amherst would be going to China. Barrow didn't think the mission would succeed, partly blaming the Chinese government – 'cowardly and jealous' – and partly because no one so far appointed, that is Amherst and Henry Ellis, his secretary, knew any Chinese – an allusion, probably, to the fact that Barrow had supported George Thomas Staunton, who did speak Mandarin, to be the ambassador.[76]

Banks probably already knew about Amherst but it was important for Barrow to tell him himself. Remembering perhaps the problems with the Macartney embassy, Banks soon started thinking about getting a naturalist appointed to the new embassy but, by the beginning of

December, he had still not found a suitable candidate.[77] Either because
Banks had no one suitable to propose or because they didn't want a
naturalist at all, no such appointment was made. Instead Banks had
to be content with the services of Clarke Abel, who was the embassy's
surgeon but was also interested in natural history.

It was a good compromise. It is very likely that Barrow who knew
everyone in the Admiralty had selected Abel. After meeting him, Banks
was satisfied that the surgeon had 'always felt an inclination to Pursue
the Science' and he was convinced that Abel would serve the needs
of natural history well.[78] Still, he did not want to let an opportunity
to collect for Kew pass by. So, Banks outlined a plan to appoint a
Kew gardener who was well versed in 'Exotic Botany' (this was not
one of Abel's strengths) to the embassy to help with the collections.
Kew would be one of the recipients of the plants and so would the
East India Company's Museum – seeds would be of no use to the
latter since they had no gardens.

Amherst immediately took Banks's suggestion to Charles Grant and
Thomas Reid at the East India Company. A few days later they agreed
to it and Reid added that the collectors should also leave one specimen
of each plant in Canton so that they could be distributed to the
Company's botanic gardens in Calcutta and Madras.[79]

By the beginning of the New Year, most of the preparations for the
voyage had been made including selecting the ship, HMS *Alceste* and
its commander, Captain Murray Maxwell. All that was missing was
the name of the gardener, but Banks had that in hand. On 2 January
1816, he wrote to James Hooper, a foreman at Kew, informing him of
his appointment to the embassy. Hooper had worked at Kew for six
years and came highly recommended by William Townsend Aiton.[80]
Banks enumerated his responsibilities: he would be Abel's assistant
and follow all his instructions, but, Banks added, since Abel knew
botany rather than gardening, it would be up to Hooper, who knew
what grew in Kew, to get the right plants – 'you are better acquainted',
Banks pointed out, 'than he is with the Plants Cultivated in Kew
Gardens & will often know by its first appearance a plant which he
ably make out by application to his books, in all Such Cases you are
Chearfully to name to him the Plants you are acquainted with in order
to Spar him the trouble & Waste of Time that would Ensue were he
Obligd to Seek them in his books.'[81] Hooper was told he was to collect
plants as dried specimens, mostly for the Company; seeds for Kew;
and living plants, also for Kew, which he would need to plant in boxes

and pots which would be kept in the plant cabin he would be taking with him to China.

Abel was well equipped for the voyage. The bill for his expenditures exceeded £550 and the single largest item was natural history books.[82] Banks expected great things from him and Hooper. Just before he left England, Banks wrote to Abel with his instructions.[83] He reminded him how little botanical knowledge had emerged from the Macartney embassy. If he was to take the same route through China as Macartney, he should collect anything at all interesting he saw: if he took a different route, he could assure Abel that 'you will Scarce meet with a vegetable that is Known to European botanists', and so he should collect everything he could. The Chinese loved flowers and Banks suggested he should fill the plant cabin for its outward voyage with roses, lilies and other elegant flowers he might find in Portsmouth and leave them in Canton as presents for the Hong merchants.[84] As Banks knew from previous experience, they would return the favour with their finest plants. He hoped the plant cabin on its return would be filled with rare azaleas and yellow and blue tree peonies. Though azaleas, especially *Azalea indica*, had been sent back before, not one of them arrived alive.[85] Banks also wanted acorns of the Chinese oak, notable for its long life; and examples of the pitcher plant, which had previously been introduced and lived for several years at Kew.

Banks couldn't emphasise enough that Abel and Hooper had a unique opportunity. For the first time since Banks had been involved in collecting Chinese plants for Kew he would have a Kew gardener returning with the plant cabin – 'the accession yours will make to the Royal Gardens will be immence', Banks stressed.

At 8 a.m. on 9 February 1816, the three-ship flotilla consisting of HMS *Alceste*, HMS *Lyra* and the East India Company ship *General Hewitt* set sail from Spithead. The instructions to Captain Maxwell on HMS *Alceste* commanded him to sail through the Atlantic, stopping at Madeira, Rio de Janeiro, the Cape, and then into the Indian Ocean through the Straits of Sunda, and to make a stop at Java.

In order not to alarm the Chinese authorities in Canton, it was decided that the ships sent from England would rendezvous with English ships sent from Canton at an anchorage between the island of Lamma and Hong Kong Island, both of which at the time were sparsely inhabited mostly by fishermen and their families. All the ships met there on 10 July 1816.

Barrow had been anxious about the lack of Chinese speakers in the embassy, but had he been at the anchorage that day, he would have seen how unnecessary his worries were. It was a linguistic power house. There was George Thomas Staunton, who had returned to Canton in September 1814 and had just become Chief of the East India Company factory. With him were five other people associated with the Company in Canton who could all speak and write some Mandarin: Alexander Pearson, the Company's surgeon; Francis Hastings Toone and John Francis Davis, writers; Thomas Manning, who had come to Canton in 1807, the same year as Robert Morrison, the last and most proficient Chinese speaker in the entourage, who was on a mission to convert the Chinese to Protestantism.[86] What a contrast with the situation twenty-five years earlier when Macartney could find no Briton who could speak Chinese and had had to rely on the two Chinese priests he found studying in Naples. The change in the linguistic abilities of the Anglo community in Canton was amazing. In fact, the right of the English to learn and communicate in Chinese with their Chinese counterparts, which previously the Chinese had discouraged and regulated, was one of the issues on which the embassy sought clarification.

On 13 July the ships followed the same route as the Macartney embassy towards their landing place en route to Peking. On 10 August 1816, seventy-five members of the Amherst embassy, including a dozen musicians, embarked on Chinese boats upriver destined for an audience with the Jiaqing Emperor and the delivery of the presents at the Summer Palace at Yuanmingyuan.[87] Meanwhile, the *Alceste* and the *Lyra* left for an exploratory voyage of the coast of Korea and the Ryukyu Islands in the Japan Sea, stretching from southern Japan to the coast of Taiwan.[88]

Just what went wrong is unclear.[89] Perhaps it was Amherst's refusal to kowtow; or perhaps it was Staunton's presence – the Chinese did not like him and his ability to understand their language – or perhaps it was a misunderstanding on the day of the audience as to what was to happen or something else entirely. Whatever the cause, the meeting, scheduled for 29 August, never took place and the embassy was ordered to leave the country immediately.

The diplomatic mission was a fiasco. The botanical mission, which interested Banks more, was, by contrast, a success. The entourage were returned to Canton along the same inland route that Macartney had followed, except that when they got to Yangzhou, they left the Grand Canal and travelled inland (where Macartney had not been), on the

Yangtze River – among the earliest westerners to do so – until they reached Nanchang, south of Lake Poyang, when they rejoined Macartney's route back to Canton.

It took four months to get to Canton from Peking. Despite the fact that Abel was ill for a significant period, Hooper managed to collect some entirely new plants, both as dried and living specimens and thousands of seeds for Kew.[90] Both Abel and Hooper were allowed more freedom to observe and collect than the gardeners on the Macartney embassy had been.

On 1 January 1817 the embassy reached Canton by which time the *Alceste* and the *Lyra* had returned from their exploratory mission to the Korean coast and the islands north of the Yellow Sea. Three weeks later, the *Alceste* with the returning party left Whampoa en route to England by way of Manila, which the ship reached on 3 February.

Six days later they were at sea again. The *Alceste* followed a southwesterly track through the South China Sea alongside the island of Borneo and headed for the Strait of Sunda and into the Indian Ocean. On 18 February at 7 a.m. in the Gaspar Strait, just off the southeastern part of the Sumatra coast, the ship hit a reef or a sunken rock so hard that it was immediately obvious that it was doomed. The priority had to be to save lives and essential stores, which is what Captain Maxwell did. Everything else, including the entire botanical bounty, to which Hooper had contributed at least three hundred packages of seed, went to the bottom of the sea.[91]

The captain ordered the first lieutenant to take Amherst and his party to Batavia by rowing them there in one of the ship's boats. The rest of the survivors had to wait on a nearby island to be rescued, which they were on 12 April by an East India Company ship, the *Ternate*, which Amherst had had despatched from Batavia. Eventually the company regrouped and Amherst chartered a ship called the *Caesar* to resume the homeward voyage. The ship anchored in Spithead on 15 August 1817. It had been eighteen months since the embassy had left for China.

It was probably Banks who was most disappointed by the botanic disaster. Never before had such a rich collection of Chinese plants been made. It was everything that Banks had hoped for until the shipwreck.

As for Abel and Hooper, though their entire collection was at the bottom of the sea, they did leave a small package of duplicate seeds with Staunton in Canton which did finally get to London. They both went on to better things. Hooper chose to remain in Java and was appointed the first head gardener of the newly created Buitenzorg Botanic Gardens.[92] Abel's next and last appointment was as surgeon to Lord Amherst when he became Governor-General of India in 1823.

21

1814: Accidentally in Brazil with Bowie and Cunningham

In 1814, William Townsend Aiton at Kew wrote to Banks telling him he thought it was time to revive their project of making Kew the finest garden in Europe. The war with France that had continued almost without a break since 1793 was over and intense relief was felt throughout Britain and in many parts of Europe, including Paris.[1]

As Aiton reminded Banks it had been the King's wish before he became ill to send collectors out to procure 'fresh & choice supplies of Seeds roots & plants'; but because of the war with France and the relative lack of interest of his son, George, the Prince Regent, little new had been 'added to keep up the Royal collection'.[2]

'The great deterioration of species [and] the loss of plants that no ordinary means of care & culture could obviate', Aiton added, meant that urgent action was required. Now that Napoleon had abdicated and travel was safer, there were several would-be collectors at Kew whose desire to go abroad, which had been frustrated for years, could now be satisfied.[3] Aiton was deferring, as he and his father always had done, to Banks's greater knowledge and experience, asking him for advice on where to send these men. Once he had the recommended destinations from Banks, Aiton was certain he could convince the Prince Regent to give the project 'His gracious commands'.

In concluding his letter, Aiton reminded Banks he had previously suggested the Cape and Australia as potentially rich areas for plant collecting. Just over a week later, Banks responded. He reiterated the sentiment that he had expressed on so many occasions before that nothing gave him as much pleasure as improving Kew; and that he would gladly resume his role in directing collectors and drawing up their instructions. Aiton's recollection of which destinations might prove most fruitful was correct. The southern temperate zone, Banks reaffirmed, was the best area to target: the Cape and New South Wales were not only rich in flora, but 'the Plants of both these Countries are beautifull in the Extreme and are Easily managed as they Suit the Conservatory'.[4] The time was right, Banks agreed, to resume collecting: he was sure that the Emperor of Austria, whose garden at Schönbrunn was Kew's main competitor, was thinking along the same lines. There was no time to lose. Banks would contact the governors in both the Cape and New South Wales to ensure a warm welcome for the Kew collectors.[5]

Two collectors from Kew, one to each destination, would suffice. Just over two months after Aiton and Banks first discussed sending collectors abroad, James Bowie contacted Aiton from Hitchin where he had been working on a garden owned by the local dignitary and MP, William Wilshere.[6] Bowie, the son of a London seedsman, had been employed at Kew before moving to Hitchin.[7] Allan Cunningham, the second collector, also wrote to Aiton volunteering his services a few weeks after Bowie.[8] The son of a head gardener at Marlborough House, Cunningham's education, which included the study of classics, led him initially to work at a conveyancer's office in Lincoln's Inn.[9] At some point, possibly 1808 or 1810, Cunningham came to Aiton's attention and the latter invited him, probably because of his superior education, to help him prepare the second edition of the *Hortus Kewensis*.[10] One of Cunningham's last tasks at Kew had been to deliver personally two plant cabins to the *Archipelago*, a frigate of the Imperial Russian Navy, eventually destined for the Dowager Empress, Maria Feodorovna.[11] Cunningham was twenty-three years old and Bowie perhaps a year or two older.

By the beginning of September, Banks knew that Bowie and Cunningham had been chosen and he turned his attention to getting the British bureaucracy to cooperate and support the project. The Prime Minister, Lord Liverpool, whom Banks had known since he was a child, had already learned of the plans and through George Harrison, the assistant secretary of the Treasury, had asked Banks for more information.[12] He was provided with a résumé of Kew's history and a blueprint for its future. Financial considerations were a major issue as were those of its responsibilities and governance.

Banks argued the case that while in the past Kew had been a charge on the royal purse, in future the government should take over the responsibility for sending and paying for collectors for Kew gardens.[13] Banks reminded the Prime Minister that the recent conflict with France had meant that the shipment of living plants, which he had been accustomed to organise for the King and country for a quarter of a century, had become nearly impossible with the result that collectors had not been sent out from Kew for some time. Now, with the restoration of peace, it was time to resume the botanic transfers. Rival gardens, especially the one at Schönbrunn, were setting the pace. It had always been the practice to choose collectors from gardeners trained at Kew; they were usually interested in collecting in the field, but the wages and conditions had stagnated for years. It was now time, Banks went on, to shift the expenditure to the public purse and to increase the basic salaries of collectors. Four hundred pounds a year per collector would cover all expenses, salary included.

This was, Banks assured the Prime Minister, an investment. He envisioned the government taking over Kew, 'which does honor to the Science of the Country . . . aids its Population & Enables the Sovereign & his Ministers to make acceptable Presents to Crowned heads without incurring any Expence in Providing them'. Kew collectors, Banks pointed out, 'whose Talents Enabled them to Excell in it', often built on their experiences as gardeners and went on to study botany, and thus added to the stock of scientific knowledge. That was one kind of investment, but Banks also thought that the circulation of rare and exotic plants from Kew would also benefit British commerce, and, thus, pay for itself. He was sure that visitors to the equivalent European gardens would be so struck by the novelties from Kew that they would place orders with British nurseries, so as to have them growing in their own gardens.

Banks added that he had two Kew-trained men ready to go. They were highly qualified and had the full support of William Aiton. The idea was to send them both to the Cape where they would collect in the environs of the town those local plants which, at Kew, had 'died from old age, before the means of increasing Them Could be discovered' but which were known from their entries in Aiton's *Hortus Kewensis*. These were the plants that Francis Masson had collected on his two expeditions there during the last thirty years of the previous century. Once the replacements had been found, one of the two collectors would head off to New South Wales while the other would stay behind to explore inland where 'there still Remains Vast Tracts of unexplored Countrey'. In both cases the collectors would be looking for plants that did not require special care and conditions like tropical plants, which, in the northern European climates required heated greenhouses, an expense few could afford. Temperate plants, which were often very beautiful, required at most a conservatory and some could be grown in the open (as Banks hoped those from Tasmania might).

It didn't take long for Harrison (and the Prime Minister) to be convinced by Banks's arguments.[14] Just over one week after Harrison first asked Banks for his plans for Kew, he wrote to Bowie and Cunningham telling them that they had been appointed as plant collectors, and they should expect to receive their instructions from, and their expenses approved by, Banks. They would be going to Rio de Janeiro on a naval ship, with the understanding that the British naval authorities there would facilitate their passage on to the Cape.[15]

Banks confirmed to Harrison that the Admiralty had no ships going directly to the Cape and that sailing to Rio de Janeiro was Bowie's and Cunningham's best option. Banks was sure that they would not find themselves stranded there as an 'abundance of opportunities will occur of Proceeding to the Cape of Good Hope'.[16] On 21 September 1814, Harrison learned that Sir Richard Bickerton, who was Commander-in-Chief at Portsmouth, had authorised Bowie and Cunningham to sail on HMS *Duncan*, scheduled to leave for Rio de Janeiro on 30 September.[17]

Banks had wasted no time in preparing the instructions for Bowie and Cunningham.[18] He had done this so frequently before that it must have been easy for him. At this point, the instructions dealt only with Brazil. When they reached their final destinations, they would get further specific instructions also prepared by Banks, from the governors of the respective settlements.

Banks had visited Brazil almost fifty years earlier when the *Endeavour* had anchored in the harbour of Rio de Janeiro on 13 November 1768. Then he had been frustrated by the authorities who continually stood in the way of his botanising, confining him, for most of the time, to the ship. Some sailors, who were allowed on shore to supply the ship with food and water, brought plants back for Banks and, on the one or two occasions when he managed to sneak on shore, he found interesting and largely unknown botanical specimens. Banks's own knowledge about Brazil's natural history was based on Willem Piso and Georg Marcgrave, whose co-authored book, *Historia Naturalis Brasiliae*, appeared in 1648.[19] Overall, and especially when compared to other places where the *Endeavour* had stopped, Brazil had been a disappointment.

Little had changed over the intervening years. Hardly any of the ships on which Banks had sent collectors had stopped in Rio de Janeiro on their way to the South Atlantic with the result that little had been added to the stock of Brazilian botanical knowledge or to the gardens in Kew. Banks suspected that the Brazilian naturalists were not sharing information with the British visitors, and the few plants that were sent back home went devoid of flower and fruit, parts crucial to their proper botanical identification – both Arthur Phillip and George Leonard Staunton, when they were in Rio de Janeiro, had come to this conclusion.[20]

Still, there was some room for hoping that conditions had changed. On 29 November 1807, the Portuguese royal court, fearing the rapid advance of French forces, with whom they were at war, fled Lisbon in a convoy of over thirty ships to join a Royal Navy squadron for protection for the voyage to Brazil. The size of the entire Portuguese entourage was probably about 10,000 people. The fleet arrived in the northern port of Salvador, Bahia on 22 January 1808, and were in Rio de Janeiro on 7 March, at which point the Portuguese Empire in exile established itself.[21]

With the arrival of the royal court, the population of Rio de Janeiro swelled. The British, who had played an essential role in the court's removal and transfer, exacted a price for their cooperation, effectively opening the Brazilian economy to British commerce. The number of British merchants who began to trade out of Rio de Janeiro increased markedly especially after 1810 when new commercial and navigational treaties were agreed. The presence of the Royal Navy which, from 1808 used Rio de Janeiro as the base for its South American fleet, also increased the population.[22] It wasn't all positive, however. There was a backlash against the incursion of Protestants and especially against those foreigners who were hostile to one of the country's most lucrative activities, the slave trade, which Britain had abolished in 1807.[23]

So, it is not surprising, in the light of his own negative experiences and with the new situation in Brazil, not entirely positive, that Banks had little to say to Bowie and Cunningham about what they should do in Brazil.[24] What he did suggest was that, in line with the arguments he had made to Lord Liverpool about the kind of plants that should be imported, they shouldn't bother with tropical varieties but concentrate on more temperate types. One plant, the *Fuchsia coccinea*, Banks knew came from a colder part of Brazil and this and other plants growing in that habitat would be most welcome in Britain. Banks knew from the recently published narrative of John Mawe's travels in the interior of Brazil that it was impossible to venture inland, especially to the gold- and diamond-mining districts, without permission.[25] In his instructions to Bowie and Cunningham, his only advice regarding this issue was that they should 'very particularly . . . avoid all conversations relating to the mines of Gold or diamonds and indeed all such [matters] as may render the people of the Colony doubtful of the real object'.

Bowie and Cunningham were at Portsmouth on 29 September 1814 but none of their possessions, which had been sent ahead, had arrived. HMS *Duncan*, it turned out, was not in a fit state for them to board. The next days were tense as the two collectors hoped their possessions would arrive before the ship sailed.[26] A little later, all was resolved and they boarded the ship the day before it left. Then contrary winds meant that the *Duncan* had to put into Plymouth Sound where it stayed for more than three weeks waiting for better sailing conditions. On 29 October, the anchors were raised and the ship began its passage to Rio de Janeiro.[27]

The voyage across the Atlantic was unremarkable apart from the fact that the weather at one point was so bad that they were unable to visit Madeira as planned. Almost two months later, and on Christmas Day, land at Cabo Frio, Brazil, was sighted and three days later the *Duncan* anchored in the harbour of Rio de Janeiro.[28]

For several days, Bowie and Cunningham lived on the ship, making daily visits to the residence of Percy Smythe, Lord Strangford, who had been instrumental in the transfer of the Portuguese royal family to Brazil and was now Britain's envoy to the Brazilian court. As Britain's senior diplomat in Rio de Janeiro, Strangford was supposed to arrange for the necessary papers that would allow the two collectors to travel outside of Rio de Janeiro. For days, and then weeks on end, nothing happened, so they tried to busy themselves in various ways: they familiarised themselves with the botany of those parts of the city that were within easy walking distance; and they got to know Baron Georg Langsdorff, a German physician and naturalist, who had been on a Russian voyage of exploration in the Pacific and was now, and had

been since 1813, Russia's consul-general to Brazil. Langsdorff was in
several ways very like Banks: not only was he a keen and talented
naturalist, but he supported the work and collecting expeditions of
others who came to Brazil to botanise; his home in the north of Rio
de Janeiro was like Soho Square – with its natural history library and
its open door, it was a scientific haven.[29] It was at Langsdorff's that
Bowie and Cunningham met Friedrich Sello [Sellow], a Prussian
gardener who, after studying botany in Paris, had gone to England in
1811 to continue his studies, largely in Banks's library. Sello had been
in London in January 1813, when Langsdorff was passing through on
his way to Brazil and, with his encouragement, Sello had set off for
Brazil to collect for a consortium of subscribers, possibly headed by
Banks and John Sims, a physician and botanist and editor of Curtis's
Botanical Magazine and the *Annals of Botany*.[30]

While waiting for the papers from Strangford, Bowie and
Cunningham also met Henry Chamberlain, Britain's consul-general
and he had had news for them. He told them bluntly that there was
no chance of a ship to take them to the Cape and to New South
Wales. Their best option, Chamberlain advised, was to return to
London and start again.[31] On the other side of the Atlantic and maybe
even at the same time, Banks had come to the same conclusion.[32]

It seems it was unusual for British ships to stop at both Rio de
Janeiro and the Cape regardless of whether they were naval or merchant
vessels (including whalers and sealers) or East India Company ships.
One can only suppose that Banks overlooked this in his anxiety to
get Bowie and Cunningham on their way. Once they understood they
were stuck in Brazil, they decided they might as well do some collecting
while they waited for further instructions.

By the end of February 1815, their passports and letters of introduction were ready and they made plans to visit the interior. About
a month later, on 3 April 1815, they began their botanical explorations
in an area beyond Rio de Janeiro in the direction of São Paulo, the
capital of the fourth most populous province, and, at a site over 2500
feet in elevation.[33] It was an area largely unknown to European botany
and, because of its location, Bowie and Cunningham hoped it would
produce the kind of plants Banks wanted.

In the few months they had been in Rio de Janeiro, Bowie and
Cunningham had managed to collect plants locally, and to send them
to London on a regular basis. The British Post Office ran a monthly

service to and from Falmouth and this set their pattern for writing to Aiton and Banks, sending their journals and the seeds and bulbs.[34] They had found that travel in Brazil was hard and that they had to hire mules and slaves in order to get about.

Getting to São Paulo proved particularly arduous. The road, if it could be called that, was rough and passed through several mountainous areas which the two collectors compared to 'Alpine travelling'. Often a river blocked their route and they had to resort to finding a canoe to take them and the luggage while the mules swam. There were no places to stay or eat. They had to sleep outdoors, often in cold conditions, and carry all their food with them. It took them a month to cover 250 miles – 'the Figure we cut is grotesque', Cunningham wrote to Aiton.[35]

They arrived in the city on 2 May and immediately headed to the home of an English cabinet maker where they would be staying thanks to the help of Georg Langsdorff, who provided letters of introduction. The next day they presented their letters to the Governor, the Conde de Palma, Francisco Assis de Mascarenhas, who 'enquired, in a particular manner, after the Health of the Right Hon(ble) Sir Joseph Banks, and when we informed him that Sir Joseph was in good bodily Health . . . the Conde seem'd highly pleased.' Bowie added that the Conde particularly praised 'the Important Services Sir Joseph Banks Rendered to the Whole World'. This lifted Bowie's spirits.[36]

At last their problems were over; the collecting could begin. Whenever the weather allowed, they went out botanising, both within the city's precincts and along the various roads that fanned out into the surrounding countryside. The highlight of the trip was collecting at the estate that had belonged to a previous Governor and to which the present Governor, the Conde de Palma, had granted them free access. Jaraguá, as it was called, was about fifteen miles to the northwest of São Paulo and as it was near the base of a mountain, it satisfied all of Bank's conditions. Every moment of the ten days when they made Jaraguá their base was spent botanising and came with substantial results.[37]

They were back in São Paulo on 30 June 1815 and returned to making daily excursions to areas that they had not yet explored to look for plants, but the time of their departure was quickly approaching. They finished their letters to Banks and Aiton. They planned just a quiet farewell to the Governor, but he had other plans. He invited them to a lavish dinner with several other gentlemen who had literary and scientific interests. The toasts, not surprisingly, were to the respective royal heads in Brazil and Britain; but Banks was not forgotten, hailed by the Governor as 'The Greatest Botanist in England'.[38]

On 14 August 1815, having packed up their collection to go to the port of Santos and from there by ship to Rio de Janeiro, Bowie and Cunningham left São Paulo. For the next few weeks they made their way back through familiar country botanising all the way, collecting plants that they had only sighted but not picked on their way to São Paulo. After a short stay in Itaguaí, to the west of Rio de Janeiro, they reached the city on 28 September, taking lodgings at a spot close to the mountains.[39]

Several things were becoming clear to Bowie and Cunningham. One was that Brazil was a wonderful place for collecting, especially on the higher ground where, as Banks had suggested, plants suitable for England could be found: they had already sent hundreds of seeds and had come across many new begonias, fuchsias and araucarias within the vicinity of Rio de Janeiro and now with their lodgings much closer to the higher ground, they could hope for even greater success. The second, and less positive realisation, was that the only way to ensure the best possible outcome for the seeds they had collected was to take them personally to a ship heading for London. The seeds and the living plants that they had collected in and around São Paulo, which they had sent on to Santos for shipping to Rio de Janeiro, arrived in a terrible state – it had taken three months to make the relatively short sea voyage between the two ports.[40] It was no better when they sent the seeds and plants they'd collected en route back from São Paulo by sea from Itaguaí to Rio de Janeiro. They were in such a hopeless condition that Bowie had to return to Itaguaí to replace the damaged specimens.[41] He was away from Rio de Janeiro for almost all of November and early December 1815.[42]

Most seriously of all, it seemed that, as Henry Chamberlain had warned them, they were stranded in Brazil and that their only option to get to the Cape would be to find a ship returning to London and start all over again. Banks seemed resigned to this outcome and had already asked Chamberlain to look out for a suitable plot of land where Bowie and Cunningham could keep the living plants alive until they could find a ship to take them and their plants back to England.[43] In fact, the piece of land that could serve as a garden didn't materialise until the end of January 1816 at a place called Gloria a few miles out of the city and near the sea; and it was the British merchant community rather than Chamberlain that helped out.[44] Banks was informed that they would need a year to establish the plants and that the best time to move them across the Atlantic wouldn't be until April.[45]

Over the next few months, the pair collected in the mountains surrounding Rio de Janeiro, especially in the Serra dos Órgãos, the Organ Range of mountains, to the northeast, sending seeds on the packet boats and growing plants in their garden. Chamberlain praised their diligence and success and thought that Kew would benefit if they stayed longer, especially 'now they have learnt something of the language and Manners of the People'.[46]

What Bowie and Cunningham thought about this isn't known but the truth was that no matter how successful they were, Rio de Janeiro was not where they were meant to be. Then on 22 March, the *Alceste*, carrying the Amherst embassy to China arrived. Rumours had it that the ship was supposed to call at the Cape, and Cunningham lost no time trying to contact Murray Maxwell, the captain, but it took a few days before he responded. In the meantime, Cunningham met and spent some time botanising with the ship's naturalists – Clarke Abel, the surgeon, and James Hooper, the Kew-trained gardener.[47] Then on 25 March, Cunningham finally met Maxwell, who explained that he had come to Rio de Janeiro because Lord Amherst wanted to see the city and that now that the watering and victualling had been completed in Rio de Janeiro they had no reason to stop at the Cape – the ship could go directly to China.[48] Cunningham's entry in his journal of this day – 'I remain still here' – shows his disappointment.[49] In fact, the *Alceste* did touch at the Cape on 18 April – as Bowie found out sometime before the end of July.[50]

With little prospect of leaving, Bowie and Cunningham kept exploring, one heading out into the hinterland while the other stayed behind to look after the collections. With no further news from London, from either Banks or Aiton, Bowie and Cunningham began planning another extended journey, this time to Minas Gerais, the mining district in the interior to the north of Rio de Janeiro, as Banks had recommended when he wrote to them in the previous year.[51]

Now Banks wrote to tell them that he was very pleased with their work and everything they had sent had arrived: the major disappointment was that the plants from São Paulo were in bad shape and, to minimise such loss in future, they should not venture too far from their base in Rio de Janeiro. The main point of the letter was to tell them that he had just learned that three convict ships were on their way to New South Wales and would be stopping at Rio de Janeiro. If any of them were also stopping at the Cape, then Bowie should take that ship; if not, then Bowie should proceed to New South Wales with Cunningham and make his way to the Cape from there. It was, at last, a concrete proposal and the first time that Banks had assigned the pair to their final destinations.[52]

Bowie went on his planned journey on 1 August 1816 and Banks's letter arrived one week later. Cunningham immediately packed his bags and went to find Bowie who, as he explained to Banks, was taking it 'leisurely in order the better to attend to the drying of the Seeds & Specimens'. Cunningham found him at a spot about twenty miles inland and Bowie decided to stay there working on his collection until he heard from Cunningham that the ships had actually arrived.[53]

On 17 September, the *Surry* (*Surrey*) entered the harbour having left Cork on 14 July with forty-nine Irish convicts aboard.[54] Two days later, Cunningham was all set: Thomas Raine, the ship's captain, said he had an order to take him to Sydney.[55] Bowie, meanwhile, who wanted to go with Cunningham to Sydney and botanise with him there, found that instead a passage had been arranged for him on the *Mulgrave Castle*, a merchant ship bound for the Cape.[56] It had put into Rio de Janeiro's harbour on 6 July in bad condition having struck rocks in the Cape Verde Islands. How long the repairs would take was uncertain: Henry Chamberlain, presumably in conversation with James Ralph, the ship's commander, had learned that he was intending to go to the Cape but not when he would be leaving.[57]

The *Surry*, with Cunningham on board, set sail on 25 September. Incredibly, it was shot at from the shore and had to return to port for repairs.[58] The *Mulgrave Castle* was now ready to sail. The misunderstanding that had led to the shooting was sorted out and two days later both ships left Rio de Janeiro together.[59]

Six days later, in the Atlantic, they parted.[60] Bowie went to the Cape and Cunningham to Sydney. They would never meet again.

It had only been intended as a stopover, but the long stay at Rio de Janeiro proved to be a major success. A little over a year after arriving in Brazil from London, Allan Cunningham wrote to his brother Richard, who was working at Kew gardens, that the seeds they had collected 'will be pleasing – Kew Gardens will shine in Melastoma, Malpighia, Banistera, Bignoniaceae, Gradeniae and many other new

& interesting Genera . . . If new Houses are building in Kew Gardens to contain our Treasures, one at least ought to be called Brazilian House'.[61]

It didn't happen but both Aiton and Banks shone in the reflected glory of their collectors' achievements. They had accomplished all that Cunningham had hoped they would. When Banks wrote to Cunningham in Sydney (which he had reached on 20 December 1816 in a record-breaking passage of 85 days from Rio de Janeiro), he congratulated him, especially for the examples of Epidendrum and Tillandsia – 'your bulbs also have produced some splendid flowers'.[62] On another occasion, he also praised Bowie, who had arrived safely at the Cape on 1 November 1816. 'The Garden at Kew', Banks wrote, 'is this Spring enlivened by abundance of the new plants' and, the following late summer, he commented that 'Kew Gardens have this year received much improvement by the New Plants sent home by both of you the Brazil Plants prove very Brilliant & Many of them New.'[63]

Writing from Rio de Janeiro, Cunningham told Aiton that even with the large amount of material they had collected and sent to London they had barely scratched the surface of what was there. He estimated they had sampled only about a quarter of the plants in and around the city and that it 'would occupy a Series of Years residence here to procure them, the Seasons are so variable'.[64]

Bowie, for his part, was in no doubt that he should have continued his collecting, and he expressed regret that he had been called away from his last journey into the interior.[65] He had learned much since first arriving: about how to dry specimens, how to pack them for safe road travel, and how to use local people to get further introductions – he had even learned Portuguese.

Henry Chamberlain, too, regretted that Bowie's journey had been cut short: Bowie, he told Banks, had learned how to behave in Brazil and was able to get what he wanted from the local people. No European botanist had ever visited these places before but after Bowie's departure, the field was left to Augustin François de Saint-Hilaire, who had arrived in Rio de Janeiro on 1 June 1816 with the French diplomatic mission. He had met the two English collectors before they left, and he went on to publish some of the first and finest botanical studies of the country.[66]

22

1815: Lockhart Survives the Congo

Banks was now seventy-two years old and had been involving himself in far-flung botanical adventures for more than half of his life. As he himself admitted he was tired and attacks of gout often incapacitated him. 'Age, infirmity & above all the loss of the use of my Legs', he wrote, 'have some time ago compelld to withdraw myself from public business, for which I no longer retain the activity & energy necessary to secure success.'[1]

However, when on 29 July 1815, Banks read a letter from John Barrow, he was transformed into a new man. 'Your letter has made my old blood circulate with renewed vigour. If anything will cure the gout it must be pleasure I derive from finding our Ministers mindful of the credit we have obtained from Discovery.'[2] So, with his enthusiasm restored, Banks was soon to embark on one of the most difficult and, as it turned out, last major botanical projects of his life.

Banks and Barrow had known each other for years. Barrow had done very well for himself. Born in 1764 in Ulverston, Lancashire, the son of a tanner, Barrow was largely self-taught and excelled in mathematics. He added navigation to his skills and, in his early twenties, when teaching mathematics at a private school in Greenwich, he so impressed George Leonard Staunton that he invited him to tutor his son, George Thomas Staunton.[3] When Macartney's diplomatic mission to China was organised in 1792, Staunton invited Barrow to join it to continue tutoring his son and to act as the embassy's comptroller – in charge of the presents to the Emperor and court mandarins. This was when Banks met Barrow.

In 1797, three years after the embassy returned, Barrow accompanied Macartney to the Cape, which had recently come under British control, wrested from the Dutch. Macartney had been chosen to be the colony's first British Governor and Barrow became his private secretary. The British occupation lasted until 1803, when, under the terms of the Treaty of Amiens signed by Britain and France, the colony was returned to the Batavian (Dutch) Republic, and both Macartney and Barrow went home. A year later, and through the patronage of Macartney and Henry Dundas, now 1st Viscount Melville, and First Lord of the Admiralty, Barrow was appointed to the post of Second Secretary to the Admiralty, which while answerable to the First Secretary, still carried with it the potential for the exercise of independent power.[4] By 1815, with the exception of one year, when because of a change in government he was out of the job, Barrow had been the Second Secretary for ten years and continued in that role for a further thirty years.[5]

On 18 June 1815, the French Army was defeated by a combined British and Prussian military force at Waterloo in present-day Belgium. Four days later, on 22 June, Napoleon abdicated for the second time. In August he was exiled to St Helena and a lasting peace established.

The British government, having made the Atlantic slave trade illegal in 1807, needed to find other commercial opportunities in West Africa. It was widely believed that there was a rich hinterland that could be accessed by the fabled Niger River. Arab sources and previous British expeditions had confirmed that the river rose in the Guinea highlands and flowed east, some thought into an inland sea or even as far as the Nile. Others believed that it turned south at some point emptying either into the Atlantic or joining the River Congo, a view that Barrow shared.[6]

Even during the Napoleonic Wars government officials had been thinking about sending expeditions to discover the true course of the Niger; and with peace they could put these plans into operation. The main proponents were Henry Goulburn, the Under-Secretary of State for War and Colonies, Earl Bathurst, the Secretary of State for War and Colonies, and John Barrow.[7] The idea was to organise two expeditions. One of them overland from the Atlantic coast and the other a maritime expedition up the Congo.[8] If the Niger and the Congo joined at some point, then the two expeditions would meet in the interior. Because of the maritime nature of the second expedition, Barrow was given full responsibility for it.

Barrow wanted to promote science and was determined it would be a scientific expedition. As soon as he received the go-ahead from the War and Colonial Office, he contacted Banks and explained that both Goulburn and Bathurst were very keen on the Congo expedition, and that, though he was confident that he could manage the naval aspects of the expedition, when it came to more scientific matters he would be relying entirely on Banks.[9]

Barrow's plan was 'that a good, clever, sensible Lieutenant of the Navy, or perhaps a Commander, should command the expedition'. It was to the maritime version, as he put it, of the Lewis and Clark expedition, referring to Captain Meriwether Lewis and second lieutenant William Clark's trek from St Louis to the Pacific and back in 1804 to 1806. This expedition had had some serious scientific objectives as Banks would have known. Barrow went on to write: 'you will perhaps be good enough to cast about for a proper person as a Naturalist; the more extensive his knowledge of course the more desirable.'[10]

Just as he didn't need to explain about Lewis and Clark, Barrow didn't need to tell Banks about the Congo. Banks had in his possession a chart of the river from its mouth to a point about 130 miles upstream published in 1795 and produced by George Maxwell, a British trader who knew the area well.[11] He also had copies of letters that Maxwell had sent to two correspondents in which he made the case for why he believed the Congo and the Niger were one and the same.[12]

Banks liked the sound of the expedition and agreed with Barrow that the commander should be especially dedicated to geographical discovery – someone like Bligh or Flinders. As for the naturalist, he would need more information before offering names. Barrow would have expected this, but Banks had some really surprising advice: he suggested that the river should be explored by a steamboat capable of coping with its current.[13]

In 1815, the sight of a steamboat in Britain was rare. Only seventeen had been built in the previous three years following the appearance of the first commercial paddle steamboat, the *Comet*, which ran on the Clyde between Glasgow and Helensburgh.[14] Steam engines were already in use in royal dockyards but when Banks recommended that the Admiralty should build such a vessel, there was no precedent. They had none, only sailing ships. His suggestion was nothing short of revolutionary.[15]

Banks knew a lot about steam engines as he had been using Boulton and Watts machines to drain the land on his Lincolnshire estates, and to pump water from the mines that he owned, principally in Derbyshire. Banks's association with Matthew Boulton went back almost forty years. Banks had also known John Rennie, Britain's leading civil engineer, for just as long, and had used his knowledge of steam engineering to help with drainage problems.[16]

Banks's idea was that the steamboat would travel as far up the river as it could and then be moored, with a few hands and a few cannon – the rest of the survey would be done by a lighter boat which had been towed to this point.[17] Barrow needed no convincing that Banks's idea was a good one. 'The moment you suggested a steam boat', Barrow wrote, 'I applied to Rennie, and he to Boulton & Watt.'[18]

Rennie thought that the engine, with two boilers, would weigh a little over 30 tons. Though James Watt Jr. offered an engine that was already working on the Tyne and whose owners were willing to sell, it wasn't right as it was designed to burn coal, whereas the plan was that the steam vessel would use wood as its fuel.[19] So the engine had to be built from scratch and it wasn't until near the end of December that it arrived in Deptford, accompanied by a trained technician, where the ship, carrying the appropriate name, HMS *Congo*, was, itself, under construction.[20]

The first naval steamboat was 'afloat' during the first week of January 1816 and, on the 11th, Barrow announced this important event to Henry Goulburn at the War and Colonial Office.[21] The very next day, Barrow wrote to Rear Admiral Sir Home Popham, who had earned a reputation as one of the navy's most scientific officers. He asked him to direct a series of experiments to determine how the ship's speed varied as the draught was changed by reducing and adding to its weight by means of iron ballasts, thus simulating the actual situation when provisions, instruments and people would be on board.

The experiment took place one week later with Popham and the engineer from Boulton and Watt in charge. Robert Seppings, Surveyor of the Navy, was also on board.[22] The results, Popham told Barrow, were very disappointing. In short, the *Congo*'s speed altered little as the draught decreased and, at little more than three knots per hour, was consistently less than the required seven knots per hour. Popham concluded that, for some reason, the engine was not producing enough power. The engineer thought that if the weight were reduced the speed would increase accordingly. Popham disagreed but in the interests of scientific knowledge, he gave the engineer another chance to prove that the engine was not at fault.[23]

Barrow, Popham thought, would be disappointed: 'the report . . . is not exactly such as I imagine you had a right to expect.' A few days later, on Monday, 22 January, the next experiment took place. The *Congo* hosted a large number of experts. Besides Popham and the engineer, James Watt Jr. was there, having come down from Birmingham especially for the occasion, and John Rennie was there too. This time everything that could be removed from the ship, including its sails, stores and ballast, was taken off and, still, it wouldn't go any faster than five and one-half knots per hour on a stretch of the Thames between Deptford and Limehouse.[24]

Though James Watt and his engineer were not yet willing to give up, Barrow was. He thought the sooner the engine was removed and the ship returned to a sailing vessel the better. As Barrow admitted to Popham, though enthusiastic at first he had had his doubts about the suitability of steam power for navigating the Congo. The possibility of something going wrong with the engine and the quantity of time and effort required for cutting wood were just two factors that worried him about the idea.[25]

Britain's first naval steamboat was no longer. The Boulton and Watt steam engine was on its way to its new life in Plymouth Dockyard where it was to be used to pump water.

Once it had been decided that the *Congo* would sail as a schooner, the ship was taken out on a sea trial which proved to be as unpromising and inconclusive as the steam trial. James Kingston Tuckey, who had been appointed to command the expedition on 13 September 1815, and had been present at all of the experiments, had had enough of hearing other people's opinions.[26] Popham, he thought, didn't like the ship at all, while Seppings thought it perfect. Tuckey's own opinion was that it was the best of a poor lot. 'In this sea of opinion', he remarked to Barrow, 'we may float to Eternity.'[27] He urged Barrow to put an end to the discussions and get on with preparing the *Congo* for the expedition. Tuckey did not agree with Barrow that the Niger and the Congo were the same.[28]

Tuckey, born in 1776, went to sea at an early age and joined the Royal Navy when he was seventeen years old. He did well and in 1802 he was appointed first lieutenant on HMS *Calcutta*, which had been fitted out to transport convicts and settlers for a new colony at Port

Phillip – present-day Melbourne. While there in 1803 Tuckey carefully surveyed the port and the surrounding area.[29] The *Calcutta* and Tuckey returned to England in early August 1804, but in the following year, and on different duties, a French squadron captured the ship near the Isles of Scilly and Tuckey was imprisoned in Verdun. He was not released until 1814.

Barrow had already chosen Tuckey as the most promising commander of the expedition in early August 1815 and had shared his enthusiasm with Banks. Banks knew something of Tuckey since he had read his narrative of the Port Phillip expedition in manuscript form and had seen the survey that he had produced.[30] Banks trusted Barrow's judgement about Tuckey: 'I never cultivated his acquaintance,' he admitted.

Though Banks must have been disappointed in the failure of his plan to have a steamboat lead the expedition, he had made it clear to Barrow that in this respect he relied entirely on the expertise of others: Rennie for issues of engineering and Seppings for naval architecture. 'I have no other concern in the Expedition', Banks stated to Barrow, 'than that of procuring a Naturalist,' though, as he added he couldn't help but consider the wider issues involved – the difficulty of working against the current of the Congo, the expected hostility of the local people and the sheer distance that would have to be covered.[31] He warned Barrow not to try and do the expedition on the cheap, meaning that there should always be more than one ship – the lucky escape of the *Endeavour* when it came to grief on the reef, and the mutiny on the *Bounty*, had taught him that 'it is better to adopt a Plan of sufficient extent at first; than to do it after a failure'.[32]

It was taking Banks a long time to find a naturalist. He was still searching by the beginning of December and decided to wait for the return from Paris of William Elford Leach, an assistant librarian in charge of the zoological collections at the British Museum.[33] Banks was under the impression that Barrow wanted 'to procure a man of general knowledge in natural history' rather than a botanist, and he thought that Leach knew such people better than he did.

Within a few days, Leach had provided the man.[34] His name was John Cranch, he was born in 1785, and had taught himself natural history, Latin and French, while being apprenticed to a shoemaker.[35] By the time the *Congo* expedition was being organised, he had managed to put together a natural history museum in Devon and it was through a visit there that Leach had first met Cranch.[36] Leach, seeing that

Cranch was an assiduous collector over a whole range of natural-history objects, asked him to collect for the British Museum, and from this their relationship grew. Leach recommended Cranch for the expedition; Banks immediately accepted the recommendation and Cranch thanked him profusely for arranging the appointment.[37]

There seems to have been little discussion about having a botanist on the expedition. From the beginning Barrow had thought in terms of a scientific group, consisting of a naturalist and artist – preferably the same person – a comparative anatomist and 'a collector of plants & seeds (gardiner)'.[38] Unexpectedly this was about to change.

Christen Smith, a Norwegian botanist, born in 1785, and who had studied medicine and botany at the University of Copenhagen, came into Banks's life towards the end of 1814.[39] He had come to London in July 1814 following an appointment as Professor of Botany at the newly founded Royal Frederick University in Christiana, now Oslo. His main intention was to find someone at Kew to help him establish a botanical garden in Christiana, which he did, and then he set off on a tour of Britain.

When Smith returned to London in December 1814, he met Banks and other leading botanists and soon, together with the geologist Leopold Von Buch, he decided to get his first taste of tropical botany by visiting Madeira and the Canary Islands.[40] Starting in April 1815, the pair spent two weeks in Madeira and almost six months in the Canaries, arriving back in London on 12 December 1815.[41]

Smith had intended to publish the results of his botanical excursions in London before returning to Norway, but, seeing his chance, Banks persuaded him to join the *Congo* expedition, which he did. Less than two months later, Banks got a gardener from Kew, David Lockhart, born in Cumbria in 1786, to assist the Professor, as Smith came to be called.[42]

On 7 February 1816, Barrow, with Banks's assistance, finalised the instructions for Smith and Cranch, and William Tudor. Tudor was a comparative anatomist, who had been selected to join the expedition at an early stage of its planning on the recommendation of the surgeon and Banks's personal physician, Sir Everard Home and Sir Benjamin Brodie, physiologist and surgeon.[43] On 10 February Lockhart's instructions were confirmed.

In his instructions to Smith, Banks told him that this was a once-in-a-lifetime chance.[44] There had never been a collection from this part of the world and, apart from South America, nothing at this latitude. He was to look out for anything and everything that appeared new to him. Collecting, drying and preserving botanical specimens was a major objective, and that would satisfy his academic interests; but collecting seeds and bulbs for the royal gardens at Kew was also important. It was here that Lockhart would play his part. The key thing was to try to keep everything alive as long as possible so that the moment the material arrived at Kew, the gardeners there would be able to take over caring for the living plant material. At the same time, Lockhart was to ensure that at each place and for each plant he procured seeds and bulbs; he was to note everything about their habitat and to cross-reference them with the dried specimens.[45]

In the whole history of British scientific expeditions since the *Endeavour* there had never been a ship that carried so many different scientists. In this respect, the *Congo* expedition was much more like its French counterparts where the presence of up to half a dozen scientists was not unusual. Banks may have been disappointed that the idea of a steamboat never materialised but he must have rejoiced in the attention paid to science. For that Barrow was responsible.

Tuckey had his ship, all the provisions, instruments, presents, the crew – including two men from the region of the Congo, Benjamin Benjamore and Sam Simmonds, as they were recorded in the ship's muster, who had advised on the kind of presents to take – plus the scientists. There were fifty-six men in all, ready to sail from Deptford on 16 February 1816, in company with the store ship *Dorothy*, but, as often happens with sailing ships, the weather was adverse and getting to sea proved to be more difficult than anticipated. It was not until 19 March, more than a month later and after a tedious stay in Falmouth, that the *Congo* cleared the Channel. Tuckey shaped a course for Madeira and better weather.[46]

After a few days, when the *Dorothy* rolled so heavily that most of the men aboard, including several of the scientists, were very seasick, the ships entered calmer waters and warmer surroundings. Tuckey soon passed Madeira and then the Canaries without stopping heading directly for the Cape Verde Islands, the closest point to the African coast.

It seems that Tuckey hadn't intended to stop at Cape Verde but the *Congo* was taking on water because the planks were shrinking and the caulking faulty. Tuckey knew from experience that no amount of trials in the Thames could prepare a ship for the real conditions at sea and he decided to anchor in what was then called Porto Praya, and is now Praia, the capital of Cape Verde and located on Santiago, the largest island in the group. The ships reached their anchorage on 9 April. This was the first time that Smith, who had been disappointed at not revisiting the Canaries, and Lockhart and Tudor had a chance to examine the island's natural history and they headed inland as soon as they could.[47]

Tuckey didn't want to stay any longer than necessary. The naturalists only had time to make a few observations. The caulking was done quickly and the ships raised their anchors on 11 April with the course set for Cape Palmas (at the border of modern Liberia and the Ivory Coast) and on to the Portuguese islands of Principe and São Tomé in the Gulf of Guinea, the latter coming into view on 18 May. The African coast was Tuckey's next objective and, on 3 June, they saw it for the first time at a distance of 15 kilometres. The course was now set southward, always keeping within sight of land, and anchoring whenever possible. This part of the voyage was, perhaps, the most tedious, but on 30 June the ships were approached by a boat and two canoes from one of the coastal villages. The main occupant was a merchant, dressed in a fine red waistcoat, who offered to sell them slaves, a trade which had been carried on in this part of the African coast for more than three centuries. Impressing on the merchant that the British slave trade was now illegal (the Slave Trade Act of 1807 abolished the slave trade in the British Empire), Tuckey returned to the business at hand and continued sailing southward.

On 8 July, the *Congo* and the *Dorothy* finally reached their destination, the mouth of the Congo River, and found an anchorage at a place called Shark Point, which George Maxwell, from his long experience trading on the river, had recommended as the best place to anchor.

Anxious to begin exploring the river, Tuckey removed the provisions from the *Dorothy* and put them on board the *Congo*. On 20 July, the *Congo*, taking advantage of the sea breezes that blew upstream from the Atlantic, set sail against the river's current. The scientific entourage, together with their instruments, moved to their new accommodation.

The Congo River teemed with traffic looking for trading opportunities. As they moved upstream they were able to collect. On one occasion, Smith found thirty new plant specimens. He was exhilarated by the botanical riches he saw. 'I found myself in a new world, which was before known to me in imagination only, or by drawings,' he wrote in his journal.[48]

The river was still wide, more than two miles across. Opportunities for anchoring presented themselves on both banks, often opposite villages. The current varied as the ship proceeded but was never a great impediment – the sail power was sufficient to propel them upriver. On 25 July, Sam Simmonds, one of the Congo men brought from England, was visited by his family – his brother and father came on board the ship.

A little over ten days later, on 5 August, the *Congo* had gone about as far as it could. The breezes grew more unreliable and Tuckey ordered the ship to be moored, as planned, with a skeleton crew of fifteen plus the surgeon and the purser, John Eyre, who was put in charge. Tuckey and the others continued upriver in the double boats, a long boat and punts that had been provided for just such an eventuality. Some of the boats were rowed, others had sails.

This stretch of the river had been visited by Maxwell and the places they passed were named on his chart. On 8 August Tuckey got bad news from the local people: not far ahead, they told him, the river became unnavigable because of the presence of 'a great cataract named Yellala'. Two days later, it became clear that the current was too strong for the boats to go further upriver. On 12 August, having climbed high enough to see the lay of the land and the course of the river, Tuckey decided that the boats would have to be left behind because there was no possibility of carrying them over the rough terrain. In order to continue exploring the river, the party would have to proceed on foot, with the help of guides,. Tuckey's first objective was to visit the waterfall. He took with him Smith and Tudor and fifteen other men, two interpreters and one guide.

The weather was pleasant but the going was very tough. Tudor and several others had to give up and return to the boats. Tuckey continued but, seeing that his companions were exhausted, decided to turn back. Ten hours later they arrived at a resting place where they found Tudor desperately ill with fever. The next morning, 17 August, Tudor was too weak to walk and had to be carried in a litter.

From this point it all went downhill. Tuckey was determined to find out what course the river took upstream from the waterfalls, whether it was navigable or not, and as no one was able or willing to tell him, he tried to find out for himself. He made repeated attempts pushing further and further onward but there were no pack animals and all the provisions had to be carried by the men themselves – it was more exhausting than anything they had ever experienced before.[49]

It was all too much even for Tuckey himself, and he had to turn back to the *Congo*.

The death toll was very high. John Cranch had succumbed after a few days of being ill. Then William Tudor, one of the first to get ill, died on 29 August. Many of the others followed suit. Smith died on 14 September; John Eyre, the purser, died on 27 September, and, Tuckey, one of the last to be struck down, died on 4 October. In total, 21 men died out of a total company of 54. It was a disaster.

Lockhart survived and so did Lewis Fitzmaurice, master and surveyor, who took command of the voyage following the deaths of both the captain and the lieutenant. He decided to leave the site of the tragedy and try to save those who were still alive. So he set a course straight across the Atlantic to Bahia, in northeastern Brazil, where he entrusted the sick men to a local hospital.

The *Congo* and the *Dorothy* arrived in Bahia on 29 October after a straight run of 28 days across the Atlantic.[50] They returned to Spithead on 23 February 1817 with the ship's logs and journals as well as part of the natural history and mineral collections. The tragic story of the voyage was already known, many of the details having been published in several British newspapers.[51]

Lockhart was lucky. Although he was too ill to leave Bahia with the ships, he returned to London at the end of June 1817 as a passenger on the brig *Regent*. He reported his arrival to Barrow and added that he had with him Christen Smith's handwritten journal, which the latter had entrusted to him before his death.[52]

The herbarium collected in the Congo by Smith and Lockhart arrived in London on the expedition's ships. The specimens went straight to Soho Square where Robert Brown examined them, who was now Banks's librarian, following Jonas Dryander's death in 1810. There were around 600 species and Brown could say with confidence that it was the largest collection that had ever been made in equatorial Africa. The results of Brown's investigation of the herbarium were published as an important appendix to Tuckey's narrative.[53] Brown, who was in a position to compare this herbarium with others collected in Africa, concluded that of the species he examined almost half were entirely new.[54] There were also some seeds from Santiago, Cape Verde and the Congo that Lockhart kept back, and these Banks presented to the royal gardens at Kew, the latter being the first of their kind.[55]

In his assessment of the collection, Brown praised Smith and

Lockhart though Banks appeared to have been less effusive. Nevertheless, his largesse spoke volumes.[56] In 1818, Lockhart was appointed as the first Superintendent of the Trinidad Botanic Gardens, a post for which undoubtedly Banks recommended him, and in which he remained until his death in 1845.

Epilogue

Joseph Banks died at his country home, Spring Grove, to the west of London, about ten miles from Soho Square, on 19 June 1820, aged seventy-seven, nearly five months after the death of George III. Banks was still President of the Royal Society, a position he had occupied for over forty-one years, the longest presidency in the Society's entire history.

Since 1779, Banks had often spent some time during the summer months at Spring Grove but in the last few years of his life, he began spending almost the entire summer there. At Spring Grove, with its extensive grounds, a garden, conservatories and greenhouses, Banks had carried out many horticultural experiments whose results were frequently published, particularly after 1807, in the *Transactions of the Horticultural Society*, the main publication of the learned society he had helped establish in 1804. Now, however, he spent most of his time there confined to bed suffering from the debilitating gout that had plagued him since late 1787 when he was in his early forties.

When not at Spring Grove in these last few years, Banks was at his Soho Square home – his regular annual visits to his Lincolnshire estates in September and October had been abandoned. At Soho Square he spent long periods as an invalid, but he was not alone nor without things to do. He had the pleasure of greeting some illustrious guests, especially from France: there was, for example, Joseph Gay-Lussac, the eminent chemist and physicist; the mathematician, physicist and astronomer, Dominique Arago; and, perhaps of greatest significance to Banks, because of their mutual interests, the naturalist, geographer and explorer, Alexander von Humboldt. When not entertaining distinguished visitors, Banks still continued with committee meetings that were now held at his home. He also ensured that his herbarium and library remained thriving enterprises, up-to-date and available for scholars from all over the world.

Banks's last projects turned out to be the Tuckey expedition to the Congo and the Amherst embassy to China in 1816, two very different parts of the world. This was characteristic of Banks's wide interests: two very different projects in two unconnected and distant places and both setting out from England within days of each other. Banks did lend his support to the polar expedition of John Ross and William Parry in 1818, but it was not his expedition: it belonged to John Barrow, the first of many such polar expeditions that he would organise for decades to come. Banks's own world was shrinking.

Though Banks never went to sea after 1773, almost all of his interests from then on involved ships and their captains; the oceans they crossed, the ports they visited. Central to this were the plants, particularly living ones, which were transported in their own specially designed cabins in the open air on the ship's quarterdeck, and those many gardeners, naturalists, ship's surgeons, merchants, government officials and travellers who collected and cared for them. Plants were moved in their thousands in this way and were introduced in places where they had never been seen before, both along the ship's route or at the final destination. What such plant movements did to the local ecologies, a major concern today, was not given any attention then as a successful transplantation was seen as an achievement to be celebrated.

Banks did not see his extensive botanical projects as fulfilling personal ambitions. He was often, as he himself said, trying to facilitate the desires of others and none more than that of his sovereign and friend George III, whose gardens at Kew he wanted to astonish the world. Banks's reverence for the King and his wishes may seem unfashionable and anachronistic today, but at the time they were shared, understood and lauded.

Banks's enthusiasm for increasing Britain's wealth, power and prestige continued to motivate many of his compatriots throughout the nineteenth century, but in the ways he tried to achieve this he acted more as a Georgian gentleman of science than as a Victorian imperialist.

Postscript

With the deaths of both Joseph Banks and George III in 1820 the royal gardens at Kew lost the support and interest from which they had benefitted since the 1770s. Over this period of almost fifty years, as we have seen, Banks had arranged for plant collectors to join governmental and commercial maritime expeditions throughout the world, to find and return with plants, often living examples, for Kew.

During Banks's time, the royal gardens had had no scientific basis, employed no botanists – there was no herbarium or library. All of the work was carried out by the head gardener and his staff. Kew's main purpose was to give the royal family the pleasure of seeing plants from all over the world growing in one place.

The Prince Regent, George III's son, had little interest in Kew and as George IV he chose to support other projects, especially the Pavilion at Brighton. Money for Kew became scarce. As a sign of the times, James Bowie, one of Banks's collectors who had been to Brazil and the Cape, was recalled in 1823; and Allan Cunningham, who had been with Bowie in Brazil and then relocated to Australia, was also recalled in 1830.

William IV, who succeeded his elder brother in 1830 upon the latter's death, had a fondness for Kew but, even so, expenditures on the gardens kept falling. Kew continued receiving plants from abroad, often from collectors, such as David Lockhart, who had been associated with Banks; and from other gardens in continental Europe, but these deliveries, as they were called at the time, were haphazard. The Horticultural Society of London, of which Banks was a founding member in 1804, took up the kind of collecting that Banks had organised: collectors, sponsored by the Society, went all over the world – China, North and South America, Africa and the West Indies – searching for and sending back living plants for its subscribers. Only a few went to Kew.

It seemed that Kew was destined to be abandoned.

Then, in 1837, with the death of William IV and the succession of his niece Victoria, the royal household was plunged into financial crisis. The income from Hanover ceased.[1]

A Royal Gardens Committee, set up by the Treasury, was charged with deciding the financial implications of this for all of the royal gardens. On 8 February 1838, they turned to the botanist John Lindley, the professor of botany at University College (who as a young man had worked for Banks as an assistant in his herbarium), to report on their 'Superintendence, Management, & Expenditure'.[2] When it came to Kew, Lindley was asked to consider particularly whether, in his opinion, the gardens should be maintained for the Queen and the public as a pleasure garden, or if it should be re-invented 'solely for the objects of science'.[3]

Lindley got to work a few days after receiving his instructions and a little over a fortnight later he had completed a draft report.[4] In it he made a strong case that Kew should either be abandoned altogether; or (what was clearly his preference) that it should be transferred from the royal household to the public, 'made worthy of the country, and converted into a powerful means of promoting national science'. He then spelled out the reasons why such an institution was needed, emphasising that it could act centrally to coordinate all the botanical gardens that had sprung up within the British Empire, such as Calcutta, Bombay, Sydney and Trinidad, for the exchange of botanical information and plants. The Royal Gardens Committee accepted Lindley's report and its recommendations and passed it on to the Treasury for their approval.

For the next two years leading members of the government wrangled over the decision. Some were fully behind Lindley's recommendations for the new Kew; others less so; and there was an influential contingent who simply wanted the gardens shut down. With no sign of a decision, Lindley took it upon himself to have the question of Kew raised in Parliament, on 4 May 1840. The report was published a week later. On 25 June 1840, the Treasury agreed to the proposals to transfer Kew to a public body.

Nine months later, on 25 March 1841, William Townsend Aiton, who had been Kew's head gardener since 1793, resigned his post. Now Kew's new future could be put into place.

On 3 April 1841, William Jackson Hooker, the Regius Professor of Botany at the University of Glasgow, and a friend of Banks's during the last decade of his life, took up his appointment as the botanical garden's first Director.

Like Banks, Hooker had accumulated a substantial herbarium and library and he brought them with him to form the core of a new imperial and scientific Kew such as Lindley had proposed.[5] Like Banks, Hooker soon began sending collectors from Kew to all parts of the world, eventually transforming the gardens into the institution familiar to visitors and scientists alike today.[6]

Acknowledgements

Even more than the pleasure I have had from working on this book for several years has been the delight in meeting so many people who were incredibly generous with their time in answering queries from me, greeting me in their archives and libraries, sharing their own interests with me and generally helping me out. I would like to thank them all. In Australia: Michelle Hetherington and Tony Orchard. In Belgium: Jan Vandermissen. In Brazil: Lorelai Kury. In Canada and the United States: Bob Batchelor, Kate Collins, Peter Crane, Sheila Ann Dean, Kimberley Fisher, Marnee Gamble, Cheryl Gunselman, Diana Kohnke, Paul Lovejoy, Haleh Motiey-Payandehjoo, Robert Peck, Tony Willis, Colyn Wohlmut, Hilary Dorsch Wong, and Anya Zilberstein. In China: May Bo Ching. In Denmark: Niklas Thode Jensen. In France: Georges Métailié and Kapil Raj. In Germany: Sabine Hackethal and Dominik Hünniger. In Great Britain: John Agar, Kate Bailey, Francesca Bray, Charlotte Brooks, Lynda Brooks, Lorna Cahill, Joe Cain, Neil Chambers, Nancy Charley, Isabelle Charmantier, Elaine Charwat, Sara Chiesura, Felix Driver, Cathy and Andrew Duncan, Sarah Easterby-Smith, Henrietta Harrison, Andrea Hart, David Helliwell, Julian Hoppit, Aaron Jaffer, Charlie Jarvis, Arlene Leis, Philippa Lewis, Trishya Long, Moira Lovegrove, Anne Marshall, Luciana Martins, Lorna Mitchell, John Moffett, Keith Moore, Henry Noltie, Alison Ohta, Lynn Parker, Leonie Paterson, Hellen Pethers, Suzanne Reynolds, Nigel Rigby, Edwin Rose, Benedetta Rossi, Suzanne Schwarz, Ruth Scobie, James Taylor, Adrian Thomas, Wai-hing Tsei, Pieter van der Merwe, Jacek Wajer, Dorota Walker, Ed Weech, Simon Werrett and Frances Wood. In Japan: Monotori Makino and Yoshiteru Yamamura. In Singapore: Jessica Hanser. In Sweden: Maria Asp, Moa Bergkvist, Lisa Hellman, Hanna Hodacs and Anne Miche de Malleray.

A big thank you to everyone who helped me with my research at these libraries and archives and at these institutions: Alexander Turnbull Library, Birmingham Archives and Heritage, British Library, British Museum, Cornell University, Duke University, Fitzwilliam Museum, Harvard University, Kent History and Library Centre, National Archives (Kew), Natural History Museum, Oak Spring Garden Library, Royal Botanic Gardens, Royal Society, Royal Swedish Academy of Sciences, Sutro Library, State Library of New South Wales, Toyo Bunko, University College London, University of London, University of Michigan, University of Toronto, Uppsala University, Washington State University, Wellcome Library and Yale University. I would also like to the thank the various organisers of conferences, workshops, seminar talks and lectures for inviting me and giving me the opportunity to present my ideas about Joseph Banks as they evolved; and to the many attendees whose stimulating questions and comments helped me hone my own thoughts.

It's not always that one thanks a book but in this case I wish to: David Mackay's *In the Wake of Cook: Exploration, Science & Empire, 1780–1801* was an immediate inspiration and sustained me throughout the years of researching and writing this book.

My agent and friend Will Francis has been supportive of this project from the very beginning as has Arabella Pike, my editor at HarperCollins, who read parts of the book as I completed them and all of it when it was delivered to her. I would like to thank her and her team for bringing this book to such a professional conclusion. I would also like to thank Luke Brown for his astute copyediting.

Special thanks go to Anna Backman and Malin Emara who helped me by translating important work from Swedish; and my friends who went out of their way to help me and read all or most of the book in manuscript and whose comments and suggestions have improved it well beyond my first attempt: Janet Browne, Mark Carine, Josepha Richard, Nigel Rigby and, particularly, Cordelia Sealy, whose close reading of the text resulted in avoiding several pitfalls and in making many significant and welcome improvements.

My final and biggest thanks go to my partner Dallas Sealy, who was there from the beginning, listening to stories about Banks, reading, commenting and substantially improving the text. The result would have been much poorer had it not been for her – with my deep gratitude and love.

Abbreviations

In citing the sources I have used in writing this book, I have adopted the following scheme. For manuscript material and published primary sources, I have used abbreviations as below. For all other sources, I have cited the author's surname followed by a shortened title of the work in question – the full citation appears in the bibliography.

ASB	Asiatic Society of Bengal
BAH	Birmingham Archives and Heritage
BdL	Bodleian Library
BL	British Library
CUKL	Cornell University, Carl A. Kroch Library
DURL	Duke University, Rubinstein Library
HUHL	Harvard University, Houghton Library
HRNSW	*Historical Records of New South Wales*, Sydney, Charles Potter, Government Printer, 1892–1901
I&P	*The Indian and Pacific Correspondence of Sir Joseph Banks*, Neil Chambers, ed., volumes 1–8, London: Pickering and Chatto, 2008–14
IOR	India Office Records
KHLC	Kent History and Library Centre
LS	Linnean Society
NHM	Natural History Museum, South Kensington
NLS	National Library of Scotland
NMM	National Maritime Museum
OSGL	Oak Spring Garden Library
PRONI	Public Record Office of Northern Ireland
RBGA	Royal Botanic Gardens Archives, Kew
RBGE	Royal Botanic Gardens, Edinburgh
RS	Royal Society

RSASCHS   Royal Swedish Academy of Sciences Center for
           History of Science
SC   *The Scientific Correspondence of Sir Joseph Banks*,
           NeilChambers, ed., volumes 1-6, London: Pickering
           and Chatto, 2007
SLNSW   State Library of New South Wales
SL   Sutro Library
TB   Toyo Bunko
TNA   National Archives, Kew
UCLLSC   University College London Library Special Collections
ULLSC   University of London Library Special Collections
WLAM   Wellcome Library Archives and Manuscripts
YUBRBML   Yale University, Beinecke Rare Book & Manuscript
           Library

Notes

Introduction

1 Banks to James Bland Burges, 19 January 1796, BdL, Dep. Bland Burges 21, fol. 7.

2 Unless otherwise noted I have followed Beaglehole, *The Endeavour Journal* and Carter, *Sir Joseph Banks*, for these biographical details.

3 Banks was born when the Julian, or Old Style, calendar was in operation. In some accounts, his birthdate is given as 24 February, which is the Gregorian or New Style date that came into force in 1752.

4 See Glyn, 'Israel Lyons'.

5 A very good discussion of Linnaeus's life and work can be found in Koerner, *Linnaeus*.

6 These students have variously called Linnaeus's disciples or apostles – for more on this, see Koerner, 'Purposes', Nyberg, 'Linnaeus' Apostles, scientific travel' and Nyberg, 'Linnaeus's Apostles and the Globalization'.

7 This voyage is covered in Lysaght, *Joseph Banks*.

8 For the artists, which also included a little-known Peter Paillou, assigned to represent the *Niger*'s natural history collection, see Lysaght, *Joseph Banks*. For more on Ehret, see Calmann, *Ehret* and on Parkinson, see Carr, *Sydney Parkinson*. For Lee, see Easterby-Smith, *Cultivating*.

9 The story of the Royal Society's involvement in the project to observe Venus's track is told in Carter, 'The Royal Society' and Cook, 'James Cook'. The voyage of the *Endeavour* has, naturally, attracted much attention. In my story of the expedition I have, unless otherwise noted, relied on a number of these, especially: Carter, *Sir Joseph Banks*, Beaglehole, *The Endeavour Journal*, vol.1 and Williams, *Naturalists*, ch. 4.

10 The 'Memorial' is in RSA, RS Misc. MSS V 39 and has been reproduced in Chambers, *Endeavouring Banks*, pp.29–30. Maskelyne's rectangle was first described in a paper by Thomas Hornby, the Professor of Astronomy at the University of Oxford – Williams, 'The *Endeavour* Voyage', p.4.

11 Beaglehole, *The Journals*, vol.1, p.511.

12 The Royal Society's involvement in this astronomical project is discussed in

several sources, particularly Beaglehole, *The Journals*, vol.1, Woolf, *The Transits* and Wulf, *Chasing*.

13 Carter, 'The Royal Society', p.251. The Council of the Royal Society recorded the King's consent to the expedition on 24 March 1768 – Beaglehole, *The Journals*, vol.1, p.513. For the correspondence relating to the King's consent and the Admiralty's responses, see Knight, 'H.M. Bark Endeavour', p.292.

14 Knight, 'H.M. Bark Endeavour', pp.296–7.

15 Beaglehole, *The Journals*, vol.1, p.511.

16 The biographical details of James Cook come from Beaglehole, *The Life*.

17 Though Cook and Banks were in the same harbour on 27 and 28 October 1766, there is no evidence that they met on this occasion – see Carter, *Sir Joseph Banks*, p.36.

18 Williams, 'The *Endeavour* Voyage', p.5.

19 On Wallis's expedition, see Cock, 'Precursors', Williams, '"To Make Discoveries"' and Patel, *Exploration*. For Dalrymple's terminology, see Dalrymple, *An Account*.

20 The story of Wallis's discovery of the island is told in Salmond, *Aphrodite's*.

21 Patel, *Exploration*, pp.xxiii–xxiv.

22 Beaglehole, *The Journals*, vol.1, p.513 and Knight, 'H.M. Bark Endeavour', p.299.

23 Beaglehole, *The Journals*, vol.1, p.134 and pp.cclxxix–cclxxxi.

24 Ibid., p.cclxxx.

25 Beaglehole, *The Endeavour Journal*, vol.1, p.22 and Beaglehole, *The Journals*, vol.1, p.514.

26 Carter, *Sir Joseph Banks*, p.32.

27 Ibid., pp.55–6.

28 Ibid., p.27.

29 See ibid., pp.58–64 and Cook, 'James Cook', pp.47–8.

30 Stephens to Cook, 22 July 1768, TNA, ADM 2/94.

31 I take these details from Duyker, *Nature's Argonaut*.

32 This and the following information about Solander's relationship with Banks and events leading up to plans about the scientific part of the *Endeavour*'s voyage is taken from Banks to Johan Alströmer, 16 November 1784, *SC*, document 533.

33 Solander's version of events was far less dramatic. As he put it to Linnaeus in a letter to him of 1 December 1768: '. . . a young rich gentleman named Joseph Banks decided to go too for the sake of Natural History; I have been well acquainted with him for many years and he proposed to me to go with him at his expense. I thought that such an offer and such an opportunity should not be turned down' – Duyker and Tingbrand, *Daniel Solander*, letter 96, 1 December 1768.

34 Duyker and Tingbrand, *Daniel Solander*, letter 90, 24 June 1768.

35 See Carter, *Sir Joseph Banks*, p.70. Ellis's 19 August 1768 letter to Linnaeus is in LS, The Linnean Correspondence, L4101.

36   Banks to William Perrin, 16 August 1768, *I&P*, vol.1, document 13.

37   James Codd to Banks, 18 August 1768, *I&P*, vol.1, document 14.

38   Beaglehole, *The Journals*, vol.1, pp.cclxxix–cclxxxi.

39   Banks to James Douglas (Lord Morton), 1 December 1768, *I&P*, vol.1, document 19.

40   Banks's list has been transcribed from his manuscript journal and printed in Beaglehole, *The Endeavour Journal*, vol.2, pp.281–9. See also Solander to Linnaeus, 1 December 1768, in Duyker and Tingbrand, *Daniel Solander*, letter 96.

41   *Lloyd's Evening Post*, 23–25 May 1768, p.502.

42   Beaglehole, *The Endeavour Journal*, vol.1, p.350.

43   Ibid., p.258.

44   Ibid., p.308.

45   Beaglehole, *The Journals*, vol.1, p.cclxxxi.

46   These instructions can be found in Beaglehole, *The Journals*, vol.1, pp.cclxxxii–cclxxxiv.

47   Beaglehole, *The Life*, p.224. Cook's journal entry, which explains how he came to this decision, is in Beaglehole, *The Journals*, vol.1, pp.272–3. Cook does not mention Banks as being involved in the decision-making process but reading Banks's account of it makes one suspect that he might have been involved – see Beaglehole, *The Endeavour Journal*, vol.2, p.38.

48   Tasman did not know that Van Diemen's Land was an island. A contemporary map of New Holland, drawn by Jacques-Nicolas Bellin, a French hydrographer, showed a coastline between the southern edge of Van Diemen's Land stretching to the north to about the southern coast of New Guinea. On the map, along the representation of the coastline, Bellin noted 'I suppose that Van Diemen's Land could be joined to the land of St. Esprit' which was discovered, in the early seventeenth century, by the Portuguese navigator, Pedro de Queirós, and which he placed near the southern part of present-day Cape York, Australia – the Bellin map is reproduced in Clancy, *The Mapping*, p.93.

49   Carter, *Sir Joseph Banks*, p.73. The map, which was enclosed with Dalrymple's, *An Account*, showed the coastline of Van Diemen's Land petering out at some northern point, after which the map went blank until the southern coast of New Guinea. For a discussion of the contents of the *Endeavour's* on-board library, see Carr, 'The Books'.

50   Beaglehole, *The Endeavour Journal*, vol.2, p.77.

51   Edwards, *The Journals*, pp.138–9.

52   Beaglehole, *The Endeavour Journal*, vol.2, p.78.

53   Carter et al. 'The Banksian'.

54   Morgan, 'From Cook'.

55   For more on Batavia, see Blussé, *Visible Cities*.

56   For example, there was one short article based on a letter that was received in London from Sydney Parkinson, which also mentioned that the ship was

returning with 'a vast number of plants and other curiosities' – see *The London Evening Post*, Tuesday, 14 May to Thursday, 16 May 1771.

57   Beaglehole, *The Endeavour Journal*, vol.2, p.232.

58   Cook's charts have been reproduced in David, *The Charts*.

59   *General Evening Post*, Saturday, 3 August 1771.

60   Carter, *Sir Joseph Banks*, p.59.

61   See John Ellis to Linnaeus, 10 May 1771, LS, The Linnean Correspondence, L4505.

62   *General Evening Post*, Monday, 12 August 1771. Pringle's biographical details come from Kippis, *Six Discourses*.

63   Carter, 'The Royal Society'.

64   See examples of the conversations as reported in letters from Pringle to Albrecht von Haller in Berne, published in Sonntag, *John Pringle's Correspondence*.

65   Carter, *Sir Joseph Banks*, p.96.

66   *Craftsman or Say's Weekly Journal*, Saturday, 3 August 1771.

67   *London Evening Post*, Thursday, 29 August 1771.

68   The ship returned by way of the Orkney Islands. The voyage is the subject of the authoritative Agnarsdottír, *Sir Joseph Banks*.

69   Though they make only a small appearance in this book, it is important to remember that Sara Sophia and Dorothea played a large role mediating between the circulation of knowledge and sociability at the Soho Square home. They were also great collectors in their own right. No one has done more to illuminate this neglected but very important area than Arlene Leis – her publications on these topics include 'Cutting', 'Displaying', 'A Little Old-China' and 'Sarah Sophia Banks'. See also Leis, 'Sarah Sophia Banks: Femininity'. Eagleton, 'Collecting' focuses on Sarah Sophia's extensive coin collection.

70   Chambers, 'Letters'.

71   Joppien and Chambers, 'The Scholarly Library'.

72   Officially, Solander was Keeper of the Natural History Department of the British Museum and had a residence there. His relationship with Banks was not, it seems, founded on a formal basis: Solander helped Banks with his herbarium and library and many other things as well and kept an eye on Soho Square when the Banks household removed to Revesby in September and October, but he was not, as Jonas Dryander would become in 1782, Banks's librarian. For his assistance, Banks granted Solander an allowance – £200 annually. For all this see Duyker, *Nature's Argonaut*.

73   The catalogue was published between 1796 and 1800.

74   Dawson, *The Banks Letters*, p.xxviii. How many of that number has survived is difficult to say but a figure of 20,000 or more would not be too far off the mark. The collections with the largest concentration of surviving correspondence are: the Sutro Library, the British Library, the Natural History

Museum, the Royal Botanic Gardens, the State Library of New South Wales and Yale University. A more detailed inventory can be found in Carter, *Sir Joseph Banks (1743–1820: A Guide)*, pp.31–49. There are, in addition to the letters, many slips of paper, comprising lists, notes, etc. These are sometimes tucked into the correspondence but the largest number on a diverse range of subjects are in the Sutro Library.

75 Banks to Hamilton, 4 December 1778, *SC*, document 146. The relevant quote is at the very beginning of the book.

76 Unless noted otherwise the following is based on Desmond, *Kew*, chs.1–5 and Berridge, *The Princess's Garden*.

77 Bute was Prime Minister briefly from May 1762 to April 1763. His role as a botanic adviser is discussed in Miller, 'My Favourite', Phillips and Shane, *John Stuart* and Taylor, *Passion*. There is a list of the plants Bute donated to Kew in Phillips and Shane, *John Stuart*, p.67. For London's nurserymen, see Easterby-Smith, *Cultivating*.

78 For Aiton, see Pagnamenta, 'The Aitons'. For Miller, see Le Rougetel, *The Chelsea Gardener*.

79 Desmond, *Kew*.

80 See Hadlow, *The Strangest Family*, pp.215–16.

81 *Gentleman's Magazine*, May 1793, p.390.

82 In his history of Kew, Ray Desmond states that Banks was in this role by 1773, but provides no evidence for it – see Desmond, *Kew*, p.89.

83 Banks to Campo d'Alange, 10 April 1796, BL, Add MS 56299, p.29.

84 See Bladon, *The Diaries* and Banks to Unknown Correspondent, 23 February 1789, *The Gentleman's Magazine*, August 1820, p.99.

1 1772: Masson Roams the Atlantic

1 Pringle to Haller, 28 November 1768, in Sonntag, *John Pringle's Correspondence*, p.120. Evidence that Pringle and the King met privately in 1763 comes in a letter from Pringle to Haller, 11 June 1763, in Sonntag, *John Pringle's Correspondence*, pp.55–6.

2 Jarrell, 'Francis Masson' notes that Masson came to Kew in 1771 but does not provide the evidence for this.

3 Banks Memorandum, [1782], NHM, DTC, vol.2, pp.213–20. A draft of this memorandum is in WLAM, MS 5218, fol.20.

4 Masson, *Stapeliae*, p.vi.

5 For a sense of the enormity of the task, see Diment et al., *Catalogue*. See also Lack and Ibañez, 'Recording Colour'.

6 See Edwards, 'Sir Joseph Banks' for the botany of the *Endeavour* voyage.

7 Beaglehole, *The Endeavour Journal*, vol.2, p.255.

8 Britten, 'Some Early Cape', p.47.

9 See Aiton, *Hortus Kewensis*, passim. For Miller, see Le Rougetel, *The Chelsea Gardener*.

10 Carter, *Sir Joseph Banks*, p.28. See also Dandy, *The Sloane Herbarium* and Britten, 'Some Early Cape'.

11 For more on these collectors, see Exell, 'Pre-Linnean Collections', Britten, 'Some Early Cape'. For a discussion of the publications, see Gunn and Codd, *Botanical Exploration*.

12 For the Cape's shipping profile, see Ward, 'Tavern', Worden, 'VOC Cape Town' and Worden et al., *Cape Town*. For the English East India Company, St Helena was more important as a stopping place for returning ships than was the Cape – see McAleer, 'Looking East' for more about St Helena.

13 Beaglehole, *The Voyage of the Endeavour*, pp.456–66.

14 This fact and the following is revealed in a letter from Oldenburg to Peter Johan Bergius, a student of Linnaeus's and a Stockholm-based physician and botanist – see Oldenburg to Bergius, 8 March 1773, RSASCHS, Bergianska brevsamlingen, vol.16, pp.465f. I would like to thank Maria Asp for sending me the transcript of the letter and Anna Backman for providing me with a translation of it. Little is known of Oldenburg, who was born in Stockholm in 1740 and died in Madagascar in 1774. See de Kock and Krüger, *Dictionary*, vol.II, p.524 and Gunn and Codd, *Botanical Exploration*, p.319.

15 About one thousand specimens that Oldenburg collected at the Cape came into Banks's possession – see *The History of the Collections Contained in the Natural History Departments of the British Museum*, vol.I, London, 1904, p.171.

16 See Reverend William Sheffield to Banks, 1 April 1772, *I&P*, vol.1, document 72.

17 Sandwich to Cook, 5 May 1772, ADM 2/97.

18 Beaglehole, *The Life*, pp.294–5.

19 Worden, *Cape Town*, pp.50–1, 60.

20 Masson, 'An Account'. The Khoikhoi were referred to at this time as Hottentots.

21 Beaglehole, *The Voyage of the Endeavour*, p.464.

22 There is no doubt that Masson would have sent William Aiton a list of what he had collected. This, and many other valuable documents, including letters, were, unfortunately, destroyed – see Desmond, *Kew*, p.190.

23 Unless otherwise noted, I have taken the information concerning Thunberg and his time at the Cape from Skuncke, *Carl Peter Thunberg* and Karsten, 'Carl Peter Thunberg'.

24 Masson, 'An Account', p.278.

25 Ibid., pp.297–8.

26 Karsten, 'Francis Masson', p.179 doubts that it was a stapelia.

27 See Britten, 'Lady Anne Monson' and Noltie, 'John Bradby Blake'. Banks and Solander both knew Lady Monson.

28 Thunberg, *Travels in Europe*, vol.2, p.132.

29 For Sparrman's time at the Cape, see Karsten, 'Sparrman' and Beinart, 'Men, Science'.

30 Sparrman, *Voyage*, vol.I, pp.80–4.

31 Karsten, 'Francis Masson', p.204.

32 On 27 March Sparrman wrote to Thunberg, whom he knew, saying that Masson was on the point of leaving for home – Sparrman to Thunberg, 27 March 1775, quoted in Karsten, 'Sparrman', p.56.

33 I have not found a precise date nor how he got home. On the other hand, there were at least four East India Company ships at anchor in Table Bay when Sparrman wrote his 27 March 1775 letter to Thunberg, and Masson could have gone home on any one of these, all of which had arrived in London by late August.

34 Skuncke, *Carl Peter Thunberg*, p.87.

35 Duyker and Tingbrand, *Daniel Solander*, p.358.

36 Gough to Tyson, 6 March 1776, Nichols, *Literary Anecdotes*, vol.VIII, p.618. For more on Gough and his circle, see Sweet, 'Antiquaries'.

37 Tyson to Gough, 4 May 1776, in Nichols, *Literary Anecdotes*, vol.VIII, p.620. Tyson knew a fair amount of natural history. He had botanised with Israel Lyons (who had lectured Banks when he was at Oxford) – Nichols, *Literary Anecdotes*, vol.VIII, p.208.

38 Masson to Linnaeus, 28 December 1775, Linnean Society, The Linnean Correspondence (online site), L5170.

39 It was listed as being introduced in 1773 by Banks in Aiton's, *Hortus Kewensis*, vol.II, p.55. There is some debate as to who gave the plant this name, Banks or Aiton – see Moore and Hyppio, 'Some Comments' and Britten, 'The History', p.5.

40 Karsten, 'Carl Peter Thunberg', p.126. Thunberg had come across a variety of *Strelitzia* near Plettenberg Bay. Britten, 'The History', p.5, states that Masson did a drawing of this plant.

41 Franz Bauer, one of Banks's most dedicated botanical artists, produced a volume of drawings of species of *Strelitzia* in 1818 – see Bauer, *Strelitizia Depicta*.

42 For the reference to *Strelitzia* as being Banks's favourite, see Carter, *Sir Joseph Banks*, p.304. The quote is from Carter, 'Sir Joseph Banks and the Plant Collection', p.346. I would like to thank Ekaterina Heath for sharing a chapter of her PhD with me which looks at the gardens at the Pavlovsk Palace, St Petersburg, where the Kew plants went – see also Heath, 'Joseph Banks'.

43 Banks Memorandum, [1782], NHM, DTC, vol.2, p.214.

44 In his memorandum, Banks explicitly states that in 1776 it was Pringle who 'again petitioned his Majesty . . . to Mr Masson's again undertaking an extensive Plan of operations' – Banks Memorandum, [1782], NHM, DTC, vol.2, p.214.

45 See Britten, 'R. Brown's List' and Beaglehole, *The Endeavour Journal*, vol.2, pp.159–65.

46 Masson to Linnaeus, 6 August 1776, LS, The Linnean Correspondence, L5229. The date of 5 June as Masson's arrival in Madeira comes from Banks Memorandum, [1782], NHM, DTC, vol.2, p.219.

47 There is some evidence that in addition to his time-consuming work he was in poor health. See Kippis, *Six Discourses*, pp.lvii ff.

48 Banks Memorandum, [1782], NHM, DTC, vol.2, p.216.

49 Masson to Banks, 28 July 1776, *I&P*, vol.2, document 164 and Masson to Banks, 8 August 1776, *I&P*, vol.2, document 168.

50 There is a large literature about botanical exploration on Madeira and the Canary Islands: see, for example, Dandy, *The Sloane Herbarium*, for an overview of how the material came to Sloane; for British collectors on the Atlantic Islands in general, see Francisco-Ortega et al., 'Early British'; for Madeira, see Britten, 'R. Brown's List', Francisco-Ortega et al., 'Plant Hunting' and Francisco-Ortega et al., 'The Botany'; for the Canaries, see Francisco-Ortega and Santos-Guerra, 'Early Evidence' and Frade and Menéndez, *El Descubrimiento*. I would like to thank Charlie Jarvis for alerting me to this latter reference.

51 See Aiton, *Hortus Kewensis*.

52 The following is based on a series from letters from Masson to Banks in SLNSW, Sir Joseph Banks Papers, Series 13.02–13.15 plus the addition of Masson to Banks, 4 February 1777, *SC*, document 94.

53 Murray to Banks, 14 July 1777, BL, Add MS 33977, p.82.

54 I would like to thank Mark Carine for telling me that Masson collected plants on São Jorge – personal communication, 17 December 2017.

55 I have taken the itinerary from Banks Memorandum, [1782], NHM, DTC, vol.2, p.219.

56 A sense of this can be gleaned from his account of the Azorian island of São Miguel, in a letter to Aiton and which was read to the Royal Society by Banks on 2 April 1778 and published as Masson, 'An Account of the Island of St. Miguel'.

57 Unless otherwise noted, this section is based on the letters Masson sent to Banks which are in SLNSW, Sir Joseph Banks Papers, Series 13.16–13.19.

58 O'Shaughnessy, *An Empire Divided*.

59 Banks Memorandum, [1782], NHM, DTC, vol.2, p.215.

60 Masson to Banks, 26 July 1780, SLNSW, Sir Joseph Banks Papers, Series 13.18. By 'philosopher' (sometimes referred to as 'natural philosopher'), Masson meant what we would now call a 'scientist' – that term did not come into use until after 1834.

61 Banks Memorandum, [1782], NHM, DTC, vol.2, p.215. This hurricane, which is estimated to have killed over 20,000 people in the Caribbean, is still considered the deadliest in the history of the Atlantic.

62 Banks Memorandum, [1782], NHM, DTC, vol.2, p.216.

63 Masson to Linnaeus the younger, 12 December 1778, LS, The Linnean Correspondence, L5637.
64 Masson to Banks, 1 February 1779, SLNSW, Sir Joseph Banks Papers, Series 13.15.
65 Carter, *Sir Joseph Banks*, p.38.
66 For more on De Visme's Lisbon garden and the one he developed north of the city in Sintra, see Luckhurst, 'Gerard De Visme'.
67 Masson to Banks, 17 March 1783, SLNSW, Sir Joseph Banks Papers, Series 13.20.
68 De Visme to Banks, 3 December 1784, *SC*, document 540. Charles Murray made a similar observation about Masson – Murray to Masson, 20 September 1779, BL, Add MS 33977, fol.109
69 Banks to Giovanni Fabbroni, 4 February 1785, *SC*, document 557. In this letter Banks notes that 'Masson is lately returnd from Barbary'. According to Karsten, 'Francis Masson', p.9, Jonas Dryander, Banks's librarian following Solander's death in 1782, produced two lists, one of plants Masson collected in North Africa, and the other, of plants he collected in Portugal, Spain and Gibraltar, but I have not been able to locate them.
70 Unless otherwise noted the documentation for Masson's time at the Cape is contained in the letters, all but one of them from Masson to Banks, in SLNSW, Sir Joseph Banks Papers, Series 13.22–64.
71 The story behind this is told in Karsten, 'Francis Masson', pp.286–90.
72 Masson to Thunberg, 21 March 1793, UUL, Thunberg Collection, G 300 s, transcribed and printed in Karsten, 'Francis Masson,', pp.301–3.
73 Masson's receipts are in SL, Sir Joseph Banks Collection, A 1:21–8 and RBGA, JBK/1/4, fol.332. Banks's letter in which he disapproved of Masson's excursions is Banks to Masson, 3 June 1789, SLNSW, Sir Joseph Banks Papers, Series 13.40.
74 Masson to Thunberg, 21 March 1793, UUL, Thunberg Collection, G 300 s, transcribed and printed in Karsten, 'Francis Masson', p.302.
75 France declared war on the Netherlands early in 1793 and, in return, Britain launched an attack on Dutch bases. The Cape fell to a British invasion force on 16 September 1795 – Potgieter, 'Admiral'. For his collection of living plants and his garden, see Masson to Thunberg, 29 November 1795, UUL, Thunberg Collection, G 300 s, transcribed and printed in Karsten, 'Francis Masson', pp.307–8. See also Banks's comments in Banks to Campo d'Alange, 10 April 1796, BL, Add MS 56299, p.30. For Masson's arrival in London, see Banks to John Koster (Liverpool merchant), 19 November 1795, *SC*, document 1340, in which he comments that he had arrived three months earlier from the Cape.
76 Britten, 'Francis Masson', p.121.
77 Masson, *Stapeliae Novae*, London, 1796.
78 Banks to Koster, 19 November 1795, *SC*, document 1340.
79 Banks to Campo d'Alange, 10 April 1796, BL, Add MS 56299, p.30.

80   Letters of introduction and passports, 30 July and 3 August 1797, BL, Add MS 56299, pp.32–3. The botany of Canada was not as well-known in Europe as its zoology – see Binnema, *Enlightened Zeal* and Houston, Ball and Houston, *Eighteenth-Century Naturalists*.
81   The ordeal is described in Masson to William Townsend Aiton, 1 January 1798, RBGA, Record Book 1793–1809, pp.84–5.
82   Masson's itinerary can be constructed from Masson to Banks, 18 October 1798, RBGA, Record Book 1793–1809, pp.77–8.
83   Much more about this group, who were called United Empire Loyalists, can be found in Jasanoff, *Liberty's Exiles*.
84   For Hamilton, see his entry by Bruce G. Wilson, *Dictionary of Canadian Biography*, online at *www.biographi.ca/en/bio/hamilton_robert_5E.html*. A few years earlier, the recently appointed Lieutenant Governor of Upper Canada, John Graves Simcoe, and his wife Elizabeth, were also the guests of Hamilton and probably for the same reason as Masson's stay – see Innis, *Mrs. Simcoe's Diary*, p.77.
85   York's population at the time was no more than 250 people. Montreal's population was around 9000 while that of Lower Canada was more like 200,000.
86   Masson to Banks, 17 October 1798, SLNSW, Sir Joseph Banks Papers, Series 13.65. Banks had been trying to grow wild rice in England, and with some success – see Zilberstein, 'Inured to Empire'.
87   For more on the North West Company, see Campbell, *The North West Company*.
88   The canoe, paddled for the most part by *voyageurs*, Quebec French-speaking men, was the means of transport throughout the trading system. Huge freight canoes, capable of taking loads, primarily beaver pelts, up to 4 tons in weight, and paddled by ten men, made their way from Grand Portage through the Great Lakes (Superior, Huron, Erie and Ontario) and finally down the St Lawrence River to Montreal.
89   See Masson to Banks, 5 November 1799, SLNSW, Sir Joseph Banks Papers, Series 13.66.
90   Masson to Banks, 12 January 1800, SLNSW, Sir Joseph Banks Papers, Series 13.67.
91   Masson to Banks, 1 November 1800, SLNSW, Sir Joseph Banks Papers, Series 13.68.
92   This is what Banks wrote to Masson and which Masson remarked upon in his return letter – Masson to Banks, 1 November 1800, SLNSW, Sir Joseph Banks Papers, Series 13.68. Unfortunately there is no sign of Banks's letter to Masson.
93   Banks to Masson, 11 February 1802, BL, Add MS 33981, p.1. In 1798, Banks voiced his disappointment to James Edwards Smith that Masson wasn't going far enough west, that is, beyond the confines of Upper Canada: Banks may

have been thinking of this when remarking that Masson was playing it safe – see Banks to Smith, 9 December 1798, *SC*, document 1493.

94 Masson to William Townsend Aiton, 14 May 1805, RBGA, Record Book 1793–1809, p.238.

95 See Masson to Aiton, 17 October 1805, RBGA, Record Book 1793–1809, p.243; Masson to Aiton, 26 October 1805, RBGA, Record Book 1793–1809, p.244 and Banks to Masson, 15 July 1805, SLNSW, Sir Joseph Banks Papers, Series 13.69.

96 Gray to Banks, 18 March 1806, BL, Add MS 33981, pp.240–2.

2 1779: Return to Botany Bay by Way of Southwest Africa

1 This is clear from a search across the newspapers collected in *17th–18th Century Burney Collection Newspapers*. For more on the public interest in Tahiti, see Maxwell, 'Fallen Queens' and Russell, 'An "Entertainment"'. See also Scobie, 'To Dress'.

2 The description of New Holland can be found on pp.109–24. Typically, Tobias Smollett's posthumously published review of this book concentrated almost exclusively on Tahiti – see Smollett, 'A Journal'. See note 22 below regarding the likely authorship of this book.

3 The Admiralty had placed a ban on publications other than its official account of the voyage, which it commissioned Hawkesworth to produce.

4 Parkinson, *A Journal*.

5 For a discussion of the politics of the Royal Society, see Miller, 'Into the Valley', Miller, 'Sir Joseph Banks' and Chambers, 'Letters'.

6 *Journal of the House of Commons*, vol.36, p.846. For more on hulks as supplements for prisons, see James, '"Raising"'.

7 For the wider context of transportation, see the following good sources: Maxwell-Stewart, 'Convict Transportation', Anderson and Maxwell Stewart, 'Convict Labour' and Christopher and Maxwell-Stewart, 'Convict Transportation'.

8 Ekirch, *Bound for America*. For a good account and further reading into the convict experience in colonial America, see Christopher, *A Merciless Place*, ch.1.

9 Frost, *Botany Bay*, pp.63–4. Solander visited the ship to inspect its sanitary conditions because he was considered a reliable expert witness of shipboard health having sailed with Cook on the *Endeavour* – Cook had not lost a single man to scurvy or other diseases on long sea voyages – Foley, 'The British Government', pp.72–5.

10 Foley, 'The British Government', p.78.

11 *Journal of the House of Commons*, vol.37, p.307. The best account of both committee hearings and the parliamentary discussions leading up to their

formation is Foley, 'The British Government'. Frost, *Botany Bay* and Oldham, *Britain's Convicts* are vital additions. Bunbury's report was laid before the House on 1 April 1779.

12 *Journal of the House of Commons*, vol.37, pp.310–11.

13 Cook, the only other authoritative voice, was away on his third voyage to the Pacific. Unknown to anyone beyond the Pacific, he was already dead, having been killed in Hawaii on 14 February 1779.

14 Banks's evidence appeared in the *Journal of the House of Commons*, vol.37, p.311. A typographical error was introduced onto this page making it seem that the date was 10 April, but, in fact, it was 1 April, as were all of the dates in this part of the *Journal*. It has been transcribed and published, with the incorrect date, in *I&P*, vol.1, document 194, see below.

15 Banks to Bunbury Committee, 10 April [1 April] 1779, *I&P*, vol.1, document 194.

16 Banks did not mention the Endeavour River area in his assessment of New Holland, which he wrote in his journal following the entry for 31 August 1770. He did, however, write that the land he saw as the ship made its way along the coast was barren, the soil sandy and very light – Beaglehole, *The Endeavour Journal*, vol.2, pp.111–37.

17 The evidence from these witnesses can be found in the *Journal of the House of Commons*, vol.37, pp.311–13. For a discussion, see Christopher, *A Merciless Place*.

18 *Journal of the House of Commons*, vol.37, p.314.

19 These are discussed in Frost, *Botany Bay*, Foley, 'The British Government', Oldham, *Britain's Convicts* and Gillen, 'The Botany Bay'.

20 I have taken these biographical details from Frost, *The Precarious Life*, pp.1–4.

21 Beaglehole, *The Endeavour Journal*, vol.2, p.86.

22 Matra is now generally acknowledged to be the author of the anonymous, *A Journal of a Voyage round the World*, published at the end of September 1771. It carried the publisher's effusive dedication to Banks and Solander. See Frost, *The Precarious Life*, p.6.

23 On Matra's visits to Banks and Solander, see Frost, *The Precarious Life*, p.71. Just when and how Matra and Banks became closely associated is unclear but Matra had been in contact with Banks not long after the *Endeavour* returned because, in a letter, he reminds him that in 1772 he was aware that Sarah Sophia was collecting coins and that Banks had just added examples from the East Indies to her collection – see Matra to Banks, 18 February 1779, BL, Add MS 33977, pp.88–9.

24 Frost, *The Precarious Life*, p.83.

25 Ibid., pp.84–5.

26 Giordano, *The Anonymous Journal*, p.x.

27 Matra to Banks, 28 July 1783, *I&P*, vol.2, document 19.

28 Foley, 'The British Government', pp.131–2. For more on Young and his

subsequent involvement in the debate over transportation, see Frost, *Dreams*.

29 Matra Proposal, *I&P*, vol.2, document 21, also in *HRNSW*, vol.1, pt.2, pp.1–6. On the need to find a home for those Americans who sided with the British, see Jasanoff, *Liberty's Exiles*.

30 Matra Proposal, *I&P*, vol.2, document 21, also in *HRNSW*, vol.1, pt.2, pp.1–6.

31 Matra Proposal, *I&P*, vol.2, document 21, supplementary comments, also in HRNSW, vol.1, pt.2, pp.6–8.

32 Beattie, *Crime*.

33 Frost, *Botany Bay*, pp.83–4.

34 The story of these troubled voyages is told in fascinating detail in Ekirch, 'Great Britain's Secret'.

35 Frost, *Botany Bay*, p.106. Christopher, *A Merciless Place*.

36 *Journal of the House of Commons*, vol.40, pp.954–9.

37 Ibid.

38 See Frost, *Botany Bay*, pp.97–8. Young's plan is in *HRNSW*, vol.1, pt.2, pp.10–13.

39 Matra's testimony can be found in Frost, *The Precarious*, pp.118–22.

40 Banks to Beauchamp Committee, 10 May 1785, *I&P*, vol.2, document 69.

41 Ibid.

42 Ibid.

43 Foley, 'The British Government', pp.167–8.

44 Frost, *Convicts*, pp.40–1.

45 Call's New South Wales plans are discussed in Frost, *Convicts*, pp.19–28.

46 Frost, *Botany Bay*, pp.109, 192.

47 *Journal of the House of Commons*, vol.40, pp.1163–4.

48 For a discussion of possible reasons for the government's interest in south-western Africa see Foley, 'The British Government', pp.169–77 and Frost, *Convicts*, pp.41–4.

49 Frost, *Botany Bay*, p.126.

50 These instructions are in Admiralty to Thompson, 15 September 1792, TNA, ADM 2/1342.

51 Banks to Aiton, 29 August 1785, HUHL, MS Hyde 10 (19).

52 Blagden to Banks, 10 September 1785, RBGA, Banks Correspondence, JBK/1/3, fol. 203. For Graefer, see Coats, 'Forgotten'.

53 David Lewis to Banks, 16 November 1786, SLNSW, Sir Joseph Banks Papers, Series 23.29.

54 Blagden to Banks, 10 September 1785, RBGA, Banks Correspondence, JBK/1/3, fol. 203.

55 Banks to Blagden, 15 September 1785, *SC*, document 598.

56 Blagden to Banks, 10 September 1785, RBGA, Banks Correspondence, JBK/1/3, fol. 203.

57 Dryander to Banks, 12 September 1785, *SC*, document 597.

58 Blagden to Banks, 15 September 1785, *SC*, document 599.

59   Ibid.

60   Ibid., document 600.

61   'Heads of Instructions to be given by Captain Thompson of His Majesty's Ship Grampus to Mr Antoni Pantaleon Howe [sic]', TNA, ADM 2/1342, filed under Captain Edward Thompson.

62   TNA, ADM 36/10702.

63   Thomas Boulden Thompson's parentage is uncertain. Some accounts state that Edward Thompson was his father, others that he was his uncle – see his entry in the *Oxford Dictionary of National Biography*.

64   Christopher, 'From', p.761.

65   TNA, ADM 55/92, 'NAUTILUS: Journal kept by Captain T B Thompson'. There is more about the voyage of the *Nautilus* in Christopher, 'From', Kinahan, 'The Impenetrable Shield' and Mackay, *In the Wake*, pp.32–6.

66   Thompson to Philip Stephens [Admiralty], 15 August 1785, TNA, ADM 1/2594, filed under Thomas Boulden Thompson.

67   Some idea of the government's commitment to Das Voltas Bay can be gleaned from Gillen, 'The Botany Bay', pp.751–2.

68   Gillen, 'The Botany Bay', p.753.

69   TNA, T 1/639, paper 2176. This is printed in *HRNSW*, vol.1, pt.2, pp.14–20.

70   See *HRNSW*, vol.1, pt.2, pp.20–4.

71   See Frost, *The First Fleet* and Gillen, 'The Botany Bay', p.76.

72   I have focused on the convict problem in this chapter because that is how Australia came back into Banks's life. Academics have long debated whether the decision for Botany Bay was only about convicts or whether it also had imperial dimensions to it – a good overview of the arguments can be found in Frost, *Botany Bay*.

73   Howe's seed catalogue is in RBGA, Banks Correspondence, JBK/1/4, fol.231.

3 1780: The First Circumnavigation of Archibald Menzies

1   See Montagu to Banks, 10 January 1780, *I&P*, vol.1, document 203, in which he first told Banks about Cook's death. The news of what happened in Hawaii and subsequent events were contained in a letter written by Captain Charles Clerke, who had assumed overall command of the expedition, to the Admiralty dated 8 June 1779 from the anchorage in the Kamchatka Pensinsula. The letter was entrusted to the care of Magnus von Behm, commander of the Russian territory, who was on his way overland to St Petersburg. He delivered the letter to Sir James Harris, the British Ambassador to Russia, who forwarded it to the Admiralty. Clerke's letter is reprinted in Beaglehole, *The Journals*, vol.III, pp.1535–40. On von Behm, see O'Grady-Raeder, 'Major von Behm' and Crownhart-Vaughan, 'Clerke in Kamchatka'. Clerke died at sea within sight of Kamchatka on the ships' return from the Bering Sea on 22 August 1779. His final written communication may well

have been a deeply warm letter he had written to Banks on 18 August 1779 – *I&P*, vol.1, document 198.

2 Montagu to Banks, *I&P*, vol.1, document 213. For Montagu's role in the Cook voyages, see Rodger, *The Insatiable Earl*.

3 King to Banks, 12 October 1780, *I&P*, vol.1, document 214.

4 Edwards, *The Journals*, 30 March 1778, p.540.

5 This quote appears in Igler, *The Great Ocean*, p.105. A full description of the sea otter by Steller can be found in Steller, *Journal*, pp.142–8. See also Williams, *Naturalists*.

6 Pallas lived in St Petersburg after 1767, having been invited by Catherine II to become a professor at the Academy of Sciences. He and Banks were good friends: they began exchanging natural history specimens in 1779. Pallas's journey through Russia was published in St Petersburg in three volumes between 1771 and 1776 under the title *Reise durch verschiedene Provinzen des Russischen Reichs*. The description of the trade at Kyakhta can be found in vol.3, pp.136–42. On Pallas, see Jones, 'Peter Simon Pallas'. See also Bockstoce, *Furs and Frontiers*, pp.103–14

7 For discussions of the Russian–Chinese fur trade, see Gibson, *Otter Skins*, Ravalli, 'Soft Gold', Jones 'Sea Otters', Jones, A "Havoc"', Zilberstein, 'Objects of Distant Exchange' and Stolberg, 'Interracial Outposts'.

8 *I&P*, vol.2, document 21.

9 King, *A Voyage*, p.437.

10 Ibid., p.441.

11 Ibid., p.438.

12 Williams, *Captain Cook's Voyages*, p.xxxiii.

13 Galois, *A Voyage*, pp.299, 348.

14 Etches to Banks, 14 March 1785, *I&P*, vol.2, document 66.

15 The report to Sandwich is at NMM, Sandwich Papers, F/5/38 and is discussed in King, 'The Long', pp.10–12.

16 For details of this visit, see Skuncke, *Carl Peter Thunberg*. For Banks's interests in Japan, see also Chambers, 'Sir Joseph Banks, Japan'.

17 Banks to Thunberg, 17 June 1785, *SC*, document 585.

18 Etches to Banks, 14 March 1785, *I&P*, vol.2, document 66.

19 See 'Agreement with David Nelson', 26 April 1776, *I&P*, vol.1, document 156.

20 Dixon to Banks, 27 August 1784, NHM, DTC, vol.4, pp.47–8.

21 Banks to Dixon, 29 August 1784, NHM, DTC, vol.4, p.49. Banks also mentioned that he had heard a rumour that the American Congress was considering just such an expedition.

22 Galois, *A Voyage*, p.301, King, '"The Long"', pp.6–7 and Mackay, *In the Wake*, p.61.

23 'Additional Proposals Relative to the establishing a Trade between the North West Coast of America – and the Coast of Asia, the Japanese Islands &c.', (29 April 1785) BL, IOR, H/494, fol.365.

24 Galois, *A Voyage*, p.7. In a later letter to Banks, Etches explained that Portlock and Dixon intended selling furs in Japan but the attempt was abandoned because the time of the year was wrong – see Etches to Banks, 23 July 1788, *I&P*, vol.2, document 230.

25 Mackay, *In the Wake*, p.64.

26 See Portlock, *A Voyage*, pp.6, 69.

27 This was not the first British venture to trade in sea otters. In April 1785, John Henry Cox, a merchant involved in private trade in Canton, and James Hanna, who was in command of the ship, set sail from Canton for the Pacific Northwest. In the following year, another venture was started by the Bengal Fur Company, based in Calcutta. A good overview of British activity in the Pacific Northwest–China trade is provided by Wilson, 'King George's Men'. See also Lamb and Bartroli, 'James Hanna', Pierce, *A List*, Nokes, *Almost a Hero* and Gough, *The Northwest Coast*. It was also not the first time that Banks's advice was sought about trade in sea otter. Before Etches came along, Banks had already been contacted by James Strange, an East India Company employee in Madras, who was impressed by King's account of the Pacific Northwest fur bonanza; and his partner, David Scott, a highly successful private merchant based in Bombay and soon to be a director of the East India Company, seeking his advice as to how to make their expedition, planned to leave from Bombay, as scientific as possible. See Strange to Banks, 3 December 1785, *I&P*, vol.2, document 79; Scott to Banks, 1785, *I&P*, vol.2, documents 64 and 65; Ayyar, Hosie and Howay, *James Strange's Journal*; and Gough, *India-Based Expeditions*.

28 Portlock, *A Voyage*, pp.7–8.

29 Though Etches didn't know it, in July 1786 the *King George* and the *Queen Charlotte* had reached their destination, Cook's River (now Cook Inlet, in Alaska), and began trading, working their way down the coast towards Nootka Sound.

30 Galois, *A Voyage*, pp.2–4. See also Howay, *The Journal*, pp.xi–xiii.

31 Galois, *A Voyage*, p.77.

32 Menzies to Banks, 30 May 1784, NHM, DTC, vol.4, fol.29.

33 The oldest surviving letter is dated 17 April 1766 but from its contents it is clear it is not the first one – see Hope to Banks, 17 April 1766, RBGA, JBK/1/2, fols.1–2.

34 For further information about Hope, see Noltie, *John Hope*.

35 Noltie, *John Hope*, p.26.

36 I take this from McCarthy, *Monkey Puzzle Man*, pp.29–30 and especially, n.30, p.31.

37 See Gwyn, *Frigates*.

38 Menzies to Banks, 30 May 1784, NHM, DTC, vol.4, fol.29.

39 Ibid.

40 Menzies to Banks, 2 November 1784, RBGA, JBK/1/3, fol.175.

41 See Menzies to Banks, 28 January 1786, RBGA, JBK/1/3, fol.221 and Menzies to Banks, 8 June 1786, RBGA, JBK/1/4, fol.234.

42 See Menzies to Banks, 8 June 1786, RBGA, JBK/1/4, fol.234. Banks had endorsed the letter with this date. For the Company's permission to Etches, see Galois, *A Voyage*, p.8 and p.302, n.70.

43 Menzies to Banks, 21 August 1786, NHM, DTC, vol.5, fol.43.

44 Menzies to Banks, 7 September 1786, *I&P*, vol.2, document 91. There is another version of this story, which appears in *I&P*, vol.2, p.127, n.1 and in Mackay, *In the Wake*, p.68 and p.80, n.61. According to these sources, Banks had already recommended a certain John Melvill to the post of surgeon but, when Melvill fell seriously ill in August, he pulled out leaving the vacancy to Menzies. The evidence for this assertion is a letter from Melvill to Banks, 21 October 1786, RBGA, JBK/1/4, fol.248. There is, however, no mention in the letter of an August date; nor is there any indication that Melvill was leaving his post or that he was a surgeon. On the contrary, from the internal evidence, Melvill was embarking on a long sea voyage for health reasons, which is what some of the country's most eminent physicians, whom he had consulted, had recommended. It was only when his condition got severely worse that he abandoned the idea altogether. Melvill states in the letter that he had already met Menzies in his capacity as the expedition's surgeon. My interpretation of the letter is that Melvill was not a surgeon but was a gentleman whom Banks knew. In late September Etches told Banks that there was room on the ship for a gentleman should he know someone – see Etches to Banks, 29 September 1786, *I&P*, vol.2, document 93. It is very likely, I would argue, that Banks passed Melvill's name on to Etches with a letter of introduction from Lord Mulgrave – Melvill to Banks, 21 October 1786, RBGA, JBK/1/4, fol.248.

45 Hope to Banks, 22 August 1786, *I&P*, vol.2, document 89.

46 Carter, *Sir Joseph Banks*, p.222 speculates that Menzies met Banks at Soho Square between 21 and 28 August but provides nothing more than that.

47 Menzies to Banks, 7 September 1786, *I&P*, vol.2, document 91; Etches to Banks, 29 September 1786, *I&P*, vol.2, document 93 and Dryander to Banks, 28 September 1786, *SC*, document 678.

48 Galois, *A Voyage*, p.78.

49 Menzies to Banks, 16 November 1786, *I&P*, vol.2, document 94.

50 Ibid.

51 Galois, *A Voyage*, p.78.

52 Ibid., p.332, n.50; Sloan [sic], 'An Account' and Melvill to Banks, 21 October 1786, RBGA, JBK/1/4, fol.248. Banks's description of the plant can be found in Beaglehole, *The Endeavour Journal*, vol.1, pp.216–17.

53 Menzies to Banks, 11 February 1787, *I&P*, vol.2, document 109.

54 Ibid.

55 Menzies to Daniel Rutherford, 12 September 1789, RBGE, GB235 RUT.

56 See *https://en.wikipedia.org/wiki/Duke_of_York_(1780_ship)*, accessed 12

September 2016; and Galois, *A Voyage*, p.305, n.106. Menzies did not know that Hope already had a specimen of *Wintera aromatica*, which he had received from John Byron, commander of HMS *Dolphin* during its circumnavigation between 1764 and 1766 – see Hope to Linnaeus, 9 May 1767, The Linnean Correspondence, at *http://linnaeus.c18.net/* Letter/L3908, accessed 8 September 2016 and Menzies to Daniel Rutherford, 12 September 1789, RBGE, GB235 RUT.

57   Menzies to Banks, 16 November 1786, *I&P*, vol.2, document 94.

58   Galois, *A Voyage*, p.100.

59   Ibid., pp.11–12.

60   For this interesting story, see Galois, *A Voyage*, p.200 and the n.107 and 108 on pp.373–4. See also Fullagar, *The Savage Visit*, pp.165–7.

61   Galois, *A Voyage*, pp.15–17.

62   Groves, 'Archibald Menzies' (1992), pp.4–5. If Menzies kept a journal, it has not survived and neither has a plant list. What we know about Menzies's botanising on the Colnett expedition is the result of painstakingly going through the herbaria where his specimens still survive. That is what Eric Groves did and which he reported in this essay.

63   Groves, 'Archibald Menzies' (1992), pp.5–7.

64   Galois, *A Voyage*, pp.17–18.

65   Unfortunately for Colnett and the venture, the whole time that they were in the Pacific Northwest, there were often many other traders around, so alluring was the potential wealth of furs. There was one ship, commanded by an English captain, Charles William Barkley, but flying the colours of Austria, who happened to have reached Nootka Sound before Colnett and scooped the finest furs; and then there was George Dixon in the *Queen Charlotte*, Etches's first venture, now on his second trading expedition. See Galois, *A Voyage*, pp.11–12 and Gough, *The Northwest Coast*. Banks and Etches knew about Barkley's expedition – see Etches to Banks, 30 July 1788, *I&P*, vol.2, document 232. For Barkley, see *https://en.wikipedia.org/wiki/Charles_William_Barkley* and Hill and Converse, *The Remarkable World*.

66   Galois, *A Voyage*, p.18.

67   See Galois, *A Voyage*, pp.18–19 and respective notes for this episode. See also Galloway and Groves, 'Archibald Menzies', p.12.

68   McCarthy, *Monkey Puzzle Man*, p.65.

69   The estimate is in Groves, 'Archibald Menzies' (1992), p.24.

70   Menzies to Banks, 21 July 1789, *I&P*, vol.3, document 26.

71   Groves, 'Archibald Menzies' (1992), p.25.

72   Menzies to Daniel Rutherford, 19 October 1789, RBGE, GB235 RUT.

4 1782: The Brothers Duncan in Canton

1 The important standard text for the introduction of Chinese plants into Europe is Bretschneider, *History*. See also Dumoulin-Genest, 'L'Introduction' and Kilpatrick, *Gifts*, pp.95–101.

2 Russian traders were the exception. See Fletcher, 'Sino-Russian' and Perdue, 'Boundaries and Trade'.

3 The mechanism of the Canton System has recently been described in Van Dyke, *The Canton Trade* and also in Carroll, 'The Canton System'.

4 For a description of life in the factories during the trading season see Downs, *The Golden*. See also Carroll, '"The Usual"', Conner, *The Hongs* and Farris, 'Thirteen Factories'. For the commercial world of the Hong merchants, see Wong, *Global Trade*.

5 Van Dyke, 'The Hume Scroll'.

6 There were exceptions to this. Puankequa I, a leading Hong merchant, for example, spoke Spanish (he had commercial dealings in the Philippines) and some pidgin English I would like to thank Josepha Richard for pointing this out to me. On the subject of linguists, see Van Dyke, *The Canton Trade*, especially pp.77–93 and the Special issue 'Chinese Pidgin English: Texts and Contexts' of the *Hong Kong Journal of Applied Linguistics*, vol.10 (2005). There were, however, exceptions to the general rule that Europeans did not learn or know Chinese, but they were few in number. Two notable British exceptions were James Flint, who was in Canton from the 1730s to the early 1760s – see Farmer, 'James Flint', and Golden, 'From the Society of Jesus'; and Thomas Bevan on whom see Morse, *The Chronicles*, vols.I and II. For exceptions in the Swedish factory community and other examples of evading restrictions, see Hellman, *Navigating*.

7 For daily life in the foreign communities in Canton and Macao, see Hellman, *Navigating*, especially ch. 5.

8 Falconer to Banks, 16 April 1768, in Chambers, *I&P*, vol.1, document 6.

9 For an overview of the European understanding of Chinese botany in this period, see Métailié, *Science and Civilisation*.

10 Estimates of the number of Jesuits in China put the figure at near one thousand before 1800. See Dehergne, *Répertoire*, Witek, 'Catholic Missions', Mungello, *Curious* and Brockey, *Journey*.

11 For Boym, see Golvers, 'Michael Boym'. Another important early title was Athanasius Kircher's *China Monumentis . . .*, 1667, which drew heavily on Boym's book. For Kircher, see Findlen, *Athanasius Kircher*. For Du Halde and the importance of his work for the circulation of knowledge of China in eighteenth-century Europe, see Foss, 'A Jesuit Encyclopedia'. On the *Lettres édifiantes* see Zhesheng, 'The "Beijing Experience"'. Cams, 'The China Maps' studies the maps that were published in Du Halde's book. For the *Mémoires*, see Dehergne, 'Une grande collection'.

12 I have relied on Haberland, *Engelbert Kaempfer* for the biographical details.

On Kaempfer's years working for the Dutch East India Company, see Van Gelder, '*Nec semper*'. See also Kowaleski-Wallace, 'The First Samurai'.

13   On Linnaeus and Chinese plants, see Cook, 'Linnaeus'.

14   Muntschick, 'The Plants', p.81.

15   For the importance of illustrating plants as a means of circulating botanical information, see Fan, *British*, Tobin, 'Imperial Designs' and Kusukawa and Maclean, *Transmitting*.

16   For Blake's work, and for his relationship to Whang at Tong see Goodman and Crane, 'The Life' and Goodman and Jarvis, 'The John Bradby Blake'. Whang at Tong came to London in 1774 and stayed for a few years. He was introduced, by Banks, to a number of prominent naturalists, for whom he helped make sense of their Chinese objects, and to the Fellows of the Royal Society. Whang helped arrange the Duchess of Portland's impressive porcelain collection at her country home Bulstrode. Whang's movements are covered in detail in Goodman and Crane, 'The Life'. Lady Dorothea Banks also had a porcelain collection at Spring Grove. Her collection is discussed in Leis, '"A Little Old-China"' and Newport, 'The Fictility'.

17   For the history of Sloane's collection, see Delbourgo, *Collecting*. On Petiver's collection and other important examples of Chinese plants in Britain, see Jarvis and Oswald, 'The Collecting Activities' and Stearns, 'James Petiver'. For Kaempfer and Sloane, see Brown, 'Engelbert Kaempfer's Legacy', Hinz, 'The Japanese Collection', Brown, 'Kaempfer's Album' and Dandy, *The Sloane Herbarium*.

18   As quoted in Knox, 'The Great Pagoda', p.26. See also the extended discussion of the changes that Chambers made to Kew in Desmond, *Kew*, pp.44–63. Chambers knew what a pagoda looked like first-hand: he had been to Canton himself on two occasions, 1743–4 and 1748–9.

19   John Duncan's posting in Canton dated from 18 June 1783. See Kilpatrick, *Gifts*, pp.95–6, Pritchard, 'The Crucial Years', p.179 and BL, IOR, G/12/61.

20   Cranmer-Byng and Wills, 'Trade and Diplomacy', pp.236–7.

21   Pigou and J. Duncan to Banks, 31 May 1782, *I&P*, vol.1, document 240. The two men had also left instructions with a local resident to collect the seeds and fruits of plants that were not yet in season and that half of those were to go to Banks and the other half to Gerard de Visme, Banks's friend of more than fifteen years, in Lisbon.

22   Pigou and J. Duncan, 31 December 1783, *I&P*, vol.2, document 37. Wilson had sailed to China in command of the *Antelope Packet* on the seldom used western route, that is across the Pacific from Cape Horn. Unfortunately, the ship wrecked on 10 August 1783 on a reef on one of the eastern outer islands of the Palau group in the western Pacific. Thanks to the good will of the local people, Wilson and his crew were able to build a vessel to take them back to Macao. In addition to caring for Pigou and Duncan's collection, Wilson was also taking Lebuu, the son of the region's leader, to visit England. The *Morse* arrived on the English coast on 14 July 1784 but Lebuu, aged

twenty years, only lived for another five months before succumbing to smallpox. The story of Wilson, the *Antelope*, and Lebuu has been told many times. See Nero, Thomas and Newell, *An Account* and Fullagar, *The Savage*, pp.156–65.

23 Pigou and J. Duncan, 31 December 1783, *I&P*, vol.2, document 37.

24 J. Duncan to Banks, 1 December 1786, *I&P*, vol.2, document 96.

25 For likely sources of Chinese plants in Canton, see Richard and Woudstra, 'Thoroughly'. This may be the same as the Fa-tee nursery, but it isn't called that in the sources I have used. See Fan, *British*, Le Rougetel, 'The Fa Tee' and 'Father's Journals', Bryant P. Tilden Papers, Phillips Library, Peabody Essex Museum – I would like to thank Josepha Richard for this reference.

26 SL, Sir Joseph Banks Collection, C 1:11. Also transcribed in *I&P*, vol.4, document 323 and in Kilpatrick, *Gifts*, p.115.

27 Banks, 'Hints', *I&P*, vol.3, document 304 and J. Duncan to Banks, 23 January 1791, *I&P*, vol.3. document 124.

28 J. Duncan to Banks, 4 April 1787, *I&P*, vol.2, document 123.

29 This is clear from the surviving copy letter book of Alexander Duncan. I would like to thank Cathy and Andrew Duncan for permission to look at this very valuable document.

30 J. Duncan to Banks, 20 March 1787, *I&P*, vol.2, document 117.

31 Banks, 'Hints', *I&P*, vol.3, document 304.

32 J. Duncan to Banks, 20 December 1787, *I&P*, vol.2, document 197.

33 Kincaid to Banks, 28 June 1788, RBGA, JBK/1/4, fol.305.

34 Crawford, *Roll*, p.30.

35 A. Duncan to Banks, 1 February 1788, *I&P*, vol.2, document 208.

36 Banks to Morton, 24 June 1788, *I&P*, vol.2, document 222.

37 J. Duncan to Banks, 7 March 1789, *I&P*, vol.2, document 278 and A. Duncan to Banks, 12 December 1789, *I&P*, vol.3, document 63.

38 A. Duncan to Banks, 6 December 1788, *I&P*, vol.2, document 261.

39 A. Duncan to Banks, 12 December 1789, *I&P*, vol.3, document 63.

40 For Consequa, see Kilpatrick, *Gifts*, pp.189–90 and Anon, 'Biography of Consequa'. A variety of *Wisteria sinensis* is named after him. For the gardens of Hong merchants in general see Fan, *British*, pp.31–5 and Richard, 'Uncovering the Garden'.

41 A. Duncan to Banks, 12 December 1789, *I&P*, vol.3, document 63.

42 This James Lind, born in 1736, should not be confused with the other James Lind, his cousin, born in 1716, who was famous for his *Treatise on the Scurvy*.

43 Lind to Banks, 23 October 1779, *SC*, document 169; Banks to Blagden, 16 June 1782, *SC*, document 259; and Lind to Banks, 15 July 1784, *SC*, document 503.

44 Lind's abilities in Chinese and his wide interests in China, especially, its music, come out clearly in the exchange of letters he had with Charles Burney, the musician and composer, in the 1770s – see Ribeiro, *The Letters*,

pp.172–80 and Lind to Burney, 11 August 1774, YUBRBML, OSB, MS 3, Box 12/884.

45 Lind to Banks, 19 October 1786, *SC*, letter 692.

46 The manuscript is in NHM, Lind MSS. The letter explaining the purpose of the document is Lind to Banks, 25 January 1790, *SC*, letter 969.

47 Banks's copy of the Chinese herbal was the Studio Three Joys Reprint Bencao Gangmu. I thank Frances Wood for this information. On the *Bencao gangmu* see Métailié, 'The *Bencao gangmu*', Métailié, 'Des mots', Métailié, 'The Representation' and Nappi, *The Monkey*. It is interesting that the two source materials that Charles Darwin relied on for his knowledge of Chinese natural history were the *Bencao gangmu* and the *Mémoires concernant l'histoire, etc. des Chinois*. For a discussion of Darwin's Chinese sources, see Pan, 'Charles Darwin's'. Du Halde's volumes relied on *Bencao gangmu* for some of its information about Chinese medicinal plants – see Fan, *British*, p.95. On the history of the *Hortus Kewensis*, see Britten, 'The History'.

48 I have not been able to find the Book and assume that it has not survived. What I know of it I have taken from its description in Dryander, *Catalogus*, vol.3, p.183. Banks explained the system of crosses in his 'Hints on the Subject of Gardening Suggested to the Gentlemen who attend the Embassy to China', which is catalogued as MS 115, Linnean Society, London and has been transcribed and published in *I&P*, vol.3, document 304.

49 A. Duncan to Banks, 12 December 1789, *I&P*, vol.3, document 63.

50 A. Duncan to Banks, 23 January 1790, *I&P*, vol.3, document 70 and A. Duncan to Banks, 8 February 1790, *I&P*, vol.3, document 71. The plants' numbers (the Book's page numbers) ranged from number 9 to number 83.

51 A. Duncan to Banks, 12 December 1789, *I&P*, vol.3, document 63.

52 Bretschneider, *History*, p.198.

53 A. Duncan to Banks, 8 February 1790, *I&P*, vol.3, document 71.

54 A. Duncan to Banks, 26 November 1791, *I&P*, vol.3, document 216 and Banks, 'Hints', *I&P*, vol.3, document 304.

55 Duncan later excused himself to the Superintendent of the Calcutta Botanic Garden for his lack of botanical knowledge – 'the want of Linnaeus's sexual system deprived me of the scientific phraseology', he noted – see Duncan to Richard Hall, President of the Select Committee, 30 October 1796, SL, Sir Joseph Banks Collection, C 1:10.

56 A. Duncan to Banks, 8 February 1790, *I&P*, vol.3, document 71. The other letters in which Duncan complained about the Chinese are in *I&P*, vol.3, documents 153 and 216.

57 A. Duncan to Banks, 25 December 1793, *I&P*, vol.4, document 105.

58 John Livingstone, a surgeon in Canton for almost a quarter of a century, wrote in 1819 that in his experience only one out of the one thousand plants shipped from Canton to London survived – see Livingstone, 'Observations', p.427. This was clearly an exaggeration. The naturalist John Ellis, who knew more than most people about transporting seeds and plants and who had

written a very valuable pamphlet on the subject, thought that the figure was more like one in fifty – Ellis, *Directions*, p.1.

59 There is a good discussion of the perils of shipboard life for a plant leaving Canton for England in Fan, *British*, pp.36–9. See also Rigby, 'The Politics'. There were attempts to reduce plant mortality by designing better containers and providing best-practice advice – see Ellis, *Directions* and, generally, Laird and Bridgman, 'American Roots' and Parsons and Murphy, 'Ecosystems'.

60 Glegg to Banks, 4 December 1787, BL, Add MS 33978, pp.157–8.

61 Banks knew that the captain had been entrusted with a plant – see J. Duncan to Banks, 20 March 1787, *I&P*, vol.2, document 117.

62 For Slater's life and activities, see Kilpatrick, *Gifts*, pp.118–29. Main's time in Canton is recounted in Main, 'Observations' and Main, 'Reminiscences'.

63 A. Duncan to Banks, 7 March 1794, *I&P*, vol.4, document 118.

64 Main to Banks, 23 September 1794, *I&P*, vol.4, document 145.

65 Main to Banks, 22 October 1794, *I&P*, vol.4, document 150. Main's own account of the return voyage was published many years later as Main, 'Reminiscences'.

66 Kilpatrick, *Gifts*, p.128.

67 H. Andrews, *The Botanists Repository*, London, 1797, vol.VI 'Plate CCCLXXIII: Paeonia Suffruticosa'.

68 A. Duncan to Banks, 15 January 1795, *I&P*, vol.4, document 164.

69 Banks to A. Duncan, 29 May 1796, SL, Sir Joseph Banks Collection, C 2:46.

70 Ibid.

71 Ibid. The gardener Banks had in mind was Peter Good who had just returned from Calcutta with a cargo of living plants. See Chapter 14 for this story.

72 A. Duncan to Richard Hall, President of the Select Committee, 30 October 1796, SL, Sir Joseph Banks Collection, C 1:10.

73 See Duncan, *A Family Called Duncan*, p.6. The last surviving letter to Banks is A. Duncan to Banks, 15 June 1796, *I&P*, document 256. Like most Europeans living in Canton, Alexander Duncan did business on the side – an account book of his dealings has survived and is in a private collection.

74 RS, EC/1798/02.

5 1786: The Madras Naturalists and Dreams of Oaxaca

1 Unless otherwise noted, the following is based on Jensen, 'Negotiating People', the latest and fullest study of Koenig.

2 Duyker, *Nature's Argonaut*, ch.3.

3 For Solander's time in London before he joined Banks on HMS *Endeavour*, see Duyker, *Nature's Argonaut*, chs.3–8.

4 There is some evidence that Banks's decision to go to Iceland was influenced in part by Solander who knew of Koenig's previous expedition to the island to collect plants – see Agnarsdóttir, *Sir Joseph Banks*, p.8.

5   See Gross, 'Background'.

6   On Tranquebar and natural history see Jensen, 'The Tranquebarian Society',
     Jensen, 'Making it in Tranquebar' and Berg, 'Useful Knowledge', pp.132–4.

7   For the development of Madras, see Neild, 'Colonial Urbanism', and
     Nightingale, 'Before Race'. These, as well as other, studies often rely heavily
     on Love, *Vestiges*, an encyclopaedic history of early Madras. A good general
     and popular account of the history of the East India Company can be found
     in Keay, *The Honourable Company*. More academic treatments are in Bowen,
     The Business, Bowen, McAleer and Blyth, *Monsoon Traders* and Chaudhuri,
     The Trading World. For a view of Madras from the transactions of a single
     private merchant, see Hanser, 'Mr. Smith'.

8   For Lady Monson, see Britten, 'Lady Anne Monson'. See also Noltie, 'John
     Bradby Blake'. Solander knew Lady Anne by 1762. She had met Banks by
     the time he and Solander were ready to sail on HMS *Endeavour* – see
     Duyker, *Nature's Argonaut*, pp.60, 90.

9   Koenig to Solander, 24 January 1774, in Rendle, 'John Gerard Koenig',
     pp.147–9.

10   Duyker, *Nature's Argonaut*, pp.63–71. Solander was appointed to the British
     Museum in 1763 and, a year later, he was elected a Fellow of the Royal
     Society and met Banks for the first time.

11   Koenig to Solander, 24 January 1774, in Rendle, 'John Gerard Koenig', p.148
     – I thank Henry Noltie for alerting me to this source. See also Solander to
     Banks, 22 August 1775, *I&P*, vol.1, document 149, which confirms that
     Koenig's plants had become incorporated into Banks's herbarium.

12   See Furber, 'Madras in 1787' for a description of the world around the
     Nawab.

13   For the Nawab's unfortunate financial relationship with the Company, see
     Phillips, 'Private Profit'.

14   BL, IOR, H/360, pp.459–62.

15   See George Wombwell to Banks, 31 March 1779, *I&P*, vol.1, document 192.
     The key words in this letter, which told Banks that Koenig had been
     appointed, are 'agreeable to his Request'. See also Patrick Russell to Banks,
     26 December 1784, *I&P*, vol.2, document 63 in which he mentions 'the
     employment you had so great a share in procuring for him'.

16   See the letters to Solander between 1774 and 1782 in NHM, Kønig MSS
     (part of the Banksian MSS) and those to Banks in *I&P*, vol.2, documents
     20 and 60.

17   Patrick Russell to Banks, 9 July 1785, *I&P*, vol.2, document 73. Given how
     he felt about Solander, it is likely that Koenig would have left him his
     possessions, but Solander died in 1782.

18   The following section is based on Hawgood, 'The Life' and Hawgood,
     'Alexander Russell'.

19   Russell to Banks, 27 December 1782, *I&P*, vol.1, document 267. See also
     Russell, *A Continuation*, p.xi.

20 Russell to Banks, 12 March 1786, *I&P*, vol.2, document 83. Banks was of the opinion that Koenig was a great naturalist: 'Rarely again will the Company Meet with a Servant So well Qualified to do them Essential Service', he told Thomas Morton, Secretary to the East India Company Court of Directors. He also added that to replace him, the Company should look to young men 'in the North of Europe Universities' who had been educated in all the branches of natural history, as was common there – Banks to Morton, 22 February 1787, *I&P*, vol.2, document 111. What Banks meant here were Scandinavian, primarily Swedish, universities. Even though Morton told Banks that the Council in Madras had appointed Russell to the vacancy left by Koenig's death, Banks persisted in trying to find someone in Sweden – Morton to Banks, 12 March 1787, *I&P*, vol.2, document 115. The person Banks wanted to go to India and whom he asked was Olof Swartz, a pupil of Linnaeus's son, who had been working on a study of West Indian plants and had just been studying Banks's collections in Soho Square. Despite Banks's pleas, Swartz decided to remain in Sweden – see Swartz to Banks, 29 August 1787, *SC*, document 766, Swartz to Banks, 4 January 1788, *I&P*, vol.2, document 202, and Banks to Swartz, 28 February 1788, *SC*, document 827.

21 Unless otherwise stated, this section is based on Robinson, *William Roxburgh*, chs.1,2.

22 Roxburgh to Banks, 8 March 1779, *I&P*, vol.1, document 191.

23 Ibid. Banks's letter to Roxburgh does not seem to have survived but we know it was dated 25 March 1778 from what Roxburgh told Banks in his letter to him. Roxburgh's letter is in BL, Add MS 33977, pp.93–5. The transcribed letter in *I&P*, vol.1, document 191, has Banks's letter incorrectly dated as 1776.

24 BL, IOR, H/360, p.461.

25 For the medical establishment in Madras at the time, see Chakrabarti, 'Networks of Medicine' and Chakrabarti, '"Neither of Meate"'.

26 Crawford, *Roll*, vol.1, entry 125. See also Sweet, 'Instructions', p.412.

27 Love, *Vestiges*, p.336.

28 Anderson to Banks, 3 December 1786, *I&P*, vol.2, document 97.

29 For an account of the development of European knowledge of cochineal, see Kellman, 'Nature, Networks, and Expert Testimony' and Ratcliff, *The Quest*, ch.3.

30 See Lech et al., 'Identification' and Serrano, 'Analysis'. See also Phipps, *Cochineal Red*.

31 For the early history of cochineal and other red dyestuffs in Europe, see Marichal, 'Mexican Cochineal', Lee, 'American Cochineal' and Munro, 'The Medieval Scarlet'.

32 See Phipps, *Cochineal* and Padilla and Anderson, *A Red*.

33 For a modern interpretation of this phenomenon, see Moran, 'Interactions'. I thank Mark Carine for this reference.

34 See Réaumur, *Mémoires*, vol.4, pp.87–111 for his discussion about cochineal. The comparison between the wild and the domesticated insects appears on pp.90–1. The French dye master Jean Hellot reiterated Réaumur's verdict in his famous *L'Art de la teinture des laines et des étoffes de laine*, Paris, 1750, p.279.

35 Anderson to Banks, n.d. (but probably 24 December 1786), *I&P*, vol.2, document 101.

36 Banks to Anderson, 22 May 1787, *I&P*, vol.2, document 132.

37 In 1786, England imported over 200,000 pounds weight of cochineal from Cadiz – see SL, Sir Joseph Banks Collection, EN 1:37. In value terms, cochineal represented either the second or the third most important import from Cadiz, surpassed only by silver in all years and wine in some years – see Lario de Oñate, *La colonia*, pp.184–93.

38 Morton to Banks, 13 April 1785, *I&P*, vol.2, document 67.

39 See Mackay, *In the Wake*, pp.168–98 and Damodaran, Winterbottom and Lester, *The East India Company*, especially the chapters by Kumar and Damodaran.

40 See the correspondence: Banks to Yonge, 15 May 1787, *I&P*, vol.2, document 130; and Banks to Dundas, 15 June 1787, *I&P*, vol.2, document 137.

41 The Company asked Banks to review the case for a botanic garden put forth by Robert Kyd, who became its first superintendent – some of the correspondence, covering the years 1786 to 1788, is in *I&P*, vol.2, documents 84, 86, 90 and 258. For more on the Calcutta Botanic Garden see Thomas, 'The Establishment'. Banks's plans for introducing tea to India are outlined in Banks to Jenkinson, April and May 1788, *I&P*, vol.2, documents 214 and 216 respectively.

42 For more on de Menonville, see Saraiba Russell, 'En Búsqueda', Edelstein, 'Spanish Red', Greenfield, *A Perfect Red*, pp.213–34 and McClellan, *Colonialism and Science*. De Menonville's book was translated into English and published in John Pinkerton's, *A General Collection of the Best and Most Interesting Voyages and Travels*, vol.13, London, 1812.

43 For the history of Sloane's important publication, see Delbourgo, *Collecting*. Delbourgo points out that this is the only engraving that pictured a plantation – p.58.

44 Banks's notes on de Menonville can be found in SL, Sir Joseph Banks Collection, EN 1:49, 52–5.

45 In 1794, Jose Antonio Alzate y Ramirez, a Mexican author who wrote widely on many areas, including science, published his observations on cochineal cultivation in his *Gazeta de Literatura de México*, vol.3, numbers 26–33. He had written the piece in 1777. See Adank, 'The "Memoria"'. Alzate y Ramirez's articles have recently been republished – Sánchez Silva and de Avila Blomberg, *La Grana*. For more on the *Gazeta*, see Clark, 'Lost in Translation'.

46 Morton to Banks, 23 June 1788, *I&P*, vol.2, document 221.

47 Anderson to Banks, 20 October 1788, with enclosure of Kyd to Anderson,

29 September 1788, *I&P*, vol.2, document 252; and Anderson to Kyd, 5 March 1789, in Anderson, *Letters on Cochineal Continued*, pp.21–2.

48 Petrie to Banks, 29 May 1787, *I&P*, vol.2, document 133.

49 For a full list of the published letters, which after 1788 contained correspondence on cochineal with a great many others scattered throughout the Asian world, see Williams, 'Scale Insects'.

50 Banks to Anderson, 6 June 1790, *I&P*, vol.3, document 95.

51 Banks's draft instructions, which he delivered to Thomas Morton on 29 March 1788, are in SL, Sir Joseph Banks Collection, EN 1:49. They were published in Anderson, *Letters on Cochineal Continued*, pp.9–10.

52 Banks to Morton, 17 November 1791, *I&P*, vol.3, document 210.

53 Besides keeping up a lively correspondence with many others on the subject of cochineal, Anderson was in close touch with his East India Company superiors in Madras to whom he reported his developments. These are scattered in the East India Company papers: see especially BL, IOR, P/240, 241 and 242, searching for Anderson, Cochineal and Nopalry in each year's index.

54 Cochineal was one among many other of Anderson's projects. See Berg, 'Passionate Projectors' for Anderson's unsuccessful sericulture project.

55 Donkin, 'Spanish Red', p.49 and BL, IOR, F/4/78, p.81.

56 Banks to Morton, 17 November 1791, *I&P*, vol.3, document 210. For Banks's close reading of the sources and his conclusions, see *I&P*, vol.2, document 200.

57 'Calcutta Botanic Garden Plants', November 1789, *I&P*, vol.3, document 57. See also Kyd's Comments on Banks's Report on the Company's Calcutta Garden, *I&P*, vol.3, documents 40, 41.

58 Banks to Morton, 17 November 1791, *I&P*, vol.3, document 210, referring to an anonymous extract from a letter in Kyd to Nathaniel Smith (Director, the East India Company), 23 March 1791, *I&P*, vol.3, document 157.

59 Banks's draft instructions are in SL, Sir Joseph Banks Collection, EN 1:49. See Banks to Devaynes, 11 March 1790, *I&P*, vol.3, document 77.

60 Banks to Devaynes, 11 March 1790, *I&P*, vol.3, document 77.

61 Banks to Morton, 17 November 1791, *I&P*, vol.3, document 210 and Banks to Baring, 15 June 1792, *I&P*, vol.3, document 280.

62 Baring to Banks, 15 June 1792, *I&P*, vol.3, document 279. In 1787, the same year that Anderson contacted Banks about his cochineal discovery, Baring and his counterpart at Hope & Co, a powerful Amsterdam-based bank, attempted, though without success, to monopolise the European cochineal market – see Buist, *At Spes*, pp.431–51.

63 Banks to Baring, 15 June 1792, *I&P*, vol.3, document 280.

64 Baring to Banks, 26 June 1792, *I&P*, vol.3, document 284.

65 Baring to Banks, 16 August 1792, *I&P*, vol.3, document 298.

66 Banks's first reaction is contained in Banks, 'Private Memorandum', 20 August 1792, *I&P*, vol.3, document 305. The change of heart is noted in Banks to Baring, 21 August 1792, *I&P*, vol.3, document 306.

67 Banks to Baring, 21 August 1792, *I&P*, vol.3, document 306. Banks's reference of the situation in 1790 is to the letter he wrote to William Devaynes, Chairman of the Court of Directors, on 11 March 1790 – *I&P*, vol.3, document 77.

68 Bartlet to Banks, 21 July 1791, SL, Sir Joseph Banks Collection, EN 1:62.

69 See Quintanilla, 'Mercantile Communities' and Quintanilla, 'The World'.

70 Banks to Bartlet, 14 January 1793, *I&P*, vol.4, document 19.

71 Banks to Inglis, 18 August 1796, *I&P*, vol.4, document 269.

72 Banks thought that Bartlet may have died but this was not the case. He died on 24 January 1800 on St George's Caye, in present-day Belize – see Usher, *Memorial Inscriptions* and Springs, 'St. George's Caye'.

73 Sproat to Banks, 5 March 1802, BL, Add MS 33981, pp.6–7. Little is known about Sproat, but see Dawson, 'The Evacuation', pp.78–81 and Dawson, 'Robert Kaye'.

74 Banks to Sproat, 27 August 1802, *I&P*, vol.6, document 74.

75 For an overview of cochineal and the East India Company, see Frey, 'Prickly Pears'.

6 1786: The First and Second Fleet

1   Frost, *Arthur Phillip*, p.55.

2   Ibid., pp.59–65.

3   Ibid., pp.82–3. See also Marques, 'Escola de homens'.

4   Pembroke, *Arthur Phillip*, pp.41–2, 49–50.

5   Quoted in Frost, *Arthur Phillip*, p.83.

6   Frost, *Arthur Phillip*, p.131. See also Knight, *Britain Against Napoleon*.

7   This is the conclusion that Frost comes to in *The First Fleet*.

8   Howe to Sydney, 3 September 1786, *HRNSW*, vol.1, pt.2, pp.22–3.

9   See Frost, *Arthur Phillip*, pp.132–42 for this.

10 Phillip's thoughts on the nature of the settlement can be found in *HRNSW*, vol.1, pt.2, pp.50–4. The dating of this document in this printed version is very late February or very early March 1787. Frost, *The First Fleet*, p.30, convincingly argues that the document is more likely to have been written in early October 1786 just after Phillip had received his commission.

11 The original lists are in SL, Sir Joseph Banks Collection, SS 1:48, 49 and TNA, T1/639, fols.253–5 and are reproduced in Frost, *Sir Joseph Banks*, pp.5–11.

12 I would like to thank Ben Breen for sharing his ipecacuanha references with me.

13 Lee, 'Ipecacuanha'.

14 Phillip to Banks, 31 August 1787, *I&P*, vol.2, document 155. The plant is currently known as *Carapichea ipecacuanha*.

15 Phillip to Banks, 31 August 1787, *I&P*, vol.2, document 155.

16 Frost, *The First Fleet*, p.173.

17   Masson to Banks, 13 November 1787, SL, Sir Joseph Banks Collection, EN 1:36.
18   Phillip to Banks, 9 November 1787, *I&P*, vol.2, document 176.
19   As quoted in Frost, *The First Fleet*, p.174.
20   Frost, *The First Fleet*, p.175.
21   Frost, *Arthur Phillip*, p.148.
22   Phillip to Sydney, 15 May 1788, *HRNSW*, vol.1, pt.2, p.136.
23   On the statistics of the First Fleet, see Gillen, *The Founders* and Frost, *The First Fleet*. On the kangaroo, see Phillip to Banks, 2 July 1788, *I&P*, vol.2, document 225.
24   See Phillip's letters to Banks in *I&P*, vol.2, documents 134, 155, 225, 227, 231, 246, 254 and 255. The names of the plants that were growing in Kew and were brought back on First Fleet ships can be found in Aiton's *Hortus Kewensis*, second edition.
25   Phillip to Sydney, Nepean and Stephens, July 1788, in *HRNSW*, vol.1, pt.2.
26   Phillip to Sydney, 12 February 1790, *HRNSW*, vol.1, pt.2, p.295.
27   Phillip's despatches arrived in London on 25 and 26 March 1789 – see Frost, *Sir Joseph Banks*, p.17.
28   Sydney to Lords of the Admiralty, 29 April 1789, in *HRNSW*, vol.1, pt.2, p.230.
29   Nash, *The Last Voyage*, p.xxi.
30   Banks to Nepean, 27 April 1789, *I&P*, vol.2, document 283.
31   Banks to Grenville, 7 June 1789, *HRNSW*, vol.1, pt.2, pp.247–8.
32   Riou to Banks, 3 June 1789, *I&P*, vol.3, document 5.
33   Ibid.
34   Ellis, *Directions*, and Duhamel du Monceau, *Avis*. See also Allain, *Voyages*, Laird, 'American Roots', Parsons and Murphy, 'Ecosystems', Rigby, 'The Politics' and Nelson, 'From Tubs'.
35   Banks to Grenville, 7 June 1789, *HRNSW*, vol.1, pt.2, p.248.
36   A plan for this coach can be found in SL, Sir Joseph Banks Collection, G 1:5.
37   Banks, 'Notes on the plant coach for the *Guardian*, 5 June 1789', in SL, Sir Joseph Banks Collection, G 1:9, also in Frost, *Sir Joseph Banks*, p.21.
38   Maddison and Maddison, 'Spring Grove'.
39   Banks to Riou, 5 June 1789, *I&P*, vol.3, document 6.
40   Riou to Banks, 6 June 1789, *I&P*, vol.3, document 7.
41   Grenville to Phillip, *HRNSW*, vol.1, pt.2, p.262.
42   Smith, 'Receipt', 6 June 1789, *I&P*, vol.3, document 8.
43   Banks to Riou, July 1789, SLNSW, Sir Joseph Banks Papers, Series 35.10, also in Frost, *Sir Joseph Banks*, pp.24–5.
44   Banks to Smith, July 1789, *I&P*, vol.3, document 16.
45   Ibid.
46   Riou to Banks, 15 July 1789, *I&P*, vol.3, document 21.
47   SL, Sir Joseph Banks Collection, G 1:16 and Smith to Banks, 20 July 1789,

I&P, vol.3, document 25. For more on Ronalds, see Ronalds, 'Ronalds Nurserymen'.

48 Riou to Banks, 15 July 1789, *I&P*, vol.3, document 21.
49 Riou to Banks, 24 August 1789, *I&P*, vol.3, document 34.
50 Dickson, *HMS Guardian*, pp.7–8 and Nash, *The Last Voyage*, pp.xxiii–xxiv.
51 Smith, 'Report of the State of the Plants &c', SL, Sir Joseph Banks Collection, G 1:21.
52 Smith to Banks, 9 December 1789, *I&P*, vol.3, document 61.
53 Ibid.; Riou to Banks, 10 December 1789, *I&P*, vol.3, document 62 plus notes attached. On Gordon in the Cape see Cullinan, *Robert Jacob Gordon*.
54 Nash, *The Last Voyage*, pp.33–4.
55 Riou to Stephens, 22 February 1790, *HRNSW*, vol.1, pt.2, p.310.
56 The names of the 61 who stayed with the *Guardian* were printed in the *London Gazette*, 7–10 August 1790, p.502.
57 Riou to Stephens, 2 May 1790, *HRNSW*, vol.1, pt.2, pp.336–9.
58 Nash, *The Last Voyage*, pp.202–3, 232–3.
59 Bateson, *The Convict Ships*, p.31.
60 Nash, *The Last Voyage*, p.xxxi. Six arrived in London on 23 April and six later. The other three, apparently, remained at the Cape.
61 Nash, *The Last Voyage*, p.199.
62 Riou to Stephens, 15 April 1791, TNA, ADM 1/2395.
63 Masson to Banks, 7 March 1791, SLNSW, Sir Joseph Banks Papers, Series 13.50; and Tripp to Banks, 22 May 1791, BL, Add MS 33979, p.98.
64 Banks to Riou, 15 May 1791, *I&P*, vol.3, document 163.
65 Banks to Dundas, 30 April 1794, *I&P*, vol.4, document 125.

7 1787: Anthony Pantaleon Hove in Gujarat

1 Desmond, *Dictionary*, p.358, refers to him as 'Anton Pantaleon Hoveau'.
2 RBGA, 'Seeds Sent Home by Howe Rec'd May 3 1786', JBK/1/3 f.231.
3 *Hortus Kewensis*, vol.2, pp.422, 428. They were first described in 1787, the year after they were collected, by the French botanist and close friend of Banks, Charles-Louis L'Héritier de Brutelle.
4 Riello and Roy, *How India* – for an overview of the Indian cotton industry from the perspective of the Indian Ocean, see Parthasarathi, 'Cotton Textiles', and from a European perspective, see Riello, 'The Globalization'.
5 There is no shortage of works on the British cotton industry in particular and the Industrial Revolution in general. Approachable and recent works on cotton, which include discussions of the British scene, include Beckert, *Empire*, Riello, *Cotton* and Lemire, *Cotton*. For the Industrial Revolution, a popular account is Weightman, *The Industrial Revolutionaries*. Mokyr, *The Enlightened Economy*, places the Industrial Revolution firmly in the context of the Enlightenment.
6 Farnie, 'The Role of Merchants', p.29 makes the point about Britain: 'From

1784 it became the largest producer of cotton goods in Europe. From 1785 it became an exporter of cotton yarn to the Continent. From 1787 its foreign trade became the largest in the world.'

7   In the previous year, supplies from the West Indies accounted for 45 per cent of the import of raw cotton into Britain. The Levant, or the region including the countries bordering the eastern Mediterranean and including present-day Iraq and Syria, and Brazil accounted for most of the rest of the supplies (imports at this time from the United States were negligible) – see Wadsworth and Mann, *The Cotton Trade*, pp.183–92 and Edwards, *The Growth*, pp.250–1.

8   See Edwards, *The Growth*, pp.75–6.

9   10 March 1786, TNA, BT 6/140, p.17.

10 TNA, BT 5/4, p.192.

11 For Drinkwater, see Bennett, 'Alignments' and Chaloner, 'Robert Owen'. For the Board of Control, see Foster, 'The India Board', Sutherland, *The East India Company* and Bowen, *The Business*, pp.53–83. On Dundas, see Furber, *Henry Dundas*.

12 NLS, MS 1064, pp.3–13.

13 The meeting's minutes for this day are in TNA, BT 5/4, pp.196–200.

14 TNA, BT 5/4, pp.198–9.

15 Nepean to Banks, 22 February 1787, *I&P*, vol.2, document 112. It is frustrating that Nepean wrote another letter to Banks on the same subject but it is undated and, therefore, cannot easily be fitted into the chain of events. With this letter, Nepean informed Banks that Hawkesbury was well-disposed to the plan that the three men had sketched out and that Hawkesbury was going to discuss it with the Prime Minister, William Pitt – see Nepean to Banks, n.d., BL, Add MS 33982, p.327. I am of the opinion that this letter was subsequent to the one he wrote on 22 February. My interpretation of what happened contradicts that of Mackay, *In the Wake*, p.146, who states that it was Banks who first suggested the idea to Hawkesbury.

16 A scattering of letters from Banks to Hawkesbury before 1787 can be found in BL, Add MS 38218–38446.

17 Nepean to Banks, 22 February 1787, *I&P*, vol.2, document 112. Banks confirmed that Nepean recommended Hove because of 'his good conduct in a former occasion' – see Banks to Hawkesbury, 26 November 1788, WLAM, MS. 5219, p.19. The story of Hove in India has been told in Mackay, *In the Wake* and in Berg, 'Useful Knowledge'. My account differs from theirs. The minutes are in TNA, BT 5/4, p.226.

18 Nepean to Banks, n.d., BL, Add MS 33982, p.327.

19 The earliest date that confirms this statement is 30 March 1787 – see TNA, BT 6/246.

20 Banks's draft is in RS, MM/6/61 and a final, neat, copy is in TNA, 6/246. The former has been transcribed and published in *I&P*, vol.2, document

104 – there the document is given a provisional date of 7 January 1787 but this is clearly incorrect.

21 Banks to Hove, [7 January 1787], *I&P*, vol.2, document 104 and TNA BT 6/246. For more on the cotton industry, from growing to manufacture, in Gujarat, see Berg, 'Craft and Small Scale', Nadri, 'The Dynamics' and Nadri, *Eighteenth-Century Gujarat*.

22 This was revealed many years later in Banks to Thomas Knight, 27 November 1812, NHM, DTC, vol.XVIII, p.189. See Moore, *The Knife Man*, for more on Hunter.

23 For all of Banks's adult life the British East India Company was the world's largest trading and territorial business. Founded in 1600, with headquarters in Devonshire Square, London, the Company had its main interests in India, southeast Asia, and Canton and owned the island of St Helena in the Atlantic.

24 See TNA, BT 6/246 which has the instructions sent from London to Bombay.

25 See Bowen, 'British Exports of Raw Cotton'.

26 Gupta, 'Competition' and Riello, 'The Indian Apprenticeship'.

27 Some of the Polish letters have survived. One is in TNA, BT 6/246 and another is in WLAM, MS. 5219. The English translations can also be found there.

28 Hove to Banks, 13 April 1787, *I&P*, vol.2, document 125.

29 Banks to Hawkesbury, 1 September 1788, TNA, BT 6/246. An earlier draft has been transcribed and published in *I&P*, vol.2, document 239.

30 Pagès, *Voyages autour du monde*. Banks revealed this to a meeting of the Board of Trade on 16 January 1790 – see BL, Add MS 38392, p.17.

31 This was also revealed by Banks at the same meeting as noted above. Forster published the account of his adventures as *A Journey from Bengal* in 1790. Forster knew that Hove had arrived safely in Bombay – see Forster to Banks, 28 August 1788, *I&P*, vol.2, document 238.

32 Banks to Hawkesbury, 20 April 1787, TNA, BT 6/246.

33 The following account is mostly drawn from Hove's journal, which was published as *Tours in Guzerat* in 1855. The original manuscript account is in BL, Add MS 8956. Extracts from the journal, concentrating on descriptions of the Gujarat cotton area, are in BL, IOR, H/374. When and by whom they were made is unknown.

34 Hove to Banks, 28 September 1787, WLAM, MS. 5219, pp.21–2v.

35 Hove to Banks, 2 February 1788, *I&P*, vol.2, document 209. One version of this letter in English is in TNA, BT 6/246 (but dated 10 March 1788), followed by a French translation and the original Polish one. Another English version, dated 2 February 1788, also with a French translation and the Polish original, is in WLAM, MS. 5219. It is this one that was transcribed and published in *I&P* cited above.

36 Hove to Banks, 2 February 1788, TNA, BT 6/246. This is marked as being a copy. The original is likely to be the one in WLAM, MS. 5219, pp.32–3v.

37 WLAM, MS. 5219, 17 March 1788, p.42.

38 Banks to Hawkesbury, 1 September 1788, TNA, BT 6/246. There is a slightly different copy in WLAM, MS. 5219, dated 31 August 1788 and most likely the draft version – it has been transcribed and published in *I&P*, vol.2, document 239.

39 Banks to Hawkesbury, 31 August 1788, *I&P*, vol.2, document 239. The next day, Banks reminded Hawkesbury that Hove's destination was decided on the basis of little more than 'a passage from a book which Mr. Dundas read to their Lordships of the Committee of Privy Council' – Banks to Hawkesbury, 1 September 1788, TNA, BT 6/246.

40 Steele to Banks, 14 June 1787, Add MS 33978, p.131. For more on Millington, see *www.ucl.ac.uk/lbs/person/view/2146644237* accessed on 28 October 2017, and Mackay, *In the Wake*, pp.152–5.

41 Kyd to East India Company, 1 June 1786, *I&P*, vol.2, document 86.

42 Banks's letter has not survived but Jones's reply has – see Jones to Banks, 25 February 1788, *SC*, document 826. Not much of the Banks–Jones correspondence has survived. What there is has been discussed in Cannon, 'Sir William Jones'. For more on Jones and the circulation of knowledge in eighteenth-century Bengal, see Raj, *Relocating*, pp.95–138.

43 Hawkesbury to Banks, 9 September 1788, *I&P*, vol.2, document 240.

44 Banks to Hawkesbury, 20 September 1788, *I&P*, vol.2, document 243 and Hawkesbury to Banks, 27 September 1788, WLAM, MS. 5219, p.17.

45 Banks to Hove, 30 September 1788, *I&P*, vol.2, document 249.

46 Hove to Banks, 10 January 1789, WLAM, MS. 5219, p.37.

47 Hove, *Tours*, p.195.

48 Hove to Banks, 15 September 1789, WLAM, MS. 5219, p.85.

49 Banks to William Fawkener, 28 August 1789, TNA, BT 6/246. A complete list of all the plants and seeds can be found in TNA, BT 6/246 and WLAM, MS. 5219, pp.91–7.

50 A first batch of cotton seeds sent by Hove in 1788 were in the hands of the governors of several West Indies islands by early 1789 – see Minutes Board of Trade, 21 March 1789, BL, Add MS 38391, p.253. See also Minutes Board of Trade, 14 January 1790, BL, Add MS 38392, pp.12–13. What happened to the seeds and to what extent they were cultivated is unclear. Cotton output in the West Indies did not keep pace with British cotton production and in 1801, the United States became Britain's main cotton supplier, as it would be for some time to come – see Jaquay, 'The Caribbean Cotton', p.84, Riello, *Cotton*, pp.198–210 and Beckert, *Empire*, pp.98–135.

51 Banks's note in WLAM, MS. 5219, pp.102–3.

52 Aiton to Banks, n.d. [likely 15 September 1789], WLAM, MS. 5219. pp.68–9.

53 Dryander to Banks, 20 September 1789, *SC*, document 946.

54 The minutes of the meeting are in BL, Add MS 38392, pp.13–23 and in *I&P*, vol.3, document 69.

8 1787: Mr Nelson's Unfortunate *Bounty* Voyage

1 Beaglehole, *The Endeavour Journal*, vol.1, p.252.
2 Piercy Brett, one of Anson's lieutenants, had drawn a scene which included the breadfruit tree in the foreground – this appeared as one of the plates in Anson's narrative of his voyage. Georg Eberhard Rumpf, when he was a merchant working for the Dutch East India Company on the island of Ambon in eastern Indonesia, had also described breadfruit in his posthumously published *Herbarium Amboinense*, Amsterdam, 1741–55. Banks had a copy of Dampier's and Anson's narratives and Rumpf's herbal – see also Ferrer-Gallego and Boisset, 'The Naming'.
3 Beaglehole, *The Endeavour Journal*, vol.1, p.252. For a discussion of Dampier and Anson and breadfruit, see Smith, 'Give Us'.
4 Beaglehole, *The Endeavour Journal*, vol.1, pp.341–5.
5 Several scholars have discussed the symbolic language surrounding breadfruit: see, for example, Bewell, 'Traveling Natures', DeLoughrey, 'Globalizing', Smith, 'Give Us' and Newell, *Trading*.
6 Beaglehole, *The Endeavour Journal*, vol.1, p.341.
7 Ibid., vol.2, p.330.
8 For what is known about Morris, see Waters, *The Unfortunate*.
9 Morris to Banks, 17 April 1772, *I&P*, vol.1, document 78.
10 Hawkesworth, *An Account*, vol.II, p.197.
11 Long, *History*, p.905.
12 Rauschenberg, 'John Ellis' and Ellis, *Directions*.
13 Ellis, *A Description*.
14 ULLSC, West India Committee Papers, M915, reel 1. For the activities of the West Indies' interest in British politics, see Penson, 'The London', Hall, *A Brief History*, McLelland, 'Redefining the West India' and Franklin, 'Enterprise and Advantage'.
15 Hall, *A Brief History*, p.5.
16 Bligh, *The Log*, p.21.
17 Ibid., p.18.
18 For the role of the West India Committee during the period of the American War of Independence, see O'Shaughnessy, 'The Formation'.
19 Higman, *Jamaica*, p.267.
20 The two men exchanged seeds for the duration of their acquaintance. See Hall, *Planters* and the Hinton–Banks correspondence in RBGA, JBK/1/3, 4 and 5.
21 East to Banks, 19 July 1784, *I&P*, vol.2, document 48.
22 See Sheridan, 'The Crisis', p.627. DeLoughrey, 'Globalizing' takes a different view of why planters wanted to introduce breadfruit for their slave population citing the fear of insurrection.
23 Wallen to Banks, 23 September 1784, *SC*, document 515.
24 Wallen to Banks, [1785], *SC*, document 544.

25 Wallen to Banks, 6 May 1785, *SC*, document 578.

26 Yonge to Banks, 29 June 1786, *I&P*, vol.2, document 88.

27 East to Banks, 24 April 1785, RBGA, JBK/1/3, 196 and East to Banks, 19 August 1786, RBGA, JBK/1/4, 239. See also Carter, *Sir Joseph*, p.218.

28 See Sheridan, 'The Crisis', p.632 and Sheridan, 'Captain Bligh', p.38.

29 ULLSC, West India Committee Papers, M915, reel 3.

30 For the ships of the First Fleet at Portsmouth in February 1787, see Frost, *The First Fleet*, pp.129–39. Banks's instructions to Pitt are contained in Banks to Pitt, February 1787, *I&P*, vol.2, document 114. Banks also advised on the best route the ship should take to Tahiti and then to the West Indies.

31 TNA, HO 42/11, pp.328–9.

32 *HRNSW*, vol.1, part.2, p.55.

33 Lee to Banks, 24 April 1776, *I&P*, vol.1, document 154.

34 Nelson to Banks, 26 April 1776, *I&P*, vol.1, document 155 and 'Agreement with David Nelson', 26 April 1776, *I&P*, vol.1, document 156.

35 'Receipt' and 'Statement of Money', 1779, SL, Sir Joseph Banks Collection, SS 1:87 and 89, respectively, and Banks to Nelson, 12 May 1787, *I&P*, vol.2, document 129.

36 Clerke to Banks, 23 November 1776, *I&P*, vol.1, document 171.

37 Anderson to Banks, 24 November 1776, *I&P*, vol.1, document 172.

38 Cook to Banks, 26 November 1776, *I&P*, vol.1, document 173.

39 St John, 'Biography'.

40 Carter, *Sir Joseph Banks*, p.214. SL, SS 1 has several documents pertaining to Nelson's stay in Plymouth during which time he spent money he didn't have.

41 Note, 22 March 1787, SLNSW, Sir Joseph Banks Papers, Series 45.04.

42 Banks's letter to Hawkesbury, dated 30 March 1787, is in *I&P*, vol.2, document 118 and the one to Mulgrave is in TNA, HO 42/11, pp.210–13. I am using the former source for this section of the text.

43 The letter is in TNA, ADM 1/4152, paper 68. See also Knight, 'H.M. Armed', pp.183–4.

44 Details of the steps leading to the launch of the new ship are given in Knight, 'H.M. Armed', McKay, *The Armed Transport* and in Mackay, *In the Wake*.

45 Mackay, 'Banks, Bligh', pp.70–1.

46 Ibid., and Bligh, *A Voyage*, pp.1–2.

47 See, for example, the instructions to the Officers of Deptford Yard, 30 May 1787, in TNA, ADM 3/103.

48 Mackay, *In the Wake*, p.134.

49 Bligh to Banks, 6 August 1787, SLNSW, Sir Joseph Banks Papers, Series 46.02.

50 There has been a lot written on Bligh. The most comprehensive account is Mackaness, *The Life*. The most recent treatment, and the one on which I have relied, is Salmond, *Bligh*. In the notes to the latter and in Alexander,

The Bounty, the interested reader can find most anything they might wish to know about Bligh's life.

51   For information on Campbell, see Byrnes, 'Emptying'.

52   Salmond, *Bligh*, p.109.

53   Bligh to Banks, 6 August 1787, SLNSW, Sir Joseph Banks Papers, Series 46.02. According to Harold Carter, Bligh first heard that new employment in the navy would be coming his way from Campbell in May 1787, but he provides no evidence of this letter – see Carter, *Sir Joseph Banks*, p.220.

54   See Duncan Campbell to Dugald Campbell, 30 August 1787, *Convict Transportation and the Metropolis*, reel 3. Unfortunately, Campbell did not go into details. For some time, Campbell had a greater presence in the story of the *Bounty* thanks to George Mackaness's assertion in his authoritative biography of Bligh, that Campbell owned the *Bethia*, which became the *Bounty* – see Mackaness, *The Life*. p.37. That assertion was challenged and corrected several years after Mackaness's book appeared when it was discovered that the *Bethia* belonged to a company called Wellbank, Sharp and Brown – see Knight, 'H.M. Armed Vessel', pp.185–6.

55   Carter, *Sir Joseph Banks*, p.220.

56   For the solicitations, one from Mannheim and the other from Lausanne, see Petry to Banks, 12 June 1787, BL, Add MS 8096, p.498 and Reynier to Banks, 15 August 1787, BL, Add MS 8096, p.500. Banks may have appointed Brown in July 1787 – see Note, ca July 1787, SLNSW, Sir Joseph Banks Papers, Series 47.03, but certainly no later than 25 August 1787 – see Note, Nelson and Brown, TNA, ADM 1/4152, paper 84a. There is very little known about William Brown. There is, however, some confusion about Brown mostly because at one time there were two William Browns on the *Bounty*. One of them was a second lieutenant who had been recommended to Banks by Lord Gainsborough as a very fine sailor – see Gainsborough to Banks, 12 August 1787, SLNSW, Sir Joseph Banks Papers, Series 72.063 (there is an enclosed letter from William Brown). Though he seems to have joined the ship, at the last minute, Lord Howe, the First Lord of the Admiralty, reassigned him to another vessel – see Bligh to Banks, 5 November 1787, *I&P*, vol.2, document 175. As an example of the confusion, see Alexander, *The Bounty*, p.56 where the Brown referred to is the second lieutenant, and not, as is written there incorrectly, the gardener.

57   Banks to Nelson, 20 August 1787, *I&P*, vol.2, document 150.

58   Ibid.

59   The plan had been for the *Bounty* to carry 600 pots but 800 arrived at the dock. Nelson decided to take them all even though he felt that some of them would be damaged. As it turned out only a few broke and the rest were all used – see Nelson to Banks, 18 December 1787, *I&P*, vol.2, document 194. The presents for the Tahitians are listed in TNA, ADM 1/4152, p.84a. The definition of toey comes from Salmond, *Bligh*, p.116.

60 For the brewing of hostilities between Britain and the Netherlands, see Rose, 'Great Britain' and Whatmore, *Against War and Empire*.

61 The correspondence relating to the delays and the frustrations can be found in *I&P*, vol.2, documents 159–96.

62 Bligh to Banks, 17 December 1787, *I&P*, vol.2, document 193 and Bligh to Banks, 19 December 1787, *I&P*, vol.2, document 196.

63 See 7–10 January 1788, Bligh's *Bounty* log at *www.fatefulvoyage.com/logbook/logbookHome.html*.

64 See 13 April 1788, Bligh's *Bounty* log at *www.fatefulvoyage.com/logbook/logbookHome.html*.

65 See 17 April 1788, Bligh's *Bounty* log at *www.fatefulvoyage.com/logbook/logbookHome.html*.

66 Bligh, *A Voyage*, p.39.

67 Ibid., p.49.

68 See 30 August 1788, Bligh's *Bounty* log at *www.fatefulvoyage.com/logbook/logbookHome.html*.

69 Edwards, *The Journals of Captain Cook*, p.447. Taking seeds and bulbs from one part of the world and sowing them in another was not such an unusual practice – see Craciun, 'The Seeds'.

70 The following section of this chapter is based on the daily remarks in Bligh's log, at *www.fatefulvoyage.com/logbook/logbookHome.html*.

71 For a modern discussion of the botany of breadfruit, see Jones et al., 'Beyond the Bounty'.

72 Bligh to Banks, 13 October 1789, *I&P*, vol.3, document 48.

73 The story of the mutiny has been told many times. A succinct version of events can be found in Salmond, *Bligh*, pp.211–15. A longer and more detailed account can be found in Alexander, *The Bounty* and in Dening, *Mr Bligh's*. For a study of the wider context of the mutiny, see Lincoln, 'Mutinous'.

74 Bligh to Banks, 13 October 1789, *I&P*, vol.3, document 48 and Smith and Thomas, *Mutiny and Aftermath*, p.54.

9 1790: The Second Circumnavigation of Archibald Menzies

1 Banks to Nepean, 27 August 1789, *I&P*, vol.3, document 35.

2 For this part of the chapter, I am relying on the account presented in Mackay, *In the Wake*.

3 For more detail on the British whaling industry in the South Seas, see Clayton and Clayton, *Shipowners*, Stackpole, *Whales & Destiny* and Newton, *A Savage History*.

4 Samuel Enderby & Sons to Banks, 29 August 1788, *I&P*, vol.2, document 237.

5 The *Sappho*, belonging to Ogle & Co, had left England in 1788. See *www.*

bswf.hull.ac.uk/fass/bswf.aspx accessed on 29 September 2016 and Clayton, *An Alphabetical List*.

6   See Mackay, *In the Wake*, pp.42–3.

7   King, 'George Vancouver', p.7.

8   Dixon to Banks, 20 October 1789, *I&P*, vol.3, document 51.

9   The original intention was to fit the *Discovery* with the frame of a small sailing boat – a shallop – to be put together if and when necessary but after it was found that it took up too much room, the plan was scrapped in favour of having two ships in the expedition – see Lamb, *A Voyage of Discovery*, p.22.

10   Banks to Thunberg, 19 March 1790, *I&P*, vol.3, document 81.

11   Mackay, *In the Wake*, p.44 and King, 'George Vancouver', p.7.

12   The letters to the Duke of Leeds referred to here are printed in Burges, *A Narrative*, pp.1–10.

13   More details about the events at Nootka Sound can be found in Clayton, *Islands*, Cook, *Flood Tide* and Gough, *The Northwest*, especially ch.9. Colnett's own version of events appeared in his journal – see Howay, *The Journal*.

14   Mackay, *In the Wake*, pp.87–92.

15   Menzies to Banks, 4 April 1790, *I&P*, vol.3, document 85 and Mackay, *In the Wake*, p.91.

16   TNA, HO 42/16, pp.86–8.

17   The Memorial is in TNA, HO 42/16, pp.98–116.

18   See Gough, *The Northwest*, Cook, *Flood Tide* and Norris, 'The Policy'. See also the wider international context as it is explored in Frost, 'Nootka Sound'.

19   Mackay, *In the Wake*, pp.44–5.

20   Lamb, *A Voyage of Discovery*, p.27.

21   Bown, *Madness*, pp.88–9.

22   Portlock to Banks, 26 December 1790, *I&P*, vol.3, document 117.

23   Mackay, *In the Wake*, p.45. It isn't clear whether the change in destination preceded or followed the change in command. Certainly, Roberts still believed he was going to go on the original voyage but on another ship – see Roberts to Stephens, 24 December 1790, TNA, ADM 1/2395 and Roberts to Stephens, 23 June 1791, TNA, ADM 1/2395. Menzies thought so too – see Menzies, 'Journal', BL, Add MS 32641, p.1. Roberts never went to the South Atlantic but to the Baltic instead as part of a task force to survey entrances into the Baltic in case of a real naval threat from Russia and Sweden – see King, 'George Vancouver', p.18 and Frost, 'Nootka Sound', p.119. Instead of Roberts, James Colnett went to the South Atlantic and further to satisfy the whaling interests which had not been forgotten. After his release from detention at San Blas, Mexico, in May 1790, Colnett returned to Nootka Sound with the *Argonaut* and then sailed to Canton to sell his furs. He was back in London in April 1792. Several months later, in early January 1793, he was back at sea now in command of the *Rattler* which sailed on a mission to find an island in the southern Pacific

which could be used as a whaling base – the Pacific had by now become the southern whaling fisheries' main destination. He returned in November 1794. See Mackay, *In the Wake*, pp.49–52 and Colnett, *A Voyage to the South Atlantic.*

24 Menzies, 'Journal', BL, Add MS 32641, p.2.

25 Menzies to Banks, 1 March 1791, *I&P*, vol.3, document 147.

26 Menzies to Banks, 15 December 1790, *I&P*, vol.3, document 114.

27 Banks to Nepean, 15 December 1790, *I&P*, vol.3, document 115. Menzies had received his warrant as a surgeon two days earlier, on 13 December 1790 – Keevil, 'Archibald Menzies', p.799.

28 Menzies, 'Journal', BL, Add MS 32641, p.3.

29 Portlock to Banks, 26 December 1790, *I&P*, vol.3, document 117.

30 Banks to Nepean, 1 January 1791, *I&P*, vol.3, document 120.

31 These instructions are in TNA, HO 28/8, pp.17–24.

32 The surveying instructions can be found in SLNSW, Sir Joseph Banks Papers, Series 60.05 where they are identified, by Banks, as being Rennell's. Another set, without attribution to Rennell, is in TNA, HO 42/18, pp.168–77.

33 Banks to Nepean, 20 February 1791, *I&P*, vol.3, document 136.

34 These were the same trade goods that Menzies had recommended to Banks when he was asked in April 1790. The final list can be found in TNA, HO 28/8, pp.27–8.

35 Admiralty to Vancouver, TNA, ADM 2/1344, also printed in Lamb, *A Voyage of Discovery*, p.1615.

36 The instructions can be found in Banks to Menzies, 22 February 1791, *I&P*, vol.3, document 139.

37 King, 'George Vancouver', p.7. The plant cabin, though agreed upon, was not yet on the *Discovery* in early October 1789 – at that time Menzies was anxious that Banks should be available during its construction – see Menzies to Banks, 8 October 1789, *I&P*, vol.3, document 47.

38 Menzies, 'Journal', BL, Add MS 32641, p.3. As late as late February 1791, the plant cabin was unglazed – 50 panes had been ordered – see Menzies to Banks, 24 February 1791, *I&P*, vol.3, document 142. A week later they arrived – Menzies to Banks, 1 March 1791, *I&P*, vol.3, document 147.

39 The plan of the plant cabin is in SLNSW, Sir Joseph Banks Papers, Series 60.12. The photograph of a replica, built in 1986, is printed and discussed in Groves, 'Archibald Menzies'.

40 Vancouver, *A Voyage*, p.xxii.

41 Banks to Menzies, 22 February 1791, *I&P*, vol.3, document 139.

42 Menzies to Banks, 1 March 1791, *I&P*, vol.3, document 147 and Menzies to Banks, 20 March 1791, *I&P*, vol.3, document 156.

43 Menzies to Banks, 20 March 1791, *I&P*, vol.3, document 156.

44 The last time that Banks asked for the documents was on 13 May 1793 – Banks to Nepean, 13 May 1793, *I&P*, vol.4, document 54.

45 Banks to Menzies, 10 August 1791, *I&P*, vol.3, document 187.

46 Ibid.

47 Menzies to Banks, 1 March 1791, *I&P*, vol.3, document 147.

48 Lamb, *A Voyage of Discovery*, p.310.

49 Menzies to Banks, 5 May 1791, SLNSW, Sir Joseph Banks Papers, Series 61.08.

50 Menzies to Banks, 10 August 1791, *I&P*, vol.3, document 188.

51 Lamb, *A Voyage of Discovery*, p.337.

52 Menzies, 'Journal', BL, Add MS 32641, 30 September 1791. Parts of this journal, which covers the period December 1790 to 16 February 1794, have been transcribed and published. Where a published version exists, I cite it in the forthcoming notes; in the absence of such publications, I cite the manuscript. The Linnean Society also has a copy of the journal in manuscript form – see LS, Ms 418.

53 Menzies, 'Journal', BL, Add MS 32641, 2 October 1791. See McCarthy, *Monkey Puzzle Man*, p.91 for a list of the new species of *Banksia* Menzies with which Menzies has been credited.

54 Further details of Menzies's stay in western Australia can be found in Groves, 'Archibald Menzies's Visit'.

55 Menzies, 'Journal', BL, Add MS 32641, 4 November 1791.

56 Menzies, 'Journal', BL, Add MS 32641, 17 and 18 November 1791.

57 Lamb, *A Voyage of Discovery*, p.396.

58 Ibid., p.393.

59 Menzies, 'Journal', BL, Add MS 32641, 31 December 1791.

60 Menzies, 'Journal', BL, Add MS 32641, 8 and 9 January 1792 and Lamb, *A Voyage of Discovery*, p.440.

61 Wilson, *Hawaii*, p.19.

62 Menzies, 'Journal', BL, Add MS 32641, 18 January 1792.

63 Menzies to Banks, 1–14 January 1793, *I&P*, vol.4, document 13.

64 Menzies, 'Journal', BL, Add MS 32641, 2 March 1792.

65 Menzies, 'Journal', BL, Add MS 32641, 4 March 1792.

66 Wilson, *Hawaii*, pp.39, 43.

67 For a discussion of Vancouver's surveying methods, see David, 'Vancouver's Survey Methods'.

68 The species has been identified as *Valerianella congesta*.

69 Newcombe, *Menzies' Journal*, p.18. Vancouver, who was with Menzies at the time, was also deeply impressed. His words were: 'Our attention was immediately called to a landscape, almost as enchantingly beautiful as the most elegantly finished pleasure grounds in Europe' – Lamb, *A Voyage of Discovery*, p.513.

70 Newcombe, *Menzies' Journal*, pp.31, 54.

71 Lamb, *A Voyage of Discovery*, pp.94–5.

72 Ibid., pp.652–3. Dening, *The Death of William Gooch* tells the full story.

73 Newcombe, *Menzies' Journal*, p.103.

74 Ibid., p.124. Menzies added that this was only one-third of the number of

ships that had already visited during the year. As it turned out, 1792 was the year in which the presence of British vessels on the Pacific Northwest was at its maximum: within a few years they and ships from other nations would withdraw from the trade leaving the field entirely to the Americans – see Cook, *Flood Tide*, Appendix E.

75 Menzies to Banks, 1–14 January 1793, *I&P*, vol.4, document 13.

76 Menzies to Banks, 26 September 1792, *I&P*, vol.3, document 311.

77 A list of the plants Menzies collected throughout his time in the Pacific Northwest (not including Alaska) can be found in Newcombe, *Menzies' Journal*, pp.132–50.

78 Eastwood, 'Menzies' California Journal', p.289.

79 Menzies to Banks, 1–14 January 1793, *I&P*, vol.4, document 13. The list of seeds, some from western Australia, some from the Pacific Northwest and some from California, is in SLNSW, Sir Joseph Banks Papers, Series 61.15. On the protracted negotiations between Vancouver and Bodega y Quadra, see Lamb, *A Voyage of Discovery*, pp.98–110 and Cook, *Flood Tide*, ch.9. For more on Bodega y Quadra, see Tovell, *At the Far Reaches* and Tovell, 'The Other Side'. Broughton sailed to San Blas in Mexico with Bodega y Quadra and from there crossed by land over to Veracruz from where he sailed eventually to London. For Broughton's journey across Mexico, see Mockford, 'The Journal'.

80 Menzies to Banks, 14 January 1793, *I&P*, vol.4, document 20.

81 Menzies and Vancouver differed on the exact number of each – see Eastwood, 'Menzies' California Journal', p.291 and Lamb, *A Voyage of Discovery*, p.787.

82 Wilson, *Hawaii*, pp.76–7, 84.

83 Groves, 'Archibald Menzies', p.89.

84 Menzies to Banks, 23 May 1793, *I&P*, vol.4, document 57.

85 For Menzies's collecting activities before September, see Groves, 'Archibald Menzies', pp.89–91.

86 Olson, *The Alaska Travel Journal*, p.60. Menzies understood about acclimatisation. For another example of 'inuring' plants, one directly involving Banks, and the contemporary understanding of this process, see Zilberstein, 'Inured to Empire'.

87 Olson, *The Alaska Travel Journal*, p.64. The plant has been identified as *Clintonia uniflora*, a species of the lily family – see Groves, 'Archibald Menzies', p.91.

88 Menzies, 'Journal', BL, Add MS 32641, 5 October 1793.

89 Groves, 'Archibald Menzies', pp.92 and 112, n.60.

90 Eastwood, 'Menzies' California Journal', p.338.

91 Ibid.

92 Ibid., p.339. The cattle were entirely new to these islands. While the ships were in Hawaii during the winter of 1794, the island's first calf was born – Wilson, *Hawaii*, p.145.

93 Menzies to Vancouver (and Banks), 18 November 1793, *I&P*, vol.4, document 96.

94 Ibid.

95 Wilson, *Hawaii*, p.161.

96 Menzies to Banks, 6 February 1794, *I&P*, vol.4, document 116 and Lamb, *A Voyage of Discovery*, p.1154. The *Daedalus* was back on the English coast at the end of April 1795.

97 Menzies's description of the journey to the summit can be found in Wilson, *Hawaii*, pp.176–99. Using a portable barometer he had been given at the Cape by Colonel William Paterson, Menzies estimated the height of the volcano at just 41 feet short of its official height. The ascent of Mauna Loa had been attempted when Cook was at the same anchorage in late January 1779. The party interestingly included George Vancouver. The attempt was then unsuccessful – see St John, 'Biography'.

98 Lamb, *A Voyage of Discovery*, p.1225.

99 Vancouver described the moment of discovery and replaced the word 'river' with that of 'inlet' – Lamb, *A Voyage of Discovery*, pp.1242–3. For the history of the search for the temperate northwest voyage until Vancouver, see Williams, *Arctic* and Williams, *The British Search*.

100 Menzies to Banks, 8 September 1794, *I&P*, vol.4, document 140.

101 Lamb, *A Voyage of Discovery*, p.1367.

102 Olson, *The Alaska Travel Journal*, p.196.

103 Groves, 'Archibald Menzies', p.96.

104 Menzies to Banks, 8 September 1794, *I&P*, vol.4, document 140.

105 Lamb, *A Voyage of Discovery*, pp.1421–2.

106 For the remaining part of this voyage I have relied on the account as given in Lamb, *A Voyage of Discovery*, pp.183–207.

107 Nelson, 'Archibald Menzies's Visit to Isla'.

108 Nelson and Porter, 'Archibald Menzies on Albemarle'.

109 Menzies to Banks, 26 March 1795, *I&P*, vol.4, document 171.

110 Menzies to Banks, 28 April 1795, *I&P*, vol.4, document 178.

111 Menzies to Banks, 14 September 1795, *I&P*, vol.4, document 196.

112 Ibid.

113 Ibid.

114 Menzies to Banks, 4 October 1795, *I&P*, vol.4, document 200. Menzies had collected five live specimens. One went to Banks and the other four to Kew – see Groves, 'Archibald Menzies', p.114, n.109. See also the discussion in McCarthy, *Monkey Puzzle Man*, pp.192–3.

115 For Menzies's collections from the Pacific Northwest and California that featured as growing in Kew up to 1812, see Groves, 'Archibald Menzies', pp.98–9.

116 Menzies to Banks, 4 October 1795, *I&P*, vol.4, document 200.

117 Some of these can be found in *I&P*, vol.4, documents 198, 199, 200, 203–5, 207 and 209. Vancouver's despatches to Nepean are in TNA, ADM 1/2629.

118 Nepean to Banks, 26 October [1795], *I&P*, vol.4, document 207.

119 Menzies to Banks, 14 September 1795, *I&P*, vol.4, document 196.

120 Under Banks's direction, Menzies was supposed to write up and publish his journal before Vancouver did. For a number of reasons this never happened. Apart from specific parts of it that have been published, the worked-up copy as a whole remains in manuscript form. Vancouver's multi-volume work came out in 1798, a few months after his death. For a discussion of this, see Lamb, 'Banks and Menzies' and Groves, 'Archibald Menzies'.

10 1791: The Gardeners of the *Providence*

1 Unless otherwise noted, I am basing this account of the voyage of *Bounty*'s launch and the subsequent journey from Timor to England on Salmond, *Bligh*, Alexander, *The Bounty*, Bligh, *A Voyage* and Rutter, *Bligh's Voyage*.

2 Bligh to Banks, 13 October 1789, *I&P*, vol.3, document 48.

3 Ibid.

4 Probably malaria which was endemic in the region.

5 Bligh to Banks, 13 October 1789, *I&P*, vol.3, document 48.

6 Bligh arranged with the Batavian authorities to find places on other Dutch East India Company ships to get the other men back to Britain.

7 An early account appeared in the *General Evening Post*, 16–18 March 1789.

8 See Alexander, *The Bounty*, pp.170–3.

9 Bligh to Banks, 24 October 1790, *I&P*, vol.3, document 109.

10 Ibid.

11 Banks to Chatham, 10 December 1790, *I&P*, vol.3, document 113.

12 Bligh to Banks, 20 December 1790, *I&P*, vol.3, document 116.

13 East to Banks, 6 October 1790, *I&P*, vol.3, document 105.

14 Ibid.

15 See RBGA, JBK/1/6, f.182. The letter is undated but from internal evidence, it must have been before 15 December 1790, the date on which Banks replied to East and in which he must have told him that a second bread-fruit expedition was being planned. Two days later, on 17 December, Banks wrote to William Eden (Baron Auckland), Britain's Ambassador to the Netherlands, that the King had already ordered a ship to be prepared to return to Tahiti for the breadfruit plants – Banks to Eden, 17 December 1790, BL, Add MS 34434, pp.335–6. Yonge's letter has never been incorporated into the breadfruit story because it has been incorrectly bound into the volume containing Banks's correspondence from 1797. The incorrect 1797 date is repeated in *I&P*, vol.4, document 298. In his *In the Wake*, David Mackay states (p.137) that when Banks wrote to Chatham on 10 December 1790, in which he asked for Chatham's support in promoting Bligh, he referred to a second breadfruit expedition. This is not what the letter says. Banks simply states that with this promotion and a return to full health, Bligh would be ready when 'he is called upon

to depart again' (Banks to Chatham, 10 December 1790, *I&P*, vol.3, document 113).

16   Yonge to Banks, n.d., *I&P*, vol.4, document 298.
17   Mackaness, *Fresh*, p.15.
18   SLNSW, Sir Joseph Banks Papers, Series 53.02.
19   Wiles to Banks, 16 January 1791, *I&P*, vol.3, document 122.
20   Banks to Wiles, 27 January 1791, *I&P*, vol.3, document 129.
21   Ibid.
22   Wiles to Banks, 9 February 1791, *I&P*, vol.3, document 133.
23   For the Perry company and Blackwall, see *www.british-history.ac.uk/survey-london/vol43-4/pp553-565#h3-0009*, accessed 18 January 2016.
24   Bligh to Banks, 23 February 1791, *I&P*, vol.3, document 140.
25   Chatham to Banks, 9 March 1791, *I&P*, vol.3, document 152.
26   Banks to Chatham, 7 March 1791, *I&P*, vol.3, document 149.
27   These orders are detailed in TNA, ADM 2/267.
28   15 July 1791, TNA, ADM 2/121, pp.475–6.
29   See 29 May 1792, Bligh's log at *www.fatefulvoyage.com/providenceBligh/index.html*. Here Bligh clearly states that in Tahiti he had '12 Pine Apple' planted. He had also brought tree seedlings with him and these are referred to as 'firs'. Bligh felt that the Tahitians valued the latter the most as 'they are likely to produce them plank and Masts'. The invoice for the 'queen pines' is in SL, Sir Joseph Banks Collection, G 1:18. For the cultivation of pineapples in Britain, see Levitt, '"A Noble Present"' and Cole, 'The Cultural History'. On the wider cultural networks of pineapple, see Beauman, *Pineapple*, Okihiro, 'Of Space/Time' and Gohmann, 'Colonizing'. For more on the nursery of Ronalds, see Ronalds, 'Ronalds Nurserymen'.
30   TNA, HO 28/62 and SL, Sir Joseph Banks Collection, G 1:18.
31   Mauritius was Banks's suggestion as the ship might be able to purchase or barter for valuable spice trees. Banks wanted to have Bligh take trees to Norfolk Island and Sydney as he had planned for HMS *Guardian*, but which never happened. See 18 April 1791, TNA, HO 28/8, pp.62–3.
32   15 July 1791, TNA, ADM 2/121, pp.475–80.
33   18 April 1791, TNA, HO 28/8, pp.62–3.
34   Salmond, *Bligh*, p.346.
35   Wiles's biographical information can be found on the website *www.anbg.gov.au/biography/wiles-james-1768-1851*, accessed on 12 December 2015. SLNSW, Sir Joseph Banks Papers, Series 53.02. See also Jacob Reynardson to Banks, 8 September 1794, SLNSW, Sir Joseph Banks Papers, Series 53.23. Samuel Reynardson, a local dignitary, Chancery clerk and a Fellow of the Royal Society since 1742, had created a classical garden beginning in 1732 with the help of the astronomer and garden designer Thomas Wright. See Phillipson, *Holywell* for Reynardson and McCarthy, 'Thomas Wright's "Designs for Temples"' and McCarthy, 'Thomas Wright's Designs for Gothic'. Reynardson's election to the Royal Society is in RS, EC/1741/17.

36 The Kew connection is stated in Salmond, *Bligh*, p.333.

37 SLNSW, Sir Joseph Banks Papers, Series 53.03. For a biography, such as it is, of William Pitcairn and a short description of his garden in Islington, see Rembert, 'William Pitcairn'.

38 For Brass, see Keay, 'Botanical', Coats, *Quest*, pp.245–7 and Fox, *Dr. John Fothergill*. Fothergill had earlier tried to form a syndicate with Banks and Pitcairn to collect plant specimens in the West Indies – see Fothergill to Banks, 1 July 1777, SLNSW, Sir Joseph Banks Papers, Series 72.051.

39 SLNSW, Sir Joseph Banks Papers, Series 49.07. An earlier draft of the instructions was written by Banks himself – Series 49.06. The final version is in TNA, HO 28/8, pp.54–5.

40 Banks to Wiles, 25 June 1791, *I&P*, vol.3, document 168.

41 Ibid.

42 The plans for the *Providence* are in NMM, ZAZ6585-6586 and the plans for the *Bounty* are in ZAZ6664-6668 and ZAZ7848. The extra purchases made for the voyage are in SLNSW, Sir Joseph Banks Papers, Series 53.09.

43 Schreiber, *Captain Bligh's*, p.16.

44 Portlock had been in command of the *King George*, the first fur-trading voyage in which Banks had been involved and which sailed from England for the Pacific in September 1785 – see Chapter 3 for more details.

45 See Bligh's letters to Philip Stephens at the Admiralty in TNA, ADM 1/1507 and his letters to Banks, recounting the delays, especially, 28 July 1791, *I&P*, vol.3, document 182.

46 Estensen, *Matthew Flinders*, pp.8–13.

47 Bligh to Banks, 30 August 1791, *I&P*, vol.3, document 192.

48 Ibid.

49 Bligh to Banks, 13 September 1791, *I&P*, vol.3, document 198.

50 Bligh to Banks, 7 November 1791, *I&P*, vol.3, document 206.

51 Bligh to Banks, 24 November 1791, *I&P*, vol.3, document 214.

52 Wiles and Smith to Banks, 28 November 1791, *I&P*, vol.3, document 217 and Bligh to Banks, 17 December 1791, *I&P*, vol.3, document 220.

53 Bligh to Banks, 24 November 1791, *I&P*, vol.3, document 214.

54 Bligh thought he'd be under way by 6 December, but he had not recovered as much as he had hoped and stayed on – Bligh to Banks, 17 December 1791, *I&P*, vol.3, document 220. Bligh was ready to go on 18 December – Bligh to Banks, 18 December 1791, *I&P*, vol.3, document 221.

55 Bligh, *The Log of H.M.S. Providence*, Remarks, 10–11 February 1792.

56 Bligh, *The Log of H.M.S. Providence*, Remarks, 22 February 1792.

57 Schreiber, *Captain Bligh's*, p.71. This entry was written later, in February 1797.

58 The list of presents can be found in TNA, ADM 1/1507. Bligh personally chose them. The bill came to £290.

59 Unless otherwise indicated, all of the information regarding the breadfruit arrivals and events surrounding their care and movements, is taken from

Bligh's log as transcribed at the Fateful Voyage website, *www.fatefulvoyage. com/providenceBligh/index.html*. This is a faithful and easy-to-use transcription of the original logs which are in TNA, ADM 55/152–153. A facsimile reproduction of the original logs has been published as Bligh, *The Log of H.M.S. Providence*.

60 The following description, unless otherwise stated, can be found in the log's entry for 10–13 and 14–16 July 1792, transcribed at the Fateful Voyage website, *www.fatefulvoyage.com/providenceFlinders/index.html*.

61 Wiles and Smith to Banks, 17 December 1792, *I&P*, vol.4, document 7.

62 Flinders's log at *www.fatefulvoyage.com/providenceFlinders/index.html*.

63 Bligh to Banks, 16 December 1792, *I&P*, vol.4, document 6.

64 This and the following detail comes from 1st Lieutenant Francis Bond's log of 24 April and 18 June 1792, transcribed at the Fateful Voyage website, *www.fatefulvoyage.com/providenceBond/index.html*.

65 Entry for 18 June 1792 at website, *www.fatefulvoyage.com/providenceBligh/ index.html*.

66 Wiles and Smith to Banks, 17 December 1792, *I&P*, vol.4, document 7.

67 Bligh to Banks, 2 October 1792, *I&P*, vol.3, document 313.

68 Wiles and Smith to Banks, 17 December 1792, *I&P*, vol.4, document 7.

69 Ibid.

70 10 November 1792, Bligh's log at *www.fatefulvoyage.com/providenceBligh/index. html*.

71 Bligh to Banks, 2 October 1792, *I&P*, vol.3, document 313.

72 Bligh to Banks, 16 December 1792, *I&P*, vol.4, document 6.

73 All of the information is from Bligh to Banks, 16 December 1792, *I&P*, vol.4, document 6.

74 Wiles and Smith to Banks, 17 December 1792, *I&P*, vol.4, document 7.

75 Ibid. and Wiles to Banks, 22 January 1793, *I&P*, vol.4, document 21.

76 The phrase appears in the 19–26 December entry in Bligh's log as transcribed at the Fateful Voyage website, *www.fatefulvoyage.com/providenceBligh/index. html*.

77 Flinders's log for 9 January 1793, as transcribed at the Fateful Voyage website, *www.fatefulvoyage.com/providenceFlinders/index.html*.

78 Bligh's log for 31 December 1792, as transcribed at the Fateful Voyage website, *www.fatefulvoyage.com/providenceBligh/index.html*.

79 Bligh's log for 11 January 1793, as transcribed at the Fateful Voyage website, *www.fatefulvoyage.com/providenceBligh/index.html*.

80 Wiles to Banks, 22 January 1793, *I&P*, vol.4, document 21.

81 Bligh to Banks, 23 January 1793, *I&P*, vol.4, document 23.

82 Bligh to Banks, 27 January 1793, *I&P*, vol.4, document 25.

83 Bligh's log for 6–10 February 1793, as transcribed at the Fateful Voyage website, *www.fatefulvoyage.com/providenceBligh/index.html*.

84 Bligh's log for 11 January 1793, as transcribed at the Fateful Voyage website,

www.fatefulvoyage.com/providenceBligh/index.html. See also Sheridan, 'Captain Bligh'.

85 This and the following information is taken from Bligh's log for 11 March–2 June 1793, as transcribed at the Fateful Voyage website, *www.fatefulvoyage.com/providenceBligh/index.html*.

11 1791: 'An Intertropical Abode': Afzelius in Sierra Leone

1   See Oldfield, *Popular Politics* and Stott, *Wilberforce*.

2   For more on Thornton, see Meacham, *Henry Thornton*, Atkins, 'Piety' and Stott, *Wilberforce*.

3   Wilberforce to Banks, 21 December 1791, NHM, DTC, vol.VII, p.293.

4   The story of the African Association is told in Hallett, *Records* and in Boahen, 'The African'.

5   A detailed discussion of the bill's passage is given in Braidwood, *Black Poor*, pp.225–68.

6   These are currently held in SL, Sir Joseph Banks Collection, at A 2:26 and 47.

7   Smeathman has been the subject of a number of recent studies: see Coleman, *Romantic*, ch.1, Douglas, 'The Making' and Douglas and Driver, 'Imagining'.

8   For the Aurelians, in general, see Salmon, *The Aurelian* and, specifically for Smeathman, Douglas, 'Natural History', pp.63–74.

9   The limited extent of European knowledge of Africa's natural history is discussed in Coats, *Quest*, pp.243–65.

10 For the interests of the sponsors of the expedition, see Douglas, 'Natural History', pp.74–80. See also Smeathman's own recollection of how he became involved in this venture, which can be found in Smeathman to Lettsom, 19 October 1782, in Lettsom, *The Works of John Forthergill*, pp.575–82. The Duchess of Portland, another leading naturalist and collector, joined the consortium in the summer of 1773 on the same terms as the other subscribers – Douglas, 'Natural History', p.94.

11 For Berlin, see Hamberg, 'Anders Berlin' and his entry and letters to Linnaeus in Hansen, *The Linnaeus Apostles*, vols.4 and 8. Smeathman's lack of botanical knowledge and Drury's solution are discussed in Douglas, 'The Making', pp.6–7.

12 Douglas, 'Natural History', p.97.

13 Smeathman recollected that he had sent about six hundred different species of plants from both Sierra Leone and the West Indies though he provided no further details – see Smeathman to Lettsom, 19 October 1782, in Lettsom, *The Works of John Forthergill*, p.579. Adam Afzelius (see further in this chapter) thought that Banks had 600 Sierra Leone plant species in his herbarium – Afzelius to Thunberg, 17 January 1792, UUL, MS, G 300 a.

14 For his various projects, see his entry, by Starr Douglas, in the *Oxford Dictionary of National Biography*. On termites, see Douglas and Driver, 'Imagining'. On ballooning, see Coleman, *Romantic*, pp.54–6 and Smeathman to Banks, 11 February 1784, NHM, DTC, vol.IV, pp.7–11.

15 In 1783, Smeathman put his thoughts on colonising Sierra Leone in a long letter to Dr Thomas Knowles. Knowles was a Quaker physician, one of a small group of similar-minded Quakers who, in July 1783, had set up an informal group to consider how they could press for the abolition of the British slave trade. In May 1787, this same group, together with other Quakers, joined with three Anglican members, Philip Sansom, Thomas Clarkson and Granville Sharp, to form the first public abolitionist pressure group in Britain, the Society for Effecting the Abolition of the Slave Trade – see Jennings, *The Business* and Davis, *The Problem*. It is likely that Smeathman's plan circulated in the form of this letter for some time. It was published posthumously in 1790 in *The New Jerusalem Magazine*, pp.279–94.

16 The story is told in detail in Braidwood, *Black Poor*.

17 Thompson to Banks, 23 January 1787, BL, Add MS 33978, p.97. Unfortunately no record survives of Banks's request.

18 See Braidwood, *Black Poor*, ch.4.

19 See Wilberforce to Banks, 21 December 1791, NHM, DTC, vol.VII, p.294.

20 See Wadstrom [sic], *An Essay on Colonization*, p.229 and Lindroth, 'Adam Afzelius', p.203.

21 Note by Banks, 28 December 1791, RBGA, JBK/1/5, fol.58.

22 Lindroth, 'Adam Afzelius. En Linnean', pp.2–3 (I would like to thank Malin Emara for helping me translate this article). Nyberg, 'Linnaeus's Apostles' discusses the idea of an apostle in relation to an earlier period.

23 Thunberg's stay at Soho Square is discussed in Skuncke, *Carl Peter Thunberg*, pp.177–85. For Thunberg's comments on Afzelius, see Thunberg to Banks, 15 January 1790, BL, Add MS 8097, pp.351–2. Before he left Sweden Afzelius had asked Thunberg for a letter of recommendation – see Lindroth, 'Adam Afzelius. En Linnean', p.15.

24 For a recent discussion of links between Soho Square and Sweden during this period, see Hodacs, 'Local, Universal'. Pehr Afzelius's stay in London is mentioned in Lindroth, 'Adam Afzelius. En Linnean', p.14.

25 Afzelius went to London to improve his life chances – see Adam Afzelius to Pehr Afzelius, 17 January 1792, UUL, MS, G2c for a sense of his self-appraisal. As to when Afzelius arrived at Soho Square, see Banks to Charles Louis L'Héritier de Brutelle, 25 December 1789, *SC*, document 961.

26 These details are drawn from Lindroth, 'Adam Afzelius. En Linnean', pp.15–20.

27 Banks to Swartz, 26 December 1790, SC, document 1024. Swartz, a student of Linnaeus, the younger, and an expert on ferns and West Indies plants, was a frequent correspondent – he stayed with Banks for nine months in 1786 and 1787 – see Galloway, 'Olof Swartz's Contributions'.

28 Lindroth, 'Adam Afzelius. En Linnean', p.23.

29 Staunton, *Memoirs*, p.6.

30 Bibliographical information on Hüttner can be found in Proescholdt, 'Johann Christian Hüttner'.

31 See Chapter 12.

32 Afzelius always wrote to Staunton in Latin. The progress of their field trip once Staunton had left for London is in a letter, from Afzelius to Staunton, written in Newcastle, and dated 27 October 1791 – see Staunton, *Memoir*, pp.333–5.

33 George Clifford, who died in 1760, was from a wealthy Amsterdam banking family, who were originally from Lincolnshire. He had a substantial estate near Haarlem, where he had an extensive garden and herbarium. Linnaeus described the plants growing in Clifford's estate in the justly famous *Hortus Cliffortianus* (1737), the precursor for his own publication of 1753, the *Species Plantarum*, in which his revolutionary binomial classification system was first made public – see Thijsse, 'A Contribution'. Banks thought that the auction would take place in November 1791 but Jonas Dryander, Banks's librarian, was uncertain whether Afzelius, who would be qualified to form an opinion of the herbarium's value, would be back from his field trip before then – see Dryander to Banks, 19 October 1791, *SC*, document 1064 and Banks to Dryander, 26 October 1791, *SC*, document 1069. The auction, as it turned out, was held in March 1792 on which occasion Banks purchased the herbarium – Thijsse, 'A Contribution', pp.137–8. By then Afzelius had already agreed to go to Sierra Leone.

34 George Staunton offered to take Afzelius with him to France and Italy sometime in January 1792 – see Chapter 12.

35 See Adam Afzelius to Pehr Afzelius, 17 January 1792, UUL, MS, G 2c. I would like to thank Anna Backman for translating this part of the letter for me.

36 For a recent discussion of Swedenborg's scientific interests, especially as they related to the problem of determining longitude at sea, see Schaffer, 'Swedenborg's Lunars'.

37 See Rix, *William Blake* and Paley, '"A New Heaven"' for more on Swedenborg's spirituality.

38 Lineham, 'The Origins'. For the Swedenborgians in London, see Garrett, 'Swedenborg and the Mystical Enlightenment'.

39 Coleman, *Romantic*, p.64. The final quote is from Paley, '"A New Heaven"', p.65.

40 For Afzelius's commitment to Swedenborg, see Widstrand, 'Adam Afzelius', pp.XII–XIII. The title of the Swedenborgian pamphlet is *Plan for a Free Community Upon the Coast of Africa Under the Protection of Great Britain but Intirely Independent of All European Laws and Government*, published in London in 1789. For more on Wadström and Nordenskjöld, see Ambjörnsson, '"La république"', Rönnbäck, 'Enlightenment', Paley, '"A New Heaven"',

pp.83–5, Rotberg, 'The Swedenborgian Search' and Dahlgren, 'Carl Bernhard Wadström'.

41 This was a Swedish government-sponsored plan to colonise a part of Senegal: Anders Sparrman, who had been with James Cook on his second voyage to the Pacific, went with him – see Rönnbäck, 'Enlightenment'. On returning to London, Wadström published his experiences of that visit as *Observations on the Slave Trade*.

42 His evidence can be found in *Minutes of the Evidence Taken before a Committee of the House of Commons Being a Select Committee Appointed on the 23rd of April 1790* . . . (London, 1790), on pp.18–44. Rönnbäck, 'Enlightenment' discusses the evidence. For more on Wadström as an abolitionist, see Ahlskog, 'The Political Economy'.

43 Granville Sharp had written to the black settlers in Granville Town on 16 May 1788 that a group of twelve Swedish men of distinction would soon be embarking for the colony. In the published version of an extract from this letter, the members of the party were identified only by their initials: one was A.N., another was C.B.W. and another one was A.A. – see Wadström, *An Essay on Colonization*, p.339 and Lindroth, 'Adam Afzelius', pp.195–6.

44 How much Banks knew of the Swedenborgians is an interesting question. Banks had at least one brush with Swedenborg himself. Banks was the pallbearer, the only non-Swedish one, at the funeral of Daniel Solander in 1782. He was laid to rest in the Swedish Church in Shadwell next to the body of Emmanuel Swedenborg. Of course, at the time, the Swedenborgian movement had not yet begun – see Carter, *Sir Joseph Banks*, p.181.

45 Banks to Thunberg, 19 March 1790, *I&P*, vol.3, document 81.

46 There are many spiritual references in the 'Plan' but it is between pp.38 and 41 that the explicit Swedenborgian connections are spelled out. Banks was one of the subscribers to Wadström's more substantial *An Essay on Colonization*, published in 1794. For more on Wadström's plans for an African settlement, see Hodgson, 'Dedicated to the Sound'.

47 Thunberg to Banks, 15 January 1790, BL, Add MS 8097, pp.351–2.

48 Banks to Thunberg, 19 March 1790, *I&P*, vol. 3, document 81. Once Afzelius learned of the mission and that its destination was to be the Pacific Northwest, he lost interest in it – see Adam Afzelius to Pehr Afzelius, 11 January 1791, UUL, MS, G 2c. I would like to thank Hanna Hodacs for providing me with this reference. Once Afzelius made this decision he was free to join the Staunton party on their summer field trip.

49 Banks to Afzelius, 2 January 1792, SL, Sir Joseph Banks Collection, A 2:39.

50 Banks to Wilberforce, 2 January 1792, NHM, DTC, vol.VIII, p.1.

51 Ibid.

52 Banks's instructions to the Company regarding Afzelius's needs are contained in Banks to Wilberforce, 2 January 1792, NHM, DTC, vol.VIII, pp.2–4. There is another copy in RBGA, JBK/1/5, fols.61–2.

53 Wilberforce to Banks, 4 January 1792, NHM, DTC, vol.VIII, p.5.

54 See Ingham, *Sierra Leone*, pp.71, 74 and Wilson, *John Clarkson*, p.85.

55 Afzelius to Banks, 2 July 1792, NHM, DTC, vol.VIII, p.29. The rain figures and the description of tornadoes are from Afzelius to Clarkson, 29 December 1792, BL, Add MS 12131, pp.20–1.

56 This section rests on Fyfe, pp.31–7 and Schwarz, 'A Just and Honourable', p.7. The story of this migration has been told many times, most recently in Schama, *Rough Crossings* and Jasanoff, *Liberty's Exiles*, where readers can find references to the standard works on the topic. The 'long' history of Sierra Leone is best covered in Lovejoy and Schwarz, *Slavery, Abolition and the Transition*.

57 Afzelius to Banks, 2 July 1792, NHM, DTC, vol.VIII, p.30.

58 Afzelius to Clarkson, 29 December 1792, BL, Add MS 12131, pp.6–7.

59 Afzelius to Banks, 30 December 1792, NHM, DTC, vol.VIII, p.141.

60 A committed Swedenborgian, Nordenskjöld had arrived in Freetown a little after Afzelius. Although he was unwell, he petitioned the colony's Council to let him travel inland to look for Swedenborg's 'true Church'. He got the permission in August but the journey went horribly wrong. He arrived back in Freetown and, several days later, on 10 December 1792, he died, aged thirty-eight. See Ingham, *Sierra Leone*, p.78, Wilson, *John Clarkson*, p.95 and Wadström, *An Essay on Colonization*, pp.229, 235–40.

61 Afzelius complained that he was already suffering from fever in August 1792 – see Kup, *Adam Afzelius*, p.2.

62 Lindroth, 'Adam Afzelius', p.200 and Afzelius to Banks, 28 October 1793, SL, Sir Joseph Banks Collection, A 2:43.

63 Wilson, *John Clarkson*, p.128.

64 Banks to Afzelius, 8 April 1793, *SC*, document 1178.

65 Banks to Wilberforce, 8 April 1793, NHM, DTC, vol.VIII, p.196.

66 Afzelius to Clarkson, 29 December 1792, BL, Add MS 12131, p.9.

67 Banks to Wilberforce, 8 April 1793, NHM, DTC, vol.VIII, p.196.

68 Afzelius to Clarkson, 29 December 1792, BL, Add MS 12131, p.15.

69 Banks to Wilberforce, 8 April 1793, NHM, DTC, vol.VIII, p.197.

70 Ibid.

71 Williams to Banks, 17 June 1793, SL, Sir Joseph Banks Collection, A 2:44. Williams signed himself 'JR'. I thank Suzanne Schwarz for helping me discover that 'JR' stood for James Rice.

72 Thornton to Nepean, 30 July 1793, TNA, CO 267/10, p.9.

73 Nepean to Banks, 16 August 1793, *I&P*, vol.4, document 77. Banks confessed to Nepean that he had 'contrived' the whole idea of sending Kew plants to Sierra Leone – see Banks to Nepean, 16 August 1793, *I&P*, vol.4, document 78.

74 Aiton to Banks, 24 August 1793, *I&P*, vol.4, document 82 and Banks to Aiton, 25 August 1793, *I&P*, document 83.

75 Banks to Aiton, 25 August 1793, *I&P*, document 83.

76 Banks to Williams, 25 August 1793, SL, Sir Joseph Banks Collection, A 2:83.

See also Williams to Banks, 30 August 1793, SL, Sir Joseph Banks Collection, A 2:45.

77 Afzelius to Banks, 28 October 1793, SL, Sir Joseph Banks Collection, A 2:43.

78 Afzelius to Banks, 11 May 1794, NHM, DTC, vol.IX, p.48. Afzelius spent almost two months sailing and staying in southern England, beginning in Portsmouth and finally departing from Plymouth – see Afzelius to James Edward Smith, 13 February 1794, LS, Correspondence of James Edward Smith, COR/1/29_1. Afzelius sailed in company with the West Indian Fleet (Britain was now at war with France).

79 Banks to Williams, 24 April 1794, SL, Sir Joseph Banks Collection, A 2:78.

80 Williams to Banks, 10 May 1794, SL, Sir Joseph Banks Collection, A 2:77. The *Harpy* was 380 tons, twice the size of the next biggest Company ship, and had a complement of 20 guns – see Misevich, 'The Sierra Leone Hinterland' and Wilson, *John Clarkson*, p.101. The *Harpy* arrived in Portsmouth on 26 May 1794.

81 Banks to Williams, 24 July 1794, SL, Sir Joseph Banks Collection, A 2:84. The correspondence detailing the arrangements are in SL, Sir Joseph Banks Collection, A 2:74,79,80,81 and 82.

82 This and the following account is based on the description found in *Substance of the Report of the Conditions of the Sierra Leone Company* (London, 1795).

83 Fyfe, *A History*, p.59, asserts that this was a renegade outfit, most likely an offshoot of the recently overthrown Reign of Terror in France.

84 Afzelius reported on the state of the garden to the Governor and Council of Sierra Leone comparing it before and after the French attack – see BL, Add MS 12131, pp.181–8.

85 Afzelius to Banks, 13 November 1794, NHM, DTC, vol.IX, p.118.

86 Once the colony began to recover from the attack, the Directors of the Company in London contacted Banks in April 1795 to see if the plant transfer might be tried again. There is no evidence that it went any further than a few exchanges of letters – these, between Banks and Williams, are in SL, Sir Joseph Banks Collection, A 3:15 and 16.

87 Afzelius to Banks, 13 November 1794, NHM, DTC, vol.IX, p.119.

88 Ibid., pp.120–1.

89 Ibid., p.123.

90 Banks to Afzelius, 17 February 1795, SLNSW, Sir Joseph Banks Papers, Series 73.047.

91 For a discussion of Afzelius's collecting activities and practice during this period, see Hodacs, 'Circulating Knowledge' and Hodacs, 'Linnaean Scholars'.

92 Afzelius to Banks, 13 July 1795, RBGA, JBK/1/6, fol.124.

93 This is clear from the entries in his journal from 17 April 1795 to 17 May 1796 (apart from a missing gap from August 1795 to end December 1795), transcribed in Kup, *Adam Afzelius*.

94 Afzelius to Banks, 4 February 1796, RBGA, JBK/1/6, fol.134 and Afzelius to Banks, 22 April 1796, RBGA, JBK/1/6/, fol.139.

95 Lindroth, 'Adam Afzelius', p.207.
96 'Passage from Sierra Leone to London', in Hansen, *The Linnaeus Apostles*, vol.4, pp.628–9.
97 RS, EC/1798/04.
98 Afzelius died in 1837, aged eighty-six.
99 Afzelius to Banks, 13 December 1799, *SC*, document 1523.

12 1792: Macartney, Staunton and the China Embassy

1 For the history of diplomatic missions to China, see Cranmer-Byng and Wills, 'Trade', Hevia, *Cherishing* and Bickers, *Ritual*. For the Russian situation, see Fletcher, 'Sino–Russian', Mosca, 'The Qing State' and Perdue, 'Boundaries and Trade'. The Dutch tried hard to convince the Emperor to allow them to have an embassy but failed – see Blussé, 'Peeking' and Kops, 'Not Such'.
2 See Carroll, 'The Canton System'.
3 Dundas had been influenced by George Smith, a private merchant operating out of Madras and who knew conditions in Canton and the reports he had written in 1783 and 1784 – see Hanser, 'Mr. Smith', pp.242–8 and BL, IOR, H/434. See also pp.117–20 for other influences on Dundas's attitude towards China. Dundas alerted Earl Cornwallis, Governor-General of India, that a diplomatic mission had the King's approval – see Dundas to Cornwallis, 21 July 1787, in Ross, *Correspondence*, p.315.
4 Dundas, in the letter quoted above, had already told Cornwallis that Cathcart was the most suitable candidate to carry out the mission.
5 Furber, *Henry Dundas*, analyses Dundas's rise to power in government. See also Fry, *Dundas*.
6 The documentation for the Cathcart mission is in BL, IOR, G/12/90. Unless otherwise noted all quotes are from this series of documents. See Pritchard, *The Crucial Years*, ch.6, for a discussion of this material.
7 See Morse, *The Chronicles*, vol.2, pp.170–1 and Stewart, 'His Majesty's Subjects'.
8 Strachan to Jervis, 21 March 1788, BL, IOR, G/12/90, p.187.
9 Cathcart to Strachan, 28 May 1788, BL, IOR, G/12/90, p.192.
10 Strachan's letter to Lord Sydney, announcing the death of Cathcart, is dated 8 October 1788 in Plymouth Sound – BL, IOR, G/12/90, p.190. Galbert died on 17 July 1788 – Strachan entered the notice in his log, a copy of which was sent to Dundas – BL, IOR, G/12/90, p.226. The original is in TNA, ADM 51/1031.
11 Charles Greville to Nathaniel Smith, 12 October 1788, BL, IOR, G/12/20, pp.582–93.
12 For this episode in the progress of the Chinese diplomatic mission, see Pritchard, *The Crucial Years*, pp.264–8.
13 There is no sign of the letter of 11 January 1791. Those dated 28 February

1791 (which mentions the January letter) and 23 December 1791 are in CUKL, MSS. DS M117, vol.2, docs.15 and 24 respectively.

14 I have used Roebuck, *Macartney*, Robbins, *Our First Ambassador*, Bickers, *Ritual* and the entry in the *Oxford Dictionary of National Biography* to sketch out this period of Macartney's life and career.

15 Quoted in Robbins, *Our First Ambassador*, pp.167–8. Macartney's wife, Lady Jane, was the daughter of John Stuart, the 3rd Earl of Bute.

16 The Literary Club is discussed in Brewer, *The Pleasures*, pp.46–51.

17 Macartney to Dundas, 4 January 1792, CUKL, MSS. DS M117, vol.3, doc.28. Macartney's letters to those involved in the preparations for the embassy, apart from Banks, are in several places: those letters actually received by the government and the East India Company are in BL, IOR, G/12/91–3; Macartney's letter books, containing copies of both incoming and outgoing correspondence, are in CUKL, MSS. DS M117; and another letter book, covering the period up to the sailing of the ships but less extensive in its range of correspondence, is in TB, Macartney Papers, MS-42.

18 Macartney to Dundas, 4 January 1792, CUKL, MSS. DS M117, vol.3, doc.28.

19 Ibid.

20 Ibid.

21 Cathcart was Sir William Hamilton's nephew. Hamilton, who was British Ambassador to Naples, an antiquarian and amateur volcanologist, is best remembered as his wife Emma's husband. He was an old friend of Banks's. Fabricius had spent a lot of time at Soho Square.

22 Banks to Macartney, 22 January 1792, SL, Sir Joseph Banks Collection, C 1:95, also reprinted in *I&P*, vol.3, document 231.

23 A copy of Macartney's plan for the embassy, in tableau form, can be found in BL, IOR, G/12/91.

24 SL, Sir Joseph Banks Collection, C 1:95, also reprinted in *I&P*, vol.3, document 231.

25 See Zeng, 'Scientific Aspects', pp.169–71. I would like to thank John Moffett who helped me confirm the title of this Chinese encyclopaedia.

26 Macartney to Dundas, 28 January 1792, CUKL, MSS. DS M117, vol.3, doc.57.

27 Apart from the entry in the *Oxford Dictionary of National Biography*, I have also benefitted from the details in Eastberg, 'West Meets East', pp.29–45 for this and subsequent biographical information.

28 Staunton, *Memoir*, pp.296–7.

29 Macartney to Staunton, December 1790, in Staunton, *Memoir*, p.330.

30 Staunton was formally admitted as a Fellow on 19 May 1789. I thank Elaine Charwat of the Linnean Society for this information.

31 Banks to Staunton, 19 June 1790, in Staunton, *Memoir*, pp.329–30. Staunton knew Banks well enough to call in to see him – Staunton to Banks, 4 July 1790, SL, Sir Joseph Banks Collection, Ag 3: 79.

32 See Staunton, *Memoir*, p.333. It may be that the business concerned the

possibility of Staunton being appointed to be President of the Select Committee of Supercargoes in Canton or a resident Consul; or preparations for the embassy – both matters were being discussed at the time and may have not been mutually exclusive (see William Richardson to Macartney, 14 December 1791, CUKL, MSS. DS M117, vol.2, doc.20 and William Richardson to Macartney, 20 December 1791, CUKL, MSS. DS M117, vol.2, doc.21; and Macartney to Staunton, 18 December 1791, in Staunton, *Memoir*, pp.337–8).

33 Staunton to Banks, 26 January 1792, *I&P*, vol.3, document 234.

34 George Thomas kept a diary of his time in Paris. The original, 'Journal of a Voyage to China' is in DURL, Staunton Papers, but it has been micro-filmed for the series *China Through Western Eyes*, Adam Matthew Publications.

35 It was probably true that no one in Britain knew Chinese, though Macartney's sneering remarks to Francis Baring, the Chairman of the East India Company, that of the supercargoes in Canton, 'not one of them has had the industry or Curiosity to acquire the Language', is not entirely accurate. Possibly some of the supercargoes in Canton in 1792 matched Macartney's description, but at least two, Thomas Bevan (who worked with John Bradby Blake on the Chinese botanical drawings) and James Flint, knew the language well – for Bevan and Flint, see Stifler, 'The Language'; see also King, 'Les premiers'. There were Chinese sailors living in the Stepney area of London and there were Chinese sailors working on East India Company ships, but these were not the right sort of people to act as interpreters and, in any case, they probably did not know Mandarin – see *Morning Chronicle*, 27 July 1782, for a report on Chinese men in Stepney. I would like to thank Aaron Jaffer for references to Chinese sailors on East India Company ships – an example from near this period is the *Rose* and the *Walmer Castle*, whose logbooks are respectively in BL, IOR, L/MAR/B/59D-F and BL, IOR, L/MAR/B/181A. See also Hellman, *Navigating* on the subject of knowing the Chinese language.

36 Macartney to Dundas, 4 January 1792, CUKL, MSS. DS M117, vol.3, doc.28.

37 Ibid.

38 Macartney to Dundas, 7 January 1792, CUKL, MSS. DS M117, vol.3, doc.29.

39 Dundas to Macartney, 8 January 1792, CUKL, MSS. DS M117, vol.3, doc.30.

40 Macartney to Dundas, 4 January 1792, CUKL, MSS. DS M117, vol.3, doc.28.

41 Macartney to Staunton, 21 January 1792, in Staunton, *Memoir*, p.340.

42 For a brief history of the Lazarists in France I found Rybolt, 'Saints' useful. For the presence and work of French missionary societies in China, see Standaert, *Handbook*, vol.1 and Tiedemann, *Reference Guide*.

43 Macartney to Staunton, 21 January 1792, in Staunton, *Memoir*, p.340. For the long friendship between Dutens and Macartney, see Dutens, *Memoirs*, vol.IV, p.229.

44 For further information on the Collegium Sinicum, see the following works: Rivinius, *Das Collegium*, Fatica, *Matteo Ripa*, Fatica and D'Arelli, *La Missione*, and Menegon, 'Wanted'.

45 For Hamilton's life see Constantine, *Fields*.

46 The Society of Dilettanti, established in 1734, was London's premier club promoting connoisseurship. It had a roster of very famous and influential people. Banks joined it in 1774 and Hamilton in 1777. See Brewer, *The Pleasures*, pp.256–60 and Redford, *Dilettanti*.

47 Constantine, *Fields*, pp.181–3.

48 For more information on Gaetano d'Ancora, see Ottaviani, 'Gaetano d'Ancora'.

49 Macartney had also asked Henry Dundas to write a letter introducing Staunton to Hamilton, which he did – see Macartney to Dundas, 28 January 1792, CUKL, MSS. DS M117, vol.3, doc.56. Staunton had this letter with him as he made his way to Naples – Staunton to Macartney, 11 August 1792, BL, IOR, Neg 4387.

50 Hamilton to Staunton, 21 February, 1792, in *China Through Western Eyes*, Part 2, Reel 27.

51 Staunton to Banks, 5 March 1792, SL, Sir Joseph Banks Collection, C 1:99.

52 Hamilton to Macartney, 3 April 1792, CUKL, MSS. DS M117, vol.4, doc.107. See also Harrison, 'A Faithful'.

53 Staunton to Macartney, 11 August 1792, BL, IOR, Neg 4387.

54 Fatica, 'Gli alunni del Collegium Sinicum di Napoli', p.542.

55 Banks to Staunton, 17 April 1792, *I&P*, vol.3, document 263.

56 Ibid.

57 Macartney to Staunton, 21 January 1792, in Staunton, *Memoir*, p.340. Gothenburg and Copenhagen were the headquarters, respectively, of the Swedish and Danish East India Companies.

58 The 'Tableau' is in BL, IOR, G/12/91, p.35.

59 Macartney to Hawkesbury, 23 June 1792, CUKL, MSS. DS M117, vol.4, doc.141.

60 Letters to and from Banks and Thomas Henry, Thomas Percival and Josiah Wedgwood are in *I&P*, vol.3; those to and from Macartney and Matthew Boulton and Samuel Garbett are in CUKL, MSS. DS M117.

61 Macartney to Boulton, 9 March 1792, BAH, MS 3782/12/93, doc.72.

62 Lightman, McOuat and Stewart, *The Circulation*, pp.3–4.

63 See the discussion in Berg, 'Macartney's Things'.

64 Staunton, *Memoirs*, p.11.

65 George Thomas Staunton, 'Journal to China', DURL, Staunton Papers.

66 Ibid.

13 1793: The Accidental Naturalist in Qianlong's Empire

1 Dundas to Gower, 8 September 1792, CUKL, MSS. DS M117, vol.5, doc.226.

2 Ibid.

3 For the second visit to China, see Lind to Banks, 23 October 1779, *SC*, document 169.

4   For a fuller description, see Chapter 4.

5   A comprehensive discussion of the key scientific and technical personnel of the embassy and Banks's participation in the choices can be found in Zeng, 'Scientific Aspects'.

6   Banks to Macartney, 29 January 1792, *I&P*, vol.3, document 235. The experiments Banks referred to included work with rockets, with signals and experiments on frogs, in conjunction with the Italian physicist Tiberius Cavallo, to determine the nature and extent of 'animal electricity'. Lind's correspondence with Banks, preserved in the Perceval Bequest at the Fitzwilliam Museum, Cambridge, shows the breadth of Lind's activities – many of these letters have been transcribed in *SC*. There is a separate correspondence between Lind and Cavallo in the BL, Add MS 22897–22898.

7   Banks to Macartney, 29 January 1792, *I&P*, vol.3, document 235.

8   Macartney to Banks, 29 January 1792, *I&P*, vol.3, document 236.

9   Macartney to Banks, 17 March 1792, *I&P*, vol.3, document 252.

10   Banks to Lind, 19 March 1792, *I&P*, vol.3, document 254.

11   Ibid.

12   Banks to Macartney, 21 March 1792, *I&P*, vol.3, document 255.

13   Banks to Macartney, 21 May 1792, *I&P*, vol.3, document 272.

14   Staunton to Banks, 23 October 1792, *I&P*, vol.3, document 319.

15   Staunton to Banks, 26 January 1792, *I&P*, vol.3, document 234.

16   Hansen, *The Linnaeus Apostles*, vol.8, p.41. For the relationship between Thunberg and Afzelius, see Skuncke, *Carl Peter Thunberg*. I would like to thank Lisa Hellman for explaining what the term 'oriental languages' meant then at Uppsala University.

17   Staunton to Banks, 9 May 1792, *I&P*, vol.3, document 268.

18   Banks to Staunton, 17 April 1792, *I&P*, vol.3, document 263.

19   Staunton to Banks, 9 May 1792, *I&P*, vol.3, document 268.

20   Banks to Swartz, 17 August 1792, *SC*, document 1130.

21   Stronach was recommended to Banks by Archibald Thomson, owner of the Mile End Nursery – see Thomson to Banks, 9 June 1792, *I&P*, vol.3, document 278.

22   Banks to Swartz, 17 August 1792, *SC*, document 1130.

23   Batteux et al., *Mémoires concernant l'histoire*.

24   Banks, 'Hints on the Subject of Gardening Suggested to the Gentlemen who attend the Embassy to China', *I&P*, vol.3, document 304. The original is in LS, MS 115. Banks reminded Staunton that it was only because he and Solander came equipped with a library on the *Endeavour* that they were able to collect and make botanical analyses on the move. Staunton did not have this luxury.

25   Banks, 'Hints', *I&P*, vol.3, document 304.

26   All of the quotes are from Banks, 'Hints'.

27   Banks selected about one-quarter of Kaempfer's drawings, which were in

the Sloane collection at the British Museum, to make the plates for the *Icones* – see Henrey, 'Kaempfer's'.

28 On Thunberg's time in Japan, see Screech, *Japan Extolled* and Skuncke, *Carl Peter Thunberg*.

29 Banks suggested that the 'gentlemen' consider 'the propriety of Erecting on the Quarter deck Either of the Lion or the Indiamen a hutch . . . to be able to transport with security the valuable acquisitions . . .' – for this, Banks provided Staunton with a plan of how this could be done. See Banks, 'Hints', *I&P*, vol.3, document 304.

30 Macartney to Banks, 16 October 1792, *I&P*, vol.3, document 317.

31 Ibid.

32 Staunton to Banks, 14 October 1792, *I&P*, vol.3, document 315.

33 Macartney to Banks, 16 October 1792, *I&P*, vol.3, document 317.

34 Gower, 'A Journal', BL, Add MS 21106.

35 Ibid. See also Macartney to Baring, 10 December 1792, TB, MS-40.

36 Macartney to Baring, 10 December 1792, TB, MS-40 and Macartney to Dundas, 10 December 1792, TB, MS-41.

37 Staunton to Banks, 14 October 1792, *I&P*, vol.3, document 315.

38 Phillip to Banks, 31 August 1787, *I&P*, vol.2, document 155.

39 Banks to Staunton, 18 August 1792, *I&P*, vol.3, document 302.

40 Ibid.

41 Staunton to Banks, 9 December 1792, *I&P*, vol.4, document 3.

42 Gillan to Banks, 11 December 1792, *I&P*, vol.4, document 4.

43 Enderby to Banks, 16 February 1793, *I&P*, vol.4, document 27.

44 Banks to Staunton, 24 February 1793, *I&P*, vol.4, document 30.

45 Ibid.

46 The drawings are in SL, Sir Joseph Banks Collection, EN 1:19, 20.

47 Carter, *Sir Joseph Banks*, pp.337–42. Banks recommended Alexander to the embassy – see Macartney to Banks, 25 June 1792, *I&P*, vol.3, document 283.

48 Banks to Gillan, c.30 April 1793, *I&P*, vol.4, document 51.

49 See Pritchard, *The Crucial Years*, pp.284–90 for the details.

50 Pritchard, *The Crucial Years*, p.315.

51 Ibid., p.316.

52 The packet with its enclosures had been left with the Dutch authorities in the middle of February by the *Wycombe*, an East India Company ship on its homeward voyage.

53 Pritchard, *The Crucial Years*, p.320.

54 See Blussé, 'Batavia', Blussé, *Visible Cities*, Tagliacozzo and Chang, *Chinese Circulations* and Lockard, 'Chinese Migration'.

55 Cranmer-Byng, *An Embassy*, pp.61–2. This text is a printed version of Macartney's unpublished journal. I have decided to use it in conjunction with the official narrative, written by Staunton, and published in 1797 – for more on Macartney's journal, see Clingham, 'Cultural Difference'.

56 This time of the year was outside the trading season, so no East India Company personnel were allowed to be in Canton.

57 The *Hindostan* was also carrying two Chinese missionaries from Naples and when the ship got near to Macao, they jumped into the sea and swam off – see BL, IOR, L/MAR/B/267G(A).

58 See Staunton's discussion of the charts, such as they were, and other texts that Gower relied on to get northwards along the Chinese coast – Staunton, *An Authentic Account*, pp.401–5.

59 Pritchard, *The Crucial Years*, p.324.

60 Staunton, *An Authentic Account*, pp.409–10.

61 Cranmer-Byng, *An Embassy*, p.70.

62 Ibid., p.71.

63 A short biography of both men is given in Cranmer-Byng, *An Embassy*, pp.326–31.

64 Cranmer-Byng, *An Embassy*, p.71.

65 Ibid. For an account of the five musicians and the music they played, see Lindorff, 'Burney'.

66 Yuanmingyuan was famously destroyed and looted by British and French troops in 1860 during the Second Opium War – see Hill, 'Collecting' for the latest discussion of this event.

67 For all of the above, the plans and the eventual timing, I have relied on Cranmer-Byng, *An Embassy*, pp.72–95.

68 Cranmer-Byng, *An Embassy*, p.95.

69 Attiret, *A Particular Account*. See also Kilpatrick, *Gifts*, pp.30, 113.

70 More on Yuanmingyuan can be found in Durand and Thiriez, 'Engraving'.

71 For the Qing hunts at Mulan, see Elliott and Chia, 'The Qing'.

72 This description of Jehol is based on the following sources: Forêt, 'The Intended', Forêt, *Mapping* and Olivová, 'A Map'.

73 Cranmer-Byng, *An Embassy*, p.106.

74 Ibid., p.112.

75 Ibid., p.113

76 Ibid., pp.117–18.

77 Ibid., p.122.

78 Staunton, *An Authentic Account*, vol.2, p.225.

79 Ibid., p.229.

80 Cranmer-Byng, *An Embassy*, p.122.

81 Staunton, *An Authentic Account*, vol.2, p.232. A transcript of Geroge III's letter to the Qianlong Emperor can be found in Morse, *The Chronicles*, vol.2, pp.244–7.

82 George Thomas Staunton, 'Journal of a Voyage to China', DURL, Staunton Papers, p.105.

83 Ibid.

84 Ibid., p.108.

85 An excellent discussion of Wanshu yuan can be found in Forêt, *Mapping*, and in Yu, 'Imperial Banquets'.

86 Cranmer-Byng, *An Embassy*, p.125.

87 Ibid.

88 Sing-songs was the name given to automaton clocks made in Europe and sold to the Chinese. As the manufactured object moved – sometimes a carriage, sometimes an animal, such as an elephant – the contraption played a tune. See the following sources for more on the sing-song trade: Needham, Ling and De Solla Price, *Heavenly*, Smith, 'The Sing-song Trade' and White, *English Clocks*.

89 Cranmer-Byng, *An Embassy*, p.125. The inner quote is to Milton's *Paradise Lost*, Book IV, p.35.

90 Cranmer-Byng, *An Embassy*, pp.132–3.

91 See Harrison, 'The Qianlong Emperor's Letter' for a discussion of this critical document.

92 Cranmer-Byng, *An Embassy*, p.340.

93 Ibid., pp.148–9.

94 Both Macartney, in his journal – Cranmer-Byng, *An Embassy* – and Pritchard, *The Crucial Years*, discuss these events. Gower's own version is in Gower, 'Journal of HMS LION 1 Oct 1792–17 Jul 1794 . . .', NMM, GOW/3 (there is another copy in BL, Add MS 21106) and in Gower to Admiral John Elliot, 24 October 1793, NMM, ELL/400.

95 For details of how the Grand Council, the body on which Sung-yun sat, worked, see Bartlett, *Monarchs*.

96 Cranmer-Byng, *An Embassy*, pp.175–6.

97 Gower to Admiral John Elliot, 24 October 1793, NMM, ELL/400. Gower mentioned his lack of these two medicines in his letter to Macartney.

98 Cranmer-Byng, *An Embassy*, p.178.

99 See, for example, Pritchard, 'The Instructions', Parts II and III. Also Macartney to the East India Company, TB, MS-40 and Macartney to Dundas, TB, MS-41.

100 The list of contributions to the debate is long indeed. Those interested in following the various interpretations would do well to begin with the most recent overview and work through the notes to the other sources. Kitson, *Forging* is the most recent book that treats the Macartney embassy in a large context of Sino-British cultural relations. Bickers, *Ritual* is a collection of essays on the embassy including contributions from Chinese scholars. Peyrefitte, *The Immobile* is the most comprehensive guide to the daily occurrences of the embassy, and is based on a wide range of primary sources, both western and Chinese – many of the Chinese sources Peyrefitte used have been translated into French and published (see Peyrefitte, *Un Choc*); Singer, *The Lion* is a fast-moving narrative but relies on fewer sources than Peyrefitte. See Williams, 'British Government' for the British reaction to Macartney's failure. As for the kowtow and gift-giving, see

Hevia, *Cherishing* and the literature that that spawned – I would like to thank Henrietta Harrison for sharing an unpublished article of hers that examines gift-giving on both sides; Harrison, 'Chinese'. On the Chinese views of British expansionism, nothing can yet compare to Mosca, *From Frontier*. That book and Mosca's articles, 'The Qing' and 'Hindustan' are certain to keep the controversy over Macartney going for some time to come.

101 All of these factors are covered in the literature cited above. It was Charles Blagden, Banks's trusted colleague and friend, who raised the issue of the Portuguese missionaries' attempts to undermine Macartney's efforts: well before the embassy set off from Portsmouth, Macartney had raised such concerns when he discussed the need for independent interpreters. See Blagden to Banks, 6 October 1794, *I&P*, vol.4, document 148.

102 'Sketch of the Expense of the Chinese Embassy upon the largest Scale', CUKL, MSS. DS M117, vol.10, document 437.

103 Banks, 'Hints', *I&P*, vol.3, document 304.

104 Banks to Macartney, 22 January 1792, *I&P*, vol.3, document 231.

105 Cranmer-Byng, *An Embassy*, p.126 and George Thomas Staunton, 'Journal of a Voyage to China', DURL, Staunton Papers, p.107.

106 Haxton's Journal appears to have been lost. This quote is from an extended extract copied by Banks from the original. The extract is in SL, Sir Joseph Banks Collection, C 1:11 and has been transcribed in *I&P*, vol.4, document 323. This quote has also been transcribed in Kilpatrick, *Gifts*, p.115.

107 Staunton, *An Authentic Account*. Almost a century later, the physician and botanist Emil Bretschneider went through the list very carefully and presented his own assessment, which was mixed. See Bretschneider, *History*, pp.156–83.

108 Staunton to Banks, 12 November 1793, *I&P*, vol.4, document 95. This letter, written in Hangzhou, would have been given to Captain William Mackintosh who was returning to his ship, the *Hindostan*, then anchored in nearby Chusan. Staunton had already emphasised to Banks that he was not a botanist, that he couldn't 'distinguish what is rare or worth sending, and what is not so': he added, as if Banks did not know, 'we every day regret that Mr. Afzelius is not with us' – see Staunton to Banks, 30 March 1793, *I&P*, vol.4, document 44.

109 Banks's admonitions to Staunton were made in writing as the former and Dryander worked through the collections. These letters were written over a period lasting six months and can be found in *I&P*, vol.4, documents 179, 181, 182 and 222.

110 Staunton to Banks, 24 January 1796, *I&P*, vol.4, document 234.

111 The instructions are in CUKL, MSS. DS M117, vol.10, document 429.

112 Staunton to Banks, 25 December 1793, *I&P*, vol.4, document 106; Gillan to Macartney, 16 September 1794 and Gillan to Macartney, 20 September 1794, in PRONI, Macartney Papers, D/572/19/173 and 175 respectively.

113 The Chinese plants are listed, as they are in the *Hortus Kewensis*, in Bretschneider, *History*, pp.191–200.

114 Banks to Staunton, 23 January 1796, *I&P*, vol.4, document 232.

14 1794: To Calcutta and Back

1 'Public Letter from Bengal', 3 September 1792, ASB, Banks MSS, letter 10 (microfilm copy deposited in Banks Archive Project, Nottingham Trent University). Bengal Public Consultations, 23 October 1793, NHM, DTC, VI, p.154.

2 This correspondence can be found in ASB, Banks MSS.

3 Banks to Ramsay, 14 May 1793, *I&P*, vol.4, document 56. For William Pitcairn's Election as a Fellow of the Royal Society, see RS, EC/1770/08/. William Pitcairn was the uncle of Robert Pitcairn, midshipman on Philip Carteret's *Swallow*, whose sighting of the island on 3 July 1767, led to it being named in his honour.

4 Smith, 'Plants on Board H.M.S. Providence', SLNSW, Sir Joseph Banks Papers, Series 52.16. See also Smith to Banks, 1 August 1793, *I&P*, vol.4, document 69. The figure of '5178' must be an error but it is a faithful transcription from the original, which is in SLNSW, Sir Joseph Banks Papers, Series 52.15.

5 Dancer to Banks, 10 January 1793, *I&P*, vol.4, document 17; Dancer to Banks, 3 April 1793, *I&P*, vol.4, document 45; Smith to Banks, 1 August 1793, *I&P*, vol.4, document 69; Aiton to Banks, 10 August 1793, *I&P*, vol.4, document 75; and RBGA, Record Book 1793–1809, pp.10–16.

6 Smith, 'Plants on Board H.M.S. Providence', SLNSW, Series 52.16. Many of the plants, including the breadfruit trees, were still prospering at Kew almost twenty years later – see Powell, 'The Voyage', pp.428–31.

7 Banks to Ramsay, 14 May 1793, *I&P*, vol.4, document 56.

8 Banks to Nepean, 15 December 1793, *I&P*, vol.4, document 104.

9 With thanks to Lynda Brooks of the Linnean Society for this information.

10 The Court of Directors informed the Governor-General of Bengal of Smith's appointment in a letter date 19 February 1794. See Court of Directors to Bengal, 2 July 1794, BL, IOR, E/4/641.

11 For the Calcutta Botanic Garden, see Thomas, 'The Establishment'.

12 Roxburgh to Banks, 1 December 1793, *I&P*, vol.4, document 98.

13 See for example, Murray to Banks, 25 April 1793, *I&P*, vol.4, document 49.

14 'Plants received from Robert Kyd', June 1790, *I&P*, vol.3, document 93.

15 Aiton to Banks, 29 May 1794, SLNSW, Sir Joseph Banks Papers, Series 20.04.

16 Kyd to Banks, 21 November 1791, *I&P*, vol.3, document 213.

17 Kyd to Court of Directors, 13 August 1787, *I&P*, vol.2, document 258 (p.365).

18 Kyd to Court of Directors, 13 June 1787, NHM, DTC, VII, pp.50–1 and Kyd to Court of Directors, 13 August 1787, NHM, DTC, VII, pp.51–3.

19 Banks to Devaynes, 3 June 1794, SLNSW, Sir Joseph Banks Papers, Series 17.01.

20 Devaynes to Banks, 3 June 1794, SLNSW, Sir Joseph Banks Papers, Series 17.02.

21 Banks to Good, 5 June 1794, SLNSW, Sir Joseph Banks Papers, Series 17.03.

22 Ibid.

23 Kyd to Banks, 21 November 1791, *I&P*, vol.3, document 213. See also Kyd to Court of Directors, 13 August 1787, *I&P*, vol.2, document 258 (p.365).

24 'Catalogue of Plants on the Company's Garden Jan 11 1790', NHM, DTC, VII, pp.65–7.

25 The bill for these purchases came to just over £50. See Banks, 'Account of Disbursements', 24 and 29 July 1794, SLNSW, Sir Joseph Banks Papers, Series 17.06 and Smith to Banks, September 1795, SLNSW, Sir Joseph Banks Papers, Series 16.11.

26 Banks to Devaynes, 6 June 1794, SLNSW, Sir Joseph Banks Papers, Series 17.04.

27 The *Royal Admiral* had left England on 30 May 1792 on a run as a convict ship to Port Jackson. Almost 350 convicts, 300 of whom were men, sailed on the ship to Port Jackson, which was reached on 7 October. After a short stay in the colony, the *Royal Admiral* continued its voyage to Canton, where the ship dropped anchor on 14 January 1793.

28 Parkins to Banks, 20 August 1793, *I&P*, vol.4, document 81. Staunton's letter to Banks, 30 March 1793, is in *I&P*, vol.4, document 44.

29 Banks to Bond, 1 August 1794, *I&P*, vol.4, document 134.

30 I would like to thank Margot Finn for alerting me to the link, *http://blogs.ucl.ac.uk/eicah/bond-family-members-in-the-east-india-company/*, which provided me with some biographical information of the Bond family.

31 Banks to Bond, 1 August 1794, *I&P*, vol.4, document 134.

32 Ibid.

33 Bond to Banks, 2 August 1794, *I&P*, vol.4, document 135.

34 Smith to Banks, 5 September 1794, *I&P*, vol.4, document 139.

35 Smith to Banks, 12 May 1795, *I&P*, vol.4, document 184. The plant list of September 1795 is in SLNSW, Sir Joseph Banks Papers, Series 16.11.

36 Smith to Banks, 12 May 1795, *I&P*, vol.4, document 184.

37 Roxburgh to Banks, 25 April 1795, *I&P*, vol.4, document 177.

38 Bond to Sir John Shore, 23 April 1795, BL, IOR, P/4/34, p.519.

39 Smith to Banks, 12 May 1795, *I&P*, vol.4, document 184 and SLNSW, Sir Joseph Banks Papers, Series 16.11. For Gordon's long interest in natural history and his explorations, over many years, in the countryside beyond Cape Town, see Cullinan, *Robert Jacob Gordon*.

40 SLNSW, Sir Joseph Banks Papers, Series 16.13.

41 SLNSW, Sir Joseph Banks Papers, Series 16.14.

42 Andrew Berry, the nephew of Dr James Anderson, tended the nopalry begun by his uncle – see Chapter 6 for more details.

43 SLNSW, Sir Joseph Banks Papers, Series 16.12. Jodrell, who was personally recommended by King George III, took up his appointment in 1787.

44 SLNSW, Sir Joseph Banks Papers, Series 16.12.

45 Sir Stephen Lushington to Banks, 28 January 1796, *I&P*, vol.4, document 235.

46 Banks to Customs House, undated, SLNSW, Sir Joseph Banks Papers, Series 73.046.

47 Banks to Roxburgh, 29 May 1796, *I&P*, vol.4, document 252. See also Banks to Customs House, undated, SLNSW, Sir Joseph Banks Papers, Series 73.046 and Banks to Smith, 26 May 1796, *I&P*, vol.4, document 251. Another plant list appears in RBGA, Record Book 1793–1809, pp.157–60.

48 It is unclear whether Smith died in Penang or whether he died en route to England from there – see Langdon, *Penang*, Book Five. I would like to thank Marcus Langdon for kindly sharing this part of his book with me.

15 1795: The Farrier's Son Finds Banks

1 Phillip to Banks, 2 July 1788, *I&P*, vol.2, document 225.

2 Phillip to Banks, 26 September 1788, *I&P*, vol.2, document 246.

3 According to Aiton's *Hortus Kewensis* the nurserymen Lee and Kennedy of Hammersmith introduced several Australian plants into Kew. See Easterby-Smith, *Cultivating*, for a discussion of London's horticultural market.

4 Bligh to Banks, 20 June 1788, *I&P*, vol.2, document 220.

5 Banks to Masson, 3 June 1789, SLNSW, Sir Joseph Banks Papers, Series 13.40.

6 Masson to Banks, 26 March 1789, SLNSW, Sir Joseph Banks Papers, Series 13.39.

7 Banks to Nepean, 7 May 1790, *I&P*, vol.3, document 91 and Burton to Banks, 26 January 1791, *I&P*, vol.3, document 125.

8 Phillip to Banks, 20 May 1792, *I&P*, vol.3, document 271.

9 Phillip to Banks, 15 October 1792, *I&P*, vol.3, document 316. See also Frost, *Arthur Phillip*, pp.217–18.

10 The *Atlantic* was one of the convict transports of the Third Fleet.

11 This report appeared in *Lloyd's Evening Post*, 31 May 1793. On the two Aborigines visitors, see Fullager, *The Savage Visit*. For more on Bennelong, see Smith, 'Bennelong' and Fullager, 'Bennelong'. Phillip had mentioned Bennelong to Banks in an earlier communication saying then that he was hoping that the Aborigine would accompany him to England – Phillip to Banks, 3 December 1791, *I&P*, vol.3, document 218.

12 *Gazetteer and New Daily Advertiser*, 12 July 1793.

13 The few details available of Bennelong and Yemmerawanne's time in and around London can be found in Brook, 'The Forlorn Hope'.

14 Banks to Dundas, 30 April 1794, *I&P*, vol.4, document 125.

15 Banks to Masson, 3 June 1789, SLNSW, Sir Joseph Banks Papers, Series 13.40.
16 Secord, 'Science'.
17 Webb, *George Caley*, p.2.
18 These letters, both dated 7 March 1795, are in *SC*, documents 1300 and 1301 respectively.
19 Uglow, *The Lunar Men*, pp.379–81.
20 Caley to Banks, 7 March 1795, *SC*, document 1300.
21 Banks to Caley, 7 March 1795, *SC*, document 1301.
22 Caley to Banks, 5 January 1798, *I&P*, vol.4, document 310.
23 Banks to Hunter, 30 March 1797, *I&P*, vol.4, document 287.
24 Caley's letter to Banks was dated 28 November 1798 but no copy of it has survived. Caley's recollection of the subsequent meeting is contained in Caley to Banks, 23 August 1798, *I&P*, vol.4, document 335.
25 Caley to Banks, 23 August 1798, *I&P*, vol.4, document 335.
26 Banks to Caley, 7 January 1798, *I&P*, vol.4, document 311.
27 Caley to Banks, 28 February 1798, *I&P*, vol.4, document 315.
28 Caley to Banks, 12 July 1798, *I&P*, vol.4, document 327.
29 Banks to Caley, 16 July 1798, *I&P*, vol.4, document 330.
30 This was a point Caley made to James Dickson, the nurseryman and friend. Caley recounted that Anton Hove was looking around for someone to accompany him to India, and had he, Caley, not been expecting to go to New South Wales, he would have jumped at the chance – see Caley to Dickson, 23 July 1798, SLNSW, Sir Joseph Banks Papers, Series 18.009.
31 Caley to Banks, 22 July 1798, *I&P*, vol.4, document 331.
32 Caley to Banks, 2 September 1798, *I&P*, vol.4, document 339.
33 Banks to Caley, 27 August 1798, *I&P*, vol.4, document 337.
34 Ibid.
35 Caley to Banks, 22 July 1798, *I&P*, vol.4, document 331.
36 Banks to Caley, 12 September 1798, *I&P*, vol.4, document 345.
37 Caley to Banks, 11 November 1798, *I&P*, vol.5, document 7.
38 Philip Gidley King to Banks, 12 November 1798, *I&P*, vol.5, document 10.
39 Banks to King, 15 May 1798, *I&P*, vol.4, document 318.
40 Ibid.
41 Philip Gidley King to Banks, 9 September 1798, *I&P*, vol.5, document 342.
42 Park to Banks, 14 September 1798, *I&P*, vol.5, document 350.
43 Dickson to Banks, 20 September 1798, *I&P*, vol.5, document 349.
44 These letters are in *I&P*, vol.5, documents 350, 351, 352, 354, 355, 356 and 357.
45 The first description is in Banks to Moss, 21 September 1798, *I&P*, vol.5, document 351; and the next two are in Banks to Moss, 28 September 1798, *I&P*, vol.5, document 357.
46 For this period of Park's life, see Duffill, *Mungo Park*.

47 Banks to Caley, 16 November 1798, *I&P*, vol.5, document 14.

48 Caley to Banks, 18 November 1798, *I&P*, vol.5, document 15.

49 SL, Sir Joseph Banks Collection, SS 1:47.

50 King to Banks, 3 May 1791, *I&P*, vol.3, document 161 and Banks to King, 15 May 1798, *I&P*, vol.4, document 318.

51 Banks to King, 15 May 1798, *I&P*, vol.4, document 318 and Banks to Aufrère, 3 July 1798, *I&P*, vol.4, document 325.

52 'Note concerning George Suttor', 11 December 1797, *I&P*, vol.4, document 306 and Banks to Aufrère, 3 July 1798, *I&P*, vol.4, document 325.

53 Banks to John King, 15 May 1798, *I&P*, vol.4, document 318.

54 See Hamond to Banks, 20 June 1798, *I&P*, vol.4, document 321 and Baron Farnborough to Banks, 22 June 1798, *I&P*, vol.4, document 322.

55 Banks to Suttor, 2 August 1798, *I&P*, vol.4, document 332.

56 Banks stressed the importance of introducing hops and brewing in general months before the collections for the *Porpoise* were made – see Banks to John King, 15 May 1798, *I&P*, vol.4, document 318. For the list of plants, see Suttor to Banks, 8 September 1798, *I&P*, vol.4, document 341 and SL, Sir Joseph Banks Collection, SS 1:20. For the supply of hops, see Rashleigh to Banks, 7 September 1798, *I&P*, vol.4, document 340.

57 Philip Gidley King to Banks, 13 September 1798, *I&P*, vol.4, document 346.

58 Philip Gidley King to Banks, 26 July 1791, *I&P*, vol.3, document 181. King was pessimistic about the future of the plants' welfare because of the lack of a special cabin in which to keep them – see Philip Gidley King to Banks, 3 May 1791, *I&P*, vol.3, document 161.

59 Suttor to Banks, 11 December 1798, *I&P*, vol.5, document 24.

60 Philip Gidley King to Banks, 6 February 1799, *I&P*, vol.5, document 31.

61 Philip Gidley King to Banks, 12 March 1799, *I&P*, vol.5, document 43.

62 Philip Gidley King to Banks, 30 April 1799, *I&P*, vol.5, document 61 and Philip Gidley King to Banks, 7 May 1799, *I&P*, vol.5, document 66.

63 Philip Gidley King to Banks, 18 June 1799, *I&P*, vol.5, document 74 and Philip Gidley King to Banks, 12 July 1799, *I&P*, vol.5, document 80.

64 Philip Gidley King to Banks, 5 September 1799, *I&P*, vol.5, document 90.

65 Philip Gidley King to Banks, 17 September 1799, *I&P*, vol.5, document 94.

66 Philip Gidley King to Banks, 6 October 1799, *I&P*, vol.5, document 99 and Webb, *George Caley*, p.12.

67 Banks to Suttor, 10 August 1802, *I&P*, vol.6, document 71.

16 1800: Caley and Moowattin

1 Banks to Macquarie, 13 May 1809, *I&P*, vol.7, document 217.

2 Caley to Banks, 15 February 1800, *I&P*, vol.5, document 120.

3 King to Banks, 15 February 1800, *I&P*, vol.5, document 121.

4 Caley to Banks, 15 February 1800, *I&P*, vol.5, document 120.

5 The details about Parramatta are taken from Webb, *George Caley*, pp.25–7. For New South Wales around the time that Caley was there and before, see Karskens, *The Colony*.

6 King to Banks, 28 September 1800, *I&P*, vol.5, document 151.

7 Banks to King, 22 June 1801, *I&P*, vol.6, document 289.

8 Caley did marry Mrs Wise soon after they returned to London together in 1810 – for what is known about Mrs Wise/Caley, see Webb, *George Caley* and *I&P*, vol.6, document 156 (n.1).

9 Caley to Banks, 12 October 1800, *I&P*, vol.5, document 153.

10 Ibid.

11 Ibid.

12 Ibid.

13 Caley to Banks, 22 December 1800, *I&P*, vol.5, document 165.

14 Caley to Banks, 25 August 1801, *I&P*, vol.5, document 306.

15 Ibid.

16 Caley to Banks, 26 May 1802, *I&P*, vol.6, document 54.

17 Else-Mitchell, 'George Caley', pp.457–8.

18 Banks to Caley, 14 August 1802, *I&P*, vol.6, document 73.

19 Ibid.

20 Ibid.

21 King to Banks, 5 June 1802, *I&P*, vol.6, document 60.

22 On the kangaroo, see Ellis, 'Tails' and Ellis, '"That Singular"'

23 Caley to Banks, 25 August 1801, *I&P*, vol.5, document 306.

24 Caley to Banks, 1 June 1802, *I&P*, vol.6, document 58.

25 Caley to Banks, 18 August 1803, *I&P*, vol.6, document 163.

26 Caley to Banks, 1 June 1802, *I&P*, vol.6, document 58.

27 Webb, *George Caley*, p.32.

28 King to Banks, October 1802, *I&P*, vol.6, document 82.

29 For the early history of the white exploration of this area, which was the home of the Gundungurra people, see Cunningham, *Blue Mountains*.

30 Caley to Banks, 1 November 1802, *I&P*, vol.6, document 83. For Barrallier's Blue Mountain expeditions, see Cunningham, *Blue Mountains*.

31 Caley to Banks, 28 April 1803, *I&P*, vol.6, document 122.

32 Banks to Caley, 8 April 1803, *I&P*, vol.6, document 107 and Banks to King, 8 April 1803, *I&P*, vol.6, document 108.

33 Caley to Banks, 13 May 1803, *I&P*. vol.6, document 127.

34 Caley to Banks, 1 November 1804, *I&P*, vol.6, document 219.

35 I have recounted Caley's Blue Mountain expedition from his own account published in Andrews, *The Devil's* and with the help of Cunningham, *Blue Mountains*. Caley's original diary is in NHM.

36 Caley to Banks, 16 December 1804, *I&P*, vol.6, document 223.

37 King to Banks, 14 January 1805, *I&P*, vol.6, document 231.

38 Caley to Banks, 18 April 1805, *I&P*, vol.7, document 8.

39 King to Banks, 20 May 1805, *I&P*, vol.7, document 23.

40 Moowattin was born in Parramatta in 1791 according to Smith, *Mari Nawi*, p.118.

41 Moowattin was now living in Caley's cottage in Parramatta – Caley to King, 25 September 1807, SLNSW, Sir Joseph Banks Papers, Series 18.067.

42 Smith, *Mari Nawi*, pp.118–20, states that Moowattin arrived in Norfolk Island on board HMS *Buffalo*, while Caley went separately on the *Sydney*. Norfolk Island, located in the Pacific 1400 kilometres from Australia's east coast, was uninhabited when Arthur Phillip chose it to be the second convict colony. Lieutenant, as he was then, Philip Gidley King led the first party to the island, which they reached in early March 1788. King remained in command of the Norfolk Island colony until he was appointed Governor of New South Wales in 1800. The island's economic mainstay was the production of flax for naval supplies.

43 Caley to Banks, 6 April 1806, *I&P*, vol.7, document 68.

44 As quoted in Webb, *George Caley*, p.175.

45 Banks to Caley, 30 August 1804, *I&P*, vol.6, document 211.

46 On Baudin, see Jangoux, *Portés* and Fornasiero, Monteath and West-Sooby, *Encountering*.

47 For the early history of Tasmania, see among many sources, Reynolds, *A History*.

48 Skiddaw was the name given to the mountain by Commodore John Hayes, born in Cumberland, when he was the first British officer to explore the Derwent River, which he also named after its Cumberland namesake, in 1793.

49 Caley to Banks, 6 April 1806, *I&P*, vol.7, document 68.

50 Ibid.

51 Ibid.

52 Smith, *Mari Nawi*, p.120, states that Moowattin and Caley arrived at Hobart separately: Moowattin on 15 November and Caley on 29 November.

53 Bligh, with support from Banks, was appointed in March 1805 and arrived in Sydney in early August 1806. For a succinct discussion of Bligh's stay and rule in New South Wales, see Salmond, *Bligh*, pp.456–64.

54 The details of the geography of the region, including the naming of the waterfall and the river, are in Caley to King, 25 September 1807, SLNSW, Sir Joseph Banks Papers, Series 18.067. A discussion of this expedition can be found in Else-Mitchell, 'George Caley', pp.509–11. The waterfall is now called 'Appin Falls' and the river 'Cataract River'.

55 Caley to Banks, 25 September 1807, *I&P*, vol.7, document 156.

56 Ibid.

57 Caley to Banks, 27 September 1807, *I&P*, vol.7, document 157.

58 Caley to Banks, 14 April 1808, *I&P*, vol.7, document 183.

59 Caley to Banks, 16 February 1809, *I&P*, vol.7, document 203.

60 Banks to Caley, 25 August 1808, *I&P*, vol.2, document 189.

61 Caley to Banks, 16 February 1809, *I&P*, vol.7, document 203.

62 Ibid.

63 Caley to Banks, 3 November 1808, *I&P*, vol.7, document 197. See also Clarke, *Aboriginal Plant*, pp.58–68 for a discussion of Moowattin's role in collecting.

64 Caley to Banks, 3 November 1808, *I&P*, vol.7, document 197.

65 Banks to Robert Peel, 16 August 1811, *I&P*, vol.8, document 40.

66 Caley's diary is in UCLLSC, MS ADD 325. I have used it and the commentary by Smith, *Mari Nawi*, pp.125–32, who used the same source.

67 Banks to Robert Peel, 16 August 1811, *I&P*, vol.8, document 40.

68 Suttor to Banks, 12 November 1812, *I&P*, vol.8, document 63.

69 See Ford and Salter, 'From Pluralism', for the case, and for its wider context and significance.

70 As quoted in Webb, *George Caley*, p.131.

17 1800: Not Since the *Endeavour*

1 Paterson, who was born in 1755 in Montrose, Scotland, was the son of a gardener, who trained in horticulture. He had visited Cape Town and the interior during the period 1777–9 collecting plants for the Countess of Strathmore. He received an army commission in 1780. The account of his travels was published in 1789. Paterson arrived in Sydney in 1791. Paterson had made friends with Solander no later than 1776; undoubtedly, he must have met Banks around this time. See Moore, *Wedlock*, pp.136–7. For Paterson's period at the Cape, see Gunn and Codd, *Botanical Exploration*.

2 Banks to Robert Moss, 28 September 1798, *I&P*, vol.4, document 357.

3 Flinders to Banks, 6 September 1800, *I&P*, vol.5, document 149.

4 For this part of Flinders's life, see Estensen, *Matthew Flinders*, chs.4–10 inclusive and Morgan, *Matthew Flinders*, chs.2 and 3. These two books are the best and most comprehensive studies of Flinders.

5 See note 6 below.

6 Hunter to Banks, 1 June 1799, *I&P*, vol.5, document 72 – 'it [New South Wales] may hereafter be found to be divided from that part which retains the Name of New Holland, some Narrow Sea, which may yet be navigable'. See also Morgan, *Matthew Flinders*, pp.47–8. Gerritsen, 'Getting the Strait' disagrees that contemporaries believed in the possibility of two land masses, but see Morgan, 'An Historical', who makes a strong case for it having been taken seriously at the time.

7 Flinders to Banks, 6 September 1800, *I&P*, vol.5, document 149.

8 Wiles to Banks, 16 March 1793, *I&P*, vol.4, document 40. At the bottom of this letter Banks had written 'delivered by a Mr. Flinders who is now on board the Bellerophon C. Pasley'. The letter arrived at Soho Square on 15 August 1793. See Flinders to Banks, 21 October 1793, *I&P*, vol.4, document 88 and Flinders to Banks, 20 March 1794, *I&P*, vol.4, document 120.

9 Banks to John King, 15 May 1798, *I&P*, vol.4, document 318.

10 Morgan, *Matthew Flinders*, p.7.
11 Sparrow, *Secret Service*, pp.252–3.
12 Otto to Banks, 9 June 1800, *I&P*, vol.5, document 140.
13 Banks to Otto, 13 June 1800, *I&P*, vol.5, document 141.
14 De Jussieu to Banks, 6 August 1800, *I&P*, vol.5, document 144.
15 Banks to de Jussieu, 10 August 1796, *SC*, document 1368.
16 See note 57 below.
17 See SLNSW, Sir Joseph Banks Papers, Series 63.01 and 63.03 – the latter is reproduced in *I&P*, vol.5, document 160 and Geeson and Sexton, 'H.M. Sloop'.
18 Mabberley, *Jupiter*, pp.36–40.
19 Correia da Serra to Banks, 17 October 1798, *I&P*, vol.5, document 4.
20 For Brown's biography, see Mabberley, *Jupiter*.
21 Dickson to Banks, 9 December 1800, *I&P*, vol.5, document 159.
22 Banks to Brown, 12 December 1800, *I&P*, vol.5, document 161.
23 Brown to Banks, 17 December 1800, *I&P*, vol.5, document 163.
24 Brown's activities in late December 1800 and January 1801 are recorded in his diary – see Mabberley, *Jupiter*, p.64. For more on Greville, see Constantine, *Fields*. In 1804, Greville, along with Banks and others, founded the Horticultural Society.
25 The correspondence with Banks is in *I&P*, vol.5, documents 204, 205 and 207.
26 Banks to Spencer, 14 December 1800, *I&P*, vol.5, document 162.
27 Biographical details for the Bauers are taken from Norst, 'Recognition' and Lack, *The Bauers*. For Jacquin, see Klemun and Huhnel, *Nikolaus Joseph Jacquin*.
28 The tour is described in Lack, *The Bauers*, pp.97–141 and in Harris, *The Magnificent*.
29 See Lack, *The Bauers*, pp.227–43. For Lambert's interest in pines, see Renkema and Ardagh, 'Aylmer Bourke Lambert'. And for Lambert's Australian connections, see Anemaat, *Natural*.
30 Lack and Mabberley, *The Flora Graeca Story*, pp.105–6.
31 Banks to Spencer, 14 December 1800, *I&P*, vol.5, document 162.
32 Ibid.
33 A fine selection of Alexander's natural history drawings can be found in BL, IOR, WD961.
34 Ibbetson, *An Accidence*, pp.vii–xi.
35 Clay, *Julius Caesar Ibbetson*.
36 'Draught of An Undertaking etc . . .', SLNSW, Sir Joseph Banks Papers, Series 63.08.
37 See Shellim, *Oil Paintings* and Eaton, 'Nostalgia' and Eaton, 'Between'.
38 Banks to Daniell, 29 April 1801, *I&P*, vol.5, document 253.
39 See Daniell to Banks, 1 May 1801, *I&P*, vol.5, document 259 and Daniell to Banks, 5 May 1801, *I&P*, vol.5, document 265.

40 Taylor, 'The Creation', p.33. See also Findlay, *Arcadian Quest* and Perry and Simpson, *Drawings*. See also Taylor, *Picturing*. I would like to thank James Taylor for discussing the circumstances of Westall's appointment with me.

41 A copy of the Admiralty contract, with Westall's name on it, can be found in BL, Add MS 32439, entry 31. The date given on the document is 29 April 1801, that is, a week before Daniell's 5 May 1801 letter to Banks.

42 Aiton to Banks (with enclosures Aiton to Wemyss), December 1800/January 1801, SLNSW, Sir Joseph Banks Papers, Series 63.06.

43 RBGA, Record Book 1793–1809, p.178.

44 Aiton to Banks (with enclosures Aiton to Wemyss), SLNSW, Sir Joseph Banks Papers, Series 63.06.

45 Aiton to Wemyss, January 1801, SLNSW, Sir Joseph Banks Papers, Series 63.06.

46 Wemyss to Aiton, 1 January 1801, SLNSW, Sir Joseph Banks Papers, Series 63.07 and Good to Aiton, 14 January 1801, SLNSW, Sir Joseph Banks Papers, Series 63.16.

47 Flinders to Banks, 29 January 1801, *I&P*, vol.5, document 187.

48 See Admiralty to Flinders, May 1801, *I&P*, vol.5, document 258. There was an earlier plan to have a 'greenhouse' built on the ship's quarterdeck to carry fruit trees – gooseberries, currants, cherries and nectarines – for the colony, but it is unclear whether it ever happened – see Flinders to Banks, 29 January 1801, *I&P*, vol.5, document 187. The plans of the *Investigator* from 1 January show clearly a structure on the quarterdeck but whether it is the greenhouse or the plant cabin is unclear – see, for example, Morgan, *Australia Circumnavigated*, p.140. At the beginning of April, before the Admiralty wrote to Flinders, Banks reminded Evan Nepean, First Secretary to the Admiralty, that the instructions should include a statement about managing the plant cabin so as not to repeat the occasion when 'the Kings interest & that of Science which were Emabrkd together in a Plant Cabbin on board Vancouvers Ship were abusd' – Banks to Nepean, [2 April 1801], *I&P*, vol.5, document 234.

49 Aiton to Wemyss, January 1801, SLNSW, Sir Joseph Banks Papers, Series 63.06.

50 Banks to Milnes, 20 January 1801, *I&P*, vol.5, document 179.

51 'Mr. Hawkins's Mineralogical instructions', *I&P*, vol.5, document 169.

52 Milnes to Banks, 4 February 1801, *I&P*, vol.5, document 190.

53 Band, 'John Allen'.

54 Maskelyne to Banks, 23 December 1800, *I&P*, vol.5, document 166.

55 Maskelyne to Banks, 24 December 1800, *I&P*, vol.5, document 167.

56 Banks to Spencer, December 1800 [sic], *I&P*, vol.5, document 168.

57 Banks to the East India Company, 24 April 1801, *I&P*, vol.5, document 248. One of these, the task of choosing a new and more appropriate name for the ship, for example, came not long after the meeting. The possible new names for this ship included *Searcher*, *Patience*, *Diligence*, and *Investigator*.

They can be found on an undated scrap of paper in Banks's hand, in SLNSW, Sir Joseph Banks Papers, Series 63.12. Mackay, *In the Wake*, p.3 asserts that Banks was responsible for the name but I have not found any evidence to support this. All that can be said for certain is that Banks had a hand in choosing the new name. Whoever decided it, the new name was already in use by 8 January 1801, at the latest – see Banks to Governor King, *I&P*, vol.5, document 173. The ship was registered officially by the name *Investigator* on 19 January 1801 – see TNA, ADM2/294, pp.247–8.

58   For the historical context of the *Investigator* as a scientific ship, see Rigby, 'Not At All' and Rigby, 'The Whole'.

18 1801: Australia Circumnavigated and Beyond

1   Banks to Spencer, December 1800, *I&P*, vol.5, document 168.

2   The *Lady Nelson* was a British-built cutter which the Admiralty agreed would sail to Port Jackson and remain there for use by the colonial authorities. It left Portsmouth on 18 March 1800 and reached Port Jackson later that year on 16 December. James Grant, the ship's commander, used the ship to make a survey of the southwestern coast of New South Wales in 1801. George Caley was a passenger on this survey. Two more surveys, conducted by John Murray, who had arrived in Port Jackson on HMS *Porpoise*, followed before the *Lady Nelson* was assigned in May 1802 to be the tender, under the command of Murray, to the *Investigator*.

3   The instructions can be found in Flinders, *A Voyage*, vol.1, pp.8–12 and reproduced in Morgan, *Australia Circumnavigated*, vol.I, pp.126–31. See also Morgan, 'Matthew Flinders'.

4   Austin, *The Voyage*, p.78, asserts Flinders had decided on this course before he left England, but he provides no evidence for it. It is clear, however, that the Admiralty disagreed with the manner by which Flinders wanted to do the circumnavigation – see Flinders to Banks, 17 July 1801, vol.5, document 293.

5   Vallance et al., *Nature's Investigator*, p.104.

6   Ibid., p.103.

7   See Vallance et al., *Nature's Investigator*, p.103 and Edwards, 'The Journal', p.52.

8   For an assessment of the flora and fauna collections during this part of the expedition, see Keighery and Gibson, 'The Flinders Expedition', and Dell, 'The Fauna'.

9   Brown to Banks, 30 May 1802, *I&P*, vol.6, document 56.

10   For the details of the encounter, see Morgan, *Matthew Flinders*, pp.89–92; Fornasiero, Monteath and West-Sooby, *Encountering*; Fornasiero and West-Sooby, 'A Cordial' and West-Sooby, 'Une expédition'.

11   See Sankey, 'The Baudin Expedition' for this later encounter.

12   Flinders did not know that Lieutenant John Murray, in command of the

Lady Nelson, had been in the same bay, named it Port King (after Governor Philip Gidley King) and claimed it for Britain in early March, almost two months earlier. King later changed the name to Port Phillip in honour of his predecessor. This is where the modern city of Melbourne now stands.

13   See Vallance et al., *Nature's Investigator*, pp.189–97 and Edwards, 'The Journal', pp.75–8.

14   Brown to Banks, 30 May 1802, *I&P*, vol.6, document 56.

15   The number of drawings is in Brown to Banks, 30 May 1802, *I&P*, vol.6, document 56 and is confirmed by a letter from Ferdinand to his brother Franz – see Norst, 'Recognition', pp.101–2. On Bauer's colour coding, see Lack and Ibáñez, 'Recording'.

16   Edwards, 'The Journal', p.79.

17   Good to Aiton, 20 July 1802, BL, Add MS 32439, p.70 and Brown to Banks, 30 May 1802, *I&P*, vol.6, document 56.

18   Brown to Banks, 30 May 1802, *I&P*, vol.6, document 56.

19   Flinders, *A Voyage*, pp.235–6.

20   Smith, *King Bungaree*, pp.27–44.

21   Flinders, *A Voyage*, p.235. Flinders made this request to King on 18 May 1802 – *HRNSW*, vol.4, p.755.

22   See Paterson to Banks, 21 July 1802, *I&P*, vol.6, document 66. Paterson was senior official in the colony and a frequent correspondent of Banks – for more about him, see Chapter 17.

23   Flinders, *A Voyage*, p.231. Flinders feared that in adverse circumstances he would be forced to throw the plant cabin and its contents overboard in order to stabilise the ship.

24   Vallance et al., *Nature's Investigator*, pp.202, 212. Brown and Good did not, apparently, rush to meet Caley – three weeks had passed without Brown going to Parramatta to see him – even though Banks had told them to make contact – Banks to Brown, 15 June 1801, *I&P*, document 288. Contact had been made by 18 June because on that day Brown and Caley walked to a spot the latter knew not far from Parramatta. See Caley to Banks, 28 April 1803, *I&P*, vol.6, document 122 and Caley to Banks, 7 August 1803, *I&P*, vol.6, document 154.

25   Vallance et al., *Nature's Investigator*, pp.201–24 and Norst, *Ferdinand Bauer*, p.103.

26   A list of the plants seen on 31 July is in Vallance et al., *Nature's Investigator*, p.233.

27   The list of living plants that went into the *Investigator*'s plant cabin on this occasion is in BL, Add MS 32439, p.78.

28   Brown to Banks, 17 October 1802, *I&P*, vol.6, document 80.

29   See Flinders to King, 18 October 1802, in Morgan, *Australia Circumnavigated*, vol.II, pp.125–7.

30   Brown's letter to Banks (17 October 1802) is in *I&P*, vol.6, document 80.

Bauer's, to his brother, is in Norst, *Ferdinand Bauer*, p.103. See also Norst, 'Recognition'.

31 Smith, *King Bungaree*, p.55.

32 See Morgan, 'From Cook to Flinders' for the navigational history of finding a safe way through the Torres Strait.

33 Veth et al., *Strangers* and Duyker, *The Dutch*.

34 See Morgan, *Matthew Flinders*, pp.117–18 and Morgan, *Australia Circumnavigated*, vol.I, pp.34–40.

35 Morgan, *Australia Circumnavigated*, vol.II, p.179.

36 Ibid., p.183.

37 Ibid., p.184.

38 See Morgan, *Matthew Flinders*, pp.130–2.

39 Flinders to Banks, 28 March 1803, *I&P*, vol.6, document 103.

40 Ibid.

41 This letter is transcribed in Morgan, *Australia Circumnavigated*, vol.II, pp.339–42.

42 Brown to Banks, March 1803, *I&P*, vol.6, document 104.

43 BL, Add MS, 32439, pp.78–9, 96.

44 Banks to Brown, 8 April 1803, *I&P*, vol.6, document 109.

45 The correspondence on which this section is based is in Morgan, *Australia Circumnavigated*, vol.II, pp.390–401.

46 Brown to Banks, 14 September 1803, *I&P*, vol.6, document 162.

47 Vallance et al., *Nature's Investigator*, p.417; Brown to Banks, 6 August 1803, *I&P*, vol.6, document 152 and Flinders, *A Voyage*, vol.2, p.296.

48 King to Banks, 20 November 1800, *I&P*, vol.5, document 156 and Brown to Banks, *I&P*, vol.6, document 152.

49 Brown to Banks, 6 August 1803, *I&P*, vol.6, document 152. Allen had originally chosen to remain in New South Wales. Why he changed his mind is unclear. Brown clearly thought little of Allen. Brown explained to Banks that Allen had decided to leave because he was worried about how much it would cost to stay.

50 The following is based on Flinders, *A Voyage*, vol.2, pp.295–321 with the assistance of Morgan, *Matthew Flinders*, pp.143–7.

51 The commander of the *Bridgewater*, Edward Palmer, later explained that he and his officers assumed that everyone on the wrecked ships had perished and continued the voyage to India – see Morgan, *Australia Circumnavigated*, vol.I, p.46.

52 Flinders named it Wreck Reef – see 'Account of the Loss of His Majesty's Armed Vessel Porpoise and Cato, upon Wreck Reef', *Sydney Gazette and New South Wales Advertiser*, 18 September 1803, p.2.

53 Flinders, *A Voyage*, vol.2, p.327.

54 Brown referred to Allen in these terms in a letter he wrote to Charles Greville – Brown to Greville, 7 August 1803, BL, Add MS 32439, p.121.

55 This part of Flinders's life, not surprisingly, has received a fair degree of

attention. See Morgan, *Matthew Flinders*, ch.12 for the latest. The most comprehensive study of Flinders on Mauritius is Ly Tio Fane Pineo, *In the Grips*. See also Carter, *Companions*.

56 The following is based on Vallance et al., *Nature's Investigator*, pp.433–581 and Lack, *The Bauers*, pp.277–90.

57 The following is from Geeson and Sexton, 'H.M. Sloop', pp.280–1.

58 Vallance et al., *Nature's Investigator*, p.582.

59 The comings and goings of this episode are detailed in Vallance et al., *Nature's Investigator*, pp.583–92.

60 King to Banks, 20 May 1805, *I&P*, vol.7, document 23. What happened to the plants in Caley's care is a bit of a mystery. Though Banks and King had hoped they would be sent soon, Caley did not let them out of his hands. The last mention in the written record of the plants is in a letter of 25 August 1808, in which Banks reproached Caley for not sending the plants on HMS *Buffalo* on which King was returning to London. The ship had arrived in November 1807 and those living plants that King had brought on his own account were in good condition – Banks to Caley, 25 August 1808, *I&P*, vol.7, document 189. Vallance et al., *Nature's Investigator*, pp.599–600 also finds the story puzzling.

61 Banks to Marsden, [25] January 1806, *I&P*, vol.7, document 55.

62 The second edition of the *Hortus Kewensis* detailed the Australian plants that had been grown from seed from Good's shipment of 1802. His name was honoured in that publication as being responsible for their introductions.

63 For Brown's publication, see Mabberley, *Jupiter*, ch.9 and for Bauer's see Lack, *The Bauers*, ch.11.

64 See Morgan, *Matthew Flinders*, ch.14 and Estensen, *Matthew Flinders*, ch.35.

65 Louis de Freycinet's 1811 map of a single land mass, which he called 'Nouvelle Hollande', was published in the second volume of the Atlas accompanying the official account of Baudin's expedition, *Voyage de Découvertes aux Terres Australes* – see Gerritsen, King and Eliason, *The Freycinet Map*. Freycinet's map frequently gave French names to places that already had English names. On this, see Fornasiero and West-Sooby, 'Naming'.

19 1803: William Kerr in Canton

1 Wills, *Mountain*, p.257.

2 Morse, *The Chronicles*, vol.II, p.322.

3 The statistics on ship movements in Canton can be found in Morse, *The Chronicles*, which is organised on a chronological year-by-year basis. For the history of the American presence in Canton, see Downs, *The Golden* and Dolin, *When America*.

4 For the French Wars (Revolutionary Wars and Napoleonic Wars) see, in general, Knight, *Britain*. For the conflict in the eastern seas, see Ward, *British* and the older account, Parkinson, *War*.

5   Bretschneider, *History*, p.204.

6   Morse, *The Chronicles*, vol.II, pp.358, 389. See also Dermigny, *La Chine*.

7   Banks to Brown, 8 April 1803, *I&P*, vol.6, document 109.

8   This is clear from a conversation that took place between the King and Banks when they discussed the gardener's salary. Banks suggested that whatever the sum, it should be backdated to 25 March 1802, the date on which it was decided to send this person from Kew to China (Aiton had first proposed this idea to Banks) – see Banks note, no date, *I&P*, vol.6, document 111 and Aiton to Banks, 19 April 1803, *I&P*, vol.6, document 117.

9   Banks to Hawkesbury, 13 September 1802, NHM, DTC XIII, pp.252–3.

10 Hawkesbury to Banks, 20 September 1802, NHM, DTC XIII, p.254.

11 Banks to Hawkesbury, 13 September 1802, NHM, DTC XIII, p.253.

12 Banks to Aiton, 21 December 1802, *I&P*, vol.6, document 90.

13 Kerr was born on 30 April 1779. This important date was not known before Jane Kilpatrick discovered it – see her *Gifts*, p.164.

14 This information is contained in a letter Kerr to Aiton, 4 March 1804, BL, Add MS 33981, pp.138–9. For more on Dickson's nursery, see Easterby-Smith, *Cultivating* and Shephard, *Seeds*.

15 Banks to David Lance, 30 August 1803, *I&P*, vol.6, document 160.

16 Bosanquet to Banks, 7 April 1803, *I&P*, vol.6, document 105.

17 Ibid.

18 The plan is outlined in a letter to Bosanquet on 8 April 1803, in BL, Add MS 33981, pp.90–1. For some reason, most of this letter has not been transcribed in *I&P*, vol.6, document 106.

19 The full letter from the Court to the 'President and Select Committee of Supra Cargoes at Canton in China', is reproduced in BL, IOR, R/10/37, dated 12 April 1803. The paragraphs relating to Kerr and Lance are in *I&P*, vol.6, document 111.

20 Banks to Kerr, 18 April 1803, *I&P*, vol.6, document 115.

21 This is the passage as it appears in Banks's instructions to Kerr in SLNSW, Sir Joseph Banks Papers, Series 20.34. I have used it rather than *I&P*, vol.6, document 115 because a crucial phrase is missing from the transcription.

22 Banks note, no date, *I&P*, vol.6, document 111.

23 This, and the following information about Lance, is from Kilpatrick, *Gifts*, pp.162–4. See also Lance's place in the network of late eighteenth-century London merchants in Cozens, 'East London'.

24 See Fry, *Alexander Dalrymple*, pp.179–80, Lamb, 'British Missions', pp.99–102, Ward, *British* and Parkinson, *War*, pp.187–220.

25 Banks to Bosanquet, 8 April 1803, *I&P*, vol.6, document 106. In a letter to Lance, Banks advises him of what plants he might look out for while in Cochin China – see Banks to Lance, 23 April 1803, *I&P*, vol.6, document 119.

26 Fry, *Alexander Dalrymple*, p.181.

27 Banks to Lance, 23 April 1803, *I&P*, vol.6, document 119.

28 Thomas Beale, a resident of Macao since 1792, had one of the most famous gardens in this part of China. It boasted twenty-five hundred plants in pots – see Fan, *British*, pp.44–5 and Kilpatrick, *Gifts*, pp.203–4.

29 Lance to Banks, 18 April 1803, *I&P*, vol.6, document 116.

30 Parkinson, *War*, pp.196–200.

31 Lamb, 'British Missions', pp.101–2. For Lance's mission and its aftermath, see The Anh, 'L'Angleterre'.

32 See Kilpatrick, *Gifts*, pp.165–7.

33 Kerr to Aiton, 4 March 1804, BL, Add MS 33981, pp.138–9.

34 This is how Kerr referred to this place where plants could be bought. It is unclear whether this is the famous Fa-tee nursery – see Fan, *British*. I would like to thank Josepha Richard and Winnie Wong for pointing this out to me.

35 Kerr to Aiton, 4 March 1804, BL, Add MS 33981, pp.138–9.

36 Ibid.

37 Ibid.

38 This quote and the following information comes from 'Reply to the Honble Courts Letter of the 12th April 1803' in BL, IOR, G/12/145, pp.183–5.

39 For more on the Hong merchants, especially their financial dealings with the East India Company, see Chen, 'The Insolvency'. I would like to thank Josepha Richard for alerting me to this source.

40 The first mention of this cabin is in Banks to Lance, 23 April 1803, *I&P*, vol.6, document 119, in which Banks remarks that 'If in fitting the Plant Cabbin Glass should become an object . . .', Lance could achieve the same light effect if he used only two or three panes instead of having it all glazed. The Committee's reply to the Court states that the cabin was put together in Canton.

41 The use of plants as part of or paving the way for diplomatic exchanges was well-known to Banks. One of his major triumphs in this area was arranging and sending a living plant collection as a gift from George III to Empress Catherine II of Russia in 1795. On this, see Carter, 'Sir Joseph Banks and the Plant' and Heath, 'Sowing the Seeds'. For a French example of plant diplomacy, see Easterby-Smith, 'On Diplomacy'. See also Genest, 'Les plantes'.

42 Kerr to Aiton, 4 March 1804, BL, Add MS 33981, p.138.

43 Kilpatrick, *Gifts*, p.168.

44 BL, IOR, G/12/145, p.184.

45 Kerr to Aiton, 4 March 1804, BL, Add MS 33981, p.138.

46 BL, IOR, G/12/145, p.145.

47 Britain declared war on France on 18 May 1803.

48 I have based this account of the battle on several sources: Parkinson, *War*, pp.221–35, and Gillespie, 'Sir Nathaniel'.

49 Allen to Banks, 6 August 1804, *I&P*, vol.6, document 194.

50 RBGA, Record Book 1804–26, p.8. This plant does not appear to have been

on Banks's wish list as were the first twenty-two plants: in the Kew incoming book, the number of crosses as they appeared in the Book of Chinese Drawings, was reproduced – see for example, Goodman and Jarvis, 'The John Bradby Blake', p.267.

51 Banks to Kerr, April 1805, *I&P*, vol.7, document 3.
52 Carter, *Sir Joseph Banks*, p.407. The description 'rich' is how Aiton described the collection. See Aiton to Kerr, undated, RBGA, Record Book 1793–1809, p.255.
53 The full list of plants is given in Bretschneider, *History*, p.190. The quote is from Aiton to Kerr, undated, RBGA, Record Book 1793–1809, p.255.

20 1812: And Still Not First-Hand

1 See Kilpatrick, *Gifts*, pp.149–51.
2 Bretschneider, *History*, pp.211–12.
3 The plant cabin was made at Kew and delivered to the ship, along with the plants – see TNA, LS 10/5.
4 Banks to Kerr, April 1805, *I&P*, vol.7, document 3.
5 The list of boxes and their contents, as they were made by William Aiton, Kew's head gardener, and placed on the *Hope*, can be found in *I&P*, vol.7, document 2, dated 23 March 1805.
6 Banks to Kerr, April 1805, *I&P*, vol.7, document 3.
7 Banks to Puankhequa, July 1805, *I&P*, vol.7, document 28.
8 Ibid.
9 Kerr to Banks, 24 February 1806, *I&P*, vol.7, document 59 and Pendergrass to Banks, 31 August 1806, *I&P*, vol.7, document 116.
10 Puankhequa to Banks, 28 February 1806, *I&P*, vol.7, document 60.
11 Ibid.
12 RBGA, Record Book 1804–26.
13 Banks to Kerr, April 1805, *I&P*, vol.7, document 3.
14 Kerr to Banks, 24 February 1806, *I&P*, vol.7, document 59. Without knowing that Kerr had already done this, Banks wrote again on 6 May 1806, reiterating that the presence of a Chinese gardener on board a returning ship was absolutely essential to the success of the voyage – Banks thought that Kerr could find such a person in Puankhequa II's employ. Whether A Hie worked for Puankhequa II is unknown – see Banks to Kerr, *I&P*, vol.7, document 79.
15 BL, IOR, L/MAR/B/168H. In the Kew account books, his name was written as 'Au-Hey' – TNA, LS 10/5, p.118.
16 Kerr to Banks, 24 February 1806, *I&P*, vol.7, document 59.
17 Pendergrass to Banks, 31 August 1806, *I&P*, vol.7, document 116.
18 Aiton to William Price, 21 November 1807, BL, Add MS, 33981, p.261.
19 Desmond, *Kew*, p.122. In his only surviving comment about A Hie, Banks simply said that he 'has amusd us much & Somewhat instructed us' – see

Banks to Staunton, March 1807, *I&P*, vol.7, document 140. His name does not appear on the *Hope*'s outward muster as clearly as it did on the homeward leg. It may be there but there is no designation 'Chinese gardener' to help distinguish one name from another among the Chinese aboard the ship. Whoever kept the muster typically wrote the same or nearly the same name for every Chinese person on board. See BL, IOR, L/MAR/B/168J.

20 TNA, LS 10/5, p.118. Kerr saw A-hey (his spelling) after his return to China. He was working as a gardener in one of the supercargoes' gardens in Macao. Kerr didn't have a good word to say about him – see Kerr to Aiton, 4 March 1809, RBGA, Copy Letters, Kerr Papers.

21 This plant cabin, and presumably many others, had been built at Kew – see TNA, LS 10/5, p.118.

22 'Notes regarding the transport of plants on board the *Thames* East Indiaman', 9 April 1806, *I&P*, vol.7, document 74.

23 Ramsay to Banks, 7 April 1807, *I&P*, vol.7, document 72.

24 'Notes regarding the transport of plants on board the *Thames* East Indiaman', 9 April 1806, *I&P*, vol.7, document 74. Banks incorrectly refers to this as the *Bounty*'s voyage but it is clear that he meant the *Providence* since, of course, Bligh returned empty-handed, without his ship, in the former voyage.

25 'Notes regarding the transport of plants on board the *Thames* East Indiaman', 9 April 1806, *I&P*, vol.7, document 74.

26 Ibid.

27 Aiton to Banks, 13 April 1806, *I&P*, vol.7, document 77 and 'Catalogue of Plants shipped on Board the Thames East Indiaman for Canton China from His Majesty's Botanic Garden Kew 11th Apl, 1806', RBGA, Outwards Book, 1805–, p.29.

28 Aiton to Kerr, undated, RBGA, Record Book 1793–1809, p.256.

29 Manning to Banks, 24 July 1806, *I&P*, vol.7, document 100.

30 Banks to East India Company, 18 March 1810, *I&P*, vol.7, document 242.

31 The details of these collections in 1809 and 1810 are in RBGA, Record Book 1804–26.

32 Banks to Kerr, 23 April 1810, *I&P*, vol.7, document 247.

33 Banks to Kerr, 30 June 1810, *I&P*, vol.7, document 255.

34 This account is based on Kilpatrick, *Gifts*, pp.183–4.

35 Kilpatrick, *Gifts*, p.183. Thanks to Kate Bailey for confirming the relationship between the two John Reeves.

36 Reeves was a distinguished barrister. He was elected a Fellow of the Royal Society in 1790 and in the following year was appointed chief judge for the British colony of Newfoundland in 1791 but did not reside there for more than a year. On returning to London in 1792, he was instrumental in forming an influential anti-republican association, which attracted widespread support (the 'Association for Preserving Liberty and Property Against Republicans and Levellers', was founded at the Crown and Anchor Tavern on the Strand). There is some question about Banks's involvement in it. He certainly had

a copy of its resolution at his home (SL, Sir Joseph Banks Collection, CR 1:58). On the association and Reeves, see Gilmartin, 'In the Theater' and the references therein, especially Duffy, 'William Pitt'. An expert on policing, in 1800 Reeves was appointed the king's printer and in 1803, he became the superintendent of aliens, a police position charged with monitoring aliens in Britain – for this see Dinwiddy 'The Use' and Emsley, 'The Home Office'.

37 The wide topics that Reeves and Banks discussed in their letters can be gleaned from their correspondence held at SL, Sir Joseph Banks Collection.

38 See Reeves to Staunton, 4 July 1792, *China Through Western Eyes*, part 2, reel 27.

39 Staunton, *Memoirs*, p.11.

40 The following account is taken from Eastberg, 'West Meets East' and Staunton, *Memoirs*.

41 See modern commentaries on this undertaking by St André, 'Travelling' and Ong, 'Jurisdictional Politics'.

42 Banks to George Thomas Staunton, 17 March 1812, *I&P*, vol.8, document 53.

43 See ibid.

44 See Kerr, 'The Chinese Porcelain'.

45 See for example Banks to George Thomas Staunton, March 1807, *I&P*, vol.7, document 140 and Banks to George Thomas Staunton, 12 July 1809, *I&P*, vol.7, document 219.

46 See Reeves to Banks, n.d, WLAM, MS 5217/18. 'The Squire', actually Puanyouwei, was Puankhequa I's second son, while Puankhequa II was his fourth son – thanks to Josepha Richard for this information.

47 Reeves to Banks, 27 December 1812, *I&P*, vol.8, document 68. See also Richard and Woudstra, 'Thoroughly'.

48 Reeves to Banks, 27 December 1812, *I&P*, vol.8, document 68.

49 Reeves to Banks, 1 October 1814, *I&P*, vol.8, document 106.

50 Reeves to Banks, 15 January 1815, *I&P*, vol.8, document 112.

51 Reeves to Banks, 1 October 1814, *I&P*, vol.8, document 106.

52 Bretschneider, in *History*, p.257, writes 'Not a Company ship at that time sailed for Europe without her decks being decorated with the little portable greenhouses which preceded the present Wardian cases'.

53 See Reeves to Banks, n.d, WLAM, MS 5217/18.

54 Johnston to Banks, 12 January 1815, *I&P*, vol.8, document 111.

55 WLAM, MS 5217/26, 28 and 29.

56 For Reeves and his family life, see Kilpatrick, *Gifts*.

57 For more on the relationship between Kerr and Reeves with regard to Canton plant drawings, see Goodman and Jarvis, 'The John Brady Blake'.

58 Canton Committee to Court of Directors, 29 January 1804, BL, IOR, MSS Eur, D562/16.

59 Whang to Banks, 18 June 1796, *I&P*, vol.4, document 258.

60 Canton Committee to Court of Directors, 26 February 1806, BL, IOR,

MSS Eur, D562/16. The drawings are now in Library, Art and Archives, Royal Botanic Gardens, Kew.

61 Desmond, *The India Museum*.

62 'Plants contained in the plant Cabin on board the Winchelsea . . .', 1806, RBGA, Record Book 1804–26. Kerr cross-referenced the numbered drawings using the abbreviation 'C.D.' (Company Drawings).

63 For the granting of permission to copy examples of the Chinese plant drawings, see BL, IOR, B/164, 18 December 1816, p.801, and 3 January 1817, p.857.

64 The history of the Society is told in Elliott, *The Royal*. For Reeves and the Society, especially the commissioned drawings, see Fan, *British Naturalists*, pp.43–5, Bailey, 'The Reeves Collection' and Synge, 'Chinese Flower Paintings'.

65 For Reeves's plant introductions, see Bretschneider, *History*, pp.256–66.

66 Reeves to Banks, 14 November 1820, *I&P*, vol.8, document 225.

67 There is no satisfactory biography of Barrow. The best is Lloyd, *Mr. Barrow*. Despite its title, *Barrow's Boys*, Fergus Fleming's book has little to say about Barrow. There is more about Barrow further ahead in Chapter 22.

68 Staunton, *Memoirs*, pp.42–3.

69 Ibid., pp.43–4.

70 This important letter is in BL, IOR, G/12/196, pp.2–6.

71 BL, IOR, G/12/196, p.11.

72 See BL, IOR, G/12/196, pp.33–44 and BL, IOR, MSS. Eur. F.140/35. The post was first offered to John Sullivan, a Commissioner on the Board of Control. Then it was offered to Amherst who turned it down on family grounds. The offer then went to Thomas Hamilton, Lord Binning, another member of the Board of Control, who also turned it down for family reasons and then back to Amherst, who finally accepted.

73 BL, IOR, G/12/196, pp.44–7.

74 Lloyd, *Mr. Barrow*, pp.149–52.

75 The correspondence for these projects can be found in *I&P*, vol.8.

76 Barrow to Banks, 3 October 1805, *I&P*, vol.8, document 123. Henry Ellis, born in 1788, was the illegitimate son of the Earl of Buckinghamshire and had spent many years in India, where he learned many languages including Persian, Arabic and Sanskrit, but not Chinese – see entry in *History of Parliament Online*.

77 Banks to Barrow, 1 December 1815, *I&P*, vol.8, document 128. In his letter to Barrow, Banks said he was waiting for the arrival of William Elford Leach, an assistant in the Natural History Department of the British Museum, from Paris, to suggest a suitable name.

78 Banks to Amherst, 9 December 1815, *I&P*, vol.8, document 129.

79 See Amherst to Banks, 11 December 1815, vol.8, document 130; Reid to William Charles Wells, *I&P*, vol.8, document 131; BL, IOR, G/12/196, p.60; and Amherst to Banks, 14 December 1815, *I&P*, vol.8, document 132.

80 Little is known of Hooper other than what appears in Melchior Treub to William Thistleton-Dyer, 24 April 1893, in 'Early History of Buitenzorg Botanic Gardens', *Bulletin of Miscellaneous Information (Royal Botanic Gardens, Kew)* 79 (1893), pp.173–5. Aiton recommended Hooper to Banks adding that the plant collectors, Allan Cunningham and James Bowie, who knew Hooper from their time at Kew and who were now in Rio de Janeiro, 'infer favourably of this promising young man' – see Aiton to Banks, 29 December 1815, SL, Sir Joseph Banks Collection, A 5:88.

81 Banks to Hooper, 2 January 1816, *I&P*, vol.8, document 136.

82 BL, IOR, G/12/196, p.108.

83 These are in Banks to Abel, 10 February 1816 [probably not this date as the *Alceste* set sail on 9 February – a copy of the letter adds 'about' – SLNSW, Sir Joseph Banks Papers, Series 73.127], *I&P*, vol.8, document 139.

84 When the *Alceste* got to Rio de Janeiro, there were roses and rhododendrons and other (unspecified) plants on board the ship – see Allan Cunningham to W.T. Aiton, 23 March 1816, RBGA, KCL/5/1, p.108.

85 This plant is now called *Rhododendron indicum*.

86 On Davis, see Chang, *Representing*; on Morrison, see Daily, *Robert Morrison*. For the change in the linguistic abilities of the Canton English community, see Cranmer-Byng, 'The First English' and Stifler, 'The Language Students'. For the language question, see Tuck, *Britain*, pp.xxii–xxiv.

87 Many members of the embassy published their accounts of the mission: see Staunton, *Notes*, Ellis, *Journal*, Abel, *Narrative*, Morrison, *A Memoir*, M'Leod, *Narrative* and Davis, *Sketches*. Despite these accounts and a great amount of unpublished material, the Amherst embassy has not received much attention. Exceptions are Wilson, 'Mission to China', Gao, 'The Amherst Embassy', Gao, 'British-Chinese Encounters' and Kitson and Markley, *Writing China*.

88 For this part of the mission, see Hall, *Account*, M'Leod, *Narrative* and Murray Maxwell, 'A Narrative of Occurrences and Remarks made on board his Majesty's Late Ship "Alceste" . . .', BL, IOR, Mss Eur A 8.

89 For the latest on the "Amherst" embassy, see Kitson and Markley, *Writing China*.

90 For the plants observed and collected as the mission travelled from Peking to Canton, see Bretschneider, *History*, pp.225–37.

91 Abel, *Narrative*, pp.vi–vii.

92 Melchior Treub to William Thistleton-Dyer, 24 April 1893, in 'Early History of Buitenzorg Botanic Gardens', *Bulletin of Miscellaneous Information (Royal Botanic Gardens, Kew)* 79 (1893), p.174.

21 1814: Accidentally in Brazil with Bowie and Cunningham

1 Rich and Kierkuć–Bieliński, *Peace Breaks Out*.
2 Aiton to Banks, 29 May 1814, *I&P*, vol.8, document 90.
3 Napoleon ratified the Treaty of Fontainbleau, which ended his rule and prepared the way for his exile to Elba, on 13 April 1814. This story is told in MacKenzie, *The Escape*.
4 Banks to Aiton, 7 June 1814, *I&P*, vol.8, document 91.
5 Ibid.
6 Bowie to Aiton, 4 August 1814, SL, Sir Joseph Banks Collection, Bo 1:21. How word got around that Aiton was looking for plant collectors has gone unrecorded.
7 Gunn and Codd, *Botanical Exploration*, p.101.
8 Cunningham to Aiton, 27 August 1814, SL, Sir Joseph Banks Collection, Bo 1:23. Cunningham and Bowie's letters are identical suggesting, perhaps, that they already knew informally from Aiton that they were to be chosen as collectors and the letters were simply placing their desires on the record.
9 McMinn, *Allan Cunningham*, p.2.
10 Ibid., pp.3–4. McMinn gives the date as 1810/1811 while A.M. Lucas, in the entry on Cunningham in the *Oxford Dictionary of National Biography*, gives the date as 1808.
11 TNA, LS 10/5, pp.265, 267. Cunningham was paid for doing this in the final quarter of 1815.
12 Lord Liverpool, Robert Banks Jenkinson, was the son of Charles Jenkinson, a close friend of both Banks's and George III's. He held various important appointments, including Secretary to the Treasury, Lord of the Admiralty and President of the Board of Trade, in successive governments.
13 This and the foregoing is contained in Banks to Harrison, 1 September 1814, SG, document 2015
14 See Harrison to Banks, 2 September 1814, *I&P*, vol.8, document 97 and Harrison to Banks, 7 September 1814, *I&P*, vol.8, document 98.
15 Harrison to Bowie, 9 September 1814, *I&P*, vol.8, document 99 and Harrison to Cunningham, Orchard and Orchard, *Allan Cunningham*, letter 1/d/1.
16 Banks to Harrison, 10 September 1814, SL, Sir Joseph Banks Collection, Bo 1:22.
17 Harrison to Banks, 22 September 1814, SL, Sir Joseph Banks Collection, Bo 1:24.
18 Orchard and Orchard, *Allan Cunningham*, letter 1/e/1 – this is a transcription of the draft instructions that are in SL, Sir Joseph Banks Collection, Bo 1:25. Another draft copy of the instructions, which Banks shared with Harrison, can be found in *I&P*, vol.8, document 103.
19 Banks's experiences of Rio de Janeiro are in Beaglehole, *The Endeavour Journal*, pp.187–94.

20 Banks to Staunton, 24 February 1793, *I&P*, vol.4, document 30.

21 The story of the removal, voyage and arrival of the Portuguese royal court is told in Wilcken, *Empire Adrift* but see also Schultz, *Tropical Versailles* for a more academic treatment, especially of the Rio de Janeiro period.

22 See Gough, 'Sea Power' and Bauss, 'Rio de Janeiro'.

23 See Schultz, *Tropical Versailles*, ch.6, Manchester, *British Preeminence* and Cardoso, 'Lifting'. I would like to thank Lorelei Kury for alerting me to this aspect of Luso–British relations.

24 This and the rest of this discussion is based on the instructions to Bowie and Cunningham in Orchard and Orchard, *Cunningham*, letter 1/e/1.

25 Mawe, *Travels*.

26 Cunningham and Bowie to Aiton, 29 September 1814, Orchard and Orchard, *Cunningham*, letter 2/a/1.

27 Orchard and Orchard, *King's Collectors*, p.39.

28 Ibid.

29 Barman, 'The Forgotten'.

30 Papavero, *Essays*, vol.1, pp.56–7 and Rego et al., 'On the Ornithological'. Sello's own account of his time in London is in Geheimes Staatsarchiv Preußischer Kulturbesitz, I. HA Rep. 76, Vc Sekt. 2 Tit. XXIII Litt. A Nr.5 Bd. 1. I thank Sabine Hackethal for sending me the transcript, made by Ulrich Moritz, of this document. See also Hackethal and Tillack, 'Im Auftrag'.

31 Cunningham and Bowie to Aiton, 12 February 1815, Orchard and Orchard, *Cunningham*, letter 2/a/4.

32 Banks to Cunningham and Bowie, 10 June 1815, Orchard and Orchard, *Cunningham*, letter 2/a/8.

33 Cunningham and Bowie to Banks, 29 March 1815, Orchard and Orchard, *Cunningham*, letter 2/a/6. For São Paulo, see Marcilio, 'The Population', p.63.

34 Orchard and Orchard, *King's Collectors*, p.65.

35 Cunningham and Bowie to Aiton, 10 July 1815, Orchard and Orchard, *Cunningham*, letter 2/a/11. The daily journal compiled by both of the journey to São Paulo is in Orchard and Orchard, *King's Collectors*, pp.77–138.

36 Orchard and Orchard, *King's Collectors*, p.93.

37 Ibid., pp.107–11.

38 Ibid., p.120.

39 Ibid., p.159 and Orchard and Orchard, *Cunningham*, letter 2/a/16.

40 Orchard and Orchard, *King's Collectors*, pp.182, 196.

41 Ibid., pp.179–80.

42 Ibid., pp.180–96.

43 Banks to Chamberlain, 10 June 1815, *I&P*, vol.8, document 119.

44 Orchard and Orchard, *King's Collectors*, pp.238–9. For British merchants, see Llorca-Jaña, 'British', Cardoso, 'Lifting' and Manchester, *British Pre-Eminence*.

45 Cunningham and Bowie to Banks, 22 December 1815, Orchard and Orchard, *Cunningham*, letter 2/a/20.

46 Chamberlain to Banks, 27 December 1815, *I&P*, vol.8, document 134.

47 Allan Cunningham to Richard Cunningham, 23 March 1816, Orchard and Orchard, *Cunningham*, letter 2/b/3.

48 Orchard and Orchard, *King's Collectors*, pp.257–8 and Allan Cunningham to Richard Cunningham, 23 March 1816, Orchard and Orchard, *Cunningham*, letter 2/b/3.

49 Orchard and Orchard, *King's Collectors*, p.258. In a subsequent letter to Aiton, Cunningham simply stated that he regretted that no ships were expected to go to the Cape – Cunningham to Aiton, 30 March 1816, Orchard and Orchard, *Cunningham*, letter 2/a/30.

50 Bowie to Aiton, 24 July 1816, Orchard and Orchard, *Cunningham*, letter 2/a/38.

51 Cunningham and Bowie to Banks, 22 June 1816, Orchard and Orchard, *Cunningham*, letter 2/a/34 and Banks to Cunningham and Bowie, 10 June 1815, Orchard and Orchard, *Cunningham*, letter 2/a/8.

52 Banks to Cunningham and Bowie, 1 June 1816, *I&P*, vol.8, document 142.

53 Cunningham to Banks, 29 August 1816, Orchard and Orchard, *Cunningham*, letter 2/a/39.

54 The ship's history on the run to New South Wales can be found in Bateson, *The Convict Ships*, pp.195–8.

55 Orchard and Orchard, *King's Collectors*, p.336.

56 Allan Cunningham to Richard Cunningham, 26 August 1816, Orchard and Orchard, *Cunningham*, letter 2/b/6.

57 Orchard and Orchard, *King's Collector*, p.313. For Ralph and the *Mulgrave Castle*, see the entry on the website *FriendsConvictShip.com*.

58 Cunningham to Aiton, 27 September 1816, Orchard and Orchard, *Cunningham*, letter 2/a/44.

59 Bowie to Banks, 26 September 1816, Orchard and Orchard, *Cunningham*, letter 2/a/42.

60 Orchard and Orchard, *King's Collector*, p.355.

61 Allan Cunningham to Richard Cunningham, 7 February 1816, Orchard and Orchard, *Cunningham*, letter 2/b/2.

62 Cunningham to Banks, 2 January 1817, Orchard and Orchard, *Cunningham*, letter 3/a/2 and Banks to Cunningham, 13 February 1817, Orchard and Orchard, *Cunningham*, letter 3/a/4.

63 Banks to Bowie, 26 March 1817, SL, Sir Joseph Banks Collection, BG 1:57 and Banks to Bowie, 24 September 1818, SL, Sir Joseph Banks Collection, BG 1:64.

64 Cunningham to Aiton, 29 August 1816, Orchard and Orchard, *Cunningham*, letter 2/a/40.

65 This and this paragraph is based on Bowie to Banks, 12 November 1816, SL, Sir Joseph Banks Collection, BG 1:54.

66 Chamberlain to Banks, 3 October 1816, *I&P*, vol.8, document 148. For Saint-Hilaire, see Kury, 'Auguste' and 'Botany'; Lamy et al., *Auguste de Saint-Hilaire*. Saint-Hilaire had left Brest on 1 April 1816 in company with the members of the diplomatic mission.

22 1815: Lockhart Survives the Congo

1 See Robert Martyn to Banks, 23 December 1814 and Banks to Martyn, 6 February 1815, both in SL, Sir Joseph Banks Collection, A 5:71. Martyn was a colonel in the Austrian Army. Martyn had wanted Banks's support in convincing Parliament to examine the results of a law suit which he thought had gone against him and his family unfairly. Martyn had contacted Banks because his brother, Lieutenant John Martyn, had died while accompanying Mungo Park on his expedition in 1805 to determine where the Niger River ended. Martyn was under the impression that the African Association, of which Banks was a leading member, organised the expedition and it was for this reason that he was seeking Banks's support.

2 Banks to Barrow, 30 July 1815, NMM, LBK/65/2, p.2 and Barrow to Banks, 29 July 1815, NHM, DTC, vol.XIX, pp.167–8.

3 I have taken the biographical information from Lloyd, *Mr. Barrow*.

4 At the time that Barrow was appointed Second Secretary, William Marsden held the post of First Secretary. Marsden had worked for the East India Company and became an accomplished orientalist and a Fellow of the Royal Society. He was a friend of Banks's. From 1807, when John Wilson Croker accepted the post, the First Secretary to the Admiralty had to be an MP. See Barrow, *An Auto-Biographical*. Increasingly in the nineteenth century, the Second Secretary became a civil service administrative post – I would like to thank Pieter Van der Merwe for this information.

5 During this period in the Admiralty, Barrow carved out for himself a role as promoting naval exploration, cartography and hydrography on a scientific basis, especially with regard to the Arctic – Lloyd, *Mr. Barrow* covers this part of Barrow's life as does Fleming, *Barrow's Boys*.

6 The problem of the Niger, and the various theories that were put forth on where it went, have been well discussed in the literature on exploration: see, for example, Bovill, *The Niger* and, more recently, Withers, 'Mapping', Lambert, *Mastering*, especially ch.1, Lockhart and Lovejoy, *Hugh Clapperton* and Bassett and Porter, '"From the Best Authorities"'.

7 Goulburn to Barrow, 28 July 1815, NMM, LBK/65/2, p.1 and Goulburn to Barrow, 2 August 1815, NMM, LBK/65/2, pp.3–4.

8 The overland expedition was led by Major John Peddie and Captain Thomas Campbell. It is described in Mouser, *The Forgotten*. This is also the only source that details the relationships between this and the Congo expedition.

9 Barrow to Banks, 29 July 1815, NHM, DTC, vol.XIX, p.167.

10 Ibid., pp.167–8.

11 Banks's copy of Maxwell's chart, 'A New Survey of the River Congo on a large Scale', is in SL, Sir Joseph Banks Collection, A 5:73.

12 The letters are: Maxwell to William Keir, 20 July 1804, BL, Add MS 37232, pp.52–3 and Maxwell to Park, 12 October 1804, BL, Add MS 37232, pp.56–9. They have been transcribed and are, respectively, in NHM, DTC, vol.XV, pp.7–9 and pp.148–56. The most important part of Maxwell's letter to Keir was published in Park, *Travels*, pp.clxxv–clxxviii; Maxwell's letter to Park, and Park's previous letter to Maxwell, were published in an edited form in Brown, 'Account of the Correspondence', pp.107–14.

13 Banks to Barrow, 30 July 1815, NMM, LBK/65/2, pp.2–3.

14 Williams and Armstrong, "One of the Noblest", p.19. For the story of the *Comet*, see Osborne, *The Ingenious* and Ransom, *Bell's Comet*.

15 See Macleod et al., 'Making Waves'.

16 See, for example, Robinson, 'Sir Joseph Banks and the East Fen', Robinson, *Sir Joseph Banks*, and Hoppit, 'Sir Joseph'.

17 Barrow to Banks, 16 August 1815, NHM, DTC, Vol.XIX, p.179.

18 Barrow to Banks, 8 August 1815, NHM, DTC, Vol.XIX, p.173.

19 Watt to Barrow, 2 August 1815, NMM, LBK/65/2, pp.21–3.

20 Watt to Barrow, 20 December 1815, NMM, LBK/65/2, pp.55–7. The ship was given its name in early October 1815 – Barrow to Banks, 3 October 1815, *I&P*, vol.8, document 123. There is a sketch of the paddlewheel engine for the *Congo* in Smith, *A Short History*, p.51.

21 Tuckey to Barrow, 21 December 1815, NMM, LBK/65/2, p.62 and Barrow to Goulburn, 11 January 1816, TNA, CO 267/43.

22 For more on Seppings in the navy, see Wright, 'Thomas Young'.

23 Popham to Barrow, 19 January 1816, NMM, LBK/65/2, pp.72–4.

24 Popham to Barrow, 22 January 1816, NMM, LBK/65/2, pp.81–3. See also 'Nautical Experiment', *Morning Post*, 23 January 1816.

25 Barrow to Popham, 29 January 1816, NMM, LBK/65/2, pp.85–7. This was not, of course, how he reacted to Banks when he first suggested the idea.

26 TNA, ADM 1/2616, 13 September 1815.

27 Tuckey to Barrow, 24 January 1816, NMM, LBK/65/2, p.99.

28 Tuckey's thoughts about the Niger and the Congo are laid out in Tuckey to Barrow, 30 August 1815, NMM, LBK/65/2, pp.23–30.

29 For the Port Phillip settlement, see Shaw, 'The Founding' and Tipping, *Convicts*.

30 Barrow to Banks, 5 August 1815, NHM, DTC, vol.XIX, p.169, Banks to Barrow, 6 August 1815, NHM, DTC, Vol.XIX, p.170 and Banks to Francis Hartwell, 26 August 1805, *I&P*, vol.7, document 34. Banks had some correspondence with Tuckey in 1805 and also had a copy of his 'Memoir of a Chart of Port Philip [sic]', which is in SLNSW, Sir Joseph Banks Papers, Series 35.29.

31 Banks to Barrow, 6 August 1815, NHM, DTC, vol.XIX, p.170.

32 Banks to Barrow, 12 August 1815, NHM, DTC, vol.XIX, p.174.

33 Banks to Barrow, 2 December 1815, NMM, LBK/65/2, p.55.

34 Barrow to Banks, 6 December 1815, SL, Sir Joseph Banks Collection, A 5:75A.

35 Cranch's early life is documented in Monod, 'John Cranch' and in Tuckey, *Narrative*, pp.lxxi–lxxviii.

36 Tuckey, *Narrative*, p.lxxv.

37 Cranch to Banks, n.d., SL, Sir Joseph Banks Collection, A 5:75.

38 Barrow to Banks, 3 October 1815, *I&P*, vol.8, document 123. At some point in November, Barrow ran across two Scottish botanists whom he was inclined to appoint to the expedition but Banks thought they were not up to the job – he insisted that a Kew gardener would be better: see Barrow to Banks, 30 November 1815, NHM, DTC, vol.XIX, p.220 and Banks to Barrow, 2 December 1815, NMM, LBK/65/2, p.54.

39 This section is based on Von Buch, 'Biographical Memoir'. There is a biography of Smith – Munthe, *Christen Smith*.

40 Smith's period in the Atlantic islands is discussed in Sunding et al., *Diario*.

41 'Note sur la vie de Mr Chretien Smith', NHM, MSS SMI.

42 Aiton to Banks, 2 February 1816, SL, Sir Joseph Banks Collection, A 5:74.

43 Home to Barrow, n.d., NMM, LBK/65/2, pp.8–9, and Brodie to Barrow, 8 September 1815, NMM, LBK/65/2, pp.9–10.

44 Barrow's copy of the instructions to Smith are in NMM, LBK/65/2, pp.166–71; Banks's draft is in SL, Sir Joseph Banks Collection, A 5:87. There are copies in TNA, CO 267/43 and ADM 1/2617. Shorter versions were printed in Tuckey, *Narrative*, pp.xxxviii–xl.

45 Lockhart's instructions are NMM, LBK/65/2, pp.185–7; Banks's draft is in SL, Sir Joseph Banks Collection, A 5:86. There are copies in TNA, CO 267/43 and ADM 1/2617.

46 Unless otherwise noted, the following account of the voyage is based on Tuckey, *Narrative*.

47 'Professor Smith's Journal', in Tuckey, *Narrative*, pp.233–4.

48 Ibid., pp.275–6.

49 Tuckey confided this to Barrow in a letter he wrote on 20 August 1816 on the banks of the Congo at a place called Banza Cooloo – TNA, ADM 1/2617.

50 *The Times*, 4 January 1817, p.4.

51 *Hampshire Telegraph and Sussex Chronicle*, 3 March 1817, p.4. There is a list of the collection cases but without much detail – see SL, Sir Joseph Banks Collection, A 5:83.

52 Lockhart to Barrow, 29 June 1817, SL, Sir Joseph Banks Collection, A 5:84. Smith's journal, in Danish, is in BL, Add MS 32444 but was translated and published as an appendix, 'Professor Smith's Journal', in Tuckey, *Narrative*.

53 Brown, 'Observations, Systematical and Geographical, on Professor Christian Smith's Collection . . .' in Tuckey, *Narrative*, pp.420–85.

54 Brown, 'Observations, Systematical and Geographical, on Professor Christian

Smith's Collection . . .' in Tuckey, *Narrative*, p.485. The main collectors were: Smeathman, Afzelius, William Brass (one of Banks's collectors who had spent some time around 1780 on the Guinea coast), and Michel Adanson, who had collected in Senegal around 1750. For the activities of these collectors see Keay, 'Botanical Collectors'.

55 RBGA, Inwards Book, 1809–1818.
56 See Banks to Blagden, 31 March 1817, *SC*, document 2084.

Postscript

1 Since George I in 1714, the monarchs of Great Britain were also Kings of Hanover. The House of Hanover did not allow for female succession, but the British system did: consequently, the financial assistance that Hanover had contributed to the royal household for over a century ended with Victoria's accession.

2 Lindley was assisted by two gardeners, Joseph Paxton, who worked for the Duke of Devonshire; and John Wilson, who worked for the Earl of Surrey. Lindley was the first professor of botany in London, appointed to his post in 1829. The royal gardens he was to investigate were, in addition to Kew, those of Windsor, Kensington, Hampton Court and Buckingham Palace.

3 TNA, T90/189.

4 TNA, T90/190.

5 Drayton, *Nature's*, p.108, quotes Banks as having remarked to Henry Dundas in 1787 that Kew might become 'a great botanical exchange house for the empire'. This phrase does not appear in the source Drayton cites nor in any of the other surviving copies of this letter – see I&P, vol.2, document 137 and p.416. This letter, in fact, deals with the Calcutta Botanic Garden – Kew is never mentioned and I personally doubt that Banks ever said this about Kew gardens.

6 Hooker referred to the gardens as the 'Royal Botanic Gardens' in his first report in 1844 (published 7 May 1845 by the House of Commons), and again in the first edition of the *Kew Gardens; or a Popular Guide to the Royal Botanic Gardens of Kew, London, 1847*. Thank you to Kat Harrington, Mark Nesbitt and Caroline Cornish for discussing this with me.

Bibliography

(a publisher's name appears only for books after 1900)

Abel, Clarke, *Narrative of a Journey in the Interior of China*, London, 1818.

Adank, Patricia A., 'The "Memoria Sobre la Grana Cochinilla" of José Antonio Alzate y Ramirez, 1777', unpublished MA dissertation, Arizona State University, 1974.

Agnarsdóttir, Anna, *Sir Joseph Banks, Iceland and the North Atlantic 1772–1820: Journals, Letters and Documents*, London: Hakluyt Society, 2016.

Ahlskog, Jonas, 'The Political Economy of Colonisation: Carl Bernhard Wadström's Case for Abolition and Civilisation', *Sjuttonhundratal*, 7 (2010), pp.146–67.

Aiton, William, *Hortus Kewensis*, London, 1789 (second edition 1810–13).

Alexander, Caroline, *The Bounty: The True Story of the Mutiny on the Bounty*, London: HarperCollins, 2003.

Allain, Yves-Marie, *Voyages et survie des plantes au temps de la voile*, Marly-le-Roi: Editions Champflour, 2000.

Ambjörnsson, Ronny, '"La république de Dieu". une utopie suédoise de 1789', *Annales historiques de la Révolution française*, 277 (1989), pp.244–73.

Anderson, Clare and Maxwell Stewart, Hamish, 'Convict Labour and the Western Empires, 1415–1954', in Aldrich, Robert and McKenzie, Kirsten, eds, *The Routledge History of Western Empires*, London: Routledge, 2013, pp.102–17.

Anderson, James, *Letters on Cochineal Continued*, Madras, 1789.

Andrews, Alan E.J., ed., *The Devil's Wilderness: George Caley's Journey to Mount Banks, 1804*, Hobart: Blubber Head Press, 1984.

Anemaat, Louise, *Natural Curiosity: Unseen Art of the First Fleet*, Sydney: NewSouth Publishing, 2014.

Anon., 'Biography of Consequa', *Gardener's Magazine*, 11 (1835), pp.111–12.

Atkins, Gareth, 'Piety and Plutocracy: The Social and Business World of the Thorntons', in Brown, Jane and Musson, Jeremy, eds, *Moggerhanger Park, Bedfordshire: An Architectural and Social History from Earliest Times to the Present*, Ipswich: Healeys Print Group, 2012, pp.185–99.

Attiret, Jean-Denis, *A Particular Account of the Emperor of China's Garden near Pekin*, London, 1752.

Austin, K.A., *The Voyage of the Investigator 1801–1803 Commander Matthew Flinders, R.N.*, London: Angus & Robertson, 1964.

Ayyar, A.V. Venkatarama, Hosie, John and Howay, F.W., *James Strange's Journal and Narrative of the Commercial Expedition from Bombay to the Northwest Coast of America*, Fairfield, WA: Ye Galleon Press, 1982.

Bailey, Kate, 'The Reeves Collection of Chinese Botanical Drawings', *The Plantsman* (December 2010), pp.218–25.

Band, Stuart R., 'John Allen, Miner, on Board H.M.S. *Investigator*, 1801–1804', *Bulletin of the Peak District Mines Historical Society Ltd*, 10 (1987), pp.67–78.

Barman, Roderick J., 'The Forgotten Journey: Georg Heinrich Langsdorff and the Russian Imperial Scientific Expedition to Brazil, 1821–1829', *Terrae Incognitae*, 3 (1971), pp.67–96.

Barrow, John, *An Auto-Biographical Memoir of Sir John Barrow, Bart., Late of the Admiralty*, London: John Murray, 1847.

Bartlett, Beatrice S., *Monarchs and Ministers: The Grand Council in Mid–Ch'ing China*, Berkeley, CA: University of California Press, 1991.

Bassett, Thomas J. and Porter, Philip W., '"From the Best Authorities": The Mountains of Kong in the Cartography of West Africa', *Journal of African History*, 32 (1991), pp.367–413.

Bateson, Charles, *The Convict Ships 1787–1868*, Sydney: Library of Australian History, 1983.

Batteux, C., Oudart Feudrix de Bréquigny, L.G., de Guignes, J. and Silvestre de Sacy, A.J., eds, *Mémoires concernant l'histoire . . . des Chinois*, Paris, 1776–1814.

Bauer, Francis, *Strelitzia Depicta*, London, 1818.

Bauss, Rudy, 'Rio de Janeiro, Strategic Base for Global Designs of the British Royal Navy, 1777–1815', in Symonds, Craig L., Bartlett, Merrill, Bradford, James, Gillmor, Carroll, Good, Jane E., Harrod, Frederick S., Masterson, Daniel M. and Love, Robert William, Jr., eds, *New Aspects of Naval History*, Annapolis, MD: Naval Institute Press, 1981, pp.75–89.

Beaglehole, J.C., *The Endeavour Journal of Joseph Banks 1768–1771*, Sydney: Angus and Robertson, 1962.

Beaglehole, J.C., *The Journals of Captain James Cook on his Voyages of Discovery*, Cambridge: Cambridge University Press, 1955–67.

Beaglehole, J.C., *The Life of Captain James Cook*, Stanford, CA: Stanford University Press, 1974.

Beaglehole, J.C., *The Voyage of the Endeavour 1768–1771*, Cambridge: Cambridge University Press, 1968.

Beattie, J.M., *Crime and the Courts in England, 1600–1800*, Princeton, NJ: Princeton University Press, 1986.

Beauman, Fran, *The Pineapple: King of Fruits*, London: Chatto & Windus, 2005.

Beckert, Sven, *Empire of Cotton: A New History of Global Capitalism*, London: Penguin Books, 2015.

Beinart, William, 'Men, Science, Travel and Nature in the Eighteenth and Nineteenth-Century Cape', *Journal of South African Studies*, 24 (1998), pp.775–99.

Bennett, Robert J., 'Alignments, Interests and Tensions Over "Reform" in Eighteenth-Century Britain: The Manchester Committee of Trade, 1774–1786', *Northern History*, 51 (2014), pp.61–90.

Berg, Maxine, 'Craft and Small Scale Production in the Global Economy: Gujarat and Kachchh in the Eighteenth and Twenty-First Centuries', *Itinerario*, 37 (2013), pp.23–45.

Berg, Maxine, 'Macartney's Things. Were They Useful? Knowledge and the Trade to China in the Eighteenth Century', unpublished at *www.lse.ac.uk/Economic-History/Assets/Documents/Research/.../Conf4-MBerg.pdf*

Berg, Maxine, 'Passionate Projectors: Savants and Silk on the Coromandel Coast, 1780–98', *Journal of Colonialism & Colonial History*, 14 (2013), n.p.

Berg, Maxine, 'Useful Knowledge, "Industrial Enlightenment", and the Place of India', *Journal of Global History*, 8 (2013), pp.117–41.

Berridge, Vanessa, *The Princess's Garden: Royal Intrigue and the Untold Story of Kew*, Stroud: Amberley Press, 2015.

Bewell, Alan, 'Traveling Natures', *Nineteenth-Century Contexts*, 29 (2007), pp.89–110.

Bickers, Robert A., ed., *Ritual & Diplomacy: The Macartney Mission to China 1792–1794*, London: The British Association for Chinese Studies/Wellsweep Press, 1993.

Binnema, Ted, *Enlightened Zeal: The Hudson's Bay Company and Scientific Networks, 1670–1870*, Toronto, ON: University of Toronto Press, 2014.

Bladon, F. McKno, *The Diaries of Colonel the Hon. Robert Fulke Greville*, London: The Bodley Head, 1930.

Bligh, W., *The Log of H.M.S. Providence 1791–1793*, Guildford: Genesis Publications Limited, 1976.

Bligh, William, *A Voyage To the South Sea Undertaken By Command of His Majesty, For the Purpose of Conveying the Bread-Fruit Tree To the West Indies, in His Majesty's Ship the Bounty*, 1792.

Blussé, Leonard, 'Batavia, 1619–1740: The Rise and Fall of a Chinese Colonial Town', *Journal of Southeast Asian Studies*, 12 (1981), pp.159–78.

Blussé, Leonard, 'Peeking into the Empires: Dutch Embassies to the Courts of China and Japan', *Itinerario*, 37 (2013), pp.13–29.

Blussé, Leonard, *Visible Cities: Canton, Nagasaki, and Batavia and the Coming of the Americans*, Cambridge, MA: Harvard University Press, 2008.

Boahen, A. Adu, 'The African Association, 1788–1805', *Transactions of the Historical Society of Ghana*, 5 (1961), pp.43–64.

Bockstoce, John R., *Furs and Frontiers in the Far North: The Contest Among Native and Foreign Nations for the Bering Strait Fur Trade*, New Haven, CT: Yale University Press, 2009.

Bovill, E.W., *The Niger Explored*, London: Oxford University Press, 1968.

Bowen, H.V., 'British Exports of Raw Cotton from India to China During the Late Eighteenth and Early Nineteenth Centuries', in Riello, Giorgio and Roy, Tirthankar, eds, *How India Clothed the World: The World of South Asian Textiles, 1500–1850*, Leiden: Brill, 2009, pp.115–37.

Bowen, H.V., *The Business of Empire: The East India Company and Imperial Britain, 1756–1833*, Cambridge: Cambridge University Press, 2006.

Bowen, H.V., McAleer, John and Blyth, Robert J., eds, *Monsoon Traders: The Maritime World of the East India Company*, London: Scala Publishers, 2011.

Bown, Stephen R., *Madness, Betrayal and the Lash: The Epic Voyage of Captain George Vancouver*, Vancouver, BC: Douglas & McIntyre, 2008.

Braidwood, Stephen J., *Black Poor and White Philanthropists: London's Blacks and the Foundation of the Sierra Leone Settlement, 1786–91*, Liverpool: Liverpool University Press, 1994.

Bretschneider, E., *History of European Botanical Discoveries in China*, Volume I, Leipzig: K.F. Koehler Verlag, 1935.

Brewer, John, *The Pleasures of the Imagination: English Culture in the Eighteenth Century*, New York: Farrar, Straus and Giroux, 1997.

Britten, James, 'Francis Masson', *Journal of Botany, British and Foreign*, 22 (1884), pp.114–23.

Britten, James, 'Lady Anne Monson (c. 1714–1776)', *Journal of Botany, British and Foreign*, 56 (1918), pp.147–9.

Britten, James, 'R. Brown's List of Madeira Plants', *Journal of Botany, British and Foreign*, 42 (1904), pp.1–8.

Britten, James, 'Some Early Cape Botanists and Collectors', *The Journal of the Linnean Society (Botany)*, 45 (1920–22), pp.29–51.

Britten, James, 'The History of Aiton's "Hortus Kewensis"', *Journal of Botany, British and Foreign*, Supplement III, 50 (1912), pp.1–16.

Brockey, Liam Matthew, *Journey to the East: The Jesuit Mission to China, 1579–1724*, Cambridge, MA: Harvard University Press, 2007.

Brook, Jack, 'The Forlorn Hope: Bennelong and Yemmerrawannie Go to England', *Australian Aboriginal Studies*, 1 (2001), pp.36–47.

Brown, William, 'Account of the Correspondence between Mr Park and Mr Maxwell, Respecting the Identity of the Congo and the Niger', *Edinburgh Philosophical Journal*, 3 (1820), pp.102–14, 205–18.

Brown, Yu–Ying, 'Engelbert Kaempfer's Legacy in the British Library', in Haberland, Detlef, ed., *Engelbert Kaempfer: Werk und Wirkung*, Stuttgart: Franz Steiner Verlag, 1993, pp.344–69.

Brown, Yu–Ying, 'Kaempfer's Album of Famous Sights of Seventeenth Century Japan', *British Library Journal*, 15 (1989), pp.90–103.

Buist, Marten Gerbertus, *At Spes Non Fracta: Hope & Co. 1770–1815: Merchant Bankers and Diplomats at Work*, The Hague: Martinus Nijhoff, 1974.

Burges, James Bland, *A Narrative of the Negotiations Occasioned by the Dispute Between England and Spain in the Year 1790*, London, 1791.

Byrnes, Dan, '"Emptying the Hulks": Duncan Campbell and the First Three Fleets to Australia', *www.academia.edu/9379922/Emptying_the_Hulks*.

Calmann, Gerta, *Ehret: Flower Painter Extraordinary, An Illustrated Biography*, Oxford: Phaidon, 1977.

Campbell, Marjorie Elliott, *The North West Company*, Toronto, ON: Macmillan, 1957.

Cams, Mario, 'The China Maps of Jean-Baptiste Bourguignon d'Anville: Origins and Supporting Networks', *Imago Mundi*, 66 (2014), pp.51–69.

Cannon, Garland, 'Sir William Jones, Sir Joseph Banks, and the Royal Society', *Notes and Records of the Royal Society*, 29 (1975), pp.205–30.

Cardoso, José Luís, 'Lifting the Continental Blockade: Britain, Portugal and Brazilian Trade in the Global Context of the Napoleonic Wars', in Coppolaro, Lucia and McKenzie, Francine, eds, *A Global History of Trade and Conflict Since 1500*, Basingstoke: Palgrave, 2013, pp.87–104.

Carr, D.J., 'The Books that Sailed with the *Endeavour*', *Endeavour*, 7 (1983), pp.194–201.

Carr, D.J., ed., *Sydney Parkinson: Artist of Cook's Endeavour Voyage*, London: Croom Helm, 1983.

Carroll, John M., 'The Canton System: Conflict and Accommodation in the Contact Zone', *Journal of the Royal Asiatic Society Hong Kong Branch*, 50 (2010), pp.51–66.

Carroll, John M., '"The Usual Intercourse of Nations": The British in Pre-Opium War Canton', in Bickers, Robert and Howlett, Jonathan J., eds, *Britain and China, 1840–1970: Empire, Finance and War*, Abingdon: Routledge, 2016, pp.22–40.

Carter, Harold B., *Sir Joseph Banks 1743–1820*, London: British Museum (Natural History), 1988.

Carter, Harold B., *Sir Joseph Banks 1743–1820: A Guide to Biographical and Bibliographical Sources*, Winchester: St Paul's Bibliographies, 1987.

Carter, H.B., 'Sir Joseph Banks and the Plant Collection from Kew Sent to the Empress Catherine II of Russia 1795', *Bulletin of the British Museum (Natural History) Historical Series*, 4 (1974), pp.281–385.

Carter, Harold B., 'The Royal Society and the Voyage of HMS *Endeavour* 1768–71', *Notes and Records of the Royal Society of London*, 49 (1995), pp.245–60.

Carter, H.B., Diment, Judith A., Humphries, C.J. and Wheeler, Alwyne, 'The Banksian Natural History Collections of the *Endeavour* Voyage and their Relevance to Modern Taxonomy', in Wheeler, Alwyne and Price, James H., eds, *History in the Service of Systematics*, London: Society for the Bibliography of Natural History, 1981, pp.61–70.

Carter, Marina, *Companions of Misfortune: Flinders and Friends at the Isle of France, 1803–1810*, London: Pink Pigeon Press, 2003.

Chakrabarti, Pratik, '"Neither of Meate Nor Drinke, But What the Doctor Alloweth": Medicine Amidst War and Commerce in Eighteenth-Century Madras', *Bulletin of the History of Medicine*, 80 (2006), pp.1–38.

Chakrabarti, Pratik, 'Networks of Medicine: Trade and Medico-Botanical Knowledge on the Eighteenth-Century Coromandel Coast', in Bandopadhyay, Arun, ed., *Science and Society in India 1750–2000*, New Delhi: Manohar Publishers, 2010, pp.49–82.

Chaloner, William Henry, 'Robert Owen, Peter Drinkwater and the Early Factory System in Manchester, 1788–1800', *Bulletin of the John Rylands Library*, 37 (1954), pp.78–102.

Chambers, Neil, *Endeavouring Banks: Exploring Collections from the Endeavour Voyage 1768–1771*, London: Paul Holberton Publishing, 2016.

Chambers, Neil, 'Letters from the President: The Correspondence of Sir Joseph Banks', *Notes and Records of the Royal Society of London*, 53 (1999), pp.27–57.

Chambers, Neil, 'Sir Joseph Banks, Japan, and the Far East', in Farrer, Anne, ed., *A Garden Bequest – Plants from Japan*, London: The Japan Society, 2001, pp.7–14.

Chang, Dongshin, *Representing China on the Historical London Stage: From Orientalism to Intercultural Performance*, New York: Routledge, 2015.

Chaudhuri, K.N., *The Trading World of Asia and the English East India Company, 1660–1760*, Cambridge: Cambridge University Press, 1978.

Chen, Kuo-tung Anthony, 'The Insolvency of the Chinese Hong Merchants, 1760–1843', unpublished PhD dissertation, Yale University, 1990.

Ching, May-bo, 'The Story of "Whang Tong": Traces of Ordinary Chinese in 18th Century England', *Historical Review*, x (2003)/2, pp.106–24 (in Chinese).

Christopher, Emma, *A Merciless Place: The Lost Story of Britain's Convict Disaster in Africa*, Oxford: Oxford University Press, 2010.

Christopher, Emma, 'From the "Ballad–Singing" Monkey to the "Cunning Savages": The Voyage to Found a British Colony on the Orange River, 1785–1786', *South African Historical Journal*, 61 (2009), pp.750–65.

Christopher, Emma and Maxwell-Stewart, Hamish, 'Convict Transportation in Global Context, c. 1700–88', in Bashord, Alison and Macintyre, Stuart, eds, *The Cambridge History of Australia*, Volume I, Melbourne: Cambridge University Press, 2013, pp.68–90.

Clancy, Robert, *The Mapping of Terra Australis*, Macquarie Park, NSW: Universal Press, 1995.

Clark, Fiona, 'Lost in Translation: The *Gazeta de Literatura de México* and the Epistemological Limitations of Colonial Travel Narratives', *Bulletin of Spanish Studies*, 85 (2008), pp.151–73.

Clarke, Philip A., *Aboriginal Plant Collectors: Botanists and Aboriginal People in the Nineteenth Century*, Kenthurst: Rosenberg, 2008.

Clay, Rotha Mary, *Julius Caesar Ibbetson 1759–1817*, London: Country Life, 1948.

Clayton, Daniel W., *Islands of Truth: The Imperial Fashioning of Vancouver Island*, Vancouver, BC: UBC Press, 2000.

Clayton, Jane M., *An Alphabetical List of Ships Employed in the South Sea Whale Fishery from Britain: 1775–1815*, Privately Printed, 2014.

Clayton, Jane M. and Clayton, Charles A., *Shipowners Investing in the South Sea Whale Fishery from Britain: 1775–1815*, Hassobury, Hampshire: Privately Printed, 2016.

Clingham, Greg, 'Cultural Difference in George Macartney's *An Embassy to*

China, 1792–94', *Eighteenth-Century Life*, 39 (2015), pp.1–29.

Coats, Alice M., 'Forgotten Gardeners, II: John Graefer, *The Garden History Society Newsletter*, 16 (1972), pp.4–7.

Coats, Alice M., *The Quest for Plants: A History of the Horticultural Explorers*, London: Studio Vista, 1969.

Cock, Randolph, 'Precursors of Cook: The Voyages of the *Dolphin*, 1764–8', *The Mariner's Mirror*, 85 (1999), pp.30–52.

Cole, E.J., 'The Cultural History of Exotic Fruits in England 1650–1820', unpublished PhD dissertation, University of Cambridge, 2006.

Coleman, Deirdre, *Romantic Colonization and British Anti–Slavery*, Cambridge: Cambridge University Press, 2004.

Colnett, James, *A Voyage to the South Atlantic and Round Cape Horn into the Pacific Ocean*, London, 1798.

Conner, Patrick, *The Hongs of Canton: Western Merchants in South China 1700–1900, as seen in Chinese Export Paintings*, London: English Art Books, 2009.

Constantine, David, *Fields of Fire: A Life of Sir William Hamilton*, London: Weidenfeld & Nicolson, 2001.

Cook, Alexandra, 'Linnaeus and Chinese Plants: A Test of the Linguistic Imperialism Thesis', *Notes and Records of the Royal Society*, 64 (2010), pp.121–38.

Cook, Andrew, 'James Cook and the Royal Society', in Williams, Glyndwr, ed., *Captain Cook: Explorations and Reassessments*, Woodbridge: Boydell Press, 2004, pp.37–55.

Cook, Warren L., *Flood Tide of Empire: Spain and the Pacific Northwest, 1543–1819*, New Haven, CT: Yale University Press, 1973.

Cozens, Kenneth, 'East London Merchant Networks in Asia: The Fitzhugh's, 1750–1800, East India Company Agents in Macao', unpublished paper.

Craciun, Adriana, 'The Seeds of Disaster: Relics of La Pérouse', in Craciun, Adriana and Schaffer, Simon, eds, *The Material Cultures of Enlightenment Arts and Sciences*, London: Palgrave Macmillan, 2016, pp.47–69.

Cranmer-Byng, J.L., ed., *An Embassy to China: Being the Journal Kept by Lord Macartney During his Embassy to the Emperor Ch'ien-Lung 1793–1794*, London: Longman, 1962.

Cranmer-Byng, J.L., 'The First English Sinologists: Sir George Staunton and the Reverend Robert Morrison', in Drake, F.S., ed., *Symposium on Historical, Archaeological and Linguistic Studies*, Hong Kong: Hong Kong University Press, 1967, pp.247–57.

Cranmer-Byng, John L. and Wills, John E., Jr., 'Trade and Diplomacy with Western Europe, 1644–c.1800', in Wills, John E., ed., *China and Maritime Europe, 1500–1800: Trade, Settlement, Diplomacy, and Missions*, Cambridge: Cambridge University Press, 2011, pp.183–254.

Crawford, D.G., *Roll of the Indian Medical Service, 1616–1930*, London: W. Thacker & Co., 1930.

Crownhart-Vaughan, E.A.P., 'Clerke in Kamchatka, 1779: New Information for an Anniversary Note', *Oregon Historical Quarterly*, 80 (1979), pp.197–204.

Cullinan, Patrick, *Robert Jacob Gordon 1743–1795: The Man and His Travels at the Cape*, Cape Town: Struik Winchester, 1992.

Cunningham, Chris, *Blue Mountains Rediscovered: Beyond the Myths of Early Australian Exploration*, Kenthurst, NSW: Kangaroo Press, 1996.

Currey, John, *H.M. Survey Vessel Lady Nelson and the Discovery of Port Phillip*, Malvern, VIC: Banks Society Publications, 2002.

Dahlgren, E.W., 'Carl Bernhard Wadström: Hans Verksamhet för Slafhandelns Bekämpande och de Samtida Kolonisationsplanerna i Västafrika', *Nordisk Tidskrift för Bok och Biblioteksväsen*, 2 (1915), pp.1–52.

Daily, Christopher Allen, 'From Gosport to Canton: A New Approach to Robert Morrison and the Beginnings of the Protestant Missions in China', unpublished PhD dissertation, School of Oriental and African Studies, University of London, 2010.

Dalrymple, Alexander, *An Account of the Discoveries Made in the South Pacifick Ocean*, Sydney: Horden House, 1996 (first published 1767).

Damodaran, Vinita, Winterbottom, Anna and Lester, Alan, eds, *The East India Company and the Natural World*, Basingstoke: Palgrave Macmillan, 2015.

Dandy, J.E., *The Sloane Herbarium*, London: British Museum (Natural History), 1958.

David, Andrew, *The Charts and Coastal Views of Captain Cook's Voyages: The Voyage of the Endeavour 1768–1771*, London: Hakluyt Society, 1988.

David, Andrew, 'Vancouver's Survey Methods and Surveys', in Fisher, Robin and Johnston, Hugh, eds, *From Maps to Metaphors: The Pacific World of George Vancouver*, Vancouver, BC: UBC Press, 1993, pp.51–69.

Davis, David Brion, *The Problem of Slavery in Western Culture*, Ithaca, NY: Cornell University Press, 1966.

Davis, John Francis, *Sketches of China*, London, 1841.

Dawson, Frank Griffith, 'The Evacuation of the Mosquito Shore and the English Who Stayed Behind, 1786–1800', *The Americas*, 55 (1998), pp.63–89.

Dawson, Frank Griffith, 'Robert Kaye y el Doctor Robert Sproat: Dos Británicos Expatriados en la Costa de los Mosquitos, 1787–1800', *Yaxkin*, 9 (1986), pp.43–62.

Dawson, Warren R., ed., *The Banks Letters*, London: British Museum, 1958.

Dehergne, J., *Répertoire des Jésuites de Chine de 1552 à 1800*, Paris: Letouzey & Ane, 1973.

Dehergne, Joseph, 'Une grande collection: Mémoires concernant les Chinois (1776–1814)', *Bulletin de l'Ecole française d'Extrème-Orient*, 72 (1983), pp.267–98.

De Kock, W.J. and Krüger, D.W., eds, *Dictionary of South African Biography*, Cape Town: National Council for Social Research, 1968.

Delbourgo, James, *Collecting the World: The Life and Curiosity of Hans Sloane*, London: Allen Lane, 2017.

Dell, John, 'The Fauna Encountered in South-West Western Australia: Its Culinary Value and Scientific Legacy', in Wege, Juliet, George, Alex, Gathe,

Jan, Lemson, Kris and Napier, Kath, eds, *Matthew Flinders and His Scientific Gentlemen: The Expedition of H.M.S. Investigator to Australia, 1801–5*, Welshpool, WA: Western Australian Museum, 2005, pp.115–27.

DeLoughrey, Elizabeth, 'Globalizing the Routes of Breadfruit and Other Bounties', *Journal of Colonialism and Colonial History*, 8 (2008), n.p.

Dening, Greg, *The Death of William Gooch: A History's Anthropology*, Honolulu, HI: University of Hawai'i Press, 1995.

Dening, Greg, *Mr Bligh's Bad Language: Passion, Power and Theatre on the Bounty*, Cambridge: Cambridge University Press, 1992.

Dermigny, Louis, *La Chine et l'Occident: le commerce à Canton au XVIIIe siècle*, Paris: S.E.V.P.E.N., 1964.

Desmond, Ray, *Dictionary of British and Irish Botanists and Horticulturalists*, London: Taylor & Francis/The Natural History Museum, 1994.

Desmond, Ray, *The India Museum 1801–1879*, London: Her Majesty's Stationery Office, 1982.

Desmond, Ray, *Kew: The History of the Royal Botanic Gardens*, London: Harvill Press, 1995.

Dickson, Rod, *HMS Guardian and the Island of Ice: The Lost Ship of the First Fleet and Lieutenant Edward Riou 1789–1790*, Carlisle, WA: Hesperian Press, 2012.

Diment, Judith A., Humphries, Christopher J., Newington, Linda and Shaughnessy, Elaine, 'Catalogue of the Natural History Drawings Commissioned by Joseph Banks on the *Endeavour* Voyage 1768–1771, Parts I and II', *Bulletin of the British Museum (Natural History), Historical Series*, 11 (1984); 12 (1987).

Dinwiddy, J.R., 'The Use of the Crown's Power of Deportation Under the Aliens Act, 1793–1826', *Historical Research* 41 (1968), pp.193–211.

Dolin, Eric Jay, *When America First Met China: An Exotic History of Tea, Drugs, and Money in the Age of Sail*, New York: Liveright Publishing, 2012.

Donkin, R.A., 'Spanish Red: An Ethnogeographical Study of Cochineal and the Opuntia Cactus', *Transactions of the American Philosophical Society*, 67 (1977), pp.3–84.

Douglas, Starr, 'The Making of Scientific Knowledge in an Age of Slavery', *Journal of Colonialism and Colonial History*, 9 (2008), n.p.

Douglas, Starr, 'Natural History, Improvement and Colonisation: Henry Smeathman and Sierra Leone in the Late Eighteenth Century', unpublished PhD dissertation, Royal Holloway, University of London, 2004.

Douglas, Starr and Driver, Felix, 'Imagining the Tropical Colony: Henry Smeathman and the Termites of Sierra Leone', in, Driver, Felix and Martins, Luciana, eds, *Tropical Visions in an Age of Empire*, Chicago: University of Chicago Press, 2005, pp.91–112.

Downs, Jacques M., *The Golden Ghetto: The American Commercial Community at Canton and the Shaping of the American China Policy, 1784–1844*, Bethlehem, PA: Lehigh University Press, 1997.

Drayton, Richard, *Nature's Government: Science, Imperial Britain and the 'Improvement' of the World*, New Haven, CT: Yale University Press, 2000.

Dryander, Jonas, *Catalogus Bibliothecae Historico-Naturalis Josephi Banks*, London, 1797.

Duffill, Mark, *Mungo Park*, Edinburgh: NMS Publishing, 1999.

Duffy, Michael, 'William Pitt and the Origins of the Loyalist Association Movement of 1792', *Historical Journal*, 39 (1996), pp.943–62.

Duhamel du Monceau, Henri-Louis, *Avis pour le transport par mer des arbres, des plantes vivaces; des semences, des animaux et de différens autres morceaux d'histoire naturelle*, Paris, 1752.

Dumoulin-Genest, Marie-Pierre, 'L'Introduction et l'acclimatisation des plantes chinoises en France au XVIIIème siècle', unpublished PhD dissertation, EHESS, Paris, 1994.

Duncan, U.K., *A Family Called Duncan*, privately printed, n.d.

Durand, Antoine and Thiriez, Regine, 'Engraving the Emperor of China's European Palaces', *Biblion: The Bulletin of the New York Public Library*, 1 (1993), pp.81–107.

Dutens, Louis, *Memoirs of a Traveller Now in Retirement*, London, 1805.

Duyker, Edward, *The Dutch in Australia*, Melbourne: AE Press, 1987.

Duyker, Edward, *Nature's Argonaut: Daniel Solander 1733–1782*, Melbourne: Melbourne University Press, 1998.

Duyker, Edward and Tingbrand, Per, eds, *Daniel Solander: Collected Correspondence 1753–1782*, Melbourne: Miegunyah Press, 1995.

Eagleton, Catherine, 'Collecting African Money in Georgian England: Sarah Sophia Banks and her Collection of Coins', *Museum History Journal*, 6 (2013), pp.23–38.

Eastberg, Jodi Rhea Bartley, 'West Meets East: British Perceptions of China Through the Life and Works of Sir George Thomas Staunton, 1781–1859', unpublished PhD dissertation, Marquette University, 2009.

Easterby-Smith, Sarah, *Cultivating Commerce: Cultures of Botany in Britain and France, 1760–1815*, Cambridge: Cambridge University Press, 2018.

Easterby-Smith, Sarah, 'On Diplomacy and Botanical Gifts: France, Mysore, and Mauritius in 1788', in Batsaki, Yota, Cahalan, Sarah Burke and Tchikine, Anatole, eds, *The Botany of Empire in the Long Eighteenth Century*, Washington, DC: Dumbarton Oaks, 2017, pp.193–211.

Eastwood, Alice, 'Menzies' California Journal', *California Historical Society Quarterly*, 2 (1924), pp.265–340.

Eaton, Natasha, 'Between Mimesis and Alterity: Art, Gift, and Diplomacy in Colonial India, 1770–1800', *Comparative Studies in Society and History*, 46 (2004), pp.816–44.

Eaton, Natasha, 'Nostalgia for the Exotic: Creating an Imperial Art in London, 1750–1793', *Eighteenth-Century Studies*, 39 (2006), pp.227–50.

Edelstein, Sidney M., 'Spanish Red: Thiery de Menonville's Voyage a Guaxaca', *American Dyestuff Reporter*, 47 (1958), pp.1–8.

Edwards, Michael M., *The Growth of the British Cotton Trade, 1780–1815*, Manchester: Manchester University Press, 1967.

Edwards, Philip, ed., *The Journals of Captain Cook*, London: Penguin Books, 1999.

Edwards, Phyllis I., ed., 'The Journal of Peter Good: Gardener on Matthew Flinders Voyage to Terra Australis 1801–03', *Bulletin of the British Museum (Natural History) – Historical Series*, 9 (1981), pp.1–213.

Edwards, Phyllis I., 'Sir Joseph Banks and the Botany of Captain Cook's Three Voyages of Exploration', *Pacific Studies*, 2 (1978), pp.20–43.

Ekirch, A. Roger, *Bound for America: The Transportation of British Convicts to the Colonies, 1718–1775*, Oxford: Oxford University Press, 1987.

Ekirch, A. Roger, 'Great Britain's Secret Convict Trade to America, 1783–1784', *American Historical Review*, 89 (1984), pp.1285–91.

Elliott, Brent, *The Royal Horticultural Society: A History 1804–2004*, Chichester: Phillimore, 2004.

Elliott, Mark C. and Chia, Ning, 'The Qing Hunt at Mulan', in Millward, James A., Dunnell, Ruth, Elliott, Mark C. and Forêt, Philippe, eds, *New Qing Imperial History: The Making of Inner Asian Empire at Qing Chengde*, London: Routledge, 2004, pp.66–83.

Ellis, Henry, *Journal of the Proceedings of the Late Embassy to China*, London, 1817.

Ellis, John, *A Description of the Mangostan and the Bread–Fruit*, London, 1775.

Ellis, John, *Directions for Bringing Over Seeds and Plants, from the East-Indies and Other Distant Countries, in a State of Vegetation . . .*, London, 1770.

Ellis, Markman, 'Tails of Wonder: Constructions of the Kangaroo in Late Eighteenth-Century Scientific Discourse', in Lincoln, Margarette, ed., *Science and Exploration in the Pacific: European Voyages to the Southern Oceans in the Eighteenth Century*, Woodbridge: Boydell Press, 1998, pp.163–82.

Ellis, Markman, '"That Singular and Wonderful Quadruped": The Kangaroo as Historical Intangible Heritage in the Eighteenth Century', in Dorfman, Eric, *Intangible Natural Heritage: New Perspectives on Natural Objects*, London: Routledge, 2012, pp.56–87.

Else-Mitchell, R. 'George Caley: His Life and Work', *Journal and Proceedings of the Royal Australian Historical Society*, 25 (1939), pp.437–542.

Emsley, Clive, 'The Home Office and Its Sources of Information and Investigation 1791–1801', *English Historical Review*, 94 (1979), pp.532–61.

Estensen, Miriam, *The Life of Matthew Flinders*, Crows Nest, NSW: Allen & Unwin, 2003.

Exell, A.W., 'Pre-Linnean Collections in the Sloane Herbarium from Africa South of the Sahara', in Ferandes, A., ed., *Comptes rendus de la IVe réunion plénière de l'Association pour l'étude taxanomique de la flore d'Afrique tropicale*, Lisbon: Junta de Investigações do Ultramar, 1962, pp.47–9.

Fabricius, Johann Christian, *Briefe aus London Vermischten Inhalts*, Leipzig, 1784.

Fan, Fa–ti, *British Naturalists in Qing China: Science, Empire, and Cultural Encounter*, Cambridge, MA: Harvard University Press, 2004.

Farmer, Edward L., 'James Flint Versus the Canton Interest (1755–1760)', *Papers on China (Harvard University)* 17 (1963), pp.38–66.

Farnie, Douglas A., 'The Role of Merchants as Prime Movers in the Expansion of the Cotton Industry, 1760–1990', in Farnie, Douglas A. and Jeremy, David J., eds, *The Fibre That Changed the World: The Cotton Industry in International Perspective*, Oxford: Oxford University Press, 2004, pp.15–55.

Farris, Jonathan, 'Thirteen Factories of Canton: An Architecture of Sino–Western Collaboration and Confrontation', *Buildings & Landscapes: Journal of the Vernacular Architecture Forum*, 14 (2007), pp.66–83.

Fatica, Michele, 'Gli alunni del Collegium Sinicum di Napoli, la missione Macartney presso l'Imperatore Qianlong e la richiesta di libertá di culto per i cristiani cinesi (1792–1793)', in Carletti, S.M., Sacchetti, M. and Santangelo, P., eds, *Studi in Onore di Lionello Lanciotti*, Naples: Istituto Universitario Orientale, 1996, pp.525–65.

Fatica, Michele, ed., *Matteo Ripa e il Collegio dei cinesi di Napoli (1682–1869)*, Naples: Universitá degli studi di Napoli 'l'Orientale', 2006.

Fatica, Michele and Francesco D'Arelli, eds, *La missione Cattolica in Cina tra I secoli XVII–XIX*, Naples: Istituto Universitario Orientale, 1999.

Ferrer-Gallego, P. Pablo and Boisset, Fernando, 'The Naming and Typification of the Breadfruit, *Artocarpus altilis*, and Breadnut, *A. camansi (Moraceae)*', *Willdenowia*, 48 (2018), 125–35.

Findlay, Elizabeth, *Arcadian Quest: William Westall's Australian Sketches*, Canberra: National Library of Australia, 1998.

Findlen, Paula, ed., *Athanasius Kircher: The Last man Who Knew Everything*, New York: Routledge, 2004.

Fleming, Fergus, *Barrow's Boys*, London: Granta, 1998.

Fletcher, Joseph, 'Sino–Russian Relations, 1800–62', in Fairbank, John K., ed., *The Cambridge History of China*, Cambridge: Cambridge University Press, 1978, pp.318–50.

Flinders, Matthew, *A Voyage to Terra Australis*, London, 1814.

Foley, Daniel J., 'The British Government Decision to Found a Colony at Botany Bay in New South Wales in 1786', unpublished PhD dissertation, King's College London, 2004.

Ford, Lisa and Salter, Brent, 'From Pluralism to Territorial Sovereignty: The 1816 Trial of Mow–watty in the Superior Court of New South Wales', *Indigenous Law Journal*, 7 (2008), pp.67–86.

Forêt, Philippe, 'The Intended Perception of the Imperial Gardens of Chengde in 1780', *Studies in the History of Gardens & Designed Landscapes: An International Quarterly*, 19 (1999), pp.343–63.

Forêt, Philippe, *Mapping Chengde: The Qing Landscape Enterprise*, Honolulu, HI: University of Hawai'i Press, 2000.

Fornasiero, Jean and West-Sooby, 'A Cordial Encounter? The Meeting of Matthew Flinders and Nicolas Baudin (8–9 April, 1802), *French History and Civilization*, 1 (2005), pp.53–61.

Fornasiero, Jean and West-Sooby, 'Naming and Shaming: The Baudin Expedition and the Politics of Nomenclature in the *Terres Australes*', in Scott, Anne, Hiatt,

Alfred, McIlroy, Claire and Wortham, Christopher, eds, *European Perceptions of Terres Australes*, Farnham: Ashgate, 2011, pp.165–84.

Fornasiero, Jean, Montieth, Peter and Sooby-West, John, *Encountering Terra Australis: The Australian Voyages of Nicolas Baudin and Matthew Flinders*, Kent Town, SA: Wakefield Press, 2004.

Foss, Theodore N., 'A Jesuit Encyclopedia for China: A Guide to Jean-Baptiste du Halde's *Description . . . de la Chine (1735)*', unpublished PhD dissertation, University of Chicago, 1979.

Foster, William, 'The India Board (1784–1858)', *Transactions of the Royal Historical Society*, 11 (1917), pp.61–85.

Fox, R. Hingston, *Dr. John Fothergill and His Friends: Chapters in Eighteenth Century Life*, London: Macmillan, 1919.

Frade, José M. Oliver and Menéndez, Alberto Relancio, eds, *El Descubrimiento Científico de las Islas Canarias*, La Orotava: Fundacíon Canaria Orotava de Historia de la Ciencia, 2007.

Francisco-Ortega, Javier, Santos-Guerra, Arnoldo, Jarvis, Charlie E., Carine, Mark A., Menezes de Sequeira, Miguel and Maunder, Mike, 'Early British Collectors and Observers of the Macaronesian Flora', in Williams, David M. and Knapp, Sandra, eds, *Beyond Cladistics: The Branching of a Paradigm*, Berkeley, CA: University of California Press, 2010, pp.125–44.

Francisco-Ortega, Javier and Santos-Guerra, Arnoldo, 'Early Evidence of Plant Hunting in the Canary Islands from 1694', *Archives of Natural History*, 26 (1999), pp.239–67.

Francisco-Ortega, Javier, Santos-Guerra, Arnoldo, Carine, Mark A and Jarvis, Charles E., 'Plant Hunting in Macaronesia by Francis Masson: The Plants Sent to Linnaeus and Linnaeus Filius', *Botanical Journal of the Linnean Society*, 157 (2008), pp.393–428.

Francisco-Ortega, J., Santos-Guerra, A., Romeiras, M.M., Carine, M.A., Sánchez-Pinto, L. and Duarte, M.C., 'The Botany of the Three Voyages of Captain James Cook in Macaronesia: An Introduction', *Revista de la Academia Canaria de Ciencias*, 27 (2015), pp.357–410.

Franklin, Alexandra, 'Enterprise and Advantage: The West India Interest in Britain, 1774–1840', unpublished PhD dissertation, University of Pennsylvania, 1992.

Frey, James W., 'Prickly Pears and Pagodas: The East India Company's Failure to Establish a Cochineal Industry in Early Colonial India', *The Historian*, 74 (2012), pp.241–66.

Frost, Alan, *Arthur Phillip 1738–1814: His Voyaging*, Melbourne: Oxford University Press, 1987.

Frost, Alan, *Botany Bay: The Real Story*, Collingwood: Black Inc., 2012.

Frost, Alan, *Convicts and Empire: A Naval Question 1776–1811*, Melbourne: Oxford University Press, 1980.

Frost, Alan, *Dreams of a Pacific Empire: Sir George Young's Proposal for a Colonization of New South Wales (1784–5)*, Sydney: Resolution Press, 1980.

Frost, Alan, *The First Fleet: The Real Story*, Collingwood: Black Inc., 2012.

Frost, Alan, 'Nootka Sound and the Beginnings of Britain's Imperialism of Free Trade', in Fisher, Robin and Johnston, Hugh, eds, *From Maps to Metaphors: The Pacific World of George Vancouver*, Vancouver, BC: UBC Press, 1993, pp.104–26.

Frost, Alan, *The Precarious Life of James Matra*, Carlton: Melbourne University Press, 1995.

Frost, Alan, *Sir Joseph Banks and the Transfer of Plants To and From the South Pacific 1786–1798*, Melbourne: The Colony Press, 1993.

Fry, Howard T., *Alexander Dalrymple (1737–1808) and the Expansion of British Trade*, London: Frank Cass, 1970.

Fry, Michael, *The Dundas Despotism*, Edinburgh: Edinburgh University Press, 1992.

Fullager, Kate, 'Bennelong in Britain', *Aboriginal History*, 33 (2009), pp.31–51.

Fullagar, Kate, *The Savage Visit: New World People and Popular Imperial Culture in Britain, 1710–1795*, Berkeley, CA: University of California Press, 2012.

Furber, Holden, *Henry Dundas First Viscount Melville, 1742–1811: Political Manager of Scotland, Statesman, Administrator of British India*, London: Oxford University Press, 1931.

Furber, Holden, 'Madras in 1787', in Seymour, Charles, ed., *Essays in Modern English History in Honor of Wilbur Cortez Abbott*, Cambridge, MA: Harvard University Press, 1941, pp.255–93.

Fyfe, Christopher, *A History of Sierra Leone*, Oxford: Oxford University Press, 1962.

Galloway, David J., 'Olof Swartz's Contributions to Lichenology, 1781–1811', *Archives of Natural History*, 40 (2013), pp.20–37.

Galloway, D.J. and Groves, E.W., 'Archibald Menzies MD, (1754–1842), Aspects of His Life, Travels and Collections', *Archives of Natural History*, 14 (1987), pp.3–43.

Galois, Robert, ed., *A Voyage to the North West Side of America: The Journals of James Colnett, 1786–89*, Vancouver, BC: UBC Press, 2004.

Gao, Hao, 'The Amherst Embassy and British Discoveries in China', *History*, 99 (2014), pp.568–87.

Gao, Hao, 'British–Chinese Encounters: Changing Perceptions and Attitudes from the Macartney Mission to the Opium War (1792–1840)', unpublished PhD dissertation, University of Edinburgh, 2013.

Garrett, Clarke, 'Swedenborg and the Mystical Enlightenment in Late Eighteenth-Century England', *Journal of the History of Ideas*, 45 (1984), pp.67–81.

Gascoigne, John, *Joseph Banks and the English Enlightenment: Useful Knowledge and Polite Culture*, Cambridge: Cambridge University Press, 1994.

Gascoigne, John, *Science in the Service of Empire: Joseph Banks, the British State and the Uses of Science in the Age of Revolution*, Cambridge: Cambridge University Press, 1998.

Geeson, N.T. and Sexton, R.T., 'H.M. Sloop *Investigator*', *Mariner's Mirror*, 56 (1970), pp.275–98.

Genest, Marie-Pierre, 'Les plantes chinoises en France au XVIIIe siècle: médiation et transmission', *Journal d'agriculture traditionelle et de botanique appliquée*, 39 (1997), pp.27–47.

Gerritsen, Rupert, King, Robert and Eliason, Andrew, eds, *The Freycinet Map 1811*, accessed at *www.australiaonthemap.org.au/category/freycinet-map/*

Gerritsen, Rupert, 'Getting the Strait Facts Straight', *The Globe*, 72 (2013), pp.11–21.

Gibson, James R., *Otter Skins, Boston Ships, and China Goods: The Maritime Fur Trade of the Northwest Coast, 1785–1841*, Montreal, PQ: McGill-Queen's University Press, 1992.

Gillen, Mollie, 'The Botany Bay Decision, 1786: Convicts, Not Empire', *English Historical Review*, 97 (1982), pp.740–66.

Gillen, Mollie, *The Founders of Australia: A Biographical Dictionary of the First Fleet*, Sydney: Library of Australian History, 1989.

Gillespie, R. St J., 'Sir Nathaniel Dance's Battle off Pulo Auro', *Mariner's Mirror*, 21 (1935), pp.163–86.

Gilmartin, Kevin, 'In the Theater of Counterrevolution: Loyalist Association and Conservative Opinion in the 1790s', *Journal of British Studies*, 41 (2002), pp.291–328.

Giordano, Antonio, *The Anonymous Journal*, Adelaide: Antonio Giordano, 1975.

Glyn, Lynn B., 'Israel Lyons: A Short but Starry Career: The Life of an Eighteenth-Century Jewish Botanist and Astronomer', *Notes and Records of the Royal Society of London*, 56 (2002), pp.275–305.

Gohmann, Joanna M., 'Colonizing through Clay: A Case Study of the Pineapple in British Material Culture', *Eighteenth-Century Fiction*, 31 (2018), pp.143–61.

Golden, Seán, 'From the Society of Jesus to the East India Company: A Case Study on the Social History of Translation', in Rose, Marilyn Gaddis, ed., *Beyond the Western Tradition: Translation Perspectives XI*, Binghamton, NY: State University of New York at Binghamton, 2000, pp.199–215.

Golvers, Noël, 'Michael Boym and Martino Martini: A Contrastive Portrait of Two China Missionaries and Mapmakers', *Monumenta Serica*, 59 (2011), pp.259–71.

Goodman, Jordan and Crane, Peter, 'The Life and Work of John Bradby Blake', *Curtis's Botanical Magazine*, 34 (2017), pp.231–50.

Goodman, Jordan and Jarvis, Charles, 'The John Bradby Blake Drawings in the Natural History Museum, London: Joseph Banks Puts Them to Work', *Curtis's Botanical Magazine*, 34 (2017), pp.251–75.

Gough, B.M., 'India-Based Expeditions of Trade and Discovery in the North Pacific in the Late Eighteenth Century', *Geographical Journal*, 155 (1989), pp.215–23.

Gough, Barry M., *The Northwest Coast: British Navigation, Trade, and Discoveries to 1812*, Vancouver, BC: UBC Press, 1992.

Gough, Barry M., 'Sea Power and South America: The '"Brazils" or South America Station of the Royal Navy, 1808–1837', *American Neptune*, 50 (1990), pp.26–34.

Greenfield, Amy Butler, *A Perfect Red: Empire, Espionage and the Quest for the Colour of Desire*, London: Doubleday, 2005.

Gross, Andreas, 'Background and Context of the Mission in Europe. Introduction', in Gross, Andreas, Kumaradoss, Y. Vincent and Liebau, Heike, eds, *Halle and the Beginning of Protestant Christianity in India*, Halle Delhi: Verlag der Franckesche Stiftungen, 2006, pp.3–6.

Groves, Eric W., 'Archibald Menzies: An Early Botanist on the Northwestern Seaboard of North America, 1792–1794, With Further Notes on his Life and Work', *Archives of Natural History*, 28 (2001), pp.71–122.

Groves, Eric W., 'Archibald Menzies's Visit to King George Sound, Western Australia, September–October 1791', *Archives of Natural History*, 40 (2013), pp.139–48.

Gunn, Mary and Codd, L.E., *Botanical Exploration of Southern Africa*, Cape Town: A.A. Balkema, 1981.

Gupta, Bishnupriya, 'Competition and Control in the Market for Textiles: Indian Weavers and the English East India Company in the Eighteenth Century', in Riello, Giorgio and Roy, Tirthankar, eds, *How India Clothed the World: The World of South Asian Textiles, 1500–1850*, Leiden: Brill, 2009, pp.281–305.

Gwyn, Julian, *Frigates and Foremasts: The North American Squadron in Nova Scotia Waters, 1745–1815*, Vancouver, BC: UBC Press, 2003.

Haberland, Detlef, *Engelbert Kaempfer 1651–1716: A Biography*, London: British Library, 1996.

Hackethal, Sabine and Tillack, Frank, 'Im Auftrag Preußens: Friedrich Sellow in Brasilien (1814–1831)', in Kwet, Axel and Niekisch, Manfred, eds, *Amphibien und Reptilien der Neotropis. Entdeckung deutschspraiger Forscher in Mittel- und Südamerika, Martensiella*, 23 (2016), pp.64–79.

Hadlow, Janice, *The Strangest Family: Private Lives of George III, Queen Charlotte and the Hanoverians*, London: William Collins, 2014.

Hall, Basil, *Account of a Voyage of Discovery to the West Coast of Corea and the Great Loochoo Islands*, London, 1818.

Hall, Douglas, *A Brief History of the West India Committee*, Barbados: Caribbean Universities Press, 1971.

Hall, Douglas, *Planters, Farmers and Gardeners in Eighteenth Century Jamaica*, Mona, Jamaica: University of the West Indies, 1988.

Hallett, Robin, *Records of the African Association 1788–1831*, London: Thomas Nelson and Sons, 1964.

Hamberg, Erik, 'Anders Berlin – en Linnean i Västafrika', *Svenska Linnésällskapets Årsskrift*, 1996, pp.99–108.

Hansen, Lars, ed., *The Linnaeus Apostles: Global Science & Adventure*, London: IK Foundation, 2007–9.

Hanser, Jessica, 'Mr. Smith Goes to China: British Private Traders and the Interlinking of the British Empire with China, 1757–1792', unpublished PhD dissertation, Yale University, 2012.

Harris, Stephen, *The Magnificent Flora Graeca: How the Mediterranean Came to the English Garden*, Oxford: Bodleian Library, 2007.

Harrison, Henrietta, 'A Faithful Interpreter? Li Zibiao and the 1793 Macartney Embassy to China', *International History Review* 41 (2019), pp.1076–91.

Harrison, Henrietta, 'The Qianlong Emperor's Letter to George III and the Early-Twentieth Century Origins of Ideas about Traditional China's Foreign Relations', *American Historical Review* 122 (2017), pp.680–701.

Hawgood, Barbara J., 'Alexander Russell (1715–1768) and Patrick Russell (1727–1805): Physicians and Natural Historians of Aleppo', *Journal of Medical Biography*, 9 (2001), pp.1–6.

Hawgood, Barbara J., 'The Life and Viper of Dr Patrick Russell MD FRS (1727–1805): Physician and Naturalist', *Toxicon*, 32 (1994), pp.1295–1304.

Hawkesworth, John, *An Account of the Voyages Undertaken By the Order of His Present Majesty for Making Discoveries in the Southern Hemisphere*, London, 1773.

Heath, Ekaterina, 'Joseph Banks and British Botanical Diplomacy', *Australian Garden History*, 30 (2019), pp.16–19.

Heath, Ekaterina, 'Sowing the Seeds for Strong Relations: Seeds and Plants as Diplomatic Gifts for the Russian Empress Maria Fedorovna', in Milam, Jennifer, ed., *Cosmopolitan Moments: Instances of Exchange in the Long Eighteenth Century*, EMAJ Special Issue, December 2017, n.p.

Hellman, Lisa, *Navigating the Foreign Quarters: Everyday Life of the Swedish East India Company Employees in Canton and Macao 1730–1830*, Stockholm, Stockholm University, 2015.

Henrey, Blanche, 'Kaempfer's "Icones"', *Journal of the Society for the Bibliography of Natural History* 3 (1955), p.104.

Hevia, James L., *Cherishing Men from Afar: Qing Guest Ritual and the Macartney Embassy of 1793*, Durham, NC: Duke University Press, 1995.

Higman, B.W., *Jamaica Surveyed: Plantation Maps and Plans of the Eighteenth and Nineteenth Centuries*, Kingston, Jamaica: Institute of Jamaica Publications, 1988.

Hill, Beth and Converse, Cathy, *The Remarkable World of Frances Barkley, 1769–1845*, Victoria, BC: TouchWood Editions, 2003.

Hill, Katrina, 'Collecting on Campaign: British Soldiers in China during the Opium Wars', *Journal of the History of Collections*, 25 (2013), pp.227–52.

Hinz, Petra-Andrea, 'The Japanese Collection of Engelbert Kaempfer (1651–1716) in the Sir Hans Sloane Herbarium at the Natural History Museum, London', *Bulletin of the Natural History Museum London (Botany)*, 31 (2001), pp.27–34.

Hodacs, Hanna, 'Circulating Knowledge on Nature: Travelers and Informants, and the Changing Geography of Linnaean Natural History', in Mackenthun, Gesa, Nicolas, Andrea and Wodianka, Stephanie, eds, *Travel, Agency and the Circulation of Knowledge*, Münster: Waxmann Verlag, 2017, pp.75–97.

Hodacs, Hanna, 'Linnaean Scholars Out of Doors: So Much to Name, Learn and Profit From', in MacGregor, Arthur, ed., *Naturalists in the Field: Collecting,*

Recording and Preserving the Natural World from the Fifteenth to the Twenty-First Century, Leiden: Brill, 2018, pp.240–57.

Hodacs, Hanna, 'Local, Universal, and Embodied Knowledge: Anglo-Swedish Contacts and Linnaean Natural History', in Manning, Patrick and Rood, Daniel, eds, *Global Scientific Practice in an Age of Revolutions, 1750–1850*, Pittsburgh, PA: University of Pittsburgh Press, 2016, pp.90–104.

Hodgson, Kate, '"Dedicated to the Sound Politicians of all Trading Nations of Europe": Sierra Leone and the European Colonial Imagination', in Lovejoy, Paul E. and Schwarz, Suzanne, eds, *Slavery, Abolition and the Transition to Colonialism in Sierra Leone*, Trenton, NJ: Africa World Press, 2015, pp.143–62.

Hoppit, Julian, 'Sir Joseph Banks's Provincial Turn', *Historical Journal*, 61 (2018), pp.403–29.

Houston, Stuart, Ball, Tim and Houston, Mary, *Eighteenth-Century Naturalists of Hudson Bay*, Montreal, PQ: McGill-Queen's University Press, 2003.

Hove, Anton Pantaleon, *Tours for Scientific and Economical Research, Made in the Guzerat, Kattiawar, and the Conkuns, in 1787–1788*, Bombay, 1855.

Howay, F.W., ed., *The Journal of Captain James Colnett Aboard the Argonaut from April 26, 1789, to Nov. 3, 1791*, Toronto, ON: The Champlain Society, 1940.

Ibbetson, Julius Cæsar, *An Accidence, or Gamut, of Painting in Oil*, London, 1828.

Igler, David, *The Great Ocean: Pacific Worlds from Captain Cook to the Gold Rush*, Oxford: Oxford University Press, 2013.

Ingham, E.G., *Sierra Leone after a Hundred Years*, London, 1894.

Innis, Mary Quayle, ed., *Mrs. Simcoe's Diary*, Toronto, ON: Macmillan, 1965.

James, Jeff, '"Raising Sand, Soil and Gravel" Pardon Refusers On-Board Prison Hulks (1776–1815)', *Family & Community History*, 20 (2017), pp.3–24.

Jangoux, Michel, ed., *Portés par l'air du temps: les voyages du Capitaine Baudin*, Brussels: Éditions de l'Université de Bruxelles, 2010.

Jaquay, Barbara Gaye, 'The Caribbean Cotton Production: An Historical Geography of the Region's Mystery Crop', unpublished PhD dissertation, Texas A&M University, 1997.

Jarrell, Richard A., 'Francis Masson', in *Dictionary of Canadian Biography*, at *www.biographi.ca*.

Jarvis, Charles E. and Oswald, Philip H., 'The Collecting Activities of James Cuninghame FRS on the Voyage of *Tuscan* to China (Amoy) between 1697 and 1699', *Notes and Records*, 69 (2015), pp.135–53.

Jasanoff, Maya, *Liberty's Exiles: The Loss of America and the Remaking of the British Empire*, London: HarperPress, 2011.

Jennings, Judith, *The Business of Abolishing the British Slave Trade 1783–1807*, London: Routledge, 1997.

Jensen, Niklas Thode, 'Making it in Tranquebar: Science, Medicine and the Circulation of Knowledge in the Danish–Halle Mission, c. 1732–44', in Fihl, Esther and Venkatachalapathy, A.R., eds, *Beyond Tranquebar: Grappling Across Cultural Borders in South India*, Delhi: Orient BlackSwan, 2014, pp.325–51.

Jensen, Niklas Thode, 'Negotiating People, Plants and Empires: The Fieldwork of Johan Gerhard König in South and South East Asia (1768–1785)', in Hodacs, Hanna, Nyberg, Kenneth and Van Damme, Stéphane, eds, *Linnaeus, Natural History and the Circulation of Knowledge*, Oxford: Voltaire Foundation, 2018, pp.187–210.

Jensen, Niklas Thode, 'The Tranquebarian Society', *Scandinavian Journal of History*, 40 (2015), pp.535–61.

Jones, A.M.P., Ragone, D., Tavana, N.G., Bernotas, D.W. and Murch, S.J., 'Beyond the Bounty: Breadfruit (*Artocarpus altilis*) for Food Security and Novel Foods in the 21st Century', *Ethnobotany Research & Applications*, 9 (2011), pp.129–49.

Jones, Ryan Tucker, 'A "Havoc Made Among Them": Animals, Empire, and Extinction in the Russian North Pacific, 1741–1800', *Environmental History*, 16 (2011), pp.585–609.

Jones, Ryan Tucker, 'Peter Simon Pallas, Siberia, and the European Republic of Letters', *Studies in the History of Biology*, 3 (2011), pp.55–67.

Jones, Ryan, 'Sea Otters and Savages in the Russian Empire: The Billings Expedition, 1785–1793', *Journal for Maritime Research*, 8 (2006), pp.106–21.

Joppien, Rüdiger and Chambers, Neil, 'The Scholarly Library and Collections of Knowledge of Sir Joseph Banks', in Mandelbrote, Giles and Taylor, Barry, eds, *Libraries Within the Library: The Origins of the British Library's Printed Collections*, London: British Library, 2009, pp.222–43.

Karskens, Grace, *The Colony: A History of Early Sydney*, Crows Nest, NSW: Allen & Unwin, 2010.

Karsten, Mia C., 'Carl Peter Thunberg: An Early Investigator of Cape Botany', *The Journal of South African Botany*, 5 (1939), pp.1–27, 87–104, 105–55; 12 (1946), pp.127–63, 165–90.

Karsten, Mia C., 'Francis Masson, A Gardener-Botanist Who Collected at the Cape', *The Journal of South African Botany*, 24 (1958), pp.203–18; 25 (1959), pp.167–88, 283–310; 26 (1960), pp.9–15; 27 (1961), pp.15–45.

Karsten, Mia C., 'Sparrman as a Correspondent', *The Journal of South African Botany*, 23 (1957), pp.43–63, 127–37.

Keay, John, *The Honourable Company: A History of the English East India Company*, London: HarperCollins, 1991.

Keay, R.W.J., 'Botanical Collectors in West Africa Prior to 1860', in Fernandes, A., *Comptes rendus de la IVe réunion plénière de l'association pour l'étude taxonomique de la flore d'Afrique tropicale*, Lisbon: Junta de Investigaçoẽs do Ultramar, 1962, pp.55–68.

Keevil, J.J., 'Archibald Menzies, 1754–1842', *Bulletin of the History of Medicine*, 22 (1948), pp.796–811.

Keighrey, Greg and Gibson, Neil, 'The Flinders Expedition in Western Australia: Robert Brown, the Plants and their Influence on W.A. Botany', in Wege, Juliet, George, Alex, Gathe, Jan, Lemson, Kris and Napier, Kath, eds, *Matthew

Flinders and His Scientific Gentlemen: The Expedition of H.M.S. Investigator to Australia, 1801–5, Welshpool, WA: Western Australian Museum, 2005, pp.105–13.

Kellman, Jordan, 'Nature, Networks, and Expert Testimony in the Colonial Atlantic: The Case of Cochineal', *Atlantic Studies*, 7 (2010), pp.373–95.

Kerr, Rose, 'The Chinese Porcelain at Spring Grove Dairy', *Apollo*, 129 (1989), pp.30–4.

Kilpatrick, Jane, *Gifts from the Gardens of China*, London: Frances Lincoln Limited, 2007.

Kinahan, Jill, 'The Impenetrable Shield: HMS *Nautilus* and the Namib Coast in the Late Eighteenth Century', *Cimbebasia*, 12 (1990), pp.23–61.

King, F.H.H., 'Les premiers étudiants britanniques de chinois (1734–1834)', *France–Asie/Asia*, 17 (1960–1), pp.2425–39.

King, James, *A Voyage to the Pacific Ocean . . . Volume III*, London, 1784.

King, Robert J., 'George Vancouver and the Contemplated Settlement at Nootka Sound', *The Great Circle*, 32 (2010), pp.6–34.

King, Robert J., '"The Long Wish'd For Object" – Opening the Trade to Japan, 1785–1795', *The Northern Mariner/le marin du Nord*, 20 (2010), pp.1–34.

Kippis, Andrew, *Six Discourses Delivered by Sir John Pringle, Bart. When President of the Royal Society*, London, 1783.

Kitson, Peter J., *Forging Romantic China: Sino–British Cultural Exchange 1760–1840*, Cambridge: Cambridge University Press, 2013.

Kitson, Peter J. and Markley, Robert, eds, *Writing China: Essays on the Amherst Embassy (1816) and Sino–British Cultural Relations*, Cambridge: D.S. Brewer, 2016.

Klemun, Marianne and Hühnel, Helga, *Nikolaus Joseph Jacquin (1727–1817) – ein Naturforscher (er)findet sich*, Vienna: Vienna University Press, 2017.

Knight, C., 'H.M. Bark *Endeavour*', *Mariner's Mirror*, 19 (1933), pp.292–302.

Knight, C., 'H.M. Armed Vessel *Bounty*', *Mariner's Mirror*, 22 (1936), pp.183–99.

Knight, Roger, *Britain Against Napoleon: The Organization of Victory 1793–1815*, London: Allen Lane, 2013.

Knox, Tim, 'The Great Pagoda at Kew', *History Today*, 44 (1994), pp.22–6.

Koerner, Lisbet, *Linnaeus: Nature and Nation*, Cambridge, MA: Harvard University Press, 1999.

Koerner, Lisbet, 'Purposes of Linnaean Travel: A Preliminary Research Report', in Miller, David Philip and Reill, Peter Hans, eds, *Visions of Empire: Voyages, Botany, and Representations of Nature*, Cambridge: Cambridge University Press, 1996, pp.117–52.

Kops, Henriette Rahusen-De Bruyn, 'Not Such an "Unpromising Beginning": The First Dutch Embassy to China, 1655–1657', *Modern Asian Studies*, 36 (2002), pp.535–78.

Kowaleski-Wallace, Elizabeth, 'The First Samurai: Isolationism in Engelbert Kaempfer's 1727 History of Japan', *The Eighteenth Century*, 48 (2007), pp.111–24.

Kup, Alexander Peter, ed., *Adam Afzelius Sierra Leone Journal*, Uppsala: Almqvist & Wiksells, 1967.

Kury, Lorelai, 'Auguste de Saint-Hilaire, Viajante Exemplar', *Intellèctus* 2 (2003), pp.1–11.

Kury, Lorelai, 'Botany in War and Peace: France and the Circulation of Plants in Brazil (Late Eighteenth and Early Nineteenth Century), *Portuguese Journal of Social Science*, 16 (2017), pp.7–19.

Kusukawa, Sachiko and Maclean, Ian, eds, *Transmitting Knowledge: Words, Images, and Instruments in Early Modern Europe*, Oxford: Oxford University Press, 2006.

Lack, Hans Walter, *The Bauers: Joseph, Franz & Ferdinand Masters of Botanical Illustration*, London: Prestel, 2015.

Lack, H. Walter and Ibáñez, Victoria, 'Recording Colour in Late Eighteenth Century Botanical Drawings: Sydney Parkinson, Ferdinand Bauer and Thaddäus Haenke', *Curtis's Botanical Magazine*, 14 (1997), pp.87–100.

Lack, H. Walter and Mabberley, David J., *The Flora Graeca Story: Sibthorp, Bauer, and Hawkins in the Levant*, Oxford: Oxford University Press, 1999.

Laird, Mark and Bridgman, Karen, 'American Roots: Techniques of Plant Transportation in the Early Atlantic World', in Smith, Pamela H., Meyers, Amy R.W. and Cook, Harold J., eds, *Ways of Making and Knowing: The Material Culture of Empirical Knowledge*, Ann Arbor, MI: University of Michigan Press, 2014, pp.164–93.

Lamb, Alastair, 'British Missions to Cochin China: 1778–1822', *Journal of the Malayan Branch Royal Asiatic Society*, 34 (1961), pp.1–248.

Lamb, W. Kaye, 'Banks and Menzies: Evolution of a Journal', in Fisher, Robin and Johnston, Hugh, eds, *From Maps to Metaphors: The Pacific World of George Vancouver*, Vancouver, BC: UBC Press, 1993, pp.227–244.

Lamb, W. Kaye and Bartroli, Tomás, 'James Hanna and John Henry Cox: The First Maritime Fur Trader and His Sponsor', *BC Studies*, 84 (1989–90), pp.3 36.

Lamb, W. Kaye, *A Voyage of Discovery to the North Pacific Ocean and Round the World 1791–1795*, London, Hakluyt Society, 1984.

Lambert, David, *Mastering the Niger: James MacQueen's African Geography and the Struggle over Atlantic Slavery*, Chicago: University of Chicago Press, 2013.

Lamy, Denis, Pignal, Marc, Sarthou, Corinne and Romaniuc-Neto, Sergio, *Auguste de Saint-Hilaire (1779–1853) un botaniste français au Brésil*, Paris: Muséum national d'Histoire naturelle, 2016.

Langdon, Marcus, *Penang: The Fourth Presidency of India 1805–1830, Volume Two: Fire, Spice & Edifice*, Penang: George Town World Heritage, 2015.

Lario de Oñate, Maria del Carmen, *La colonia mercantil británica e irlandesa en Cádiz a finales del siglo XVIII*, Cadiz: Universidad de Cádiz, 2000.

Lech, Katarzyna, Witkoś, Katarzyna, Wileńska, Beata and Jarosz, Maciej, 'Identification of Unknown Colorants in Pre-Colombian Textiles Dyed with American Cochineal (*Dactylopius coccus* Costa) Using High-Performance

Liquid Chromatography and Tandem Mass Spectrometry', *Analytical and Bioanalytical Chemistry*, 407 (2015), pp.855–67.

Lee, M.R., 'Ipecacuanha: The South American Vomiting Root', *Journal of the Royal College of Physicians of Edinburgh*, 38 (2008), pp.355–60.

Lee, Raymond L., 'American Cochineal in European Commerce, 1526–1625', *The Journal of Modern History*, 23 (1951), pp.205–24.

Leis, Arlene, '"A Little Old-China Mad": Lady Dorothea Banks (1758–1828) and her Dairy at Spring Grove', *Journal for Eighteenth-Century Studies*, 40 (2017), pp.199–221.

Leis, Arlene, 'Cutting, Arranging, and Pasting: Sarah Sophia Banks as Collector', *Early Modern Women: An Interdisciplinary Journal*, 9 (2014), pp.127–40.

Leis, Arlene, 'Displaying Art and Fashion: Ladies' Pocket-Book Imagery in the Paper Collections of Sarah Sophia Banks', *Konsthistorisk tidskrift/Journal of Art History*, 82 (2013), pp.252–71.

Leis, Arlene, 'Sarah Sophia Banks: A "Truly Interesting Collection of Visiting Cards and Co."', in Burrows, Toby and Johnston, Cynthia, eds, *Collecting the Past: British Collectors and their Collections from the 18th to the 20th Centuries*, London: Routledge, 2019, pp.25–44.

Leis, Arlene, 'Sarah Sophia Banks: Femininity, Sociability and the Practice of Collecting in Late Georgian England', unpublished PhD dissertation, University of York, 2013.

Lemire, Beverly, *Cotton*, Oxford: Berg, 2011.

Le Rougetel, Hazel, *The Chelsea Gardener: Philip Miller 1691–1771*, London: Natural History Museum, 1990.

Le Rougetel, Hazel, 'The Fa Tee Nurseries of South China', *Garden History*, 10 (1982), pp.70–3.

Lettsom, John Coakley, *The Works of John Fothergill, M.D.*, London, 1784.

Levitt, Ruth, '"A Noble Present of Fruit": A Transatlantic History of Pineapple Cultivation', *Garden History*, 42 (2014), pp.106–19.

Lightman, Bernard, McOuat, Gordon and Stewart, Larry, eds, *The Circulation of Knowledge Between Britain, India and China: The Early-Modern World to the Twentieth Century*, Leiden: Brill, 2013.

Lincoln, Margarette, 'Mutinous Behavior on Voyages to the South Seas and Its Impact on Eighteenth-Century Civil Society', *Eighteenth-Century Life*, 31 (2007), pp.62–80.

Lindorff, Joyce, 'Burney, Macartney and the Qianlong Emperor: The Role of Music in the British Embassy to China, 1792–1794', *Early Music*, 11 (2012), pp.441–53.

Lindroth, Sten, 'Adam Afzelius: A Swedish Botanist in Sierra Leone', *Sierra Leone Studies*, 4 (1955), pp.194–207.

Lindroth, Sten, 'Adam Afzelius: En Linnean i England och Sierra Leone', *Lychnos*, 1944–1945, pp.1–54.

Lineham, Peter J., 'The Origins of the New Jerusalem Church in the 1780s', *Bulletin of the John Rylands Library*, 70 (1988), pp.109–22.

Livingstone, John, 'Observations on the Difficulties which have Existed in the Transportation of Plants from China to England, and Suggestions for Obviating Them . . .', *Transactions of the Horticultural Society of London*, 3 (1822), pp.421–9.

Llorca-Jaña, Manuel, 'British Merchants in New Markets: The Case of Wylie and Hancock in Brazil and the River Plate, c.1808–19', *Journal of Imperial and Commonwealth History*, 42 (2014), pp.215–38.

Lloyd, Christopher, *Mr. Barrow of the Admiralty: A Life of Sir John Barrow 1764–1848*, London: Collins, 1970.

Lockard, Craig A., 'Chinese Migration and Settlement in Southeast Asia before 1850: Making Fields from the Sea', *History Compass*, 11 (2013), pp.765–81.

Lockhart, Jamie Bruce and Lovejoy, Paul E., eds, *Hugh Clapperton into the Interior of Africa: Records of the Second Expedition*, Leiden, Brill, 2005.

Long, Edward, *The History of Jamaica*, London, 1774.

Love, Henry Davison. *Vestiges of Old Madras 1640–1800*, London: John Murray, 1913.

Lovejoy, Paul E. and Schwarz, Suzanne, eds, *Slavery, Abolition and the Transition to Colonialism in Sierra Leone*, Trenton, NJ: Africa World Press, 2015.

Luckhurst, Gerald, 'Gerard De Visme and the Introduction of the English Landscape Garden to Portugal (1782–1793)', *Revista de Estudos Anglo–Portugueses*, 20 (2011), pp.127–60.

Lysaght, A.M., *Joseph Banks in Newfoundland and Labrador: His Diary, Manuscripts and Collections*, London: Faber and Faber, 1971.

Ly-Tio-Fane Pineo, Huguette, *In the Grips of the Eagle: Matthew Flinders at Ile de France, 1803–1810*, Moka, Mauritius: Mahatma Gandhi Institute, 1988.

M'Leod, John, *Narrative of a Voyage in His Majesty's Late Ship Alceste to the Yellow Sea*, London, 1817.

Mabberley, D.J., *Jupiter Botanicus: Robert Brown of the British Museum*, Braunschweig: J. Cramer, 1985.

MacGregor, Arthur, ed., *Naturalists in the Field: Collecting, Recording and Preserving the Natural World from the Fifteenth to the Twenty-First Century*, Leiden: Brill, 2018.

Mackaness, George, ed., *Fresh Light on Bligh*, Sydney: D.S. Ford, 1953.

Mackaness, George, *The Life of Vice-Admiral William Bligh*, Sydney: Angus and Robertson, 1951.

Mackay, David, 'Banks, Bligh and Breadfruit', *New Zealand Journal of History*, 8 (1974), pp.61–77.

Mackay, David, *In the Wake of Cook: Exploration, Science & Empire, 1780–1801*, London: Croom Helm, 1985.

MacKenzie, Norman, *The Escape from Elba: The Fall and Flight of Napoleon 1814–1815*, Oxford: Oxford University Press, 1982.

Macleod, Christine, Stein, Jeremy, Tann, Jennifer and Andrew, James, 'Making Waves: The Royal Navy's Management of Invention and Innovation in Steam Shipping, 1815–1832', *History and Technology*, 16 (2000), pp.307–33.

McAleer, John, 'Looking East: St Helena, the South Atlantic and Britain's Indian Ocean World', *Atlantic Studies*, 13 (2016), pp.78–98.

McCarthy, James, *Monkey Puzzle Man: Archibald Menzies, Plant Hunter*, Dunbeath: Whittles Publishing, 2008.

McCarthy, Michael, 'Thomas Wright's Designs for Gothic Garden Buildings', *Journal of Garden History*, 1 (1981), pp.239–52.

McCarthy, Michael, 'Thomas Wright's "Designs for Temples"' and Related Drawings for Garden Buildings', *Journal of Garden History*, 1 (1981), pp.55–66.

McClellan, James E. III, *Colonialism and Science: Saint Domingue in the Old Regime*, Baltimore, MD: The Johns Hopkins University Press, 1992.

McClelland, Keith, 'Redefining the West India Interest: Politics and the Legacy of Slave-Ownership', in Hall, Catherine, McClelland, Keith, Draper, Nick, Donington, Kate and Lang, Rachel, eds, *Legacies of British Slave-Ownership: Colonial Slavery and the Formation of Victorian Britain*, Cambridge: Cambridge University Press, 2014, pp.127–62.

McKay, John, *The Armed Transport Bounty*, London: Conway Maritime Press, 2001.

McMinn, W.G., *Allan Cunningham Botanist and Explorer*, Melbourne: Melbourne University Press, 1970.

Maddison, R.E.W. and Maddison, Raymond E., 'Spring Grove, the Country House of Sir Joseph Banks, Bart, P.R.S.', *Notes and Records of the Royal Society*, 11 (1954), pp.91–9.

Main, James, 'Observations in Chinese Scenery, Plants and Gardening, made on a Visit to the City of Canton and its Environs, in the Years 1793 and 1794 . . .', *Gardener's Magazine* 2 (1827), pp.135–40.

Main, James, 'Reminiscences of a Voyage to and from China, in the Years 1792–3–4', *Horticultural Register*, 5 (1936), pp.62–7, 97–103, 143–9, 171–80, 215–20, 256–62, 292–7, 335–9.

Manchester, Alan Krebs, *British Preeminence in Brazil: Its Rise and Decline: A Study in European Expansion*, New York: Octagon Books, 1964.

Marcilio, Maria Luiza, 'The Population of Colonial Brazil', in Bethell, Leslie, ed., *The Cambridge History of Latin America*, Volume 1, Cambridge: Cambridge University Press, 1984, pp.37–64.

Marichal, Carlos, 'Mexican Cochineal and the European Demand for American Dyes, 1550–1850', in Topik, Steven, Marichal, Carlos and Frank, Zephyr, eds, *From Silver to Cocaine*, Durham, NC: Duke University Press, 2006, pp.76–92.

Marques, Vera Regina Beltrão, 'Escola de Homens de Ciências: A Academia Científica do Rio de Janeiro, 1772–1779', *Educar*, 25 (2005), pp.39–57.

Masson, Francis, 'An Account of the Island of St. Miguel . . .', *Philosophical Transactions of the Royal Society of London*, 68 (1778), pp.601–10.

Masson, Francis, 'An Account of Three Journeys from the Cape Town into the Southern Parts of Africa . . .', *Philosophical Transactions of the Royal Society of London*, 66 (1776), pp.268–317.

Mawe, John, *Travels in the Interior of Brazil*, London, 1818.

Maxwell, Anne, 'Fallen Queens and Phantom Diadems: Cook's Voyages and England's Social Order', *The Eighteenth Century*, 38 (1997), pp.247–58.

Maxwell-Stewart, Hamish, 'Convict Transportation from Britain and Ireland 1615–1870', *History Compass*, 8 (2010), pp.1221–42.

Meacham, Standish, *Henry Thornton of Clapham, 1760–1815*, Cambridge, MA: Harvard University Press, 1964.

Menegon, Eugenio, 'Wanted: An Eighteenth-Century Chinese Catholic Priest in China, Italy, India, and Southeast Asia', *Journal of Modern Italian Studies*, 15 (2010), pp.502–18.

Métailié, Georges, 'Des mots, des animaux et des plantes', *Extrême-Orient–Extrême-Occident*, 14 (1992), pp.169–83.

Métailié, Georges, *Science and Civilisation in China. Volume 6, Part IV*, Cambridge: Cambridge University Press, 2015.

Métailié, Georges, 'The *Bencao gangmu* of Li Shizhen – An Innovation in Natural History?', in Hsu, Elisabeth, ed., *Innovation in Chinese Medicine*, Cambridge: Cambridge University Press, 2001, pp.221–61.

Métailié, Georges, 'The Representation of Plants: Engravings and Paintings', in Bray, Francesca, Dorofeeva-Lichtmann, Vera and Métailié, Georges, eds, *Graphics and Text in the Production of Technical Knowledge in China: The Warp and the Weft*, Leiden: Brill, 2007, pp.487–520.

Meynell, Guy, 'Kew and the Royal Gardens Committee of 1838', *Archives of Natural History*, 10 (1982), pp.469–77.

Miller, David P., '"Into the Valley of Darkness": Reflections on the Royal Society in the Eighteenth Century', *History of Science*, 27 (1989), pp.155–66.

Miller, David Philip, 'Joseph Banks, Empire and "Centers of Calculation" in Late Hanoverian London', in Miller, David Philip and Reill, Peter Hans, eds, *Visions of Empire: Voyages, Botany, and Representations of Nature*, Cambridge: Cambridge University Press, 1996, pp.21–37.

Miller, David P., '"My Favourite Studys": Lord Bute as Naturalist', in Schweizer, Karl W., ed., *Lord Bute: Essays in Re Interpretation*, Leicester: Leicester University Press, 1988, pp.213–39.

Miller, David P., 'Sir Joseph Banks: An Historiographical Perspective', *History of Science*, 19 (1981), pp.284–92.

Misevich, Philip, 'The Sierra Leone Hinterland and the Provisioning of Early Freetown, 1792–1803', *Journal of Colonialism & Colonial History*, 9 (2008), n.p.

Mockford, Jim, 'The Journal of a Tour Across the Continent of New Spain from St. Blas in the North Pacific Ocean to La Vera Cruz in the Gulph of Mexico by Lieut. W.R. Broughton in the year 1793 Commander H.M. Brig *Chatham*', *Terrae Incognitae*, 36 (2004), pp.42–58.

Mokyr, Joel, *The Enlightened Economy: Britain and the Industrial Revolution 1700–1850*, London: Penguin Books, 2011.

Monod, Théodore, 'John Cranch, zoologiste de l'expédition du Congo (1816)', *Bulletin of the British Museum (Natural History) Historical Series*, 4 (1970), pp.1–75.

Moore, Harold E. and Hyppio, Peter A., 'Some Comments on Strelizia (Strelitziaceae)', *Baileya*, 17 (1970), pp.65–74.

Moore, Wendy, *The Knife Man: The Extraordinary Life and Times of John Hunter, Father of Modern Surgery*, New York: Broadway Books, 2005.

Moore, Wendy, *Wedlock: How Georgian Britain's Worst Husband Met His Match*, London: Phoenix, 2010.

Moran, V.C., 'Interactions Between Phytophagus Insects and their *Opuntia* Hosts', *Ecological Entomology*, 5 (1980), pp.153–64.

Morgan, Kenneth, *Australia Circumnavigated: The Voyage of Matthew Flinders in HMS Investigator, 1801–1803*, London: Hakluyt Society, 2015.

Morgan, Kenneth, 'From Cook to Flinders: The Navigation of Torres Strait', *International Journal of Maritime History*, 27 (2015), pp.41–60.

Morgan, Kenneth, 'A Historical Myth? Matthew Flinders and the Quest for a Strait', *Australian Historical Studies*, 48 (2017), pp.52–67.

Morgan, Kenneth, 'Matthew Flinders and the Charting of Australia's Coasts, 1798–1814', *Terrae Incognitae*, 50 (2018), pp.115–45.

Morgan, Kenneth, *Matthew Flinders, Maritime Explorer of Australia*, London: Bloomsbury, 2016.

Morrison, Robert, *A Memoir of the Principal Occurrences During an Embassy from the British Government to the Court of China in the Year 1816*, London, 1820.

Morse, Hosea Ballou, *The Chronicles of the East India Company Trading to China 1635–1834*, Oxford: Clarendon Press, 1926–9.

Mosca, Matthew W., 'The Qing State and Its Awareness of Eurasian Interconnections, 1789–1806', *Eighteenth-Century Studies*, 47 (2014), pp.103–16.

Mosca, Matthew W., *From Frontier Policy to Foreign Policy: The Question of India and the Transformation of Geopolitics in Qing China*, Stanford, CA: Stanford University Press, 2013.

Mosca, Matthew W., 'Hindustan as a Geographic and Political Concept in Qing Sources, 1700–1800', *China Report*, 47 (2011), pp.263–77.

Mouser, Bruce L., ed., *The Forgotten Peddie/Campbell Expedition into Fuuta Jaloo, West Africa, 1815–17*, Madison, WI: University of Wisconsin, 2007.

Mungello, D.E., *Curious Land: Jesuit Accommodation and the Origins of Sinology*, Wiesbaden: Fritz Steiner Verlag, 1985.

Munro, John H., 'The Medieval Scarlet and the Economics of Sartorial Splendour', in Harte, N.B. and Ponting, K.G., eds, *Cloth and Clothing in Medieval Europe: Essays in Memory of Professor E.M. Carus-Wilson*, London: Pasold, 1983, pp.13–70.

Munthe, Preben, *Christen Smith: Botaniker og Økonom*, Oslo: Aschehoug, 2004.

Muntschick, Wolfgang, 'The Plants That Carry His Name: Kaempfer's Study of the Japanese Flora', in Bodart-Bailey, Beatrice M. and Massarella, Derek, eds, *The Furthest Goal: Engelbert Kaempfer's Encounter with Tokugawa Japan*, Folkestone: Japan Library, 1995, pp.71–95.

Nadri, Ghulam A., 'The Dynamics of Port–Hinterland Relationships in Eighteenth-Century Gujarat', in Mizushima, Tsukasu, Souza, George Bryan and Flynn, Dennis O., eds, *Hinterlands and Commodities: Place, Space, Time and the Political Economic Development of Asia over the Long Eighteenth Century*, Leiden: Brill, 2015, pp.83–101.

Nadri, Ghulam A., *Eighteenth Century Gujarat: The Dynamics of its Political Economy, 1750–1800*, Leiden: Brill, 2009.

Nappi, Carla Suzan, *The Monkey and the Inkpot: Natural History and its Transformations in Early Modern China*, Cambridge, MA: Harvard University Press, 2009.

Nash, M.D., *The Last Voyage of the Guardian Lieutenant Riou, Commander 1789–1791*, Cape Town: Van Riebeeck Society, 1990.

Needham, Joseph, Wang, Ling and Price, Derek J. de Solla, *Heavenly Clockwork: The Great Astronomical Clocks of Medieval China*, Cambridge: Cambridge University Press, 2008.

Neild, Susan M., 'Colonial Urbanism: The Development of Madras City in the Eighteenth and Nineteenth Centuries', *Modern Asian Studies*, 13 (1979), pp.217 46.

Nelson, E. Charles, 'Archibald Menzies's Visit to Isla del Coco, January 1795', *Archives of Natural History*, 40 (2013), pp.149–56.

Nelson, E. Charles, 'From Tubs to Flying Boats: Episodes in Transporting Living Plants', in MacGregor, Arthur, ed., *Naturalists in the Field: Collecting, Recording and Preserving the Natural World from the Fifteenth to the Twenty-First Century*, Leiden: Brill, 2018, pp.578–606.

Nelson, E. Charles and Porter, Duncan M., 'Archibald Menzies on Albermarle Island, Gálapagos Archipelago', *Archives of Natural History*, 38 (2011), pp.104–12.

Nero, Karen L., Thomas, Nicholas and Newell, Jennifer, eds, *An Account of the Pelew Islands by George Keate*, Leicester: Leicester University Press, 2002.

Newcombe, C.F., *Menzies' Journal of Vancouver's Voyage April to October, 1792*, Victoria, BC: Archives of British Columbia, 1923.

Newell, Jennifer, *Trading Nature: Tahitians, Europeans, and Ecological Exchange*, Honolulu, HI: University of Hawai'i Press, 2010.

Newport, Emma, 'The Fictility of Porcelain: Making and Shaping Meaning in Lady Dorothea Banks's "Dairy Book"', *Eighteenth-Century Fiction*, 31 (2018), pp.117–42.

Newton, John, *A Savage History: The Story of Whaling in the Pacific and Southern Oceans*, Sydney: NewSouth Publishing, 2013.

Nightingale, Carl H., 'Before Race Mattered: Geographies of the Color Line in Early Colonial Madras and New York', *American Historical Review*, 113 (2008), pp.48–71.

Nokes, J. Richard, *Almost a Hero: The Voyages of John Meares, R.N., to China, Hawaii and the Northwest Coast*, Pullman, WA: Washington State University Press, 1998.

Noltie, Henry J., 'John Bradby Blake and James Kerr: Hybrid Botanical Art, Canton and Bengal, c. 1770', *Curtis's Botanical Magazine*, 34 (2017), pp.427–51.

Noltie, H.J., *John Hope (1725–1786): Alan G. Morton's Memoir of a Scottish Botanist*, Edinburgh: Royal Botanic Garden Edinburgh, 2011.

Norris, John, 'The Policy of the British Cabinet in the Nootka Crisis', *English Historical Review*, 277 (1955), pp.562–80.

Norst, Marlene J., *Ferdinand Bauer: The Australian Natural History Drawings*, London: British Museum (Natural History), 1989.

Norst, Marlene, 'Recognition and Renaissance: Ferdinand Lucas Bauer 1760–1826', *Australian Natural History*, 23 (1990), pp.296–305.

Nyberg, Kenneth, 'Linnaeus's Apostles and the Globalization of Knowledge, 1729–1756', in Manning, Patrick and Rood, Daniel, eds, *Global Scientific Practice in an Age of Revolutions, 1750–1850*, Pittsburgh, PA: University of Pittsburgh Press, 2016, pp.73–89.

Nyberg, Kenneth, 'Linnaeus' Apostles, Scientific Travel and the East India Trade', *Zoologica Scripta*, 38 (supplement 1) (2009), pp.7–16.

O'Grady-Raeder, Alix, 'Major von Behm und Captain Cooks letzte Tagebücher', *Jahrbücher für Geschichte Osteuropas*, 37 (1989), pp.65–72.

Okihiro, Gary Y., 'Of Space/Time and the Pineapple', *Atlantic Studies*, 11 (2014), pp.85–102.

Oldfield, J.R., *Popular Politics and British Anti-Slavery: The Mobilisation of Public Opinion Against the Slave Trade, 1787–1807*, London: Frank Cass, 1998.

Oldham, Wilfrid, *Britain's Convicts to the Colonies*, Sydney: Library of Australian History, 1990.

Olivová, Lucie, 'A Map of the Chinese Imperial Summer Resort Discovered in a Czech Museum', *Imago Mundi*, 62 (2010), pp.232–8.

Olson, Wallace M., ed., *The Alaska Travel Journal of Archibald Menzies, 1793–1794*, Fairbanks, AK: University of Alaska Press, 1993.

Ong, S.P., 'Jurisdictional Politics in Canton and the First English Translation of the Qing Penal Code (1810), *Journal of the Royal Asiatic Society*, 20 (2010), pp.141–65.

Orchard, A.E. and Orchard, T.A., *Allan Cunningham: Letters of a Botanist/Explorer, 1791–1839*, Western Creek: Privately published, 2015.

Orchard, A.E. and Orchard, T.A., *King's Collectors for Kew: James Bowie and Allan Cunningham, Brazil, 1814–1816*, Western Creek: Privately published, 2015.

O'Shaughnessy, Andrew Jackson, *An Empire Divided: The American Revolution and the British Caribbean*, Philadelphia, PA: University of Pennsylvania Press, 2000.

O'Shaughnessy, Andrew J., 'The Formation of a Commercial Lobby: The West India Interest, British Colonial Policy and the American Revolution', *The Historical Journal*, 40 (1997), pp.71–95.

Osborne, Brian D., *The Ingenious Mr Bell*, Glendaruel: Argyll Publishing, 1995.

Ottaviani, Alessandro, 'Gaetano D'Ancora fra antiquaria, filologia e storia naturale', in Mazzola, Roberto, ed., *Le scienze a Napoli tra Illuminismo e Restaurazione*, Rome: Aracne, 2011, pp.61–78.

Padilla, Carmella and Anderson, Barbara, eds, *A Red Like No Other: How Cochineal Colored the World*, New York: Skira Rizzoli, 2015.

Pagès, Pierre Marie François de, *Voyages autour du monde, et vers les deux pôles, par terre et par mer, pendant les années 1767–1776*, Paris, 1782.

Pagnamenta, Frank, 'The Aitons: Gardeners to their Majesties, and Others', *Richmond History*, 18 (1997), pp.7–19.

Paley, Morton D., '"A New Heaven is Begun": William Blake and Swedenborgianism', *Blake*, 13 (1979), pp.64–90.

Pan, Jixing, 'Charles Darwin's Chinese Sources', *Isis*, 75 (1984), pp.530–4.

Papavero, N., *Essays on the History of Neotropical Dipterology*, São Paulo: Museu de Zoologia, Universidade de São Paulo, 1971.

Park, Mungo, *Travels in the Interior Districts of Africa*, London, 1815.

Parkinson, C. Northcote, *War in the Eastern Seas 1793–1815*, London: George Allen & Unwin, 1954.

Parkinson, Sydney, *A Journal of a Voyage to the South Seas, In His Majesty's Ship Endeavour . . .*, London, 1784

Parsons, Christopher M. and Murphy, Kathleen S., 'Ecosystems Under Sail: Specimen Transport in the Eighteenth-Century French and British Atlantics', *Early American Studies*, 10 (2012), pp.503–29.

Parthasarathi, Prasannan, 'Cotton Textiles in the Indian Subcontinent, 1200–1800', in Riello, Giorgio and Parthasarathi, Prasannan, eds, *The Spinning World: A Global History of Cotton Textiles, 1200–1850*, Oxford: Oxford University Press, 2009, pp.17–42.

Patel, Sandhya, ed., *Exploration of the South Seas in the Eighteenth Century: Rediscovered Accounts*, London: Routledge, 2016.

Pembroke, Michael, *Arthur Phillip: Sailor Mercenary Governor Spy*, Melbourne: Hardie Grant Books, 2013.

Penson, Lillian M., 'The London West India Interest', *English Historical Review*, 36 (1921), pp.373–92.

Perdue, Peter C., 'Boundaries and Trade in the Early Modern World: Negotiations at Nerchinsk and Beijing', *Eighteenth Century Studies*, 43 (2010), pp.341–56.

Perry, T.M. and Simpson, Donald H., eds, *Drawings by William Westall: Landscape Artist on Board H.M.S. Investigator During the Circumnavigation of Australia by Captain Matthew Flinders, R.N. in 1801–1803*, London: Royal Commonwealth Society, 1962.

Peyrefitte, Alain, *Un choc de cultures: La vision des Chinois*, Paris: Fayard, 1991.

Peyrefitte, Alain, *Un choc de cultures: Le regard des Anglais*, Paris: Fayard, 1998.

Peyrefitte, Alain, *The Immobile Empire*, New York: Alfred A. Knopf, 1992.

Phillips, Charlotte and Shane, Nora, eds, *John Stuart 3rd Earl of Bute 1713–92*, Luton: Luton Hoo Estate, 2014.

Phillips, Jim, 'Private Profit and Imperialism in Eighteenth Century Southern India: The Tanjore Revenue Dispute, 1775–1777', *South Asia: Journal of South Asian Studies*, 9 (1986), pp.1–16.

Phillipson, Mildred, *Holywell and the Birch Reynardsons*, Sleaford, Lincolnshire: Society for Lincolnshire History and Archaeology, 1981.

Phipps, Elena, *Cochineal Red: The Art History of a Color*, New Haven, CT: Yale University Press, 2010.

Pierce, R.A., ed., *A List of Trading Vessels in the Maritime Fur Trade, 1785–1825*, Kingston, ON: Limestone Press, 1973.

Portlock, Nathaniel, *A Voyage Round the World; but more Particularly to the North-West Coast of America . . .*, London, 1789.

Potgieter, Thean, 'Admiral Elphinstone and the Conquest and Defence of the Cape of Good Hope, 1795–6', *Scientia Militaria: South African Journal of Military Studies*, 35 (2007), pp.39–67.

Powell, Dulcie, 'The Voyage of the Plant Nursery, H.M.S. *Providence*, 1791–1793, *Economic Botany*, 31 (1977), pp.387–431.

Pritchard, Earl H., 'The Crucial Years of Early Anglo–Chinese Relations, 1750–1800', *Research Studies of the State College of Washington*, 4 (1936).

Pritchard, Earl H., 'The Instructions of the East India Company to Lord Macartney on his Embassy to China and his Reports to the Company, 1792–4. Part II: Letter to the Viceroy and First Report', *Journal of the Royal Asiatic Society of Great Britain and Ireland*, 3 (1938), pp.375–96.

Pritchard, Earl H., 'The Instructions of the East India Company to Lord Macartney on his Embassy to China and his Reports to the Company, 1792–4. Part III: Later Reports and a Statement of the Cost of the Embassy', *Journal of the Royal Asiatic Society of Great Britain and Ireland*, 4 (1938), pp.493–509.

Proescholdt, Catherine W., 'Johann Christian Hüttner (1766–1847): A Link Between Weimar and London', in Boyle, Nicholas and Guthrie, John, eds, *Goethe and the English-Speaking World*, Rochester, NY: Camden House, 2002, pp.99–110.

Quintanilla, Mark, 'Mercantile Communities in the Ceded Islands: The Alexander Bartlet & George Campbell Company', *International Social Science Review*, 79 (2004), pp.14–26.

Quintanilla, Mark, 'The World of Alexander Campbell: An Eighteenth-Century Grenadian Planter', *Albion*, 35 (2003), pp.229–56.

Raj, Kapil, *Relocating Modern Science: Circulation and the Construction of Knowledge in South Asia and Europe, 1650–1900*, Basingstoke: Palgrave Macmillan, 2007.

Ransom, P.J.G., *Bell's Comet: How a Paddle Steamer Changed the Course of History*, Stroud: Amberley Publishing, 2012.

Ratcliff, Mark J., *The Quest for the Invisible: Microscopy in the Enlightenment*, Aldershot: Ashgate, 2009.

Rauschenberg, Roy A., 'John Ellis, F.R.S.: Eighteenth Century Naturalist and Royal Agent to West Florida', *Notes and Records of the Royal Society of London*, 32 (1978), pp.149–64.

Ravalli, Richard John, 'Soft Gold and the Pacific Frontier: Geopolitics and Environment in the Sea Otter Trade', unpublished PhD dissertation, University of California, Merced, 2009.

Réaumur, René Antoine Ferchault de, *Mémoires pour servir à l'histoire des insects*, vol.4, Paris, 1738.

Redford, Bruce, *Dilettanti: The Antic and the Antique in Eighteenth-Century England*, Los Angeles, CA: J. Paul Getty Museum, 2008.

Rego, Marco Antonio, Moreira-Lima, Luciano, Silveira, Luís Fábio and Frahnert, Sylke, 'On the Ornithological Collection of Friedrich Sellow in Brazil (1814–1831), with Some Considerations about the Provenance of his Specimens', *Zootaxa*, 3616 (2013), pp.478–84.

Rembert, David H., Jr., 'William Pitcairn, MD (1712–1791) – a Biographical Sketch', *Archives of Natural History*, 12 (1985), pp.219–29.

Rendle, A.B., 'John Gerard Koenig', *The Journal of Botany British and Foreign*, 71 (1933), pp.143–53, 175–87.

Renkema, H.W. and Ardaugh, J., 'Aylmer Bourke Lambert and his "Description of the genus *Pinus*"', *Journal of the Linnean Society*, 48 (1930), pp.439–66.

Reynolds, Henry, *A History of Tasmania*, Cambridge: Cambridge University Press, 2012.

Ribeiro, Alvaro, S.J., ed., *The Letters of Dr Charles Burney, Volume 1, 1751–1784*, Oxford: Clarendon Press, 1991.

Rich, Alexander and Kierkuć-Bieliński, Jerzy, *Peace Breaks Out: London and Paris in the Summer of 1814*, London: Sir John Soane's Museum, 2014.

Richard, Josepha, 'Uncovering the Garden of the Richest Man on Earth in Nineteenth-Century Canton: Howqua's Garden in Honam, China', *Garden History*, 43 (2015), pp.168–81.

Richard, Josepha and Woudstra, Jan, '"Thoroughly Chinese": Revealing the Plants of the Hong Merchants' Gardens through John Bradby Blake's Paintings', *Curtis's Botanical Magazine*, 34 (2017), pp.475–97.

Riello, Giorgio, *Cotton: The Fabric that Made the Modern World*, Cambridge: Cambridge University Press, 2013.

Riello, Giorgio, 'The Globalization of Cotton Textiles: Indian Cottons, Europe, and the Atlantic World, 1600–1850', in Riello, Giorgio and Parthasarathi, Prasannan, eds, *The Spinning World: A Global History of Cotton Textiles, 1200–1850*, Oxford: Oxford University Press, 2009, pp.261–87.

Riello, Giorgio and Roy, Tirthankar, eds, *How India Clothed the World: The World of South Asian Textiles, 1500–1850*, Leiden: Brill, 2009.

Riello, Giorgio, 'The Indian Apprenticeship: The Trade of Indian Textiles and the Making of European Cottons', in Riello, Giorgio and Roy, Tirthankar, eds, *How India Clothed the World: The World of South Asian Textiles, 1500–1850*, Leiden: Brill, 2009, pp.309–46.

Rigby, Nigel, '"Not at all a Particular Ship": Adapting Vessels for British Voyages of Exploration', in Wege, Juliet, George, Alex, Gathe, Jan, Lemson, Kris and Napier, Kath, eds, *Matthew Flinders and His Scientific Gentlemen: The Expedition of H.M.S. Investigator to Australia, 1801–5*, Welshpool: Western Australian Museum, 2005, pp.13–23.

Rigby, Nigel, 'The Politics and Pragmatics of Seaborne Plant Transportation, 1769–1805', in Lincoln, Margarette, ed., *Science and Exploration in the Pacific: European Voyages to the Southern Oceans in the Eighteenth Century*, Woodbridge: Boydell Press, 1998, pp.81–100.

Rigby, Nigel, '"The Whole of the Surveying Department Rested on Me": Matthew Flinders, Hydrography and the Navy', in Rivière, Marc Serge and Issur, Kumari R., eds, *Baudin-Flinders dans L'Océan: Voyages, Découvertes, Rencontre: Voyages, Discoveries, Encounter*, Paris: L'Harmattan, 2006, pp.259–70.

Rivinius, Karl Joseph, *Das Collegium Sinicum zu Neapel und seine Umwandlung in ein Orientalisches Institut: Ein Beitrag zu seiner Geschichte*, Nettetal: Steyler Verlag, 2004.

Rix, Robert, *William Blake and the Radical Swedenborgians*, at *www.esoteric.msu.edu/VolumeV/Blake.htm*

Robbins, Helen H., *Our First Ambassador in China: An Account of the Life of George, Earl of Macartney*, London: John Murray, 1908.

Robinson, David, 'Sir Joseph Banks and the East Fen', in Sturman, Christopher, ed., *Lincolnshire People and Places: Essays in Memory of Terence R. Leach (1937–1994)*, Lincoln: Society for Lincolnshire History & Archaeology, 1996, pp.97–105.

Robinson, David, *Sir Joseph Banks at Revesby*, Horncastle: Sir Joseph Banks Society, 2014.

Robinson, Tim, *William Roxburgh: The Founding Father of Indian Botany*, Chichester: Phillimore, 2008.

Rodger, N.A.M., *The Insatiable Earl: A Life of John Montagu, Fourth Earl of Sandwich*, London: HarperCollins, 1993.

Roebuck, Peter, ed., *Macartney of Lisanoure 1737–1806: Essays in Biography*, Belfast: Ulster Historical Foundation, 1983.

Ronalds, Beverley F., 'Ronalds Nurserymen in Brentford and Beyond', *Garden History*, 45 (2017), pp.82–100.

Rönnbäck, Klas, 'Enlightenment, Scientific Exploration and Abolitionism: Anders Sparrman's and Carl Bernhard Wadström's Colonial Encounters in Senegal, 1787–1788 and the British Abolitionist Movement', *Slavery & Abolition*, 34 (2013), pp.425–45.

Rose, J. Holland, 'Great Britain and the Dutch Question in 1787–1788', *American Historical Review*, 14 (1909), pp.262–83.

Ross, Charles, ed., *Correspondence of Charles, First Marquis Cornwallis*, London, 1859.

Rotberg, Robert I., 'The Swedenborgian Search for African Purity', *Journal of Interdisciplinary History*, 36 (2005), pp.233–40.

Russell, Gillian, 'An "Entertainment of Oddities": Fashionable Sociability and the Pacific in the 1770s', in Wilson, Kathleen, ed., *A New Imperial History: Culture, Identity and Modernity in Britain and the Empire, 1660–1840*, Cambridge: Cambridge University Press, 2004, pp.48–70.

Russell, Patrick, *A Continuation of an Account of Indian Serpents Containing Descriptions and Figures from Specimens and Drawings*, London, 1801.

Rutter, Owen, ed., *Bligh's Voyage in the Resource*, London: Golden Cockerell Press, 1937.

Rybolt, John E., 'Saints of the Vincentian Family', at *www.vincenziani.com/nuovaconferenza.htm*

St André, James, 'Travelling Toward True Translation: The First Generation of Sino–English Translators', *The Translator*, 12 (2006), pp.189–210.

St John, Harold, 'Biography of David Nelson, and an Account of his Botanizing in Hawaii', *Pacific Science*, 30 (1976), pp.1–5.

Salmon, Michael A., *The Aurelian Legacy: British Butterflies and their Collectors*, Berkeley, CA: University of California Press, 2000.

Salmond, Anne, *Aphrodite's Island*, Berkeley, CA: University of California Press, 2010.

Salmond, Anne, *Bligh: William Bligh in the South Seas*, Berkeley, CA: University of California Press, 2011.

Sánchez Silva, Carlos and de Ávila Blomberg, Alejandro, *La grana y el nopal en los textos de Alzate*, Mexico City: Archivo General de la Nación, 2005.

Sankey, Margaret, 'The Baudin Expedition in Port Jackson, 1802: Cultural Encounters and Enlightenment Politics', *Explorations*, 31 (2001), pp.5–36.

Saraiba Russell, Ángeles, 'En búsqueda de la grana cochinilla, Thiery de Menonville en Oaxaca, 1777', *Acervos*, 23 (2001), pp.13–26.

Schaffer, Simon, 'Swedenborg's Lunars', *Annals of Science*, 71 (2014), pp.2–26.

Schama, Simon, *Rough Crossings: Britain, the Slaves and the American Revolution*, London: BBC Books, 2005.

Schreiber, Roy, ed., *Captain Bligh's Second Chance: An Eyewitness Account of his Return to the South Seas*, London: Chatham, 2007.

Schultz, Kirsten, *Tropical Versailles: Empire, Monarchy, and the Portuguese Royal Court in Rio de Janeiro, 1808–1821*, London: Routledge, 2001.

Schwarz, Suzanne, '"A Just and Honourable Commerce": Abolitionist Experimentation in Sierra Leone in the Late Eighteenth and Early Nineteenth Centuries', *African Economic History*, 45 (2017), pp.1–45.

Scobie, Ruth, '"To Dress a Room for Montagu": Pacific Cosmopolitanism and Elizabeth Montagu's Feather Hangings', *Lumen*, 33 (2014), pp.123–37.

Screech, Timon, *Japan Extolled and Decried: Carl Peter Thunberg and the Shogun's Realm, 1775–1796*, London: Routledge, 2005.

Secord, Anne, 'Science in the Pub: Artisan Botanists in Early Nineteenth-Century Lancashire', *History of Science*, 32 (1994), pp.269–315.

Serrano, Ana, Sousa, Michael M., Hallett, Jessica, Lopes, João and Oliveira, M. Conceição, 'Analysis of Natural Red Dyes (cochineal) in textiles of Historical Importance Using HPLC and Multivariate Data Analysis', *Analytical and Bioanalytical Chemistry*, 401, (2011), pp.735–43.

Shaw, Alan, 'The Founding of Melbourne', in Statham, Pamela, ed., *The Origins of Australia's Capital Cities*, Cambridge: Cambridge University Press, 1989, pp.199–215.

Shellim, Maurice, *Oil Paintings of India and the East by Thomas Daniell, 1749–1840 and William Daniell, 1769–1837*, London: Inchcape, 1979.

Shephard, Sue, *Seeds of Fortune: A Gardening Dynasty*, London: Bloomsbury, 2003.

Sheridan, Richard B., 'Captain Bligh, the Breadfruit, and the Botanic Gardens of Jamaica', *Journal of Caribbean History*, 23 (1989), pp.28–50.

Sheridan, Richard B., 'The Crisis of Slave Subsistence in the British West Indies During and After the American Revolution', *William and Mary Quarterly*, 33 (1976), pp.615–41.

Singer, Aubrey, *The Lion and the Dragon: The Story of the First British Embassy to the Court of the Emperor Qianlong in Peking 1792–94*, London: Barrie & Jenkins, 1992.

Skuncke, Marie-Christine, *Carl Peter Thunberg: Botanist and Physician: Career-Building across the Oceans in the Eighteenth Century*, Uppsala: Swedish Collegium for Advanced Study, 2014.

Sloan, Hans [sic], 'An Account of the True *Cortex Winteranus*, and the Tree that Bears it . . .'. *Philosophical Transactions*, 17 (1693), pp.922–4.

Smith, Edgar C., *A Short History of Naval and Marine Engineering*, Cambridge: Cambridge University Press, 1938.

Smith, Keith Vincent, 'Bennelong Among his People', *Aboriginal History*, 33 (2009), pp.7–30.

Smith, Keith Vincent, *King Bungaree: A Sydney Aborigine Meets the Great South Pacific Explorers, 1799–1830*, Kenthurst: Kangaroo Press, 1992.

Smith, Keith Vincent, *Mari Nawi: Aboriginal Odysseys*, Kenthurst: Rosenberg, 2010.

Smith, Roger, 'The Sing-Song Trade: Exporting Clocks to China in the Eighteenth Century', *Antiquarian Horology* 30 (2008), pp.629–58.

Smith, Vanessa, 'Give Us Our Daily Breadfruit: Bread Substitution in the Pacific in the Eighteenth Century', *Studies in Eighteenth Century Culture*, 35 (2006), pp.53–75.

Smith, Vanessa and Thomas, Nicholas, eds. (with the assistance of Maia Nuku), *Mutiny and Aftermath: James Morison's Account of the Mutiny on the Bounty and the Island of Tahiti*, Honolulu, HI: University of Hawai'i Press, 2013.

[Smollett, Tobias], 'A Journal of a Voyage round the World . . .', *The Critical Review*, October 1771, pp.256–61.

Sonntag, Otto, ed., *John Pringle's Correspondence with Albrecht von Haller*, Basle: Schwabe & Co, 1999.

Sparrman, Andrew, *A Voyage to the Cape of Good Hope, Towards the Antarctic Polar Circle, and Round the World . . .*', London, 1785.

Sparrow, Elizabeth, *Secret Service: British Agents in France, 1792–1815*, Woodbridge: Boydell Press, 1999.

Springs, Lauren C., 'St. George's Caye: A Bioarchaeological Study of Eighteenth Century Belize', unpublished MA dissertation, Texas State University – San Marcos, 2012.

Stackpole, Edouard A., *Whales & Destiny: The Rivalry Between America, France and Britain for Control of the Southern Whale Fishery, 1785–1825*, Amherst, MA: University of Massachusetts Press, 1972.

Standaert, Nicolas, ed., *Handbook of Christianity in China*, Leiden: Brill, 2001.

Staunton, George, *An Authentic Account of an Embassy from the King of Great Britain to the Emperor of China*, London, 1797.

Staunton, George Thomas, *Memoir of the Life & Family of the Late Sir George Leonard Staunton, Bart.*, Havant, 1823.

Staunton, George Thomas, *Memoirs of the Chief Incidents of the Public Life of Sir George Thomas Staunton, Bart.*, London, 1856.

Staunton, George Thomas, *Notes of Proceedings and Occurrences During the British Embassy to Pekin, in 1816*, London, 1824.

Stearns, Raymond Phineas, 'James Petiver: Promoter of Natural Science, c.1663–1718', *Proceedings of the American Antiquarian Society*, 62 (1953), pp.243–365.

Steller, Georg Wilhelm, *Journal of a Voyage with Bering 1741–1742*, edited by O.W. Frost, Stanford, CA: Stanford University Press, 1988.

Stewart, Larry, 'His Majesty's Subjects: From Laboratory to Human Experiment in Pneumatic Chemistry', *Notes & Records of the Royal Society*, 63 (2009), pp.231–45.

Stifler, Susan Reed, 'The Language Students of the East India Company's Canton Factory', *Journal of the North China Branch of the Royal Asiatic Society*, 69 (1933), pp.46–82.

Stolberg, Eva-Maria, 'Interracial Outposts in Siberia: Nerchinsk, Kiakhta, and the Russo–Chinese Trade in the Seventeenth/Eighteenth Centuries', *Journal of Early Modern History*, 4 (2000), pp.322–36.

Stott, Anne, *Wilberforce: Family and Friends*, Oxford: Oxford University Press, 2012.

Sunding, Per, Santos-Guerra, Arnoldo and Hansen, Cristina S., *Diario del Viaje a las Islas Canarias en 1815: Christen Smith*, La Orotava, Tenerife: Fundación Canaria Orotava de Historia de la Ciencia, 2005.

Sutherland, Lucy Stuart, *The East India Company in Eighteenth-Century Politics*, Oxford: Clarendon Press, 1952.

Sweet, Jessie M., 'Instructions to Collectors: John Walker (1793) and Robert Jameson (1817); With Biographical Notes on James Anderson (LL.D.) and James Anderson (M.D.)', *Annals of Science*, 29 (1972), pp.397–414.

Sweet, Rosemary, 'Antiquaries and Antiquities in Eighteenth-Century England', *Eighteenth-Century Studies*, 34 (2001), pp.171–206.

Synge, Patrick M., 'Chinese Flower Paintings: An Important Purchase by the Royal Horticultural Society', *Journal of the Royal Horticultural Society*, 78 (1953), pp.209–13.

Tagliacozzo, Eric and Chang, Wen-Chin, eds, *Chinese Circulations: Capital, Commodities, and Networks in Southeast Asia*, Durham, NC: Duke University Press, 2011.

Taylor, James, 'The Creation and Reception of William Westall's Admiralty Oil Paintings Derived from his Voyage on HMS *Investigator* 1801–1803', unpublished PhD dissertation, University of Sussex, 2015.

Taylor, James, *Picturing the Pacific: Joseph Banks and the Shipboard Artists of Cook and Flinders*, London: Adlard Coles, 2018.

Taylor, Kristina and Peel, Robert, *Passion, Plants and Patronage: 300 Years of the Bute Family Landscapes*, London: Artifice Books, 2012.

The Anh, Nguyen, 'L'Angleterre et le Viêt-Nam en 1803: la mission de J.W. Roberts', *Bulletin de la Société des Études Indochinoises*, 40 (1965), pp.339–47.

Thijsse, Gerard, 'A Contribution to the History of the Herbaria of George Clifford III (1685–1760), *Archives of Natural History*, 45 (2018), pp.134–48.

Thomas, Adrian P., 'The Establishment of the Calcutta Botanic Garden: Plant Transfer, Science and the East India Company, 1786–1806', *Journal of the Royal Asiatic Society*, 16 (2006), pp.165–77.

Thunberg, Carl Peter, *Travels in Europe, Africa, and Asia, Made Between the Years 1770 and 1779*, London, 1795.

Tiedemann, R.G., *Reference Guide to Christian Missionary Societies in China from the Sixteenth to the Twentieth Century*, Armonk, NY: M.E. Sharpe, 2009.

Tipping, Marjorie, *Convicts Unbound: The Story of the Calcutta Convicts and their Settlement in Australia*, Ringwood, VIC: Viking O'Neil, 1988.

Tobin, Beth Fowkes, 'Imperial Designs: Botanical Illustration and the British Botanic Empire', *Studies in Eighteenth Century Culture*, 25 (1996), pp.265–92.

Tovell, Freeman M., *At the Far Reaches of Empire: The Life of Juan Francisco de la Bodega y Quadra*, Vancouver, BC: UBC Press, 2008.

Tovell, Freeman M., 'The Other Side of the Coin: The Viceroy, Bodega y Quadra, Vancouver, and the Nootka Crisis', *B.C. Studies*, 93 (1992) pp.3–29.

Tuck, Patrick, ed., *Britain and the China Trade 1635–1842*, London: Routledge, 1999.

Tuckey, J.K., *Narrative of an Expedition to Explore the River Zaire Usually Called the Congo in South Africa in 1816*, London, 1818.

Uglow, Jenny, *The Lunar Men: The Friends Who Made the Future 1730–1810*, London: Faber and Faber, 2002.

Usher, John Purcell, *Memorial Inscriptions and Epitaphs: Belize, British Honduras*, London: Cassell & Co., 1907.

Vallance, T.G., Moore, D.T. and Groves, E.W., *Nature's Investigator: The Diary of Robert Brown in Australia, 1801–1805*, Canberra: ABRS, 2001.

Van Dyke, Paul A., *The Canton Trade: Life and Enterprise on the China Coast, 1700–1845*, Hong Kong: Hong Kong University Press, 2005.

Van Dyke, Paul A., 'The Hume Scroll of 1772 and the Faces Behind the Canton Factories', *Review of Culture (International Edition)*, 54 (2017), pp.64–102.

Van Dyke, Paul A. and Mok, Maria Kar-wing, *Images of Canton 1760–1822*, Hong Kong: Hong Kong University Press, 2015.

Van Gelder, Roelof, '*Nec semper feriet quodcumque minabitur arcus* – Engelbert Kaempfer as a Scientist in the Service of the Dutch East India Company', in Haberland, Detlef, ed., *Engelbert Kaempfer (1651–1716): Ein Gelehrtenleben zwischen Tradition und Innovation*, Wiesbaden: Harrassowitz Verlag, 2004, pp.211–25.

Vancouver, George, *A Voyage of Discovery to the North Pacific Ocean, and Round the World*, Volume 1, London, 1793.

Veth, Peter, Sutton, Peter and Neale, Margo, eds, *Strangers on the Shore: Early*

Coastal Contacts in Australia, Canberra: National Museum of Australia, 2008.

Von Buch, Leopold, 'Biographical Memoir of the Late Christian Smith, M.D. Naturalist to the Congo Expedition', *Edinburgh New Philosophical Journal*, (1826), pp.209–16.

Wadstrom, C.B., *An Essay on Colonization*, London, 1794.

Wadsworth, Arthur P. and Mann, Julia de Lacy, *The Cotton Trade and Industrial Lancashire 1600–1780*, Manchester: Manchester University Press, 1931.

Ward, Kerry, '"Tavern of the Seas?": The Cape of Good Hope as an Oceanic Crossroads during the Seventeenth and Eighteenth Centuries', in Bentley, Jerry H., Bridenthal, Renate and Wigen, Kären, eds, *Seascapes: Maritime Histories, Littoral Cultures, and Transoceanic Exchanges*, Honolulu, HI: University of Hawai'i Press, 2007, pp.137–52.

Ward, Peter A., *British Naval Power in the East 1794–1805: The Command of Admiral Peter Rainier*, Woodbridge: Boydell Press, 2013.

Waters, Ivor, *The Unfortunate Valentine Morris*, Chepstow: The Chepstow Society, 1964.

Webb, Joan, *George Caley: Nineteenth Century Naturalist*, Chipping Norton, NSW: Surrey Beatty & Sons, 1995.

Weightman, Gavin, *The Industrial Revolutionaries: The Creators of the Modern World 1776–1914*, London: Atlantic Books, 2007.

West-Sooby, John, 'Une expedition sous haute surveillance: le voyage aux Terres Australes vu par les Anglais', in Jangoux, Michel, ed., *Portés par l'air du temps: les voyages du Capitaine Baudin*, Brussels: Éditions de l'Université de Bruxelles, 2010, pp.187–210.

Whatmore, Richard, *Against War and Empire: Geneva, Britain and France in the Eighteenth Century*, New Haven, CT: Yale University Press, 2012.

White, Ian, *English Clocks for the Eastern Markets: English Clockmakers Trading in China & the Ottoman Empire 1580–1815*, Ticehurst: Antiquarian Horological Society, 2012.

Wilcken, Patrick, *Empire Adrift: The Portuguese Court in Rio de Janeiro, 1808–1821*, London: Bloomsbury, 2004.

Williams, D.J., 'Scale Insects (Hemiptera: Coccoidea) Described by James Anderson M.D. of Madras', *Journal of Natural History*, 36 (2002), pp.237–46.

Williams, David M. and Armstrong, John, '"One of the Noblest Inventions of the Age": British Steamboat Numbers, Diffusion, Services and Public Reception, 1812–c.1823', *Journal of Transport History*, 35 (2014), pp.18–34.

Williams, Glyn, *Arctic Labyrinth: The Quest for the Northwest Passage*, London: Allen Lane, 2009.

Williams, Glyn, *Naturalists at Sea: Scientific Travellers from Dampier to Darwin*, New Haven, CT: Yale University Press, 2013.

Williams, Glyndwr, *The British Search for the Northwest Passage in the Eighteenth Century*, London: Longman, 1962.

Williams, Glyndwr, ed., *Captain Cook's Voyages, 1768–1779*, London: The Folio Society, 1997.

Williams, Glyndwr, 'The *Endeavour* Voyage: A Coincidence of Motives', in Lincoln, Margarette, ed., *Science and Exploration in the Pacific*, Woodbridge: Boydell Press, 1998, pp.3–18.

Williams, Glyndwr, '"To Make Discoveries of Countries Hitherto Unknown": The Admiralty and Pacific Exploration in the Eighteenth Century', *Mariner's Mirror*, 82, (1996), pp.14–27.

Williams, Laurence, 'British Government under the Qianlong Emperor's Gaze: Satire, Imperialism, and the Macartney Embassy to China, 1792–1804', *Lumen* 32 (2013), pp.85–107.

Wills, John E., Jr., *Mountain of Fame: Portraits in Chinese History*, Princeton, NJ: Princeton University Press, 1994.

Wilson, Dick A., 'King George's Men: British Ships and Sailors in the Pacific Northwest–China Trade, 1785–1821', unpublished PhD dissertation, University of Idaho, 2004.

Wilson, John Malcolm, 'Mission to China: Lord Amherst's Embassy 1816', unpublished PhD dissertation, Harvard University, 1978.

Wilson, Ellen Gibson, *John Clarkson and the African Adventure*, London: Macmillan, 1980.

Wilson, W.F., *Hawaii Nei 128 Years Ago*, Honolulu, HI: n.p., 1920.

Witek, John W., S.J., 'Catholic Missions and the Expansion of Christianity', in Wills, John E., ed., *China and Maritime Europe, 1500–1800: Trade, Settlement, Diplomacy, and Missions*, Cambridge: Cambridge University Press, 2011, pp.135–82.

Wither, Charles, W.J., 'Mapping the Niger, 1798–1832: Trust, Testimony and "Ocular Demonstration" in the Late Enlightenment', *Imago Mundi*, 56 (2004), pp.170–93.

Wong, John D., *Global Trade in the Nineteenth Century: The House of Houqua and the Canton System*, Cambridge: Cambridge University Press, 2016.

Woolf, Harry, *The Transits of Venus: A Study of Eighteenth-Century Science*, Princeton, NJ: Princeton University Press, 1959.

Worden, Nigel, Bickford–Smith, Vivian and Van Heyningen, Elizabeth, *Cape Town: The Making of a City: An Illustrated Social History*, Verloren: David Philip, 1998.

Worden, Nigel, 'VOC Cape Town as an Indian Ocean Port', in Ray, Himanshu Prabha and Alpers, Edward A., eds, *Cross Currents and Community Networks: The History of the Indian Ocean World*, Oxford: Oxford University Press, 2007, pp.142–62.

Wright, Thomas, 'Thomas Young and Robert Seppings: Science and Ship Construction in the Early Nineteenth Century', *Transactions of the Newcomen Society*, 53 (1981), pp.55–72.

Wulf, Andrea, *Chasing Venus: The Race to Measure the Heavens*, London: William Heinemann, 2012.

Yu, Renqui, 'Imperial banquets in the Wanshu yuan', in Millward, James A., Dunnell, Ruth, Elliott, Mark C. and Forêt, Philippe, eds, *New Qing Imperial*

History: The Making of Inner Asian Empire at Qing Chengde, London: Routledge, 2004, pp.84–90.

Zeng Jingmin, 'Scientific Aspects of the Macartney Embassy to China 1792–1794: A Comparative Study of English and Chinese Conceptions of Science and Technology in the Seventeenth and Eighteenth Centuries', unpublished PhD dissertation, University of Newcastle (Australia), 1998.

Zhesheng, Ouyang, 'The "Beijing Experience" of Eighteenth-Century French Jesuits', *Chinese Studies in History*, 46 (2012–13), pp.35–57.

Zilberstein, Anya, 'Inured to Empire: Wild Rice and Climate Change', *William and Mary Quarterly*, 72 (2015), pp.127–58.

Zilberstein, Anya, 'Objects of Distant Exchange: The Northwest Coast, Early America, and the Global Imagination', *William and Mary Quarterly*, 64 (2007), pp.591–620.

Index